SCIENCE AND SOCIETY

BOSTON STUDIES IN THE PHILOSOPHY OF SCIENCE

EDITED BY ROBERT S. COHEN AND MARX W. WARTOFSKY

VOLUME 65

JOSEPH AGASSI

Boston University and Tel-Aviv University

SCIENCE AND SOCIETY

Studies in the Sociology of Science

D. REIDEL PUBLISHING COMPANY

DORDRECHT : HOLLAND / BOSTON : U.S.A.

LONDON : ENGLAND

Library of Congress Cataloging in Publication Data

Agassi, Joseph.
Science and society.

(Boston studies in the philosophy of science ; v. 65)
'Bibliography of Joseph Agassi': p.
Includes bibliographies and indexes.
1. Science – Social aspects. 2. Science – Philosophy.
I. Title. II. Series.
Q174.B67 vol. 65 [Q175.5] 501s [500] 81–8663
ISBN 90–277–1244–1 AACR2
ISBN 90–277–1245–X (pbk.)

Published by D. Reidel Publishing Company,
P.O. Box 17, 3300 AA Dordrecht, Holland.

Sold and distributed in the U.S.A. and Canada
by Kluwer Boston Inc.,
190 Old Derby Street, Hingham, MA 02043, U.S.A.

In all other countries, sold and distributed
by Kluwer Academic Publishers Group,
P.O. Box 322, 3300 AH Dordrecht, Holland.

D. Reidel Publishing Company is a member of the Kluwer Group.

Printed in The Netherlands

Dedicated to the memory of
Michael Polanyi

EDITORIAL PREFACE

> "If a science has to be supported by fraudulent means, let it perish."

With these words of Kepler, Agassi plunges into the actual troubles and glories of science (321). The sociology of science is no foreign intruder upon scientific knowledge in these essays, for we see clearly how Agassi transforms the tired internalist/externalist debate about the causal influences in the history of science. The social character of the entire intertwined epistemological and practical natures of the sciences is intrinsic to science and itself split: the internal sociology within science, the external sociology of the social setting without. Agassi sees these social matters in the small as well as the large: from the details of scientific communication, changing publishing as he thinks to 'on-demand' centralism with less waste (Ch. 12), to the colossal tension of romanticism and rationality in the sweep of historical cultures.

Agassi is a moral and political philosopher of science, defending, disturbing, comprehending, criticizing. For him, science in a society requires confrontation, again and again, with issues of autonomy vs. legitimation as the central problem of democracy. And furthermore, devotion to science, *pace* Popper, Polanyi, and Weber, carries preoccupational dangers: Popper's elitist rooting out of 'pseudo-science', Weber's hard-working obsessive commitment to science. See Agassi's Weberian gloss on the social psychology of science in his provocative 'picture of the scientist as maniac' (437). Nevertheless, the matter of judging science is not so quickly or cleverly turned bitter, as we see in his cheerful recommendation for reform: " . . . valiant failure . . . a bold and interesting mistake" (500). Agassi defends science *and* democracy, rationality *and* intuition, conservative competition in politics *and* in science but radical competition too, genuinely and recurrently 'starting afresh' (92).

Again, as with *Science in Flux* of 1975, we are pleased to include this selection of Joseph Agassi's papers as a coherent volume of the *Boston Studies*. We are glad also to be able to append a full bibliography of Professor Agassi's published writings to 1980, 137 papers, 6 books, and some 45 reviews and other pieces. For help in editing this volume, we once again are grateful to Carolyn Fawcett for her careful and intelligent collaboration.

Boston University
Center for the Philosophy and History of Science
June, 1981

ROBERT S. COHEN
MARX W. WARTOFSKY

TABLE OF CONTENTS

PREFACE

This, my second volume of selected essays, has Michael Polanyi as its hero and target of criticism. My first volume had Popper serve the very same function. It is not surprising, perhaps, that as a former student and disciple of Popper, I commented so much on his work. In retrospect I was surprised to find how much I commented on Polanyi; I realized only then how much I was indebted to him not only as a kind elder colleague who was so gracious as to waive his seniority but also as a challenging thinker indeed. I repeatedly express my admiration for both of these thinkers as the two outstanding philosophers of the mid-century; I take their concern for the progress of science to be their major asset, and I regard their view of science as a social phenomenon as the hallmark of contemporary *avant-garde* philosophy. I cannot think of any of their contemporaries who influenced as profoundly the current climate of opinions about science and its significance and its social character, except for Sir Edward Evans-Pritchard and Robert Merton, and no doubt Polanyi has taken cognizance of their contributions.

The fundamental concern shown by both Popper and Polanyi is for science as part and parcel of our culture and of our society. Science is traditionally characterized both by its method and by its products. Popper chose to stress the method rather than the products; and he brought to the fore the social character of science by the added assumption according to which scientific method is a product of social conventions, i.e., is a social institution. Polanyi chose the social aspect of science as primary, and deemed the methods practiced by its members as secondary. The difference is subtle but of extremely divergent consequences. Suppose scientists agree somehow to change their methods most drastically. Will this kill science as we know it? Popper will say, if and only if they cease criticizing themselves and each other, then, yes, science as we know it will thereby be over. Polanyi will say, if and only if the change be so radical that continuity will no longer be preserved, then, yes, science as we know it will thereby be over. Now, clearly, neither belittles either criticism or continuity. And both recognize this fact. Indeed, Popper speaks of the continuity of the critical tradition, and Polanyi of criticism that maintains flexibility and so fosters continuity.

Were it not so difficult to distinguish Popper and Polanyi, Imre Lakatos, who was a sharp and acute thinker, would not be able to slide from being a disciple of the one to the other. I remember one evening devoted to this point at the Brandeis University Philosophy Club with the late Harold Weisberg in the Chair. Lakatos was the speaker and I was a commentator, replacing Paul Feyerabend who failed to show up. Lakatos' punchline was, but if Polanyi only could understand himself well enough, he would not fail to notice that he is a true Popperian. The remark brought the house down: Lakatos could be trusted to bring out funnily the funny side of important matters.

It was fortunate for me that though Popper and Polanyi were the two great opponents in England when I went there for my studies, they each appealed to the better side of the other, no matter what the other's philosophic stand on the matter was. Thus, though Popper preached we should engage in critical debate only with fellow rationalists, he debated with post-critical Polanyi who, though he preached we should engage in debate only with fellow main-stream thinkers, debated with heretic Popper. For, I was thus able to enjoy the benefit and the pleasure of the company of both of them. This raises a general question: is there a limit to toleration and is toleration a necessary or a sufficient condition for critical debate? Or, we can switch words and ask, is there a limit to critical debate and is it a necessary or a sufficient condition for toleration? And are these two qualities similarly related to scientific progress? If we cannot have all three, which should we give up? Any one who both appreciates and dissents from both Polanyi and Popper is bound to ask, what is the order or priority of toleration, criticism and scientific progress? And priority here may be regarded both morally and socially. I hope the present selection of essays will help support my plea for putting this concern high on the agenda, perhaps also for clarifying its background somewhat and pointing at some possible corollaries. Let this be my homage to a great thinker. I hope my harsh criticism is relevant and is thus conducive to the noble concerns he showed in all his writings.

Tel-Aviv, winter 1980 JOSEPH AGASSI

P.S.

Paul Feyerabend has recently (*Inquiry*, 1980) branded Michael Polanyi an elitist and a Stalinist on the authority of Imre Lakatos, who had used the labels 'elitist' and 'Stalinist' interchangeably. And perhaps Lakatos was an authority: not only do Feyerabend and I agree he was both an elitist and a Stalinist; Ian Hacking too, it seems, in his extended review of Lakatos's posthumous collected works (*British Journal for the Philosophy of Science*, 1980), expressed pretty much the same view. Yet there is a paradox here: should we trust an expert Stalinist as an expert, or should we distrust him as a Stalinist? (Following Lakatos and Feyerabend I consider here Zhdanovism as part and parcel of Stalinism with no specific reference to it.)

This question may be dismissed by Popper, who said, rather naively, never trust an expert anyway. It was central to Polanyi, who said, the expert is an indispensable ingredient: he is the *connoisseur* whose taste is so very much superior to that of the inexpert, that we will lose much if we fail to utilize his good offices. Nor was Polanyi insensitive to the possibility of the expert abusing his power. He declared the abuse regrettably unavoidable, but recommended it should be controlled lest the scientific tradition be disrupted. This is why, when Stalinism was rampant and its program of planned science popular in the community of science, Polanyi was far from relying on expert judgments on planned science; rather, he worked indefatiguably to organize the Congress of Cultural Freedom and the Committee on Science and Freedom. These organizations played a significant if quiet role immediately after World War II, especially in Central Europe before the formation and consolidation there of political parties proper. Whatever history will pronounce as a judgment on these organizations and their political activities, it should be noted that Polanyi was dedicated to the cause of intellectual freedom. This made him abandon his prestigious chair in physical chemistry for a chair in sociology. It might also be noted that his judgment of his misguided colleagues was benign to the end: in 1956, soon after the revelations, in the 20th Congress of the Bolshevik Party, of the Stalinist atrocities, Polanyi still viewed the acceptance of Stalin as based on the nationalization of industry and his attraction as rooted in "*the messianic claims attached to this measure by Stalin's followers*". (The emphasis is in the original.) This judgment contrasts rather sharply with George Orwell's view of the popularity of Stalin among intellectuals as power-worship proper.

Despite Polanyi's great sensitivity to human suffering, he declared the authority of the expert binding without fooling himself or his audience: he was clear about the fact that authority wields power and can abuse it. He clearly distinguished between the external authority of the politician, which he sharply opposed, and the internal authority of the scientific leadership which he endorsed. Yet he was clear about the fact that the internal authority wields power and can abuse it too. He demanded the control of power and devoted much of his career to this task. Yet, this *control has no room in his philosophy*. The reason is simple: the desired control of power is *democratic*. Nor is there a discussion of democratic control in other writings on the philosophy of science that I know of, except for a remark here and an admission there. The critical philosophy of the Popper brand is conducive to it, yet Popper's own philosophy still includes the myth of the unanimity of science and brands as pseudo-scientific all scientific disciplines where controversy is aflame. The very preoccupation of Popper with pseudo-science is elitist. This elitism did not mix well with the democratic social philosophy of Popper. Indeed, it goes better with the democratic social philosophy of Polanyi. It is therefore not surprising that Polanyi grudgingly endorsed Popper's idea of pseudo-science, though with his usual reservation that the scientific elite can overrule any of its accepted criteria. It is therefore even less surprising that the Stalinist Lakatos tried to offer an explicit elitist criterion of demarcation of science from pseudo-science (see the note by myself and Charles M. Sawyer 'Was Lakatos an Elitist?', *Ratio*, 1980) plus the demand to suppress pseudo-science, which demand will be discussed in later chapters in this volume.

The message of the present volume is this. Science will do better and be more humane if the (inner and outer) democratic controls of the commonwealth of learning improve, become more effective, and apply to wider areas. In discussing this I find it necessary to criticize both Popper and Polanyi. Unlike Feyerabend, however, I try to take note of their devotion to the cause of freedom and democracy. Their advocacy of self-censorship of the commonwealth of learning is understandable, but dangerous and contrary to the very spirit of democracy that obviously does permeate all their writings: it is more important to criticize the undemocratic tendencies present in the writings of avowed democrats like Popper and Polanyi than the very same tendencies present in the writings of advocates of suppression and violence like Lakatos and Feyerabend.

In this, I think, I am a follower more of Popper than of Polanyi, since Polanyi viewed all dissent as strife and Popper deemed critique as homage. Let my respectful dissent from and critique of both Popper and Polanyi count as my homage to these great philosophers and as my expression of admiration to the spirit of democracy which permeates their writings. On this ground I deem all my criticism of their writings internal criticism.

ACKNOWLEDGMENTS

Much of the previously unpublished material was written or revised while I was senior fellow of the Alexander von Humboldt Stiftung, resident at the Zentrum für Interdisziplinäre Forschung, Universität Bielefeld. I am grateful to them, and to the editors and publishers of the periodicals and books from which the remainder of the material here is taken, for their permission to republish. The following list of sources of the republished material indicates my indebtedness.

Chapter 2: *Philosophica* **15** (1975), 5–20.
Chapter 4: *Brit. J. Phil. Sci.* **10** (1959), 135–146.
Chapter 5: *Manuscrito* **2** (1978), 65–78.
Chapter 7: *Diálogos*, No. 35 (1980), 27–35.
Chapter 8: *Metaphilosophy* **3** (1972), 103–122.
Chapter 9: *Organon* **3** (1966), 47–61.
Chapter 11: *Scientific Information Transfer: The Editor's Role*, ed. by Miriam Balaban (Reidel, Dordrecht, 1978), pp. 133–139.
Chapter 15: *Phil. Soc. Sci.* **5** (1975), 145–161.
Chapter 16: *The Interaction between Science and Philosophy*, ed. by Y. Elkana (Humanities Press, New York, 1975), pp. 391–405.
Chapter 17: *Synthese* **26** (1974), 498–514.
Chapter 18: *Perspectives in Metascience*, ed. by J. Bärmark; Acta Reg. Soc. Sc. et Litt. Gothob, *Interdisciplinaria* **2** (1979), 13–25.
Chapter 19: *Poznań Studies in the Philosophy of Science and the Humanities* **1** (1975), 1–8.
Chapter 21: *Zeitschrift für allgemeine Wissenschaftstheorie* **8** (1977), 30–38.
Chapter 22: *J. History of Ideas* **34** (1973), 609–626.
Chapter 23: *Physis* **17** (1975), 165–185.
Chapter 24: *Organon* **8** (1971), 138–166.
Chapter 25: *Organon* **7** (1970), 117–135..
Chapter 26: *Vistas in Astronomy* **22** (1979), 419–430.
Chapter 27: *Historical Studies in the Physical Sciences* **1** (1969), 1–36.
Chapter 30: *Boston Studies in the Philosophy of Science* **11** (1974), 249–257.
Chapter 32: *Inquiry* **16** (1973), 395–406 (Universitetsforlaget, Oslo-Bergen-Tromsø).

CHAPTER 1

INTRODUCTION: SCIENCE IN ITS SOCIAL SETTING

Our recent past is strewn with landmarks commemorating recent swift cultural developments: political events like the Cold War, the Hungarian revolt, and Khrushchev's Speech culturally considered; space probes and walks and travels taken technologically, scientifically, poetically, and politically; truth in packaging and the Suez crisis in British politics and the Watergate crisis; the eruption of black liberation and women's liberation and the sexual revolution and student revolts, and the New Left enter and exit before you could assess their worth and impact on science, on campus style of life, on international politics and on the new ecological fashions; with, finally, the Viet Nam War and its impact on cultural life in general, on politics, on ideology, on the social sciences and on social philosophy, perhaps even on scientific research.

What is intriguing about this short list of impressive landmarks is not only their close proximity in time but also their cross-reference: the impact of space programs on politics and on poetry and vice versa, for example, are fairly obvious. Now, these connect with the surge of popularity of political activism on campus in diverse ways. Viet Nam replaced the Cold War because space programs made the Cold War obsolete – technologically, with skies open to both spy-satellites and nuclear armament; culturally because people learned that space travel makes the brotherhood of men a politically viable idea with a high Nielsen rating. And so circles close, and diverse factors come together into focus.

Before getting worked up about the immense focusing of diverse trends which took place in recent years, I must observe that the peak is behind us, not ahead of us: professors and students have gone back to work without having found the happy medium where genetic factors or quantized electromagnetic fields naturally reveal their political relevance without anyone trying to compute it. *Science* magazine, the official organ of the American Association for the Advancement of Science, publishes regularly pleas to limit the number of candidates for Ph.D. degrees in the sciences so as not to flood the market. And there is no anti-trust law and no organized opposition to curb this return to the Middle Ages. The prevalent philosophy of science, especially in the social sciences which should know better, is that of Michael Polanyi, especially its vulgarized version offered by Thomas S. Kuhn, of scientists as a guild of masters and apprentices and a hierarchy of leaders who rule by the means of existing conformity to fashion, which conformity makes thought-police quite redundant. All this is taken for granted amongst commentators on works of Polanyi and of Kuhn. It is time to remember Lord Snow and his *The Two Cultures and the Industrial Revolution*, those two lectures he gave only two decades ago, which made so much stir at the time.

Funny; he said scientists were progressive and artists were reactionary – natural Luddites, he called them. And then they all joined the New Left and the Great Marches, to Aldermaston in England, to Washington in the U.S.A. And now the scientists have settled for the narrow antipolitical reactionism that Snow so deplores in artists. What has happened to the artists I cannot say. If Leonard Bernstein, the Big Black Panther of yesterday, is any indicator of trends, then his return to normal in his Norton Lectures in Harvard signifies a similar settling down in the arts: business as usual. Or is it? What has come out of the melting pot of the mixmaster is still not clear; but with Watergate still having the highest Nielsen rating and with events moving still so fast, we cannot wait for sedimentation before we begin with analysis. It is not easy to observe which things keep whirling, but I think we have no choice. Let me take a few steps back and discuss Snow's lectures, and then go further back to their cultural background, before attempting to return to the present.

I will have no time to assess the historical impact of Lord Snow. He said, quite humbly, he had no original idea to offer but struck a familiar chord: this is why so much response to his lecture took place – by resonance. Indeed, even Washington tried at the time to do something against narrow professionalism and for the humanization of the sciences and the rendering of science palatable to non-scientists. Snow was a part of a trend, and it is hard to know how much to credit him. I shall regretfully not touch upon this, except to say now, Snow had an impact, and to the good in spite of his errors and shortcomings, and his good turn deserves another: we should start again before everybody is too well saddled in his super-professional niche. So much for my introduction.

1. SNOW ON THE FACTS OF THE MATTER

Snow begins with an obvious observation. I wish to stress, and sing the praise of, the very obviousness of his observation: the more obvious and readily available, the more inescapable it is; the more dangerous is, then, the strategy of turning a blind eye to it. The obvious and superficial fact is particularly obvious to Snow himself. He has two professions: he is a person holding a physics degree and employed for decades as a high government official in some scientific capacity or another; and he is the author of a well-known series of novels 'Strangers and Brothers'; he is, as he says of himself, fortunate to be standing at a vantage point, as one with a foot in each camp. Interestingly enough, this claim was hotly contested in the most vicious attack on Snow's *Two Cultures* by F. R. Leavis, the most prominent British literary critic of the age. Leavis declared Snow's novels so poor as to hardly give Snow credit for having so much as a foothold in the literary camp. But this is sheer affront: Snow does not claim to be a physicist of any distinction, and so his distinction as a writer or its absence, as the case may be, is likewise irrelevant. All that matters is that Snow frequents circles of writers and other artists as well as circles of physicists and other scientists – let us say he merely goes to meet them at cocktail parties. From this unpretentious vantage point he already could see so very clearly the already obvious simple fact; indeed, he could not have missed it even if he tried.

The fact is that a physicist living in London is more likely to contact a physicist living across the Atlantic than an artist living in London. Now I mean any contact and under any pretext. An artist in London may be a relative of a physicist in London, or a member of the same club as he, or a co-worker in a political campaign, or they may bump into each other at a publisher's cocktail party or at a social event such as the Queen's garden party. The same may hold for a physicist in England and a physicist in the United States, yet much less likely. On the purely non-professional and on the semi-professional level any Englishman is more likely to bump into another Englishman than into an American. But if they are physicists, they lead most of their active lives in almost strictly professional circles, and their professional circles are international and highly mobile. And so the balance is tipped the other way. And this is the significant fact!

I wish to stress that Snow's discussion rests on an empirical observation. I should therefore pause a minute and ask, is the observation true? How often do, for example, physicists on one given American campus come into contact with any British scientists, and how often do they come into contact with any American artists? I should submit that the question is not easy to answer, that even if you ask one physicist on the campus here to answer the question, you will need two major operations: you will need, first, to specify, perhaps with his help, what the act of coming into contact means; and you will need, second, to give him time to record all his contacts over a period – preferably the period of one year because conventions, meetings, subscriptions to concerts, and even publishers' cocktail parties and the Queen's garden parties, do go in annual cycles. And, to be really significant and to follow Snow's idea, any communication should count as a contact, including correspondence or even the act of reading a paper or a book. Now, do physicists read only physics, or do they secretly read the national weeklies and cheap novels? This is no easy matter to find out, as I am reminded repeatedly when I discover another colleague secretly hiding a television set in his bedroom – just in order to watch the news in times of crisis, and to throw a glance now and then at *Laugh-in, Maude,* and *Mannix,* of course. All this, then, requires much empirical study. This, to the best of my knowledge, has not been done. My own observations, which are much more careful and detailed than any I have noticed, or heard about, are far too crude to allow me to judge whether Snow's observation is true or not. I frankly do not know.

I have now arrived at a contradiction. I have first considered Snow's observation obvious and now I say it is problematic. Is it obvious? I have checked repeatedly with physicists and other scientists, as well as with writers and other artists, with sociologists and with bureaucrats of all sorts, whether Snow's observation seems questionable. I have met not a single person who, presented with Snow's observation, has questioned it. I have concluded that it will pass as obvious. I have myself empirically brought it into question; my statistics, crude, poor, and impressionistic as it undoubtably is, indicates that quite possibly Snow is in error: an average English physicist is quite possibly much more likely to bump into an English artist than into an American physicist. It certainly is true of many; I do not know if it is true of the average.

I must have misread Snow, then. Snow does not mean, let us correct my presentation, that counting all contact, this is true, but, let us say, counting only professional contacts. This, however, seems quite cheap: obviously, an English physicist is more likely to come into professional contact with an American physicist than with an English artist: what possible professional contact will he have with an English artist?

2. OUR FEELINGS ON THE MATTER

Before looking at a third reading of Lord Snow I wish to take my rhetorical question as a question rather than as rhetoric: what possible professional contact can there be between a physicist, or any scientist for that matter, and an artist in the same town? We consider the Renaissance, especially the early Italian Renaissance, the Golden Age, and Leonardo its great personification. We view the Renaissance man not only as a rounded man, a polymath, but an integrated man: we look with deep love and yearning at a venture which combines physics, engineering, and architecture, which combines painting, and anatomy, etc. Of course, the little word "etc." is pregnant with all sorts of connexions we assume without much knowledge about them. We frankly exaggerate and idealize. But we feel convinced that in principle we are right. When we read a chatty and light-weight work of the period, such as Giorgio Vasari's *Lives of the Artists* is, we cannot but notice the naive lack of differentiation between art, craft, technology, and on occasion – especially with Leonardo, of course – science. No doubt, the Renaissance, though an age of irretrievable innocence, in a sense still stands, perhaps for want of a better example, as closest to the ideal rounded man that warms the heart of even many hardened a professional amongst us.

Allow me to remain with this ideal a bit, before returning to Snow's sociological observation. One of the earliest examples of modern professionalism I know of is that of Michael Faraday. He was the disciple of Sir Humphry Davy, a leading chemist and a considerable poet in the age of Coleridge. He was a contemporary and friend of Sir William Grove, the judge, the lecturer on science, the author of the *Correlation of Force* of the 1840's and the inventor of the Grove battery. Maxwell's obituary notes that Grove was the last amateur physicist who made important scientific contributions. Faraday was of a similar ilk as Davy and Grove. He loved to meet with artists. He was himself a flutist and a singer. Yet he got absorbed in his work and at the peak of his career he hardly even had any recreations: he went to the annual dinners of the Royal Academy of Art, he went for a few minutes a year to a party thrown by Grove, a few private dinners, and that is all; he had no time to make use of his free tickets to the opera. He went to the zoo and to the fair only when he was too sick to work. Yet, I do not doubt, he found this absorption in work quite deplorable. Nor was he a model for all scientists. His younger contemporary, Hermann von Helmholtz, not only was a physicist, physician, and musician; but he also integrated his interests, and even opened up vistas of new studies in physical acoustics and music theory. He was, incidentally, old fashioned in his musical taste, and his acoustic theory comes to defend old-fashioned tastes of the preclassical period; Bach's *Well-Tempered Klavier* of the early eighteenth century went way beyond the overtone theory of the scale, the basic theory in Helmholtz's researches of the late 19th century. The most celebrated student of Helmholtz was Max Planck, the celebrated father of quantum theory. Well into adult life he wondered whether to become a natural scientist or a violinist; at times he also seriously contemplated economics as a career. In his brief scientific autobiography he records with amazement that he once occasioned to listen to ancient instruments and found the modern, well-tempered, harmonies pleasanter to his ear than the older purer ones – contrary to the teaching of Helmholtz, need one say.

Yet, for all this, we could consider the violinist Planck, and even the violinist Einstein, as professional scientists hardly in contact with artists. Our impression is, and I invite you to check it empirically, that Einstein was in close contact with physicists all over the world, yet hardly in contact with ordinary mortals. He played the violin and went to concerts; he talked with artists of all sorts, as even Snow observes in his *Variety of Men*. He lectured and wrote first-rate material on philosophical subjects; he was active in political causes – of Zionism and socialism, world government and world peace. Yet, his image for us is not that of a Leonardo but that of a Faraday, a man regrettably too engrossed in professional life to have much contact and meaningful encounters with ones who do not share his professional interests.

And now we are back at Snow's observation. I wish to repeat my observation about it. Were it a statistical study akin to Kinsey's report or the Neilsen ratings, it would have had a different impact on us. No doubt, we live by the Nielsen; no doubt Kinsey has changed our outlook on American sexual mores and thereby was a major factor in the development of the new sexual morality all over the post-industrial world. The force of works like Kinsey's is in their surprise value. Hence they did not hit a familiar chord, they did not create a rich resonance. Snow's observation did, as he reports in the opening of his second lecture. It was not a new fact, but forcing us to focus our eyes on an all too familiar one, that has brought about the reverberating echo.

What, then, is this all too familiar fact? It is not that all in all a scientist meets only scientists; it is not that in one's professional career one meets only professional colleagues – all too often one meets bureaucrats, other professional members of the same college or club, students, mothers, etc. The familiar fact is hard to locate because we search for its projection whereas it is in our hearts: we are sorry to be so narrowly educated. As scientists we lament the fact that we hardly ever have time to go to the opera or read John Betjeman and Robert Frost; as artists we lament the fact that we never caught up on our high-school maths, all our good intentions notwithstanding, so that we cannot even read *The New Scientist* or *Scientific American*.

But this is not Snow: it is in opposition to what he says. He says he has interviewed in his official capacities many young scientists, and he found that they had no use at all for books – any books at all! At most they read professional periodicals, that is. And he records, likewise, contempt for science which he has observed amongst artists. This contempt is well illustrated in Leavis's response to Snow's *Two Cultures* where Leavis says science is cultureless.

3. THE POLITICS OF THE MATTER

Here, then, is my third and last reading of Snow. It is not that scientists only meet scientists, or only have business with scientists; it is that they are bored by artists, just as artists are bored by scientists. Now, is that a fact? I wish to say at once I find Snow's observation on this reading no observation at all but merely a sermon and moralizing marshalled as sociology. We have crude youths, alas! in all walks of life. They are as uninterested in culture in general as in their own specific part of it. Sociological observation will easily show that just as a chemist hates music, so he hates chemistry; just as a musician may hate mathematics he may hate music; indeed, one's hatred of one's own subject is livelier and more authentic, more direct and more significant, than one's hatred of alien subjects. I submit that many musicians hate music and many physicists hate physics and so on. Now we can look at it this way. The phenomenon of anti-culture and anti-intellectualism accepts at times the comfortable excuse of professionalism – that is to say, first of all professionalism allows one to hate everything other than one's own profession. Afterwards, the professional may, and usually does, hate his own profession as an advanced case of professionalism. Finally, the professional will have special excuses for hating just his own branch and his own participation in it. And so this has nothing whatsoever to do with Snow's observation. Snow's observation, on the contrary, is intended to hold for cultured and intellectual people who are allegedly of limited interests. In fact, however, they feel unease, displeasure, dissatisfaction, frustration, guilt even, that wishing to emulate a Leonardo in their own small way they find themselves emulating a Faraday and having less and less time and presence of mind for culture in general. And so Snow is here plainly in error.

The thrust of Snow's essay, however, is not at all sociological: it is political. He blames, no doubt, both scientists and artists for their narrow professionalism, for their want of concern for others' contributions to human culture. But this is as far as his impartiality takes him. He then explains the split between scientists and artists by reference to politics; and there he declares artists to be introverts, tradition-minded, and thus conservative and natural Luddites; scientists, he says, are the very opposite: extrovert, radical, naturally future-oriented, forward looking, left-wing oriented. This sounds strange, coming as it does from Snow's pen. Readers of his novels will be familiar with characters he draws of reactionary scientists and of progressive artists. Yet, on the whole, Snow's novels conform to the general idea: his men of science as a group are progressive, and his artists and such, less so. From his *The Masters* to his *Corridors of Power* this is clear. Even his most moving novel, *The Light and the Dark*, on a philologist friend, presents the friend as one who had to make conscious efforts to overcome his Nazi sympathies in the dark days preceding World War II.

How true is this picture? We have no statistics; we have no statistics even regarding broad outlines such as, are professional organizations divided to Left and Right? Are physicists by and large crew-cut and left-wing and are historians or painters by and large long-haired and right-winged? (I know: in the seventies crew-cut meant Haldeman, the last of the Law and Order McCarthyites; but Snow wrote in another era: he belongs to the fifties.) Before answering this, let me notice one small oddity. Snow notes that his division of our society into two cultures is somewhat arbitrary, he feels that perhaps social scientists constitute a third culture. This is, indeed, Talcott Parsons's criticism of Snow. Yet, in my opinion, Snow's division into two, radical and conservative, is the deeper one. Just as he admits that some artists may be radicals and some scientists conservative, so he may notice that some social scientists go hither and some thither. The main point is, we have the extrovert who does not need tradition and who breeds science and technology, yet we also have the sensitive introvert who is tradition-minded and thus naturally conservative. What hurts Snow is the split, the inability to appreciate the other half. He feels a bridge between radical scientist and conservative artist is possible. And he feels this bridge will make the artists politically move to the Left. In the worst period of exposure of Soviet Russia as hell on earth, the years past the twentieth Congress of the Soviet Communist Party and Khrushchev's Speech revealing Stalin's atrocities, at that period Snow looked up to Russia and its educational system for a solution, for a way to bridge the two cultures, since Russia was a country preaching radicalism yet extremely conservative.

How true, then, is Snow's observation of the split? Again, I feel, as with other facts in Snow's study, we are misled. We feel that we do observe an obvious fact, yet the facts are not well known, not empirically observed, not well analyzed. But they strike a chord. Facts are in Snow's *Two Cultures* not representative of the average objective situations but of average subjective ones. We have no knowledge of the objective fact. We do not know whether art and science are tradition-bound and tradition-free respectively, whether they are, by and large, conservative and radical respectively, whether they are socially in contact with each other or isolated from each other. But let me report an observation of mine: all too readily we assent: we feel a study akin to a Kinsey report which will confirm Snow's claims would not surprise us. Kinsey's report corroborated the claim that most married Americans have sex almost exclusively with their mates, yet this was so taken for granted that no commentator, to my knowledge, has noted the fact; rather, the small deviations he reported were noted as much larger than expected. Similarly, had a Kinsey shown that most artists are conservative and have little interest in science, we would not notice it and rather be surprised at the number of radical ones. Perhaps we will explain the prevalence of radicalism among artists by reference to Viet Nam. But we would hardly bother to explain the prevalence of conservatism among them. This would seem proper. Why? Why is it proper for artists to be wedded to tradition but not for scientists? What is our view of art and science in relation to politics?

To explain all this I must very crudely and briefly outline the two mainstream philosophies that are most influential in our present culture. These are the rationalistic ideas of the 18th century, of the Age of Reason or the Enlightenment, or Radicalism, or universalism on the one hand, and of the 19th century, Romanticism, nationalism, conservatism and reaction on the other. Let me say in advance, the views I present are very oversimplified in their brevity and in my gloss over the fact that they get deformed and altered in applications; also, let me say, though today the two views are held by some thinkers or others, many hold eclectic mixtures of elements of both; only few are working towards a coherent synthesis. It is the coherent synthesis, one hopes, that may offer a better solution to the difficulties that are felt by all and were resonated when Lord Snow struck, as he says, a familiar chord. Let me begin, then, with a brief description of rationalism and of romanticism, or rather of the caricature of these views as understood by rather unsophisticated adherents – then and now.

4. THE INTELLECTUAL HISTORY OF THE MATTER: RATIONALISM

The crux of rationalism is Sir Francis Bacon's doctrine of prejudice, a doctrine which is as inconsistent as it is still very popular. Bacon asked, what brought about the decline of Antiquity and the descent of the darkness of ignorance of the Middle Ages? Unless we know, he said, we may fall into the same error as those who brought about the disaster, and thus repeat it. Indeed, his warning was even more specific. The cause of the darkness, he said, was dogmatism: the blind adherence to Aristotle. And, he said, in the modern times Copernicanism is replacing Aristotelianism as the new dogma, and unless we stop the trend we are stepping into the new epoch of darkness.

There are two methods of inquiry, said Bacon. The one is that of certainty – slow but sure. Observe facts as they are and never jump to conclusions: let the conclusions impose themselves on you with utter certainty. The opposite method is the method of hypothesis, of making conjectures and then finding observations to support them. Here comes Bacon's world-shaking psychological discovery which many psychologists consider as a new discovery made in our own century! The discovery is this: no matter how far-fetched your hypothesis is, no matter how obvious its falsehood, if you believe in it long enough you will find facts confirming it and find ways to explain away what seems to contradict it.

Sir Karl Popper holds a variation of this view. He says, look at Marx's theory of the press. It says, everything published in the press serves class struggle. Consider anti-communist propaganda in Western newspapers. Clearly it serves the capitalist interest. Take information which goes the other way, which comes, however seldom, into the Western press: it is suppressed, and admitted grudgingly when efforts fail to suppress it completely. Take, then, politically indifferent news items, such as information regarding the latest antics of the present sex-bomb of a movie queen. Clearly, here the capitalist press tries hard to take workers' minds off their own immediate interests. To sum it up, any information, for capitalism, against capitalism, or indifferent to politics, namely any information whatsoever, once published by the press, confirms the Marxist theory of the press as serving class interest, and this quite regardless of the question, is that theory true?

You may consider this example far fetched. So let me go back to an example already mentioned: Copernicanism. Copernicus repudiated the Ptolemaians for endorsing epicycles and eccentrics, yet he himself used these. Hence the same practice condemns his opponents but supports his own views! Galileo Galilei, the leading Copernican, said, he could not adequately express his deep admiration for Copernicus for the courage he showed when he dismissed the evidence from the brightness of Mars and asserted his theory to be true nonetheless.

Sir Karl Popper has quoted Galileo's passage with great approval, claiming that this refuted Bacon's inductivism. Popper thus ignores, says Paul Feyerabend, that this quotation shows that Popper's own theory of science is refuted by the same token. Feyerabend is quite right, of course. And he illustrates the fact that Bacon's observation still holds: we still tend to see facts as if they prove us right and ignore the same facts when they prove us wrong.

For my part, I think that Bacon's theory of prejudice thus illustrated is neither true nor momentous. Yet I fail to find many ideas which had such a terrific impact on the modern world as Bacon's. The whole anti-speculative school of thought, positivism, hostility to religion, to the Middle Ages, even to innocuous superstitions, which is still commonplace, all find their rationale in Bacon's doctrine of prejudice. Even Snow's works are full of it. I should only mention his *The Affair* which describes how faith in a false theory led to the readiness to forge evidence, *pro tem*, in its favor.

Now the most important corollary to Bacon's doctrine is the idea that science is dispassionate. Oh, the scientists can love science itself passionately and run excitedly all the way to their lab. But once they are in the lab, serenity must reign; love and hate only cloud judgment!

What does this do to the relation between science and art, what does this do to the relation between science and tradition, what does this do to the relation between science and politics?

Take art first. The locus classicus is the celebrated dream of Descartes, who accepted Bacon's radicalist doctrine of prejudice. In his dream Descartes saw art and science symbolized by books and he interpreted art to be conducive to science. Bacon's idea of science as dispassionate left for art no role within science. Reversing the Renaissance tradition, this view became the new commonplace. Descartes even went further and declared thinking alone to belong to the mind and all emotion, all affection, all passion, to be animal spirits, i.e., entirely bodily.

Robert Boyle, the moving spirit of the young Royal Society of London, disagreed and declared passions to be mental. Yet he too said art has a place on the walls of the telescope, not on its lenses. And John Dryden, the contemporary leading poet and Fellow of the Royal Society of London, concurred and made art secondary. The romantic reaction to all this is well expressed by a poem of Keats or of Shelley which complains that Newton debases the rainbow; many have felt these days that NASA has killed the romance that goes with the moon since time immemorial. I confess I think these complaints have a point, and the point is the one which romanticism took up with a vengeance. Before going into that, I must present Rationalism, radicalism, and their politics.

5. THE INTELLECTUAL HISTORY OF THE MATTER: ROMANTICISM

Here my job is made easy by a terrific essay called 'Rationalism in Politics' by the contemporary arch-conservative English political scientist Michael Oakeshott. Rationalism is for Oakeshott, as for Bacon, synonomous with radicalism. Tradition, says Bacon, is frought with prejudice, and prejudice clouds our vision. It follows that we can do anything better by appeal to science rather than to tradition: all we need is to start with a clean slate and go on with the business disinterestedly and unemotionally and not rushing to any conclusion. Bacon himself noted that the application of this idea to politics is political radicalism, and he suggested to keep the law out of science: science opposes dogmatism and the law frankly endorses it. In German universities, traditionally and up till today, the law is lumped together with theology, hermeneutics, and I know not what else, into the rubric "dogmatic sciences". The shock which I felt when I first encountered when hearing this expression indicates the universality of Bacon's doctrine of prejudice and the success of the Baconians in implementing his doctrine in politics – and in the following way. What Oakeshott argues is very simple. Call it science or not, call it dogmatic or not, certain things can be learned only within a living tradition, not by observing them and presenting a recipe for their reproduction. Perhaps chemists can convey knowledge thus, but cooks cannot: nobody becomes a cook by the mere study of a cookbook. And the same goes for the art of conversation and for democracy: democracy is a deep-seated tradition, not an export item.

The attempt to eradicate tradition may lead to all sorts of modernisms, in politics or in the arts, some more successful, some less successful, as future critics may judge. The attempts to eradicate political tradition are nothing short of a political revolution. Not a revolution in political thought, but in political action. And revolution, says Oakeshott, destroys traditions irretrievably. It is not clear to me why this is so. The eighteenth century did not live by all its ideas; it could maintain national borders; it could recognize governments which allowed slavery; it could likewise recognize kings while laughing at their divine rights as superstitions and prejudices. History, however, went the other way. As Bacon declared, lack of belief did bring about the collapse of institutions based on them to a sufficient extent to cause havoc. Bacon's warning became a self-fulfilling prophecy, to use Merton's term, or an Oedipus effect, to use Popper's.

Again, I say, this is a historically very significant yet over-valued idea. We know today that kingship can survive the demise of the divine right of kings, that socialist governments can live after all shred of socialism is destroyed by terror, etc. Contrary to Bacon's view, then, we do not always live by our own ideas. And the same goes for science. Bacon's claim that science has no need for and no respect for any tradition, has survived all criticism. Snow asserts it most confidently in his *The Two Cultures* a decade after Michael Polanyi has refuted it. One reason why Kuhn's ideas are so widely popular is that they are echoes of Polanyi's traditionalism within science.

The rise of the two cultures dates from the days of the political reaction of the traditionalists to the great, noble, and foolish, rationalist political act known as the French Revolution. Rationalism could be criticized in a thousand ways. Yet the historically victorious criticism was German Idealism or Romanticism which declared science and only science subject to the radicalist code, to the doctrine of prejudice. Politics, history, art, all these are different. They are culture-bound and so tradition bound. We think scientifically in a dispassionate way, with no recourse to tradition: we extrovertly observe instruments, cold and meaningless. But in politics we think historically, with our whole being, with our affections and passions, with our prejudices which gives us both our strength and our limitations. We think in social matters with our emotions and sympathy, not looking to cold machines but at real live human beings.

It is very easy to go on with this kind of chit-chat and sound convincing. You can make the experiment yourself; you can approach the same audience and the same question on two different occasions, once from the Rationalist and once from the Romantic angle. You will be convincing to some extent each time, and with a proper choice of problem you can confuse your audience thoroughly.

The locus classicus of the confusion, I do think, is Snow's *The Two Cultures*. Snow ignores the fact that Rationalism has led to two revolutions, the French and Russian, both of which have ended up in a regime of terror and a bloody counter-revolution led by monstrous tyrants. Rather, Snow describes the role the rationalist movement had in bringing literacy to adult workers even in the mid-nineteenth century England, even to Snow's own grandfather. Snow likewise ignores the fact that Romanticism brought us fascism of all sorts all over Europe, including the Third Reich. He forgets his own novel whose hero was a Nazi sympathizer until almost the outbreak of World War II. Snow's historical analysis is remarkably British, remarkably insular, remarkably confined to an experience where neither rationalism nor Romanticism brought about any internal political disaster, any large-scale internal bloodshed, where romantic and rationalist live in a peaceful coexistence, equal but separate, hostile but polite, two cultures in one. When I read Snow on English and American scientists and artists I assumed he had chosen them partly because of his own experiences as described in *The*

New Men and *Corridors of Power*, partly due to the linguistic unity of their situation. But I have changed my view of the matter: I can hardly imagine the possibility of Snow's two cultures except on the ground of a tolerant society, where some ideas of both camps are fairly disseminated and where crossing lines is not a cause for ostracism – a Western democracy, but more likely Britain or the U.S.A. What Snow noticed is that with the entrenchment of the two cultures things may become different, mainly through increasingly specialized education at an increasingly early stage. In his *Time Machine*, H. G. Wells projected current society into the future two, upper and lower classes, which become two species of animals. By no accident Snow's projected artists are Wells' projected upper-class, and vice versa, except that Snow's projection is meant as no fiction and no long term. Presumably the split between the two cultures is not only the thing of the future as well described by Wells in his *Things to Come*; it is quite common in the nineteenth century. The split is most obvious when it occurs within one man. Men of science like Helmholtz and Planck were radicalists and universalists in science but Romantic and nationalistic in politics.

Intellectually, the compromise, the eclectic position, cannot be maintained. In particular, there is always the attempt to base nationalist politics on biology or on animal psychology, thus giving irrationalism a basis in rational science and even in experiments. Somehow, today it is fashionable to make in the name of science wild assertions about man's aggressive instincts as going beyond the limits of reason, as uncontrollable by reason. And, indeed, when reason is declared to be nothing more than the brute means of survival in a Darwinian sense, as Spencer, Dewey, and others have claimed, and when self-defense among animals is called aggression and viewed as essential for survival, the conclusion seems inevitable.

Let me conclude with an outline of the new rationalism which is neither radical nor romantic but gradualist, which neither rejects nor endorses tradition but critically takes it as a point of departure, which makes truth neither the universal and easily attainable goal of all science nor the local relative accepted view of our own part of society, but as the long term goal to which all inquiry strives; which recommends reform rather than either the blind conservatism of the total revoking of tradition.

6. THE NEW RATIONALIST APPROACH

Let me take one example which I find rather puzzling, because it presents as quite novel an idea which I think all of us will consider as rather trite and rather obvious. In the preface to his *Moses* Martin Buber sums up the traditions of study of the Bible. The book, may I remind you, barely belongs to history: it was written about the time of World War II. In the preface Buber observes that most Bible studies belong to either the traditional camp which endorses the Bible to the letter or the radicalist camp which dismisses the Bible as unreliable evidence and so worthless. Rather, says Buber, we must treat it as distorted racial memory. I need not tell you that increasingly historians look at every historical document as such: they neither endorse it nor dismiss it but they try to see through it. What puzzles me is that this could be viewed as a novelty hardly a generation ago.

Let me mention another example: scientific records. Most scientific records are contemporary and of no interest to any historian. Some scientific records, however, are historical. Already Laplace was fascinated with the existing Babylonian astronomical records that were emerging in his days. But he used them merely as evidence to strengthen his claim that the sun will assuredly rise tomorrow. Had the records disagreed, he would have rejected them. Till today we normally subject records to scientific examination and no one dreams of doubting the x-ray photograph of a mummified chimpanzee simply because the historical record declares it a mummy of a human prince. But Einstein had made us wary of making too much of that. All astronomical records prior to 1900, with one possible exception, agreed with Newtonian celestial mechanics, which Einstein put to question a decade or two later. And Einstein, we know, won and declared that agreement of observation with Newton's theory nothing less than the result of inaccuracy of observation. Yes, observation was as accurate as possible, and more than any other. Laplace was right to say that measurements of the lunar orbit were the best we ever had; yet they simply were not good enough; today we allow theory and fact to improve each other alternately. The question, we may remember, when should theory overrule observation and when not, is still open. A debate about it is still raging.

So much for the post-romantic views of science; as for the post-romantic views of art, but I have no time for them. What I can say now is that the chief philosophical foundation of the two cultures are both out. Most of Snow's observations have to be restated and reexamined. This will not, in itself, solve the problem. Too many of us specialize too much, and we should invent incentives against overspecialization and against overspecialized education. At heart, I fully sympathize with Snow – or should I say resonate. In particular, I fully endorse Snow's idea that we must look at our educational system and try to reform it before the split between the cultures goes too far. But his proposed reform is an utter failure, and for an obvious reason.

Snow's target of reform is the curriculum. We should not train our students in so specialized a way as we do: we should offer a broader common basis to all. Here Snow observes, Russia leads the world. I do not know whether this is true, and I have tried to examine it whenever I meet Russian immigrants in Israel. If my observations are true, then Snow is wider of the mark regarding Russia than regarding Britain and the U.S.A., and rests on nothing but his credulity and Russian official propaganda. But let this ride. Let us briefly glance at Western attempts at curriculum reforms.

The most famous reform was general education in colleges, whether by James Bryant Conant in Harvard or elsewhere. The pattern is the same everywhere: professional departments are happy to dump their casualties, students or faculty, on the division of general education, and center their efforts on the most promising, most professional individuals. And this is reasonable. A school is judged by its most successful members, by its most successful products. And such individuals are deemed professional regardless of how broad or how narrow their education seems to be, simply on the true assumption that for the peak of success in one field some sacrifice of interests in other fields is necessary. This makes fields of inquiry distinct and isolated and barren.

Can we change the incentive system which makes our education so specialized? Can we do so within schools or must we, again, abandon them? I say "again" alluding to the fact that universities, being a medieval guild, opposed the rise of modern science, which evolved, instead, in scientific societies divorced from the universities until the twentieth century. The modernization of the world hardly touched the guild system of the universities and of the academic professions such as medicine and the law. And now that more and more professions require university education, it is no surprise that Polanyi's theory of science as a medieval guild, as a tradition carried from master to apprentice, becomes a model and the paradigm.

What we need, I think, is the destruction of the medieval system of power. We cannot allow anyone to practice medicine, but we cannot trust a master medicine man to declare his apprentice to be a qualified medicine man. Abraham Flexner discovered half a century ago that this is the source of the inferior nature of American medical practices. He felt that if medical training be confined to universities and if university medical schools of low quality could be closed, progress could be achieved. And he was right for the time. But this was the closing down of small businesses because big businesses do not cheat in the same way – it led big businesses to cheat in different ways, and in particular to create guild quotas on entry.

I think a radical reform is needed. When one wants a permit to drive one may first benefit from a driving school, but one does not need to, except where the driving testers get kick-backs from the driving schools. Yet when one wants a permit to practice medicine or the law or school-teaching one is forced to enroll in a school. And schools give not only training but also enlightenment; and students must take all or nothing. And so enlightenment is controlled by professional examiners who thus become thought-police.

It is easy to set up examination boards which are democratically instituted by states or by public bodies: all it takes is legal reforms: any professional activity that requires qualifications also requires democratically instituted examination boards to decide who is qualified. Teachers and trainers can be represented on such boards, and even teaching or training institutions. But such institutions should not, as institutions, be the ones to grant qualification-certificates.

Once degrees will not be in any way qualification-granting, they will become sheer ornaments and so harmless even if not abolished. Students will go to college either to learn skills for which they will be examined elsewhere, or to acquire education and knowledge for their own personal consumption. Such a simple reform will clear the road to combat excessive specialization. Those who would still insist on it will not be forced to specialize; and others will be able to go one way or another. And where there is a will there may be a way, and where there may be a way there is hope.

CHAPTER 2

THE PRESENT STATE OF THE PHILOSOPHY OF SCIENCE

The reason I have chosen to present this survey is that I always find it difficult to present my own views on any specific topic, and I wish to explain this difficulty somewhat. My views are those of an apostate and one from the school of Sir Karl Popper, who was for long little known and less understood yet whose popularity is now rising somewhat. I find it hard to assume my audience to be sufficiently familiar with Popper's views to enable me to proceed to criticize him without much explanation and offer modifications to his ideas as I would wish to do. I shall soon present some gross misconceptions of Popper's views which, I feel, I have to clear beforehand. But there is a more serious obstacle than mere misconceptions. Those who are familiar with Popper's views are often defenders of the majority views; this is to be expected of course, yet it causes for me no small difficulty. I am always wary of being taken as a critic of Popper from the majority's viewpoint (from the right, as it were) whereas, in fact, if I at all enter the debate between Popper and the majority I am so much on Popper's side that my disagreement with him becomes negligible. This is, of course, no enviable position either. I confess I have myself found it amusing that, say, members of different religious camps unite against the threat of agnosticism; yet, this amusement is only just when aimed at some petty politics of a disagreement; otherwise it is perfectly reasonable that Catholic and Protestant should view differences between their views as secondary compared with what they share in opposing the agnostic. This is merely a matter of a sense of proportion; and so it would hold equally for the manner two agnostics may view any differences between themselves as compared with the differences both have with any true believer; quite naturally they will oppose him first before they start opposing each other.

I do not know how much all this will be taken for granted, how much it will be opposed. History offers us contrary examples, from religious controversies, and from political campaigns, and even from scientific and philosophical schools. There is the idea, shared by many left-wing leaders, and by Freud, which sounds perfectly practical and commonsense and reasonable. It is this. We cannot change the whole world at once. And so, at the very least we must, first and foremost, be exactly clear about what we want. Not because we are going to do things right now on a grand scale and so we must do them absolutely right, but (on the contrary) because we know our limited powers and so we must at the very least try our best to control what we can, namely our own selves, our immediate environment. And so we should keep the doctrine pure. It follows from this that the person whose opinions are closest to yours is your most dangerous opponent. In particular, the dissenter in your own midst is the bitterest enemy in the eye of the leadership of your own camp; and, of course, to the dissenter the leadership looks equally treacherous and more. I shall soon offer an example of this from the history of physics.

Those who vigilantly guard the doctrine against all impurity come to a very dead end: they cannot but repeat themselves; they incessantly teach the pure milk of the true doctrine – with or without small variations, usually not in the positive teaching but in the venomous criticism on their greatest and nearest opponents, the dissenter or the establishment as the case may be. At times they apply the old doctrine to a new case, and thereby achieve minor innovation. But on the whole they are crushing bores.

Most of us need food for thought; we want new ideas now and then. But some or most of those open to new ideas are not as open as they can be: let us consider, in particular, a doctrine so revolutionary and so important that one cannot even notice it without thereby committing a major act of heresy. I have in mind such diverse innovations as Faraday's field theory or Popper's theory of science. The historical case of Faraday is tragicomic: even recognizing Faraday's ideas as interesting was a heresy, since these ideas contradicted the well-established canons of Newtonian science. A man like Faraday could do nothing, then, except lecture and write almost exclusively about his own ideas. Let us consider the young people who went to his lectures, who were as profoundly impressed by him as one might expect, who had practically no physics in high school to counter-balance his ideas. These young people, quite naturally, had the most lop-sided view of physics. They grew up ignorant of the state of the science in other countries and in their own country a generation earlier. Faraday, almost single-handed, made such a profound change, that the great revolution in nineteenth century physics – the introduction of electromagnetic field theories – went unnoticed. This is well illustrated, I think, by the following two cases. My first case is James Clerk Maxwell, the inventor of the famous electromagnetic field equations which bear his name. He came to Cambridge, England, as a young man well versed with Continental studies on electricity but utterly unaware of even the existence of ideas by Faraday. His somewhat elder contemporary, William Thomson, later known as Lord Kelvin, told him to read Faraday. Thomson himself had heard about

Faraday from a cousin of Faraday who was a substitute professor of natural philosophy in Edinburgh. My second case, and barely a generation later, we find Silvanus P. Thompson, the founder of the Institute of Electrical Engineering in London. He gasped at the fact that he had rediscovered some ideas of Faraday which he lectured on in the Institute and no one recognized them as Faraday's.

I cannot blame Faraday for his having taught his own ideas to children so efficiently, because he was so ostracized by his peers. But I do think it is a bad situation when people lose signt of even their immediate ancestry, when they can no longer survey their own specialty and notice the trend they swim in. Just as the unexamined life is not worth living, so the unexamined current is not worth drifting in.

Sir Karl Popper used to teach his own doctrine with as much force and fervour and, I suppose, due to as much isolation. Possibly his ideas bear fruit and possibly his ideas are becoming public property without proper introduction – as Faraday's ideas were before. And I confess I find quite unpleasant this possibility, however viable or remote it may be.

1. FRANK AND SURREPTITIOUS CHANGE

There is one idea which I have learned from my schooling by Popper, or perhaps it is an observation he made which has impressed me so. The school which tends to keep vigilantly the purity of the doctrine, Popper observes, keeps changing its doctrines all the same. But it does so either by producing heretics and apostates, or by surreptitious changes, namely in an underhand manner – either by changes too small to be noticed, or by reinterpretation of its terminology, or by some other subtle means. Now surreptitious changes are merely lies of a sophisticated kind, and heretics are in fact those who break from the tradition. Various schools of thought undergo these days surreptitious changes, and I think it is better if we are on our guard, because a change effected underhand is not the best planned change and it is surely one which did not enjoy the full benefit of critical examination.

Let me offer a couple of concrete historical examples. But in all fairness I owe it to my examples to warn my reader that my presentation is biased and oversimplified. It cannot but be so, since my thesis is that the examples are unnecessarily complicated and subtle – unnecessary, that is, except for the purpose of surreptitious change. Of course, I present the examples since I think they are true, but I know that people whose ideas are allegedly represented in the next few paragraphs will claim that I distort their views – and honestly so, no doubt: I do not wish to call them liars but merely muddleheads.

My first and brief example is the influence of Freud. I do not know whether orthodox Freudians exist and how orthodox, to what degree orthodox a Freudian, say Erik Erikson, considers himself to be. There is, usually, little occasion for a mature Freudian, in practice or even in research, to compare his own views with those expressed in the standard works of the Master and see how well the two agree with each other. It requires a fresh reading of a complete work or two of Freud, and there is usually not much occasion for such an excercise. The excercise naturally presented itself in recent years with the publication of a new work of the Master – the book on Woodrow Wilson by Freud and Bullitt. It was a rather embarassing experience to judge by various responses. But Professor Erikson came to the rescue: he showed the work to consist of two separate compositions: a masterpiece by the late master and something else by the late diplomat. And, need one say, when the reading became less disturbing the question of the purity of the doctrine was laid to rest too.

My second example requires an elaborate introduction – of what is known as a whole philosophical school, the Vienna Circle. Along with its derivatives in England, the United States, and elsewhere it represents a singularly barren group of people who worked between the two World Wars, produced then almost nothing, yet nevertheless managed to impress the world of learning for quite a while. The background to their philosophy was all the credentials they had: logic, empirical science, clear thinking. They repeatedly said they sided with these. There are people who still admire that group because of their devotion to the idea that clarity equals rationality equals science. That was all: naive but clear. Traditionally, rationality was identified with science. Traditionally, science was identified with empirical examination – though another tradition of science also existed. It was, therefore, quite natural to identify the rational with the scientific with the empirically verified. Clarity came into the picture with the quaint idea that verification is what gives a proposition its meaning. Hence, the unvoiced corollary goes, either a proposition is perfectly clear or not at all. The meaning of a proposition, that is, may be crystal clear if it is verifiable, and non-existent otherwise. So far so good; but the story goes further. The pinnacle of the Vienna doctrine was put in one slogan: the meaning of a proposition, said the pundits of the Vienna Circle, is its method of verification.

I have the habit of confessing at once my inability to comprehend what I say when I say something I do not comprehend. So allow me this interruption.

I repeat the slogan of the pundits of the Vienna Circle: the meaning of a proposition is its method of verification. But I do not claim to comprehend it and indeed it is my considered opinion that this is a meaningless string of words, a pseudo-proposition so-called, or, in plainer English, stuff-and-nonsense. (This, indeed, was the verdict Friedrich Waismann, one of the leading members of that group throughout its period of existence, gave years after its demise.) Of course, no one quite knows if the slogan, 'the meaning of a proposition is its method of verification', is meaningful or meaningless because we do not as yet have a sufficiently elaborate theory of meaning. We do have a theory or theories of meaning of names and of descriptive phrases. We do have rudiments of a theory of the meaning of propositions – but not enough. The Vienna Circle were quick to judge any idea meaningless when not backed by a fully fledged theory, and so perhaps by their own precepts their slogan is meaningless. But these precepts are erroneous and hence, quite possibly, their slogan, the meaning of a proposition is its method of verification, is meaningful, I do not know.

Popper thinks the Vienna Circle's slogan is meaningful, that is to say, it is either true or false; he thinks it is false. He says, first, we cannot devise a method of testing a proposition before we fully comprehend it. Moreover, says Popper, whatever proposition we can test is scientific, but some propositions are not testable yet quite meaningful. Having meaning, he goes on, is not the same as being scientific. Furthermore, he argues, empirical tests only refute scientific generalisations, they never verify them.

Somehow, before the War his Viennese colleagues staunchly ignored his idea that science is not the totality of meaningful propositions. Rather they were impressed with his idea that science is not verifiable but refutable. The result of this was a compromise, the bastard idea that though being scientific is the same as having meaning, this is not the same as being empirically verifiable but the same as empirically refutable. This bastard idea fell between two stools; it was rejected by both the Vienna Circle and by Popper. It was erroneously attributed to Popper by various authors at one time or another: The locus classicus of this attribution is Carnap's well-known *Testability and Meaning* of 1936–7.

This bastard idea is easy to refute and it was satisfactorily refuted by various authors. The refutation goes like this. Whatever makes a proposition meaningful, one thing we are sure about the meaning of a proposition: the negation of a meaningful proposition has the opposite truth-value and hence has a truth-value and is hence meaningful. Likewise, the negation of a meaningless string of words (or pseudo-proposition) is itself meaningless. Now consider a universal proposition, say (*U*) 'all men are black'; consider a basic statement, namely a possible candidate for an observational report, such as, say (*B*) 'in Philadelphia in January 1970, there was a non-black man'; consider also the negation of the proposition (*U*) 'all men are black' i.e. (*E*) 'there exists a non-black man'. Clearly, (*B*) refutes (*U*) and (*B*) verifies (*E*). Clearly (*E*) is not refutable since the universe is possibly infinite and so we cannot possibly scan it to disprove a purely existential statement. But (*U*) is refutable since it contradicts (*B*). So (*U*) is meaningful and (*E*) is meaningless by the bastard version of Popper's theory. Since (*E*) is the negation of (*U*) it is meaningful and since it is irrefutable it is meaningless – which is absurd. And we have logically refuted the view which I have called bastard and which the Vienna Circle have erroneously ascribed to Popper. When he said scientific character equals refutability they heard him say meaning equals refutability, and they refuted what they heard.

It is a strange fact, but I recommend that those interested check it to their own satisfaction. In the fifties, series of papers appeared in England about natural theology, in *Analysis*, in a volume of essays edited by Anthony Flew on linguistic analysis and natural theology. Both believers and unbelievers among the contributors shared the refuted bastard criterion of meaning and their work, naturally, is worse than nothing. The same criterion was also employed in examining such doctrines as Marxism and other versions of the doctrine of historical inevitability, and Freudianism and offshoots of it or variants of psychoanalysis. The bastard version is sometimes used more as a criterion for a theory being scientific rather than meaningful, that is to say, there are variants of the bastard version of varying degrees of bastardness, created as if especially to illustrate my view of the folly of surreptitious change in general and of the surreptitious and unhealthy increase of Popper's influence in particular: It is doubtless the least valuable part of Popper's theory which thus far has been gaining currency, and in a distorted form to boot!

If my extended example is not too much of a nuisance, I would like to continue it a little. Of late there is a further surreptitious change to the story, and it is even a little improvement in a way – but not a big one in any way. It is this: what has happened is simply that certain philosophers have rejected Popper's characterization of science as it clashes with their intuition that both a statement and its negation must be together scientific or together unscientific. It is a bit sad that this idea is presented seriously as a reasonable one, without any further inquiry about the force of such intuitions and in the face of the fact that from the beginning of philosophy till the rise of the

Vienna Circle it was taken for granted that the negation of a scientific (i.e. proven) proposition is unscientific. It is merely the idea that meaning equals scientific character which has changed these people's intuition: take away this idea and what is left is a dangling intuition.

Again we see, I hope, how important it may be to know what is the current we are drifting in if we wish to know ourselves. And so, I am coming back to my reasons for offering a survey here. So now let me begin my survey. Let me end this long preamble by reminding my reader that all surveys are biassed. The only way to remedy this is not to search for an unbiassed survey, since all humans are suspect of bias, but to create diverse biasses in the hope that biasses may cancel each other to some extent. Well, then; here we go.

2. THE VIEW EQUATING SCIENCE AND RATIONALITY

The label philosophy of science was invented by members of the Vienna Circle between the two World Wars. It comes to replace two entirely different ones, epistemology and scientific philosophy. Epistemology, or its modern variants, theory of knowledge, or *Erkenntnislehre*, deals with epistēmē, which is the opposite of doxa, namely with scientia, or knowledge, as opposed to opinion. The word epistēmē is a technical term presumably invented by Parmenides to mean knowledge in the strict sense i.e. in the sense of fully demonstrable knowledge, i.e., knowledge that is here to stay, for all times, unalterable and unshakeable. (Parmenides spoke of logon piston, i.e. of theorem.) Little reflection will show that in the ordinary use of the word knowledge may be alterable; indeed even the words proof and demonstration, and their equivalents, do not necessarily mean in ordinary discourse, once and for all times. This is a point noticed by Gomperz and by hoards of commentators since. Ordinary language philosophers who sanctify ordinary usage, go even so far as to condemn Parmenides and his followers; for my part I see no fault in Parmenides' introduction of a new use, since he – or is it Plato? – defined it impeccably. It is not the question, how we used the word "knowledge", which matters; it is the question, how alterable do we think our views are, which matters. To take an example, Sir John Herschel, Dr. William Whewell, and other spokesmen for science and for scientific philosophy in the nineteenth century, all declared sharply that Newtonian mechanics, being scientific proper or demonstrated, is not amenable to any shaking. Mechanics will never suffer the slightest modification, they said most clearly. To them, the idea that Newtonian mechanics may be an approximation to another theory was a shocking thought. One may say, for instance, that the force of gravity does not quite act at a distance, though it does indeed travel so fast that it may be so viewed when calculating mechanical predictions. This indeed happens to be the current twentieth century view; but to nineteenth century physicists this was so unthinkable, that when Faraday said it they simply plugged their ears. Herschel and Whewell thought of Faraday as of a dear friend and as a highly esteemed colleague; yet they refused to entertain his views even tentatively. Indeed, because he was so close his views were deemed ever more dangerous; too dangerous even to muse about.

All this holds not only for physics but for any theory claiming scientific status or demonstrability, even a metaphysical theory. Thus, when Solomon Maimon accepted Kant's philosophy as true but declared its status to be that of a hypothesis, he incurred Kant's displeasure. Kant had declared – and he always remained of the staunch view – that his philosophy was scientific and so it will forever remain unchanged. He later declared that Maimon was only an intellectual parasite, as Jews so often are. I must say, anyone who could get an antisemtic remark out of the paragon of Enlightenment that Kant was, must indeed have got under his skin. Perhaps Kant was not that sure after all. But he felt he had to be sure, since he wanted to be an author of a scientific philosophy.

The phrase scientific philosophy is a synonym for rational philosophy, and was chosen for two reasons. First, to intimate the doctrine identifying rationality with scientific character, and second, because the word rational or rationalist had, traditionally, two distinct meanings. One meaning is that which is exhibited in the contrast between rationalism and irrationalism, namely rationalism as the view that man can and ought to use his reason or intellect to determine his beliefs, guide his actions, etc. The other meaning is that which is exhibited in the contrast within the rationalist school between rationalist and empiricist sub-schools, namely rationalism as the view that the grounds of reason are in the intellect itself rather than in the senses. Immanuel Kant suggested that we substitute the word intellectualism for rationalism in this narrow sense of a sub-school. But his idea did not take. And so, when we want to speak of rationalist philosophy in the broad sense encompassing both Descartes and Locke, to the exclusion of the narrow sense encompassing Descartes but not Locke, it may be preferable to use the word scientific philosophy.

Now the label "philosophy of science", as in the title of Ampère's *Essai sur la philosophie des sciences* of 1834, designates faith in the rationality of philosophy. Yet, strictly speaking, obviously an unscientific philosophy or an irrationalist philosophy may indeed include a view of science, which may be called a philosophy of science. It is most regrettable that we regularly forget the philosophy of science of people like Croce and Gentile, like Sartre and Heidegger. These thinkers do have views about science which they expound in their books which are on the reading lists in courses on phenomenology and such. Yet their views about science are hardly complimentary. And so, courses in the philosophy of science only refer to scientific philosophies of science. And so, very regrettably, they have their share in the increased gulf of non-communication and lack of understanding between the neo-Hegelians, phenomenologists, and existentialists on the one hand and the positivists, pragmatists, etc. on the other.

The philosophy of most of the writers on the one hand is almost uniformly an instrumentalist or a pragmatist philosophy. This, of course, should immediately puzzle you, as I put the pragmatists as on the other hand and spoke of the gulf between the writers on the one hand and the pragmatists. Here, again, we see the strange results of having no easily available survey of the field.

The pragmatist philosophy of the phenomenologists and the others on the one hand says, we need not study science since it is deprived of all truth, of all intellectual value: it is purely of pragmatic value. Its truths are merely pragmatic truths, or its truths are only true when judged by some transient standards of truth, such as what is useful here and now. But not by standards of eternal truth, since these are reserved for metaphysics alone. Metaphysics alone, then, is verifiable (by the authority of our intuition, incidentally) and hence only metaphysics is a true science or truly scientific.

What will a pragmatist say to all this, for example James and Peirce? They cannot disagree with the part of the doctrine just expounded as far as science is concerned; but they can disagree on metaphysics, and in two ways. They can either declare the same pragmatic standard of truth to apply to metaphysics; or they can declare that there is no such thing as metaphysics. As far as I understand James and Peirce, James accepted the first alternative, Peirce the second. For James even religious or theological truths are useful; for Peirce they are meaningless or else a part of our survival mechanism and hence a part of our science. In both cases of pragmatism no eternal truths are allowed.

The philosophers on the one hand, whom I consider irrationalists, are contemptuous of science. Their reason is not so much that it is merely an instrument of survival – they do like survival. But they say that they can also offer eternal truths, that their metaphysical doctrines greatly outshine science. The pragmatists offer nothing of the kind, and thus avoid any irrational practices. They avoid all claim for finality – in science or elsewhere – and thus all dogmatism.

This raises again the question, what has happened to verification in science so firmly upheld in the nineteenth century? The answer, in brief, is, it was killed by Einstein in 1919, when he was acknowledged publicly as the one who had successfully modified Newton's views. True, the positivists either had not heard of Einstein or had not understood him; on a rare point they made a valiant last ditch effort. The great locus classicus for the twentieth century verification principle is Ludwig Wittgenstein's *Tractatus Logico-Philosophicus* of 1922. The word "verification principle", I must hasten to explain, belongs to Waismann – the one who brought Wittgenstein to the Vienna Circle; furthermore, the formulation I quoted above belongs to Schlick – the man who created and headed the Circle until he was murdered. Still, I do think the verification principle is Wittgenstein's; he used Newtonian mechanics as an example of verified theory. And this in 1922! His disciples of the Vienna Circle went on talking of verification well into the thirties. Ignoring a few other writers who failed to be influenced by Einstein, let me mention P. W. Bridgman, the Nobel laureate physicist, who in 1927, made quite a pathetic declaration. The very fact that the Einstein revolution could ever occur, he said, shows that not all had been in order in the house of science. He called for more vigilance in keeping the purity of the language of science by the device of using in science only those terms which can be operationally defined. This, said Bridgman, will assure that from now on all will be in order in science. One must define length as a certain operation of measurement and time as a different operation of measurement (of behavior of clocks), and heat as a still different operation, etc. As is well known, the concept of simultaneity was allegedly defined by Einstein in 1905, but not by any of his predecessors. Now his definition is not entirely operational as it refers to inertial systems; moreover, even the narrow constraint on Einstein's terminology as he accepted in 1905 impeded his development of his general theory of relativity so that by 1916 he gave it up. Later on Bridgman too gave up operationism – unfortunately by a surreptitious change – by adding to the operations of measurements those of pencil and paper, i.e. of thinking. Surely this will not guarantee that no revolutions may occur in science in future.

The classical theory of empirical verification crumbled – all evidence from Vienna to the contrary not withstanding. Two schools of thought are now extant. The one sticks to the empirical and gives up verification in theoretical science and the other sticks to verification and gives up the empirical nature of theoretical science. These are inductivism and conventionalism respectively.

It so happens that in those fields of science which lean more towards theory than experience – e.g. theoretical physics – people are inclined towards conventionalism and those which lean more towards experiment – e.g. zoology and botany – show preference for inductivism; molecular biologists, for example, vacillate.

I find both schools extremely narrow and I only marvel at the fact that when the verification principle crumbled, its chief victim, rational metaphysics which merely claims the status of hypothesis, was not revived. So my battle cry is, back to Solomon Maimon. But I must not be carried away; I should proceed with a survey of what is, not of what I wish to be.

3. THE SCHOOLS IN PHILOSOPHY OF SCIENCE

For those who have the preconceived notion that surveys start when the field is divided into schools and subschools which are then properly characterized – for them, then, the survey begins right now. The field of the philosophy of science is divided into the dwindling minority who study empirical meaning and whom I shall ignore from now on and the majority who study the nature of empirical science. Of these, a majority studies the nature of empirical confirmation and belongs to the inductivist school; the rest are conventionalists who try to formalize systems or to discuss the division of scientific theory into empirical content and mathematical frameworks.

If I were to recommend a reading list for those who wish to become students of the field, let me say outright, I would recommend classical works – inductivist, Francis Bacon, William Whewell – or conventionalist, Poincaré, Duhem – I will not trouble you with contemporary works in either field. If I were pressed for more modern stuff I would recommend Meyerson, Polanyi, and Popper. Let me elaborate on this paragraph and draw this survey to its conclusion.

The main fact to observe about both contemporary conventionalists and contemporary inductivists – particularly the inductivists – is that they are repeatedly prone to fall into the pitfall of old verificationism. They sometimes express an explicit verificationist doctrine, sometimes they only imply it. The reason is complex. First, consistency is generally difficult to maintain. Second, old pitfalls are generally hard to avoid; for example, former Pagans are unable to avoid importing some of their Paganism to Christianity after they adopt it. The third and more significant reason is that contemporary views – both conventionalist and inductivist – are surrogate verificationism: They come to answer the same questions which the old verificationism came to answer, and when they fail to answer these questions their advocates tend to return for a while to the good old theory.

Conventionalism employs two standards of truth, absolute and pragmatist. Conventionalists recognize two facts which look identical but are not. First, that certainty of theory cannot rest on empirical fact; and second, that science never has the last word. They ascribe to scientific theory the status of certainty – and thus of immutability – but only as part and parcel of mathematics. Physics is then viewed only as a branch of applied mathematics; physics is then a system of pigeon holes to classify facts more or less neatly. Now the neatness with which theory stores facts is not a matter of mathematics and is not unalterable, since the stock of empirically known facts is alterable with the growth of science. So the neatness is judged by the alterable standard of pragmatic truth.

All this is highly sophisticated but quite unnecessarily so. Moreover, it is unsatisfactory as it takes no account of the fact that a scientific theory is not only a pigeon-hole system – it should also be confirmed by the facts. This criticism, incidentally, is not one which I endorse but which inductivist philosophers of science launch against conventionalism. More precisely, inductivists prefer, when they can, to avoid discussing conventionalism – partly because so many theoretical scientists are conventionalists.

The inductivists operate with one concept of truth – the absolute. And they characterize the latest in science not as demonstrably true but as probably true. (The strange exception is Hans Reichenbach who suggested once that probability may be viewed as a truth-value in an infinitely-many-valued logic; it did not work.) The replacement of verification by probability is neither new nor very interesting. The idea that the latest ideas accepted in science are probable, however, is a step in the right direction, to be sure; since the latest in science is possibly true.

There are two different aspects to the theory of confirmation – qualitative and quantitative – and both can be treated loosely or precisely. Qualitatively, we relate prediction to explanation, for example. Indeed, the theory of explanation has the lion's share of the literature in the field – almost always with relation to confirmation. And quantitatively, confirmation may be a probability measure, i.e. follow the axioms of the theory of games of chance, or it may not. All this is subject to much discussion. Also, the theory of chance tells us what is the probability of one event given another, but if the other is not quite given we do not know what to do about it. Thus, if hypotheses are probable, given some evidence, the witness must be reliable and honest for sure. No witness is. And so the whole discussion is vague – the more precise it seems the more it is a waste of time to examine it carefully. The most important works in the field of confirmation are qualitatively of Hempel and quantitatively of Carnap – both members of the Vienna Circle before its demise. Hempel showed how paradoxical the qualitative theory is and Carnap did the same with the quantitative theory, yet whereas they had relinquished verificationism, they stuck to their newer views, to non-verificationist inductivism. Carnap promised a second volume to his magnum opus on probability but gave it up in the preface to his second edition. Nevertheless he went on working to his dying day with hopes to solve the problem of induction by a strong and mathematical theory of confirmation of one sort or another, where confirmation is no longer assumed to follow the rules of the theory of games of chance yet relate to them in some way or another.

The various conventionalists and inductivists labour to one end – they wish to justify science, to answer the skeptic. The skeptic is not necessarily one who denies that, say, Einstein is correct: the skeptic is willing to endorse Einstein's theory but he insists that the theory has the status of a hypothesis. Just as Solomon Maimon did with respect to Newton and to Kant. But skepticism is still the target of the majority schools. Michael Polanyi shares with these schools their concern, but views their efforts futile. He thinks the status of science is definitely higher than that of a hypothesis: once one has endorsed science one has thereby endorsed the present body of scientific opinion. And though scientific opinion is alterable, its acceptance *pro tem* is imperative. On what does Polanyi base his claim that scientific opinion is obligatory? And who does he think decides what is obligatory? This is very interesting, and I shall have to pass it by very quickly. The elders of science, says Polanyi, decide what is current scientific opinion. They feel it in their bones or in their fingertips. He illustrates all this, but he cannot defend it, since his view is that you cannot defend science by any given rationale.

And so, to use our terminology, Polanyi has a philosophy of science – indeed he wrote chiefly about science – but it is not a scientific philosophy; it rests on an intuition of the leadership and on his declaration that the leadership has authority. Thomas S. Kuhn has much enlarged on this idea of Polanyi, coupled with a Duhemian theory of pigeon-holing, etc. I must leave all this now.

The classical views of science, verificationism or any other, upheld the authority of science, but under the heading of rationality. Reason, they said, lends authority to science. When the great 20th century developments in science took place, the old views were gone. Almost all those active in the field still have the same aspirations; they still hope to reestablish the authority of science, to justify science afresh. Polanyi upholds the authority of science, without believing that reason can justify it.

Popper claims to be the exact opposite; he claims to stick to the idea that science is rational; but he has no desire to uphold any authority of science, to justify science in any manner whatsoever. And so, to return to my opening paragraph, his agnosticism and mine make us practically close allies when it comes to debates with any believer in the authority of science. But otherwise I have little to agree with him. Let me conclude by telling you what I consider his great point, which is so important that it makes any philosopher who overlooks it quite antediluvian. I do not mean that one cannot disagree with Popper on this point, but that one cannot ignore him, that he has altered our way of thinking about, or looking at, one question quite beyond reversal. The question is, how do we learn from experience? It is the one traditionally known as the problem of induction.

The greatness of Popper's view does not depend on whether one endorses it or not. He changed the problem of induction irreversibly. When classical thinkers from Hume onward asked, how do we learn from experience, they meant, how does experience guide our choice of one hypothesis rather than another? Popper said, learning from experience is learning to reject a given hypothesis. This puts into question the implicit opinion that learning is choice. If learning is not choice, it may be the increase of the field of choice, or the decrease of it. The decrease may be done by empirical refutation; perhaps in a sense the increase too.

Whether true or false, this idea is so intriguing that one cannot overlook it except at the risk of being left behind. When the majority takes such a risk, it only means that the majority may be left behind. We have historical examples, such as the electricians who were Faraday's contemporaries yet chose to ignore his ideas. I submit that there is too much parallel between nineteenth century electromagnetism and twentieth century philosophy of science to leave one complacent.

This is not to express agreement with Popper. Having opened on a note of disagreement, I feel I should say something about it now, especially since my disagreement stems from the ideas of Émile Meyerson, the modern philosopher of science on my 'must' list. Meyerson considered Science as guided by metaphysical ideas of a kind akin to Kant's regulative ideas. At times he almost suggested that science is a handmaid to metaphysics, an elaboration on it. Meyerson's ideas were taken up by historians of science, particularly A. Koyré and I. B. Cohen, perhaps also E. A. Burtt. I find Popper's theory in part able to accommodate for Meyerson's view, in part opposed to it. For example, Popper views the refutability of a hypothesis as a necessary and sufficient reason for our being interested in it, also for our taking it as a realistic – true or false – view of things. Contrary to this, we do not take the continuum theory seriously as physics, for example, merely because it opposes our atomistic metaphysics. Many testable hypotheses are of mere technological interest and are approached instrumentalistically, not realistically. Other hypotheses, though for long barely testable, are taken very seriously and attempts to render them a little more empirical are made. It is no accident that most of the luminaries in the Popper galaxy are now concerned with research programs and their metaphysical background. The problem is, how damaging is all this for Popper's original philosophy? In my own opinion this is an interesting question which is now on the agenda.

CHAPTER 3

WAS WITTGENSTEIN REALLY NECESSARY?

1. THE NEW PSEUDO-RATIONALISM

As a philosopher, I seem to qualify: I am a target for Wittgenstein's onslaught against philosophers. I wonder what am I to do about it. Wittgenstein and his followers claim that metaphysics – with which I happen to be preoccupied – is a violation of grammar and of simple rules of clear thinking. I may accept the charge and be silent about any philosophical issue. I may, on the contrary, ignore their charge and continue with my work on metaphysics, which I find very interesting. Why, at all, should I pay any attention to those who claim that my activity is pointless?

Perhaps I should listen to them because they are right. Doubtless, I should benefit if what I do up to now is pointless and if I now cease doing it and do, instead, something valuable. So I read Wittgenstein and listened to his crowd. I met disciples and admirers of Wittgenstein, important people who knew him personally and made names for themselves, less important people, and ordinary philosophy students. I invested more energy in the study of Wittgenstein and his philosophic schools than in any other, except perhaps Popper and Polanyi. I found my effort so unrewarding that I have to explain to myself what kept my interest going while it was going. In part it was a requirement of my work. I had to know what Wittgenstein and diverse Wittgensteinians said in order to do my work while I was a student: As Polanyi observes, you must be familiar with current orthodoxy. As Popper observes, you need not endorse it. I also mixed with English philosophers in England in the fifties, and there and then Wittgenstein was the paradigm of Philosophy. And so we naturally talked about our work; I can hardly report adequately the degree of astonishment that my interlocutors were filled with, when they realized that I sincerely disagreed with the Wittgensteinian principles which they took for granted while they did their work. And they took for granted that anything worthwhile said by a non-Wittgensteinian at the time or earlier is, in reality, only an analysis and so in full accord with Wittgenstein's dictum that philosophers have nothing to say; that meaning equals use; that ordinary use as well as scientific technical use are binding; that there is no other use or usage; that hence philosophy is already dead.

What do you do when you meet someone who takes for granted something very contrary to what you take for granted? You ask. You request. For, he takes something for granted, meaning he feels no need to air it. But you, in the desire to establish contact with him, may request that he tell you. Of course, you may be lucky and he may want you to tell him your story, and then neither need request, but rather you exchange information. But, as if I need say, I was never that lucky. Not with disciples of Wittgenstein, that is. Or, to be precise, once or twice something like an exchange of information with a Wittgensteinian happened to me, but the person who was starting to listen to me soon ceased and developed a great distaste for my philosophy or for me as a philosopher. One who has not experienced this can hardly appreciate the force of Polanyi's claim that current opinion is imposed on potential dissenters, even in science, by the force of a threat of isolation. For, I think it is commonplace that, when people of different backgrounds meet, some transfer of information must occur in order to facilitate communication; and this transfer may go two ways or one way; and this depends on consent or on force. What I found extraordinary in my own case with the current philosophical opinion was my inability to appreciate anything to do with it – with the two schools of Wittgenstein – except some very marginal points. In particular, I could not see why my activity is pointless and theirs not.

Philosophy is an activity, said Wittgenstein, not a doctrine. But how can we find what activity is worthwhile, what not, unless we have a doctrine? Do we have a doctrine about intellectual activity in general, applicable to the activity called philosophy? I could not get a clear yes-or-no either from Wittgenstein's works, or from other books. I asked people and found that I thereby annoyed them. Do we have a description of these activities? No. We do have paradigms for them but these change fast. How did one keep contact with things? How did one know what was the new paradigm? Once John Austin came out of his Oxford den and read a paper in London. Great excitement. Some said here is the new way of doing philosophy, others reported that even in Oxford itself hostility to Austin was not small. No one is immune to the possibility of demotion. The question was asked and debated: how prominent was Wittgenstein himself in English philosophy or even in his own school?

In his essay on the social function of the absurd Ernest Gellner explains that giving up one's ability to think for one's own self is the supreme expression of loyalty, and some exclusive clubs will not accept less – and to this end they demand that their members believe in absurds, just like in the final scenes of George Orwell's famous novel *1984*. Gellner has published a detailed and devastating study of the philosophy then in fashion in England and offered a socioanalysis of it, namely of the social role philosophers expect to play in contemporary British society. For my part, I do not wish to pretend that I can have any rational exchange with them either: they consider my philosophy pointless, and I return the compliment. Now, not debating with them and not offering a social analysis of their views, what else can I do with them? I suppose I can offer an analysis of the role of positivism in intellectual society. Whereas Gellner studies the different role of this philosophy in society at large, I confine myself to its role in intellectual society, including scientific society, thus having to examine more closely its intellectual defects.

But let me offer the simplest analysis first. As I have mentioned, the Wittgensteinian movement is a rather exclusive club: it is not easy to enter. The criteria of entrance are not clear, except that they include the acceptance of a positivist doctrine or two, and the readiness to follow the fashion of how to do philosophy. Now, perhaps I felt rejected by them. And so in a pure retaliatory mood I shall describe them as intellectual rejects, as rejects of the scientific and the literary cultures – the two Wittgensteinian schools as rejects of the supposed two cultures!

In a more serious vein, I wish to present Wittgensteinianism as intellectually marginal. And I shall begin by illustrating what I mean by intellectual marginality in discussing the stature of traditional theology in the contemporary intellectual world. To avoid all surprise, I shall offer my conclusion first: my analogy between all sorts of Jewish, Christian, or Moslem, ancient, medieval and modern theology, and all sorts of positivism, classical and, as modern, Wittgensteinian and Polanyite, is almost complete: theology and positivism are both versions of irrationalism, and of a very special kind, namely, pseudo-rationalism. I do reject the Popperian view of pseudo-rationalists as a fifth column in the rationalist camp; I view them as rather the fringe of that camp, as the ones who did not quite make it, as the rabble and the mixed multitude of the intelligentsia, yet at times they may win a central position, theologians and positivists alike. For some thinkers pseudo-rationalism, in the guise of theology or of positivism, is really necessary; when they are too unreligious for theology, they can accept positivism as seemingly even more rational a stance than theology, and when they are tired of positivism they go back to religion. The best of them share both religion and positivism, and leading among these is Michael Polanyi. In brief, my objection to both theology and positivism is moral: rationality entails free decisions, yet they present rationality as a *post hoc* justification of a miserable state of affairs which is a perferred state for no better reason than that it has already been chosen.

Thus, the young Wittgenstein opts for science, the old Wittgenstein for commonsense, and Polanyi for science, religion and commonsense.

The crux of theological issues, as most theologians admit, is the very legitimacy of theology, the question, what place can it occupy? Not only positivists declare in their anti-metaphysical fervor that there is no place for theology; even some writers who have engaged in metaphysical discourses to great extents, even when investigating the nature of God with quite some passion, are frankly too hostile towards theology for a compromise. Will any decent philosopher declare Descartes, Spinoza, Leibniz and Kant theologians? Clearly, the answer is an empathic no. The clearest expressions of anti-theology are two, from intellectual freedom and from moral freedom. And they are closely connected. Moreover, the better theologians, or perhaps, since they are the better they are sometimes called religious philosophers, for example, Jacques Maritain, fully recognized both arguments and their interconnectedness. The one is an argument against traditional intellectual authority, the other is an argument against traditional moral authority. The one argument is universalistic, the other is individualistic. The one argument rebels against the arbitrariness, the accidentality, of one's tradition as one finds it, of one's birth into the one tradition rather than the other. The other argument declares the individual, rather than society – any society – to be the master of his decisions and his fate.

Examples abound. Descartes says, whatever I happen to think for no better reason than that I have been born a Frenchman rather than a Chinaman, I do not want to think. This at once leads one to wonder, what happens to the Frenchman's religion as opposed to the Chinaman's? But Descartes is reticent, as most seventeenth century philosophers were. Dr. Isaac Watts, the eighteenth century philosopher, educationist, publicist, poet, once very famous but now forgotten, held a different position. He was religious and even wrote hymns. (His hymns are all forgotten even by hymn-book compilers except two or three of them.) He said, how regrettable it is that so many people are Christian for no better reason than that they were born into this faith rather than into any other; and he thought he could rationally defend his peculiar religion. Now I do not wish to argue the merit or defect of Watts' position. I merely use it as an example. In principle it shows the universalist view that arguments have their validity regardless of the bonds of tradition. A little later Dr. Joseph Priestley, the Unitarian priest, better known for his discovery of oxygen, offered to argue amicably with the Jews about matters religious; his concern was not the Jews, nor whether he could convert them; his concern was to examine his own claim that his religion was universalist, not based on the force of a tradition. So much for the rational side of theology, which was a mere episode in the history of theology.

The moral aspect of theology has its locus classicus in Immanuel Kant's *Religion Within The Limits of Reason Alone*. I am responsible for my religious decisions, says Kant, and on principle. For, should God Himself reveal Himself to me and tell me whatever he tells me, it would still remain my decision to consider whether God is good or evil, and thus whether obeying His command be right or wrong. This passage and a more dramatic one by John Stuart Mill, received a great poignancy in the post World War II era, when the excuse, I was only – only! – obeying commands, was used by every condemned Nazi imaginable. Existentialist philosophy denounced this argument. A new kind of theology developed, with which Bonhöffer's name is most closely associated. a theology which paradoxically enough deprives God not only of His power to order us around, but perhaps even of His mere existence. This anti-theology is no more theology than the anti-Christ is Christ. Take from theology the claim for God's authority over our reason, and it is theology no longer. Søren Kierkegaard, who admired Abraham, the Knight of Faith, because he violates his own moral convictions at God's command to sacrifice his son, Kierkegaard is more of a theologian than Bonhöffer.

I take for my witnesses the leading Christian theologians, such as Karl Barth, Paul Tillich, and Jacques Maritain, and indeed no less a figure than the doyen of Jewish theologians, Moses Maimonides, to declare that reason is limited, that men must lean on some authority, that the meaning of life cannot be found in reason or by reason and must therefore by supplemented from the outside. It is, of course, most rational to supplement reason if reason is so limited and in need of a supplement.

2. THE TECHNIQUES OF PSEUDO-RATIONALISM

Religion and science were at odds ever since Greek and Jew met. The conflict led some people to oppose rationalism, science, and even technology. It made others welcome technology and declare science to be nothing but technology, or even rational technology, but not cosmology, and not a rationalist-universalist way of life. The universality of science was viewed by these not as the immediate corollary of any universalism, but simply the result of the fact that nature is the same everywhere. Already in late antiquity a shift was effected. The sentiment of early antiquity was best expressed by Protagoras, who observed that the laws of nature are the same everywhere but not the laws of man. He meant to slight human law and deprive it of its power; in the

Hellenistic world this same observation was twisted around: it was taken to be a matter-of-fact, not a dismissal of all tradition. This was an expression of pluralism and so, also, of my own tradition and its laws, of my own life-form, to use Wittgenstein's expression. And, indeed, the greatest complaint against Jews, that they were particularists, soon gave way to the greatest complaint against Christians, that they were disturbing the peace among the Gods. And under this pressure the particularist claims for superiority, Jewish and more so Christian and Moslem – we shall see the same in positivism soon – did require a rational excuse, i.e. a universalist excuse, a rationale. It could not be done. It could only be faked, as in the last resort the particularists had to take resort to some particular claim for some privileged access to the truth.

Here I wish to offer, then, a view of the hallmark of pseudo-rationality: it fits well Popper's theory of conjectures and refutations, and it fits well Popper's theory of reinforced dogmatism. Hence, incidentally, Popper is in error; but I deal with this error of his elsewhere.

The history of the arguments for the existence of God and the history of the arguments for the meaninglessness of all propositions outside science and logic, are both instances of series of conjectures and refutations. The history of theology runs over centuries, and that of modern positivism over mere decades. Propounders of a theological thesis, say, the Absolute is Perfect, offer some arguments in favor of the thesis, which have some sense in them, and so are not deprived of all initial plausibility. And they invite any old opponent to poke holes into their arguments. And they do so in good faith, or at least did so in the past. Indeed, when one of their arguments was refuted they gave it up and the call went out for a newer and better argument, for a newer and better theory, for a newer and better theodicy, to replace whatever criticism had demolished. Moreover, this was repeatedly reflected in their teaching, since regularly the flock had to learn the new arguments, and the new scholarly techniques. These practices, for all I know, still go on. At times Church leadership welcomes quite opposing philosophies on no better ground than that each of them supports the faith; when questioned, they display complete and quite candid open-mindness: let the better theory win, they say.

What they are for is the faith. What arguments and theories and other rational material they amass, simply serves as a buffer between the doubting faithful and the enemy at the gate. The rational material is the reinforcement of the faith. Though this process looks more like tirelessly pouring sand into a sieve to keep things going, rather than like sitting pretty behind reinforced concrete, it still is, technically, the same technique as the one described by reference to reinforced concrete as reinforced dogmatism. For, such activity can only support the faith, not undermine it.

And so we see the function of theology. It serves those who waver, those for whom autonomy and freedom of thought are attractive enough to cause them a worry, but not more; the one who wavers is neither faithful enough nor faithless enough; his expression, his intellectual expression of his wavering, is his very theological activity. There is a huge and ever expanding corpus of theological theory and argument, and the argument is, indeed, to and fro, and so within its limits perfectly rational; but the staying within its limits is not in the least rational: it is dogmatic. Since the rational activity of theological debate helps keep the faithful irrationally within the folds, its being limited to that end makes it irrational and thus pseudo-rational. And so, even though theological doctrine does exist, significantly enough theology is not a doctrine but an activity. This, of course, is a modification and a paraphrase of Wittgenstein's statement, one which he kept while changing from the one school to the other, namely, philosophy is not a doctrine, but an activity.

Thus, pseudo-rationality is perfect rationality but within limits; dogmatically imposed limits. It serves those who are in conflict about rationality in that it allows them to have their cake and eat it too.

The question is, what is the faith? The pseudo-rationality of the religious is clear enough; he believes in the religion of his fathers, be it Judaism or Islam, be it Marxism, be it the Israeli or the Iranian mixture of both. But what is the faith of the Wittgensteinian? That there is no philosophy? That sounds odd; very odd.

A faith is sociologically describable as that which keeps the crowd together, the excuse one has to offer when one joins the herd. The contents of the faith may fit the practices of the herd well, or even, as in Russia and in Israel, conflict with them. Or it may be quite inconsistent, as American Puritanism-cum-utilitarianisms obviously is. Nevertheless, or just as well, even, the faith is the psychological and intellectual justification of the individual's dependence on his crowd; it is the best social cement.

This is the classic sociology of faith as expounded, for example, by Émile Durkheim. I cannot accept it, at least not fully, since it cannot apply to the faith of the rebellious prophets, of people like Amos who wanted no cement and like Jeremiah who preached despair and capitulation to the enemy. Nevertheless, it obviously does apply to conformists, especially those who do not have in them the courage and resources for that autonomy for which the prophets are so celebrated, and for those who use religion as means for conformity, such as the Spanish Inquisition and the Iranian establishment – and even the Israeli establishment whose violence is still checked.

Now, paradoxically perhaps, one of the most famous cases of a lack of intellectual and/or moral resources for autonomy, is that of the one who wavers between autonomy and heteronomy. The most famous escape from the pains of intellectual and/or moral wavering is a leap of faith to over-commitment, known philosophically as the existentialist engagement or commitment, and known in the vernacular as fanaticism. Hence at times a fanatic minds less what is the object of his commitment than that the commitment holds and relieves him of his daily decision problems. Fanatics are people weak yet demonstratively aggressive, doubting yet of vociferously confident expression, unable to decide yet armed with prearranged decisions and decision procedures. Oddly, they can switch allegiance from one doctrine or faith to another, but provided the switch is rapid enough. The very rapidity of the switch may puzzle an onlooker, yet speed is of the essence: the fanatic suffers too much from doubt to let it linger long.

Fanaticism, thus, is the absence of intellectual and/or moral autonomy, namely dependence, coupled with an appeal for excessive and clear-cut rationality. I must stress that dependence and reason, when they come together, create a painful conflict; with the insistence on simple crystal-clear rational method there is conflict no longer. Now, often we tend to confuse fanaticism with closed-mindedness. For example, we often consider as fanatics those fundamentalists who opposed the teaching of Darwinism in school. This is a serious error. The fundamentalists have no conflict. They have no attitude towards Darwinism, and wish to know nothing about it, since – and this is all they need know – it conflicts with some Biblical tenets. They find no point in discussing the arguments against Darwinism either. Suffice it for them to read the Word of God. Closed-minded they are – because they are dependent. But there is only determination, no conflict, and so no fanaticism, in their conduct. It must be noted that in the Scopes trial, the so-called Monkey-trial, in Tennessee just within living memory, it was Darwinism's invasion into the Bible Belt, not any fundamentalist crusade, that brought the clash about. There is no need to pile on the charge of closed-mindedness the charge of fanaticism. Moreover, the fanatic, in his very excessive closed-mindedness, in his very excess, betrays his conflict, a conflict rooted in his possession of some ability to change his mind; and in a rare case indeed he does change his mind: rapidly.

Now the most paradoxical fanaticism is an anti-fanatic fanaticism, pro-science fanaticism. It is, of course, what G. E. Moore called a pragmatic paradox, the excitement of the one who tells you excitedly that there is nothing to get excited about, the tense insistence on being very calm and relaxed, the intense and dramatic insistence with which G. E. Moore finished one of his most spectacular lectures, thrusting his hand up and declaring that it is bland, common knowledge that I am in possession of my own hand. The intellectual foundation for fanaticism is already laid down by the justificationist tradition of science. Both Bacon and Descartes – or was it Galileo – fought against intellectual tyranny in an effort to preach autonomy, yet Bacon openly preached the tyranny of the senses, and Descartes the tyranny of intuition (or of logic or what have you). Yet, clearly, unlike the modern philosophy of the commitment – to religion or to commonsense or to any other view – the justificationist is not a fanatic: he is still looking for the criterion of rationality; he does not employ a given criterion in a manner that forces decisions in him. Yet justificationists still insist: the most important problem in philosophy is the problem of choice: which hypothesis should I believe in? No doubt, justificationists do take something for granted: that such a criterion does exist. Why should there be a criterion? They will not say. They repeatedly declared this to be the central problem of the philosophy of science, and they would not say why. It is possible to hold that a criterion exists by mistake, or as a dogma; but it is not fanaticism unless one uses a given criterion right and left with no hesitation and with no criticism.

Let us push matters further. Is justificationism true? Should we seek the criterion of choice of a belief? Is there such a criterion? If in fact we have no such criterion, then there seems to be no need for it, since science is doing fairly well one way or another. This is the chief point stressed by Michael Polanyi. If we do have a criterion, can we observe it? If we can observe it, then sociologists of science should settle matters for us very nicely, as sociologists are adept at making explicit criteria implicit in social practices. If, however, we cannot observe the criterion, how do we know it exists at all? If we cannot observe that it exists, then we should conclude, perhaps, that it does not exist. If we can observe that it exists, how come it is so elusive? The existence of a criterion, the argument goes, is evidenced by the existence of an observed consensus among men of science. What is it based on in fact?

And so, it turns out, the philosopher should find the rational of Professor Tom or Professor Dick in this university or that, and thereby justify their faith in Einstein or in Darwin or in Crick, even though, admittedly, these esteemed scientists themselves need not be able to articulate their own defense. Just like in theology, then, so in philosophy of science, we need the justification *post hoc*! The trouble with all *post hoc* arguments is that they are defenses of the *status quo*. If so, then their prime assumption is that the *status quo* is satisfactory. But then, make this assumption explicit, and you need no further defense! This is the claim of Michael Polanyi and of the older Wittgenstein. Popper, by distinction, says, defending the *status quo* leads to stagnation.

The popularly endorsed positivistic demand from science to settle all its disputes fairly rapidly either leads to the claim that all disputes are scientifically settleable and the rest is bunk, or to the claim that when a dispute rages outside the limits of science anything goes. Wittgenstein endorsed the former view when young and the almost latter when old. Anything goes for the old Wittgenstein, provided it is part and parcel of a life-form, of a mode of being that makes my decisions for me.

The old Wittgenstein rather refrained from explicit reference to science. And so his natural complement is Michael Polanyi. Polanyi is the one who justifies the consensus among scientists by its very existence: to try to articulate its reasons is futile, he said. The consensus, added Thomas Kuhn, is the assent to a paradigm. This goes very well with Wittgenstein's justification of commonsense as a life-form; science has paradigms, ordinary life has life-forms, both express themselves implicitly in the language in use. Kuhn unifies the Vienna and Oxford schools; and quite unwittingly! (J. W. N. Watkins observed that mediaeval theology is paradigmatic but modern theology is a multiple paradigm; Kuhn did not respond to this in his reply to Watkins.)

To do the young Wittgenstein justice, I must draw attention to his subtlety. Old Maimonides and new Comte and Marx, presented rationalists as ones who recognized no limits to science: all questions are scientifically answerable, and in principle. Wittgenstein's famous *dictum* (*Tractatus*), the puzzle does not exist, seems to mean the same. The Vienna Circle read him so, and agreed. Thus, Rudolf Carnap, for example (*Pseudo-Problems*), said, the problem of life may be a question proper and as such answerable, or it may be a deep sense of loss and a need to readjust, and therefore no question, properly speaking, but a task. For my part I think Wittgenstein showed a higher degree of sophistication than his Vienna disciples. He not only noted the limits of science; he showed how much there is beyond that limit, in the realm of the mystical, as he called it. Yet, he said, language is limited just like science – this was his chief dogma, his picture theory of language; and so, he said, we have no means by which to ask the question about the mystical. Now, by language we mean here not the language of poetry but of science. Add to these two languages diverse languages, indeed diverse sublanguages known as language-games, and hey presto! you have made room here for the old

Wittgenstein and for all the religious positivists from St. Robert Cardinal Bellarmino to Pierre Duhem.

Ernest Gellner has generalized this point. For he has noted already that, anti-religious as positivism tends to be, it finally goes very well with religion. I think he could say more here: as in the teaching of Malinowski, anti-religious positivism opposes only religious doctrine, not ever religious ritual. In the positivistic sense, then, anti-religious positivism was not quite anti-religious in the first place. But positivism may well go with religious doctrine, as it was with Duhem. Yet religion plus positivism, that is to say religious doctrine proper plus science, is very different from any other religion, and most particularly from religion plus theology. There are all kinds of religion, speaking from the viewpoint of the rationalists. There is the naive or primitive or prerational religion. There is the closed-minded religion, of those who only know that criticism exists and is dangerous, but no more. Then there are those who waver and the fanatics who need badly a rational justification of what they rightly suspect is irrational. And they find it in theology or in positivism.

3. THE INEFFECTIVENESS OF PSEUDO-RATIONALISM

I am now near my conclusion. My thesis was that positivism is anti-theological as it is a competitor with it for the same function. I concluded that in some sense positivism is not anti-religious, and can very well be pro-religious. One more corollary, and a rather surprising one, is that likewise positivism is not anti-metaphysical, all appearances to the contrary notwithstanding. Let me argue this final point now. When a positivist hears of a metaphysical doctrine conducive to science, say Greek atomism, or Spinoza's parallelism, he remains curiously unperturbed. I have observed this phenomenon often, in old books, in new positivistic publications, and in live positivists. I was always greatly puzzled by this. I found this an almost incredible dogmatism, which does not suit parties not given to so flagrant an exhibition of dogmatism. (I already spoke of the rational attitude of positivists to criticism.) They said, when asked, whatever is O.K. for science is O.K. for us; we are not against science but for it and so also against obscurantism. They even showed readiness to see a difficulty here, but felt clearly that a Democritus was not a St. Augustine and on this they were quite insistent.

Rationalists often mistakenly identified theology with obscurantism. This explains why the positivist's demand for the utmost clarity is taken for granted, all arguments against this excessive demand notwithstanding. It is therefore understandable that they attack obscure metaphysics as disguised theology – not quite accurate but accurate enough – and lumped clear and rational metaphysics with science – again not quite accurate but accurate enough. Hence, positivists oppose only the obscure kind of metaphysics, which is often lumped with theology.

Positivism, then, may be viewed as an infant disease of the rationalist movement, to use Lenin's terminology, and a left-deviation as opposed to theology's right deviation. And so, as an infant disease, positivism is hardly necessary but also hardly avoidable. Except that Lenin's terminology of infant diseases and left and right deviations is highly positivistic, with the idea that you can depend on the orthodoxy which is smack in the middle between all the deviations, and which is adult enough for you to rely on. Whereas, what seems to me the most obvious is that self-reliance, intellectual and moral autonomy, is the proper starting point. Perhaps even only that kind of moral autonomy that entails intellectual autonomy, is the only proper starting point; the only proper Archimerean point. Otherwise, nothing is easier than to identify rationalism with positivism and both with science, to admit that science is limited and to add to science religious ritual and even religious doctrine without theology. This will be the defense of the *status quo* in science and in religion, yet without any proper justification – as in the philosophy of Michael Polanyi.

Was Wittgenstein helpful in bringing this point home? Perhaps; I cannot judge. But if he did, he did so by his very preaching of a perverse heteronomy. The problem now is, how can we get out of the infant disease fastest, with the least side-effects, and with the least damage. I suppose the best answer is, by discussing autonomy and its problems. But, of course, we may just as well use alternative doctrines that serve the same function, the function of keeping rationalism in its infantile disease. Now that Wittgenstein's philosophy is coming out of fashion, Polanyi's philosophy is coming into fashion, especially in Kuhn's variant. And it, too, defends science and the *status quo* and unanimity and commensense – scientific commensense, that is. Superior as Polanyi's philosophy is to Wittgenstein's, if we allow ourselves to be seduced by it and repeat Wittgenstein's pseudo-rationalism, then the whole episode of the Wittgenstein fashion will have been utterly useless. One way or another, we had better eschew the idea of unanimity and begin with the moral autonomy of the individual.

CHAPTER 4

EPISTEMOLOGY AS AN AID TO SCIENCE

> ... that duty to science, which consists in the enunciation of problems to be solved....
>
> Faraday

Gerd Buchdahl's main point has become somewhat clearer to me after long discussions with its author, with Ernest Gellner, and with J. W. N. Watkins. I have ultimately assumed that Buchdahl's thesis is that epistemological discussion is entirely useless for scientific research, a thesis very much in tune with the philosophy now in vogue. I shall try to criticise this thesis of Buchdahl's, and to argue that in the historical case which he chose to discuss as an illustration of his point – Dumas's views of atomism – epistemological criticism might have been of great use as a tool for solving scientific problems. The point of view from which my criticism is launched is that of Popper, as I shall explain in the body of my discussion.

Indeed, Braithwaite in the last and most prominent sentence of his *Scientific Explanation* anticipates my reply to Buchdahl's thesis. For there he expresses his views on the possible influence of his epistemology on scientific research as follows:

> If this clarification has the secondary effect of encouraging scientists to construct deductive systems and to use theoretical concepts freely, it will, I believe, assist the progress of science as well as the better understanding of what science is doing.

1. DIVORCING SCIENCE AND PHILOSOPHY

If I understand Buchdahl rightly, his point is this. Discussions of scientific problems cannot gain but can only lose from epistemological arguments. An epistemological argument is one which applies equally to all hypotheses as such, while a genuinely scientific argument concerns a given hypothesis or a given group of hypotheses. Of course, nothing prevents anyone from being interested in both epistemology and science. Yet one should not, he holds, use arguments from one field in the other. The pointlessness of doing this may be well exemplified by the case of a scientific discussion concerning the choice between two specified hypotheses. Any epistemological criticism of one of them – being general in nature – may be equally applicable or equally inapplicable to both, and is thus useless for the purpose of that discussion and even confuses the issue. Therefore we should accept as a general rule not to mix science with epistemology. I shall take this maxim to be what Buchdahl recommended, although he himself seems to violate it – a fact which I shall ignore.

Buchdahl will readily admit that some scientists who fell prey to this confusion were occasionally stimulated by it. Yet this is a mere psychological fact or a historical curiosity. The logic of his case is that an argument quite clearly belongs either to science, or to epistemology, or to neither, but not to both. Science *qua* science is not concerned with epistemological generalities.

This is my summary of Buchdahl's position. It is surely a very important position, if only as a variant of the ideology behind the increasing specialisation of the present day. This position – epistemology for epistemology's sake – seems to me to imply that epistemology is barren. (Some might feel that had they accepted Buchdahl's view they would have dropped their interest in epistemology.) Yet I see no reason to accept Buchdahl's grim judgment. First of all, it is very difficult to judge epistemology as such since we do not know its future contributions, much less their impact on scientific research. Moreover, the history of scientific research is full of examples of developments which were made possible through epistemological discoveries. I would mention the examples of Galileo, Boyle, Faraday, and Einstein. However, my present aim is to explain the inadequacy of Buchdahl's suggestion that scientists should go on doing research without paying attention to epistemological arguments.

Any activity, be it scientific research or not, is based on the choice of an aim, an appraisal of the situation, and a decision as to the behaviour most likely to forward that aim in the given situation. The aim, the appraisal of the situation, and the decision are all necessary if the behaviour is to constitute rational activity. A critical attitude towards an activity implies readiness to accept criticism concerning any part of it. The criticism may be moral, i.e. concerning the desirability of the aim, or factual, i.e. concerning the correctness of the appraisal of the situation, and the adequacy of the decision. Now, when the activity in question happens to be scientific research, or the advancement of learning, the criticisms, concerning aims, circumstances and method, belong partly to ethics and mainly to what is called epistemology, using that word either in its traditional meaning, or in the meaning given to it by Buchdahl. Therefore Buchdahl's claim that epistemology cannot affect scientific research would render scientific behaviour uncritical and merely habitual. It would also deprive its practitioners of their personal responsibility. However, since we usually suspect that scientific research *is* rational, we may well suspect that Buchdahl's judgment is false, since it implies that epistemological criticism is barren.

It is quite true that for the purpose of choosing a given hypothesis, and for other purposes, some epistemological considerations may be irrelevant. Yet the very idea that research involves a choice of a hypothesis is in itself an epistemological idea. Buchdahl tells the investigator to go on doing scientific research and ignore epistemology, yet in research he includes the choice of hypotheses; which is a specific epistemological theory. Consider an investigator who has a different epistemological theory; say, one who believes that research does not involve any choice of hypotheses or any choice at all (since choice contains an element of arbitrariness), but rather verification, or proof, or the compulsion of scientific statements. It would be foolish to tell this person to go on trying to verify his hypotheses. It would be much more reasonable to advise him to stop after a while and ask himself why all his attempts to verify his hypotheses have so far ended in failure, and whether these failures are not inevitable.

Buchdahl would have told a verificationist like Dumas, who worried about this inability to verify any of his hypotheses, not to consider such epistemological problems, but rather to go on doing research. Yet by research Dumas understood trying to verify hypotheses. Therefore the only way to put him right would have been to advance epistemological criticisms, namely to explain to him the impossibility of verification. Yet according to Buchdahl's position, such a criticism would have been redundant, since, according to him, everybody understands perfectly well the meaning of scientific research.

In this respect Buchdahl's view is very similar to that of Wittgenstein, as Gellner has pointed out to me. Wittgenstein asserted in his *Tractatus* that his own epistemological theory – good old verificationism – was self-evident and unassailable ('it *shows* itself') in a manner which makes it both impossible and unnecessary to state it. Now, the standard criticism of Wittgenstein's view, first advanced by Russell, is this: Wittgenstein's epistemology entails its own meaninglessness; yet he managed to state it, and even to claim for it the status of unassailable truth; which is absurd. Buchdahl's epistemology can be viewed as watered down Wittgensteinianism. For him, epistemology is not inferior to science with respect to meaning; it is equally meaningful, yet equal-but-separate. Instead of Wittgenstein's thesis that epistemology is meaningless, Buchdahl advocates the thesis that epistemology is pointless at least as far as science is concerned. Yet the same criticism which can be levelled against Wittgenstein's doctrine of meaninglessness, namely that since it renders itself meaningless it is absurd, can be applied to Buchdahl's doctrine of pointlessness, namely that since it renders itself pointless it is absurd. For if all epistemology is pointless, so is Buchdahl's epistemology. Therefore, if we accept Buchdahl's suggestion that we go on doing research without paying heed to epistemology, we should have to ignore his own epistemology too, and go on doing research in the way it was done by most thinkers from Galileo to Niels Bohr, namely, by mixing science with epistemology. Thus Buchdahl's view is inconsistent.

On this point Buchdahl's view resembles that of Polanyi, as Watkins has pointed out to me. Polanyi thinks that research is not based on an explicit epistemological theory; indeed, not even on an expressible one. The epistemology implicit in research is that which can be transmitted only by personal contact between master and apprentice. Therefore there is no room for epistemological theory. Perhaps I can follow Polanyi's version more easily than that of Wittgenstein because in condemning the existing epistemological theories as such Polanyi rules out the possibility of any alternative to them, while Wittgenstein condemns the existing epistemological theories as such while advocating one of them – verificationism. Moreover I accept Polanyi's view that research is a traditional activity. I also follow him in holding that the scientific tradition is not completely describable in human language, simply because no fact – not even the apparently simplest fact – is completely describable in human language. Yet just as we try to describe facts about cabbages and kings, we may try to describe facts about the scientific tradition. Popper attempts to describe it as the tradition of criticism – of clashes of opinions. The criticism in question may relate to scientific or to epistemological statements. The question of when a particular criticism is scientific and when it is epistemological is an interesting and important question which Popper ventured to answer, yet it is not the major question. The main aspect of Popper's viewpoint, at least to me, is his advocacy of the critical attitude as such. Criticism is always welcome, whether it is labelled 'scientific' or not.

Polanyi's position – which is to me very stimulating though uncritical, or as he calls it 'post-critical' – rests on the following point. What Kant calls tact, what he calls personal knowledge, and what may be called a sense of discrimination, is something one learns by joining the tradition – something learned not as one learns the Latin word for 'tiger', but as sculpture and painting are learned. This point seems to me to be true and very important.

Yet this personal knowledge which we learn in the scientific tradition concerns, among other things, the choice of what should be criticised next and the ability to criticise it. If Polanyi held his view of personal knowledge consistently, he would have said that it is a matter of personal knowledge to judge when to blend science and epistemology, and in what proportion. Furthermore, what I particularly do not accept is Polanyi's idea that one needs personal knowledge in order to judge the fruits of other people's personal knowledge. True, some sense of discrimination is always necessary; yet we often need not have very much sense of discrimination in order to be able to appreciate critically the fruits of the labour of a person with the highest degree of ability.

This is the reason, it seems to me, why it is so easy to be wise after the event. If we do not wish to be wise after the event we should not blame Dumas (whose ideas Buchdahl discussed) because he did not criticise his own verificationism, though we may appreciate all the more the important contributions of his friend Faraday which were made possible by such criticism. Faraday had more sense of discrimination and more critical ability than Dumas, but this only became obvious after the success of his criticism. Faraday's superior ability should not be used against Dumas, especially since Dumas also was critically minded.

I say all this in order to stress that I do *not* think that in general epistemological criticism is more important or less important than scientific criticism. If I say that I would rather have criticised Dumas's verificationism than his atomism, I say so for two very specific reasons. First, Dumas himself succeeded in criticising his atomism, but failed to criticise his verificationism. Secondly, it was this failure, this insistence on verificationism, which got Dumas bogged down; he was intimidated by it; he considered it disreputable to deal with unverifiable hypotheses. Indeed, his friend Faraday also got bogged down at the same time, and for similar reasons. Only after ten years of disappointment did Faraday start to develop his criticism of verificationism and thus got new wings. But until Popper, epistemologists continued to preach verificationism to the scientist. When I say that Dumas might have advanced his faculty for research by criticising his own epistemological views and thus freeing his imagination, I think this merely because I have benefited – I hope – from the fruits of later criticism, by Faraday and by the followers of his tradition – in which I include Einstein and Popper. I think that Buchdahl has also somehow benefited from this tradition, and he has used its fruits in his quasi-definition of the 'scientific level'. (He does not define it as the domain of certainty!) However, he takes the fruits of later criticism of verificationism so much for granted that he regards those pronouncements of Dumas which he quotes, as irrelevant to research and as mere confusions.

Quite unintentionally, I am sure, he exhibits that attitude of superiority which some epistemologists show towards some serious scientific work, thereby causing an unhealthy hatred of epistemology. Dumas was not confused, though like all humans he committed some errors, including his belief in verification. If he mixed science with epistemology – not enough for my taste, and too much for Buchdahl's – all we can do as historians is to explain his behaviour critically and respectfully or leave it alone.

2. THE NEED TO DEMARCATE

Before trying to explain Dumas's verificationism on alternative lines, I wish to add, somewhat parenthetically, a few words on Buchdahl's criticism of Popper.

Buchdahl claims that he accepts Popper's view about the aim of science being the search for the hidden reality only after having demoted it to the 'scientific level'. On the 'epistemological level' he doubts it. This puzzled me greatly. For both Popper and Mach have theories about the aim of science and the content of hypotheses. This is why I naïvely considered both Popper's and Mach's theories to be epistemological. Now Buchdahl puts Mach's theory on the epistemological level and Popper's theory on the scientific level. This, Watkins has remarked, would imply that only Mach's epistemology must be barren but not Popper's, which being on the scientific level can have an impact on research. To this I must say that double injustice is done to Mach. First, Buchdahl overlooks the fact that Mach's epistemology also could have – nay, did have – a profound impact on research. Secondly, Mach and Popper are opponents and therefore Mach should have equal chance of being considered and perhaps accepted. Mach says that scientific theories do not mean what they seem to mean but rather that they are highly condensed reports of observations, while Popper says that they do indeed mean what they seem to mean, for they are conjectural assertions about the (unobserved) structure of the world. It is therefore unfair to Mach to tell students of scientific problems that they should ignore Mach and take Popper for granted. Yet Buchdahl puts Mach's theory on the epistemological level, Popper's on the scientific level, and tells people to ignore the epistemological level. This, obviously, amounts to telling them to ignore Mach's theory and to accept Popper's alternative uncritically. I believe it is more rational and more just to argue about the alternatives critically than to reconcile them by delivering one to the scientist and the other to the epistemologist.

3. SCIENCE AND PHILOSOPHY INTERTWINED

I shall not consider here the reasons why verificationism was accepted in the first place. I shall only discuss the reason for its persistence although it was refuted again and again.

The way to dodge refutation, which Popper calls the conventionalist stratagem, was beautifully exemplified by Polanyi, who said something like this. A theory is scientific – I should say critical or rational – if when it is refuted it is considered to be false. Once the person who applies the theory (not the theory itself) is blamed for the false prediction which he deduced from the theory, that theory ceases to be scientific. This is a nice example of the conventionalist stratagem – of the method of having a scapegoat ready in order to save the theory. The Baconian conventionalist stratagem which I shall now discuss is a particularly clever one; for it is itself an attack on the conventionalist stratagem, as I shall now explain.

It was Telesio who was the first among the moderns to claim that the senses cannot mislead us. His friend Patricio criticised him by claiming that we do not see facts as they are but rather as interpreted by our theories. This criticism was fully accepted by Galileo, who subsequently rejected the authority of the senses, and by Bacon, who did not. Applying the conventionalist stratagem, he said that when someone sees facts not as they are but as interpreted by his theories, then it is not the fault of his senses but his own fault. It is his own *fault*, because he does so in order to save his theories from refutation. Observational errors always stem from observations containing some theoretical ingredients. Such observations are bound to mislead us, for the rôle which theory plays in interpreting observation is that of distorting sensations in order to fit them into the preconceived theoretical scheme so as to save that scheme. The observer must so act because he is simply unable to admit error. He is unable to see facts as they are, being bound to look for confirmations, in order to try to flatter himself that his hypothesis was correct. He will thus become prejudiced or superstitious, since he will not be able to see the evidence against his theory.

Today when so many of us so readily admit error it is a little difficult to understand Bacon's view and its impact. But it might be helpful to mention two examples of the behaviour which Bacon described and which were very widespread in his day. The examples are dogmatism and neuroses. We know that dogmatic people cannot but see facts as wonderful confirmations of their views. We know that neurotic people, when their neuroses are challenged, may distort the facts in a most fantastic way, rather than get rid of their neuroses or reject their own rationalisations of their neuroses. According to Bacon, all hypotheses sooner or later turn into neurotic obsessions and into dogmas.

The reason why Bacon's doctrine of prejudice was so popular, in spite of its great disregard of man's critical faculty, is that in many ways it was amply confirmed by experience. Since error was identified with prejudice people could either deny that a theory was erroneous, or declare it to be a prejudice. When Lavoisier's theory was successfully criticised by Davy, great natural philosophers like Gay-Lussac refused to declare that it was an error because they knew that Lavoisier was not prejudiced. Thus, they stuck to their error and became prejudiced, as Bacon predicted, just because they accepted Bacon's theory. (This may be an instance of what Popper calls the Oedipus effect.) People who accepted Davy's criticism of Lavoisier declared Lavoisier's theory to be a prejudice. Even such appreciative people as Dr. Thomas Thomson and the young Faraday declared Lavoisier's theory to be a prejudice. Such declarations were, of course, based on Bacon's doctrine; but they were soon taken as evidence for it. Later many historians who rightly praised Lavoisier's theoretical contributions made no allusion to the fact that he had also made some mistakes. Accepting Bacon's doctrine, they thought it would be derogative to mention Lavoisier's errors. Thus, through the acceptance of Bacon's doctrine the picture of the great heroes of science became Baconian: they were painted as unerring supermen – as indeed they would have to be if Bacon's theory were true.

Another cause of the confirmation of Bacon's doctrine is this. A theory – say, Lavoisier's – is refuted by a new fact – like the existence of combustions without oxygen. After the refutation of the theory the refuting fact becomes plainly obvious and even commonplace – like the combustion of metals in sulphur, or of hydrogen in chlorine. How is it then that the great experimenters of the past – such as Lavoisier – failed to see such simple facts? Answer: because they were blinded by their own theories. Ergo, they were prejudiced.

Incidentally, all this is a very good example which itself confirms Bacon's view; for the sheer acceptance of it made people see the history of science in its light. Yet this example also shows that Bacon's theory did not solve the problem of dogmatism. Bacon himself still believed that some kind of confirmation is possible, instead of seeing, as Galileo[1] did, that what matters is not confirmation but criticism. Bacon's antidogmatic theory was soon confirmed and became *the* dogma and therefore was hardly ever criticised. It ceased to be the 'idol of the market place' only after Einstein's theories were taken seriously. It was later restated in various forms, say, by Mannheim and by Warnock who recently attributed it to Wittgenstein.

I would briefly mention Popper's criticism of Bacon's doctrine. First of all, none of us can be free of prejudice. Science is made by people, who, being human, have their own prejudices. These need not be impediments, as Bacon thought; they can be useful. Even if one person is prejudiced by a theory, says Popper, that theory may be a challenge for another to try to criticise it, and perhaps to discover the refuting instance. The refuting instance would not have been discovered at all but for the existence of that theory, and of people who are prejudiced by it, and the subsequent attempt to criticise it. Perhaps the prejudiced person will stick to his view in spite of refutations, but these may be taken more seriously by other people. Perhaps, as Planck said, only the next generation might reconsider the situation and take the criticisms more seriously; yet in one way or another the criticisms may be taken seriously and thus bring progress.

Although I reject Bacon's doctrine of prejudice my contention is that it is a very plausible doctrine, which the traditional natural philosophers took very seriously. These facts, I think, help to explain their attitude. Their acceptance of Bacon's doctrine explains their effort not to go beyond sense experience – their effort not to be committed to false views. This is why I think that I could have argued with Dumas on the lines which Buchdahl rejects.[2] The point I should have made to Dumas is that he could not escape from error, and that error need not be so dangerous as it seemed to him; that his atomism, though erroneous, was not disastrous (as he thought it was, as it is clear from Buchdahl's quotations) since it was criticised and refuted. I believe that realising this he might have been less worried, and less paralysed by the fear or error. This is just one example of the possibility that epistemological criticism may have a liberating effect. Indeed, although Faraday remained to the last a Baconian to some degree, he was liberated by his discovery that even the most cautious man is not exempt from error, and that research cannot progress without error. He was the first to rebel against the Baconian tradition by publishing systematically his own errors even after he had refuted them experimentally. He warned people not to accept his speculations but to use them critically, to try to test them experimentally. All he claimed was that his own views, limited though they might be, were not refuted while those of the majority were. Faraday's free style of writing was such a novelty that years after his death Maxwell commented on it with a great sense of surprise, admitting its helpfulness to the reader, and yet proposing to follow Ampère's Baconian style, rather than Faraday's. Fortunately, Maxwell himself did not follow his own proposal and a whole school of free expression of thought developed, culminating, I believe, with Einstein. In 1905 Einstein published two theories, one of which asserts that light consists of continuous waves, the other that light consists of discrete quantities of energy. Einstein knew very well that for mere logical reasons at least one of these theories was in error. Yet he published both, leaving it to criticism to sort out the true from the false.

It is easy now, after Einstein's new tradition and after Popper's rationalisation of it, to look back at Dumas and to claim that his worries were due to a confusion of science with epistemology. By the same token one may pooh-pooh even the famous 'crisis in physics' as mere epistemology which has nothing to do with science as we know it now (as Buchdahl does by implication). Yet this 'crisis' was the great intellectual event of liberating us from verificationism – the new Renaissance – which opened the way for our new epistemology. True, no epistemology would have solved all problems unless it is a mock-epistemology. Science is trial and error, and in order to progress one has to try. The new epistemology would not have solved Dumas's problems; yet it would have helped him. Admittedly, freedom of thought and readiness to accept criticism are always characteristics of the scientific tradition. Yet Faraday's view that science should be in a permanent crisis is still a novelty. And Popper's idea that if one is successfully criticised one need not be discouraged but should be grateful and try again – this idea, too, I am afraid, is still far from being very popular. Not only could it have helped Dumas – it can still be of service to those who have not yet considered it. In this respect the scientific tradition (which I greatly admire) seems to me to be in need of reform.

NOTES

[1] Galileo also looked for experimental evidence deciding for his Copernicanism. Yet what he looked for was a clear-cut *crucial* experiment, and he surely thought that criticism is more important than supporting evidence, much as he would have liked to have both.

[2] This is a reference to Buchdahl's claim that 'it would be somewhat of a joke' to tell Dumas that verificationism is false.

CHAPTER 5

EXTERNALISM

It is a theory of any field which tells us what is its internal logic and hence, what is its internal history. What is not internal is, then, external. E.g., military incentive is external to both science and art but not to logistics. Whereas purely external study of science, such as the studies of citations of one author by others, is plainly silly, the purely internal study of science is possible but incomplete.

1. THE PURELY EXTERNAL

Internalism and externalism concern chiefly history, but also sociology and social history, and such, chiefly of science, but also of art and other culture activities. Thus we can have the internal/external history/sociology of science/art, etc. Let me first explain the terms "internal" and "external", regarding which I think confusion abounds. It is hard to speak of the current usage when the terms are so young, and I suggest my observation more as a clarification than a factual report: anyone who wishes to offer different terms or meanings should feel free to do so; my purpose is not to fix terminology but to clarify.

Whereas an internal history of a science, say, concerns itself only with that science, external history concerns itself with the interactions that science has with other sciences, with the arts, or any other areas of human activity, intellectual or social, cultural or material. It is essential to notice that whereas internal history ignores external factors, external history does not ignore internal factors. Let us label that external history that ignores internal factors "purely external" so as to keep out much of today's confusion.

An example, and a simple and obvious one, is Flora Masson's life of Robert Boyle which, for one reason or another is well known among scholars concerned with her hero even though in the whole volume she barely touches on the fact that Boyle was a leading philosopher and scientist of the seventeenth century, and says nothing of the contents or value of his intellectual product. The book cannot but seem history frustrating to a historian of science, but he knows this fact in advance and if he reads her book nonetheless, it is because he expects to enjoy it and learn something from it, etc. A slightly less pure external and much more delightful example is the life of John Stuart Mill by Ruth Borchard, called, very aptly, *John Stuart Mill, The Man*. Here we do learn, and quite a lot, about Mill's intellectual activities, but strictly to the extent that the author feels unable to portray the man without reference to them.

It is hard to imagine a purely external history of an art, whether painting or music; or even one which uses internal factors sparingly and for specific reasons. It is equally hard to imagine a history of the Church of Rome with no, or almost no, reference to Church doctrines. Yet such things do exist. Many social and political histories include almost purely external histories of the Church of Rome, even of the Reformation and the Counter-Reformation, where terms like "the Eucharist" are unexplained yet the fact is noted that a controversy regarding the Eucharist was central to the Reformation as seen by many participants in the rise of Protestantism and the ensuing Religious Wars. Similarly, there are many histories, political and even social, which discuss the difference between *musica da camera* and *musica da chiesa* from many angles without even bothering to explain that church music is distinguished from chamber music first and foremost by the presence of an organ in one place but not the other. They likewise discuss the rise of the chamber orchestra and of the grand orchestra without a hint at the artistic value of a grand orchestra and what it can do that a chamber orchestra cannot do. Nor is this an easy matter, considering the fact that some of Schubert's last chamber music, chiefly for string, foreshadow, in their monumental style and vastness, some of the most remarkable achievements of the grand orchestra though not, of course, its timbre. Alternatively, social historians can discuss the traditional presence of the organ in churches and nowhere else with no reference to its use as a peculiar artistic instrument, and linguists may notice that choirs sing *a capella* because chapels employed choirs but courts rarely did, and then hardly without any accompaniment (which is the meaning of *a capella*, namely, choir with no accompaniment even when performance takes place in a concert hall).

The external history of science can be exciting. People totally blind to all mathematics and totally ignorant of Newtonian mechanics can be told, for example, of Newton's life and deeds, of the usefulness of Newtonian mechanics, of Newton's influence on Ben Franklin who flew a kite in a storm and miraculously lived to tell the tale and build the famous lightning rod. You need know nothing at all about electricity to understand the usefulness of the lightning rod. You need know no organic chemistry and nothing about atoms and molecules to learn about artificial dyes and their disastrous effect on the indigo industry in India, about rubber boots and rain-coats and the rubber-plantations in Malaya, about the debt motor cars have or do not have to theoretical physics or chemistry, and about their effect on sprawling suburbia and the sexual revolution.

I would be omitting a most novel and exciting idea if I left the purely external without mentioning big science and the citation index. Big science differs from little science in purely external aspects. It is research performed by large teams costing a lot of money. This makes it subject to much political interest, internal as well as external to the teams involved. One important question the student of the purely external aspect of big science can hardly ignore, he feels (he ignores it anyway, we shall see at once), is the question, what and how much do research teams produce (apart from the obvious political and social intrigue, bitterness, and similar social factors that the sociologists of science assiduously ignore). There are lovely remarks to quote when one comes to this question, outstanding among them is that of Jerome Wiesner, former president of that citadel of big science, the Massachusetts Institute of Technology, to the effect that the United States Federal Government's support of research and development in the post World-War II years has led to a blooming of science to the extent the like of which only Renaissance Italy witnessed regarding the arts. I do suspect Jerome Wiesner deeply regrets this foolishness: as Brian Aldiss says, foolish ideas never die.

The purely external sociologists of science – I am almost finished with the purely external – have invented a purely external mean of measuring whatever scientific research – big or small – produces or may produce, without opening such internal questions as what *does* research produce? let alone, is it of any value? More precisely, the father of the expression *Big Science, Little Science*, Derek J. de Solla Price, the author of a very famous best-sold worst-read book by that name, has invented the technique. It is known as citation index. A fellow whose production is higher is oftener cited, and a head of a team is cited with the precious "*et al.*" that goes for the whole team and is the hallmark of big science. Citation indexing has been criticized on the ground that one may cite authors in order to pay them homage for their fine work and one may cite authors in piety, in mock-piety, in an attempt to win easy acceptance of one's paper for publication by editors who are friends and relations of the cited, in an attempt to find a mock-opponent to a rather too-well-accepted thesis one wishes unnecessarily to defend, in an act of self-advertising, or in an act of sheer defiance. Since these are so very clearly external criteria, the advocates of purely external sociology and social history of science may look for some external means, or for minimal internal means, to distinguish between these various categories of citation. Hence, accepting this criticism, the purely external sociologists of science may still go on – even with renewed vigor. In any case, clearly, their venture will not thereby be stopped.

And, to conclude this section, there is much incentive for the purely external: it is the shelter of ignorance, and as long as there are many sociologists let loose in the market, there will be some effort to impress the public by purely external means: all one has to do is make the purely external sociology of science a subject of its own, opaque to the non-specialists, and then all one could do in order to examine the question, do members of the sub-disciplines devoted to this subject produce anything? will be the application of a purely external analysis of the sub-profession. If this happens it will take a Hercules to change the trend. My aim is to stop the torrent, to fight the windmills in the hope of preventing them from producing too much wind before it is too late.

2. INTERNALISM

It is not easy, perhaps not even possible, to say what is internal other than that the internal excludes the purely external. Internal occupation with a science is tough-and-no-nonsense. We must cut out all history, all philosophy, all methodology, all biography, and all personal concerns, before a field becomes entirely confined to the internal and thereby crystallizes to the purist's heart's content. Usually, even the very toughest tough-and-no-nonsense militant allows himself an external comment here and there, historical, biographical, at times even political or merely personal. But this is done very unobtrusively, takes practically no time, and is in the nature of a word or preamble, a comic relief, a marginal comment.

It is an important social fact that if you button-hole a tough-and-no-nonsense scientist who is also a regular professor and ask him how he and/or his colleagues in his department decide what to teach, you will not get satisfaction. He may offer generalities, explain the rationale behind the most elementary compulsory courses, refer to the highly specialized ones as depending on the qualifications of his colleagues now teaching in the department, of the need for certain techniques – mathematical, laboratory, field-work – but if you are able and willing to question details he will soon lose his temper with you. In part this is so because he has enough of this topic which fills many frustrating debates in what seems endless and futile departmental meetings. In part, he loses his temper for the same reason that makes the departmental meetings painful too. It is this. Scientists go by public opinion and by political considerations – both internal and external, of course – so that the tough-and-no-nonsense science teacher can teach as he does only after he stops questioning the framework and content of the course and simply follows the syllabus as closely as possible. Internality, then, is defined by the syllabus!

This definition of the internality of science teaching by the syllabus is poor chiefly because the syllabus is easily swayed by external causes. But let us deny this. Let us now, for the moment, assume all syllabi to be one hundred per-cent scientific, determined by merely internal causes. The question remains, nevertheless, is the choice of each item on the syllabus and of its details internal or external? How does the department decide to teach this year course x and not course y? Suppose the choice is also purely internal. By what criteria? And how does the department know the criteria?

The historian of a science can try to evade the question and teach, or write papers about, or research, the history of the tough-and-no-nonsense course; still better, he can do so with respect to items shared by all courses on the topic around the globe: for example, Heisenberg's matrix mechanics.

So far so good. But once we deviate from the present-day tough-and-no-nonsense syllabus, we are at sea. In particular, we simply cannot take last century's tough-and-no-nonsense syllabus, which might not include Maxwell's equations and Mendel's genetics, and Kierkegaard and Marx and other items by now most often obligatory and quite standard; we cannot swallow syllabi which might include models of the aether and pre-Weierstrass calculus and evolution permitting no large mutation and the value theory of commodities and racism of all sorts and plain anti-sex.

What for the historian determines what is internal is no doubt quite different from what for the educator determines the same question. For, the historian's main tool for deciding internality is his view on scientific method, whereas what determines for a science educator's what is internal is largely public opinion and at times his own judgment and/or taste. Leaving the science educator for now I shall enumerate the attitudes towards scientific method open to the historian of science who wishes to decide what is internal.

First, he may take the scientific public opinion of the period he studies to determine what is the science of that period. This is almost never the case for reasons explained in the last two paragraphs. A glaring example will be phlogistonism, anathema to most historians of science, dead or alive, yet fervently defended and vociferously admired by all or almost all chemists for the generation or two when it was the vogue.

The most popular, still, is the inductivist view – the view of science as a collection of raw, objective data, plus the collection of theories firmly based on them by induction – which theories usually were and are scientific ever since they were successfully deduced (or adduced or induced or abducted) from the facts. Those who follow this view try first and foremost to offer as tough-and-no-nonsense a history of science as they know how. They willingly ignore at first all social and philosophical and other background, even personal opinions and idiosyncrasies of the great scientists of the past: they feel that after they did the real work of the historian, the purely internal history of science, which gives the field its very right to exist, then and truly then can they add the icing on the cake, or the jam on the bread-and-butter rather, and add some conspicuous external detail so as to evolve in their studies ever so slight a human touch, akin to the most celebrated human touch in one of the most famous tough-and-no-nonsense textbooks of the century, to wit a photograph of Feynman playing bongo-drums attached to his most esteemed, most tough-and-no-nonsense textbook on quantum theory.

But not all historians of science want the foundation of the history of a science to be the facts of the internal history of that science – not all of them are inductivists. Some expressly question those statements of facts which inductivists and others feel must be uncontested, such as that Joseph Priestley discovered oxygen. For example, Thomas S. Kuhn's celebrated *The Structure of Scientific Revolutions* (which advocates the *vox populi, vox dei* theory and so ought to admit phlogistonism as scientific but does not). Can we say that Priestley discovered oxygen if he did not know of oxygen, particularly if he died while persisting in his refusal to admit the very existence of oxygen? Surely his mere holding a container filled with almost only oxygen hardly counts: every untrained person holds a container with quite some oxygen or with mainly nitrogen. If you say that little oxygen – 20% or so – is not enough, then think of the 80% nitrogen in the air which most people cannot avoid handling even with no knowledge of chemistry at all. Moreover, it is obvious that many early chemists must have held almost pure oxygen in hand quite unsuspecting, yet they knew nothing of oxygen.

There are the conventionalists who view science as a system of mathematical and conceptual frameworks which serve as pigeon-holes to classify facts in them. For them the concept of oxygen, for example, is essential for the so-called discovery of oxygen; and so they view the history of this discovery in a very different way. Clearly, for them the internal history of science conforms to their view of science, its nature, its method. Clearly, what they consider internal does not quite coincide with the inductivists' view.

Rather than discuss these two methodologies, or the late comer but by now increasingly popular one due to Sir Karl Popper – of science as a series of conjectures and refutations, or of science as series of conjectures, corroborations, and refutations (depending on how you read Popper) – let me mention two points on the relation between one's view of scientific method and one's view of what is quite definitely internal to science. First, the connexion is not always simple. There are many attempts to formulate views of scientific method, older and newer. To what extent each of the allegedly new views of science is really new and can be applied to the construction of an internal history of science remains to be seen; but, clearly, it is not *a priori* necessary that any view of scientific method should offer us a view of the internal history of science: a view of science as the national ideology of the strongest nation, or as expressing the class-struggle, or as the views having the highest survival value for the species or for the society which holds it – these are some views blocking all possible presentation of any purely or nearly purely internal history. And there are views which permit some degree of internalism but forbid pure internalism. My second point is to the contrary. Many historians now take it for granted that at any time any existing (established or nearly established) field of science poses to its students a few key questions. They likewise believe that researchers in a given field of science are supposed to answer these key questions and examine their answers (first as tentative answers and then as putatively true ones). And, finally, some of these historians believe that there is logic to this procedure. Now, clearly, one may be excited about this logic and attempt to find the rules of this logic in given historical cases and then examine one's finding by attempting a historical reconstruction of that case and of other cases. This is what is now – but not historically – viewed as purely internal history par excellence. Purely internal history of science is that history which follows the inner logic of the growth of scientific knowledge.

To this kind of exercise, to a historical reconstruction of the inner logic of a given case history, whatever it is claimed to be and however incomplete it is claimed to be, to this no student of the history of science will offer any objection. The inductivist historian of science will start his historical reconstruction from facts rather than from questions, and his reconstruction may be criticized by his non-inductivist colleagues on this very account. But his predilection for a logical reconstruction, i.e. his decision to stay strictly within internal history of science, will be questioned by no one, contested by no one, criticized by no one.

What, then, is the debate between externalists and internalists? First, there are the purely externalists whose work is criticized (e.g. in the previous section) as worthless, parasitic, and decidedly on the side of the status quo. Second, and not in parallel, there is the purist internalist. He is contemptuous of all external history, and sees it as mere embellishment on internal history. Obviously, the ordinary externalist historian of science, or the undefined one, will not quarrel with the pure internalist. Oddly, the internalist will not quarrel even with the pure externalist, since they have no concern in common. What then are the internalist and the externalist quarreling about?

I do not know, and, in particular, since there exists a classic and famous debate on the question between Thomas S. Kuhn and Imre Lakatos (*Boston Studies in the Philosophy of Science*, Volume VIII) which I do not pretend to be able (or willing) to fathom, I must be wary on this point. Let me, then, make a simple conjecture. The desideratum of the conjecture is that it present the internalist and the externalist views as conflicting. I am far from agreeing about this desideratum. For all I know possibly there is no real disagreement and the parties are merely competing – perhaps they are merely trying each to get more attention, public support, etc., for the kind of things each of them is doing. Nevertheless, I shall cling to the desideratum – perhaps because I hold a beautiful hypothesis that conforms to it, perhaps because I am eager to utilize this hypothesis in some historical context I wish to reconstruct, and perhaps because I like to believe contestants, as I do, when they say that they do have a genuine disagreement.

Well, then. In my opinion the externalist affirms, and the internalist denies, that some patently non-scientific factors are very helpful, useful, perhaps even essential, to the successful rational reconstruction of science. To put it differently, I suppose that the question they disagree about is this. Can there be a complete, or a self-contained, internal history of science?

Anyone dissenting from this hypothesis will do well if he says whether he endorses my desideratum to view the conflict as real and take it from there. For the time being, however, I wish to present the internal logic of externalism, and I shall also offer sketches of historical reconstructions of case histories of externalism in the history of science.

3. IS COMPLETE INTERNALISM POSSIBLE?

Let me expand on my conjecture. There is much criticism of the purely external on this account: it is sooner or later bound to miss the internal altogether and without noticing the fact. But there is no criticism of either the purely internal or of the external. Criticism abounds, but it is specific.

That is to say, whatever study one produces, by whatever approach, can be and is, judged and valued on its merits and criticized for its defects. Hence, there is and there need be no quarrel where one might expect it, between historians of diverse styles, concerning these diverse styles. But historiographers, namely philosophers debating the question, how can or should a good history be written? can survey the possibilities and ask, is it at all possible to write a purely internal history of science? And on this, theoretical question, they may – and indeed, do – hold different views.

So much for my conjecture. Let me now attack the question, and from the broadest angle open to me. Let me start, then, with the most crucial question regarding a culture – of any culture, of any aspect of culture, of any subspecialty of it, science, art, mores, religion, or anything else. The most crucial question is, does it at all have, or at least seem to have, an inner logic? What, then, is inner logic and how is it ever exhibited? I do not know and I have found practically nothing said about it, even though it is constantly in the air these days. I suppose inner logic is the same as rationality, namely a given goal, given circumstances, and attempt to progress towards the given goal. In science the goal is the truth, the whole truth, and nothing but the truth. What is the goal of this or that other part or aspect of culture is harder to specify. Now, rather fortunately, the function of the goal is to provide certain criteria for progress, for achievement, for any sort of judgment. And so, an inner logic may perhaps – I am not clear in my own mind about it – be criteria in action, so to speak. Internal criteria in action will be inner logic, and some activities have them but not all.

As I said before, some philosophers deny that internalism is at all possible in the history of science – e.g. some Marxist philosophers. That is to say, they deny that the criteria for judging scientific excellence belong to science. In other words they deny the very autonomy of science. This need not be a matter for concern – it all depends on the situation. Let me illustrate this with a very simple example not from science or its history but from science education. The reader may remember the frustrated science educator who, in truth, has no knowledge of any clear-cut criteria by which to decide the curriculum, though he has some, regarding the most elementary level and the most special and important cases. Now, compare him with the vocational trainer or educator. He has much less of a problem on his hands since, plainly, his criteria are external and given to him by the profession and by the market situation. Similarly, if the 'vulgar' Marxists are right, then scientific research is centered round questions important to the economy and so researchers should know what scientific problems are important and what not.[1]

I have thus far presented classical inductivism, with its view of science as the autonomous search for truth, as internalist par excellence, and classical Marxism, or a variant of it, with its view of science as a part of the social process (class-struggle) as the opposite, the denial that internalism can be kept pure. This picture is false and needs correction. Indeed, I wish to declare that as far as the disagreement could be put as plainly as that, there was hardly any interest in the debate, that interest flared up because matters did get out of hand – and in quite a fascinating manner, may I add.

The schema for things going out of hand is, I think, simple: you have a theory about what is internal history and how it works; you apply the theory; it applies with difficulty; you introduce external factors to overcome the difficulty. Query: is this move quite kosher? or should your theory be rejected? The answer hinges on the question, is your theory introduced as purely internalist? The answer to this hinges on, can there ever be a purely internalist theory? If we say, there cannot be a purely internalist history, then your rescue operation looks more kosher than if we say, there can be a purely internalist history. End of schema.

Examples. In the inductivist tradition it must be admitted that science and its history are one, since theories based on facts have, according to classical inductivism, to emerge from them. Laplace noticed that the history of astronomy as it really happened deviates from his inductive reconstruction. He explains it by an external factor: it took time for the inductive method proper to spread and be employed in all the necessary rigor. Second example from inductivism. Bacon said, no selection of subjects, or interests, of questions is permissible, since the selection is based on theory and either theory is based on facts which are thus unselected or it precedes facts and is thus a prejudice. Historians must proceed with some order from the start. So they cannot but bring in a non-intellectual selection-criterion, such as the incentives economic and technological needs offer scientific research. This view is somewhat Marxist, perhaps pseudo- or crypto-Marxist; I cannot say. I have discussed both these examples in my *Towards an Historiography of Science* in 1963 (facsimile reprint 1967; Wesleyan University Press).

A different example. The conventionalist-instrumentalist historians of science are disturbed by realistic moves in science such as scientists' search for an aether in physics, atomism, radical theory, valence theory, and the like, in chemistry, and classical value theory in economics. Now, clearly, some such ventures can be viewed as external to the history of science, since realism is, no doubt, a metaphysical theory, not a scientific one. Still further, some realistic ventures were deeply misguided, such as the search for an aether.

And yet, some such searches were epoch-making: indeed, but for the realist search for an aether Michelson would have never devised his celebrated experiment which opened the way to the Einsteinian revolution in physics. The historians who are conscientious instrumentalists will have to concede the point and allow for it somehow. The way was shown already by Pierre Duhem, the father and doyen of instrumentalist history of science, in his celebrated *The Aim and Structure of Physical Theory* (1905): realism entered science through external factors, such as Galileo's quarrel with the Church of Rome, and of course, in addition to the damage it did, it also did some good, but anyway it is plainly false; there is no aether, and Maxwell's equations are merely this: mathematical equations for the coordination of facts.

Popper's theory of science as conjectures and refutations was open to criticism from the very start. Popper conceded at the very end of his *magnum opus*, his *Logik der Forschung* of 1935, that though the most testable hypotheses should be taken up first, at times quite untestable – metaphysical – ones are taken up, such as atomism, and then rendered testable. He dismissed such cases as external; it matters not what factor stimulates a scientist to imagine a testable hypothesis, be it metaphysics, be it facts, or be it a cup of coffee, as Poincaré put it, the stimulus is external.

It was Émile Meyerson who disagreed with Poincaré on this very issue early in the century. Metaphysics, he said, stimulates more than one scientist, more than on occasion, and in a systematic manner; it is not akin to inspiration from coffee. This was taken up by Alexandre Koyré who studied in the 1930's and 40's the systematic impact of Platonic metaphysics on Renaissance thinkers and by I. B. Cohen who showed (1956) that Ben Franklin's researches were carried out along the lines of Newtonian metaphysics. Prior to Koyré, and seemingly with no connection with Meyerson, E. A. Burtt published similar views (1924). I remember well, when I discussed all this in the tenth international congress for the history of science in 1962, the ensuing debate concerned one question only: why not incorporate those metaphysical views that have systematically influenced scientific research into the body of science proper? I could not see the force of the question then, and said so. I do now, perhaps: the question was, do I deny that a purely internalist history of science is possible? I can now say, for my part I am indifferent to this question. I wish to explain why and with this close the present discussion.

The problem on which the debate rages, we remember, is, can there be a comprehensive internalist history of science? For my part, I think nothing comprehensive is ever possible. The importance of the question, we remember, stems from the recognition that a certain view of (the history of) science is inadequate plus the historian's willingness to broaden, rather than change, his view by adding (allegedly) external factors. This sounds apologetic, and the question was, is it apologetic or is it permissible? For my part, I find the question not very pressing: the worst thing about any apologetic writing is that it is boring. To make one's work interesting one only need write it from different points of view as openly as one can.

The question, can there be a complete internalist historiography? is metaphysical. Even if there existed one, it may lie far in the future. What then is a historian to do in the meantime? He can use any theory he knows, and when they all fail he should do as he pleases – on the condition that he take his reader into his confidence, of course.

A historian of science may be suspicious of all that. Why should he employ a defective theory? he may ask. But he can hardly do without a theory of art or science or religion, without knowing what art, science, or religion is. Which, of course, amounts to a theory. But there is an answer to this: rather than be an internalist, and a purist one at that, we may go to the opposite direction and make our histories as inclusively externalist as reasonable, so as to accommodate as many possible theories of art, science, or religion, as the case may be. If I understand correctly Roger Hahn's review of my *Towards an Historiography of Science* (in the *Journal of the History of Philosophy*, 3 1965), then I think this is his proposed solution. I think it is far from being indifferent to any views of science: I think it is mildly inductivist.

Before bringing this study to its natural close I wish to mention two very important writers on the internalist-externalist controversy in art, E. H. Gombrich and I. C. Jarvie, both of whom are, incidentally, disciples of Popper. They both stress the artist's inner logic, and say that the problems an artist faces, the situations within which he operates, are dictated by the general setting within which he finds himself. In other words, we cannot comprehend an artist, or a scientist, etc., much less his inner logic, unless we know quite a lot about his background; even such internal problems as, what idiom did he accept from his schooling, what he borrowed from diverse sources, and what idiom he invented – these are highly significant for any internal history, yet often they concern the interface of the internal and the external.

I think this point, obvious as it is, has been overlooked by many students of René Wellek's view of literary history: Wellek insists that criteria are internal, so that though "nobody should be forbidden from" doing external history, "the objection" to externalism is that through it "external criteria are introduced" (René Wellek and Austin Warren, *Theory of Literature*, 1924, opening to Chapter 2; see also the introduction to Part III on "the extrinsic approach"); yet he also thinks pure internalism – of the kind popularly ascribed to him – is impossible (see especially chapters 5 and 9): "literature occurs only in a social context." In other words, I understand him to say, the question of autonomism seems to him important and he suggests that art (and, I add, science) should be viewed as autonomous. (Is religion autonomous too?) But, of course, even an autonomous field is not capable of purely internalist treatment without severe limitations. Which is not to deny, but even to stress, that purely internalist analysis of historical episodes can be very exciting.

POSTSCRIPT BY I. C. JARVIE

I too wonder what is internal logic. Is this too vague: follow the argument where it leads? But then: there is more than one internal logic (e.g. *The Republic* is not *the* debate on justice); i.e. internal logics are not natural, given, essentialist, *but* reconstructions. Internal history is a caricature of real history, says Lakatos, which perhaps strives to be, but never is, quite fully, internal. His "Proofs and Refutations" is *an* internal history, and also no history at all. Is internal history the attempt by intellectual historians to explain what happened, using a minimum of external factors?

NOTE

[1] The question, was Marx a "vulgar" Marxist is very intriguing. See the very forceful, if brief, discussion in René Wellek and Austin Warren, *Theory of Literature* (Harcourt, Brace, and Co., New York, 1942, 1947, 1949), Chapter 9, pp. 102–6, where the term is traced to the Marxist critics A. A. Smirnov and V. Grib, and where Marx himself and Marxism in general are presented as ambiguous on the issue.

Let me notice that "externalism" and "internalism" are presented in that enlightening volume as "extrinsic study" and "intrinsic study", which parts III and IV, the central parts, respectively explore.

CHAPTER 6

THE AUTONOMY OF SCIENCE

The praise of the autonomy of science seems to be something so laudable it barely needs defense. This is the sign of rapidly changing times. For, it was only in the forties, as I well remember, that the defense of planned science, i.e. government regulated and controlled science, was no less in the vogue, and was taken for granted not only by left-wing scientists. Michael Polanyi is by now one of the best known philosophers of science. He began his philosophical career when he attacked the theory of planned science during World War II, when he was a well-known scientist. He began his defense of the autonomy of science arguing from the autonomy of the scientist, and he ended by denying this autonomy. I wish to argue that Polanyi's progress is logical enough: you cannot have it both ways.

1. INTERNALISM AND EXTERNALISM AGAIN

For the philosophers of the Enlightenment scientific research, or, as they called it, philosophy, was the expression of man's moral quality of intellectual autonomy. Not accidentally they viewed the science or philosophy of man as moral science or moral philosophy, not social science or philosophy: the philosophical man is above local differences and national conflicts, they were sure. The philosophic man exhibits a philosophic style of life: he belongs not to a social stratum, nor to a profession, but to the spirit of Enlightenment which bestows on him a lofty mood.

We forget these things today. We view the visit of Sir Humphry Davy to France, the enemy country, during the Napoleonic wars, as an oddity rather than as characteristic. We view as more characteristic the fact that World War I divided scientists, natural and social, philosophers, irrationalists and allegedly rationalists, and, to topple the humiliation, even the most philosophical, most universalist of all the then existing movements, namely the socialist movement. We are reconciled to the fact that some of the best scientists of the Nazi era tolerated the regime, sympathized with it, and even defended it. Otherwise Konrad Lorenz could not have won a Nobel Prize.

Yet a new opinion is emerging, which views science as not just a set of some ideas plus some laboratory experiments; which views science as not even an isolated segment of society. Rather it is an integrated and variable segment that plays different roles, which roles are reflected within its most specific activities. As one of the initiators of the current debate against the internalist school of the history of science which studies its development in intellectual isolation, I wish to state clearly that I too reject any general demarcation of science just because not only the world of Davy differs from the world of Lorenz; even the scientific world of the one differs from that of the other. We can still make comparisons and analogies, point at advantages and disadvantages, and also outline some historical continuities wherever we think they exist, and discontinuities otherwise. The anti-internalist view is winning in the discussion of the nature and history of art; it is time it also gained strength in the parallel discussion of science. The strongest internalist argument regarding science as opposed to art is that whereas art is subjective, science is objective. This is now *passé*. Nothing is purely subjective and nothing is purely objective; the division between subjective and objective is a charming eighteenth century rationalist hope fostered by nineteenth century romantic thinkers, and so it is very widespread. Art historians and aestheticians increasingly reject both subjectivism and internalism in art, and some philosophers of science do the same regarding science. Often this leads to the rejection of objectivity in science or at least to its relativization in the theory of relative truth. Yet relative truth, though still popular enough to be endorsed by Kuhn and his hoards of followers, for example, is really not worth discussing: at best the theory of relative truth is relatively true; the theory of absolute truth, then, is absolutely true.

Internalism, observes Ernest Gellner, is enjoying great popularity. In art, science, even religion, these days internalism is defended by various thinkers, from the Wittgensteinians to the scientific Catholics, to the existentialists and the New Left. Yet the best among the internalists is Michael Polanyi who defends internalism, as Duhem did it before him, in the name of autonomy. The autonomy of science, of art, of religion, the independence of these activities, impose on us the restraint of not imposing any one of them, or the conclusions of any one of them, on any other. It looks as if this makes the practitioner of one of these fields more autonomous, more free, since he does not have to fear pressures from any other field of activity. A scientist may rest assured, even if he does not know much about religion, that he is safe from attacks from its end, and vice versa. This freedom is seen by many as a great boon, and I have personally encountered this almost as often as I used in discussion with any professional man ideas and information rightfully belonging to another profession. The professional would respond with indignation and declare that he is not bound to be omniscient – and that I am bound to accept expert professional judgment, as he does. The autonomy of the different fields, then, certainly offers their practitioners the right to be ignorant of other fields and the right to trust implicitly the experts in these fields. But the autonomous man does not need a justification of his ignorance, nor does he want to protect it. The autonomy of the field at best has nothing to do with the enhancement of the autonomy of its practitioner.

2. SCIENCE AS A GUILD

Polanyi attacks autonomism as an impossible dream: no one can start afresh and so he must depend on others. Science starts neither in experiment as the classical empiricists have told us nor in reason as the classical intellectualists have. Science starts in freshman science courses in college. A scientist is, to begin with, an apprentice. Here we see science as an autonomous social entity, a medieval guild. Its members are not autonomous at all; they submit to the guild and its traditions and willingly become heteronomous at least for the time being. They cannot even decide to become heteronomous out of a rational debate; not only because rationality and heteronomy are antithetical, but also because the master cannot explain to the novice what he is going to get as an apprentice: the knowledge conveyed is personal and cannot be conveyed merely verbally: to learn it you must be an apprentice first. If later on you find out that you were in error, that you have chosen the wrong vocation, you cannot complain and you cannot see to it that others will avoid your mistake.

Is the master who is a member of the guild, is he perhaps autonomous? No, says Polanyi. To live in the guild you must accept its current norms, teach its current doctrines, etc. Even if your doctrine is the better one, you must teach your apprentices the current one. Polanyi says so explicitly. As a man who changed his vocation, as a man who had a superior doctrine yet taught the current one – and did not complain to the last – he is a man who lives by his teaching. As a scientist he just was heteronomous to the last.

And science whose membership live in heteronomy within autonomous professions is one where a unifying metaphysics plays a much smaller role than a science which has no boundaries and is practiced by autonomous individuals. But I must leave this now.

Is there anyone autonomous in Polanyi's guild? If we mean by autonomy a full autonomy, even the freedom to diverse intellectual activities as was done regularly by eighteenth century scientists, then I am afraid the answer must be negative. But some people do have autonomy, according to Polanyi, enough to make a revolution, i.e. to alter current views, traditions, or fashions. This makes them leaders. That is to say, they are leaders in virtue of the fact that the guild as a whole follows them; and the guild may one day just rebel and switch leadership. A revolution may constitute such a switch of leaders.

This theory is extremely broad in its application. It was meant, originally, to show the resemblance between the way of life of a scientist and those of an artist or a priest. The very words master and apprentice are used by Polanyi to stress this fact. And, of course, they may characterize other ways of life as well, for example that of crime.

Thomas Kuhn has applied Polanyi's views to the history of science, studiously avoiding almost all reference to other guilds. In a surprise move, then, J. W. N. Watkins showed that Kuhn's scheme applies to medieval theology, and Paul Feyerabend showed that it applied just as well to safe-crackers. But when we remember that Polanyi intended this view to apply to any professional way of life, to any guild-system, Watkins's and Feyerabend's discovery becomes less surprising, as it amounts to the discovery of guilds. Damon Runyon has told us that safe-cracking is a guild. Watkins and Feyerabend have shown how little sophistication is necessary to develop a guild, how little intellectual quality a scientific community needs to acquire Polanyi's and Kuhn's blessings. Guild society is, indeed, relatively primitive.

The reason that unwittingly Polanyi and Kuhn have managed to put the scientist and the safe-cracker on a par is clear: they forgot at the time all about the aspirations of scientists. Indeed, Kuhn's main point is merely that a normal scientist's aspirations are small; small as these are they may loom large for scientists with still smaller aspirations.

Polanyi's and Kuhn's theory, apart from sanctioning the professions' autonomy of and the heteronomy of most of their practitioners, is not specific enough to even distinguish a scientist from a safe-cracker. Their theory is simply poor as moral philosophy and poorer as social science. Social science must take some account of aims of members of social groups or of the functions of social organizations, or some other added factors. Otherwise sociology of science and of crime would merge.

3. AUTONOMY AND HETERONOMY AGAIN

Yet, clearly, Polanyi does offer us a sociological analysis. It is not of science, but of heteronomy. It can apply not only to science, art, and religion, but also to tribal politics. We can now either split the sociology of heteronomy into that of heteronomic art, science, religion, etc., or contrast it first with the sociology of autonomy and then split the two together. There is one advantage here: whereas heteronomy is prevalent, the traditional locus classicus of autonomy always was science. Of course, democracy is the politics of autonomy and Western art shows increasing measures of autonomy too. What characterizes autonomy, then?

The autonomic is self-sufficient, he makes his own rules and decisions, though not capriciously or arbitrarily. Now, strangely, the act of polarizing the rational and the capricious, of identifying the rational with the justified, namely given to proof, and of making the proven obligatory, overshoots its mark and makes rationality heteronomic. I need not mention Bartley and Popper here. Rather I wish to mention Walter Kaufmann's recent (1973) and exciting *Without Guilt and Justice*, which amply enlarges on this point.

What makes justification or proof heteronomic is that it provides unique solutions to given problems, which solutions are imposed on us, thus relieving us from the burden of responsibility and choice. Quite generally, any theory of this characteristic, any monism, is in principle, even if unintentionally, quite heteronomic. Now clearly, most choices are of actions or of opinions, and the problem of choice may be solved monistically and so relieve us of the responsibility of choice. This seems to leave science alone since the very task of the man of science is to tackle problems and sift opinions! Here, Popper, Polanyi, Kuhn, and Lakatos, all come to the rescue and with a monistic theory impose problems and opinions on us by monistic criteria. Polanyi and Kuhn do it crudely: accept the opinions of your peers, work on small problems imposed on you by the leadership. Popper says, since rationality is criticism, choose the most highly testable theory. Lakatos says, since rationality is problem-solving, accept the metaphysics which generates the program for research that prescribes the most progressive, the broadest, problems around.

Rationality, I think, is not polarized. The criteria we have are partial, the choices they offer are not uniquely determinable. Our choice is not clearly so arbitrary that any fashion will do – the whole world of science, says Popper, may be in the wrong; and it sometimes is, says I. And our choice is not clearly so rational that every dispute is between the rational and the irrational, the black and the white. The logic of autonomy, then, is much harder than the logic of heteronomy.

Come to think of it, even heteronomy is not total, and this is how normal scientists too may on a rare occasion make a significant contribution. But the asymmetry remains. For Polanyi and Kuhn heteronomy is cardinal; it is easy to describe indirectly: follow the leader; it is impossible to describe it directly, since to know it is to live it. But if autonomy is cardinal it is much harder to say what it is, and there is no leader to follow and no whim. But we can still say of some things what it is not. In particular, it is not what classical thinkers said it was – Polanyi is right here. And even autonomous scientists must coordinate their research. Polanyi is right in saying that scientists do usually know what problems have currency in the world of science, what not. Kuhn is right in saying that his heteronomic theory explains the coordination in science: it is, he says, imposed from above! If we wish to explain otherwise the existing coordination, then we must turn to more sophisticated sociological methods.

4. THE SOCIOLOGY OF SCIENCE AGAIN

Our study of science may be sociological in diverse ways. We may use the Weberian method and construct the ideal type of the scientist. Will that be more akin to Davy or more akin to Lorenz? Or shall we present as the core of science only what is shared by all scientists? Is there such a core? Is it not possible that tradition shifted? If there is no core of Christianity shown by the Apostles and the diverse Christian churches through the ages, as many historians of religion think, then perhaps the same holds for science. After all, the ideal type of x is the essence of x, and perhaps this does not exist!

This does not render the Weberian method invalid; we can develop and test theories of ideal type scientists, both global and local. Only, I think, all existing global ones are by now fairly well refuted; still, we can continue with local models, the Reformation Protestant, the Reformation scientist, the early Christian, the Greek scientist, etc., etc.

The second most popular method of sociology is the Durkheimian or functionalist method, made so very popular by Malinowski, Radcliffe-Brown, and Evans-Pritchard. The crux of this method is the rejection of views of natives about their own activities and institutions. This crux goes, indeed, to Hegel's theory of the cunning of history. Similarly, the functionalist theory of Polanyi regarding master and apprentice goes back to Hegel's master-slave theory. I find this idea otiose, and I particularly wish to see some truth in many scientists' profession of openmindedness and reluctance to indoctrinate their students. I find particularly dangerous Imre Lakatos' view that we can just as well forget all what scientists say about their traditions and aspirations. I am not forgetting, though, Einstein's terrific joke, his mock-inductivist and mock-functionalist remark, do not listen to what scientists say they do, look at what they do. Indeed, I cannot but admit that scientists may be inconsistent, as when preaching positivism while engaging in speculative metaphysics, and that scientists may preach peace and wage war. Here, I think, Einstein's joke becomes an apt counsel. But to conclude, like all functionalists and Lakatos too, that what scientists say about science is pointless, is just too much.

A third, and more general sociological method is that of the logic of the situation, which analyzes people's conduct in the light of the aims and their physical *cum* institutional conditions. It is easy to show the superiority of this method over the other two by pointing at their obvious defects which it does not suffer from.

Consider ideal types first. It is no criticism of the ideal type theory to say that, try hard as they may, not all members of a given movement manage to come equally close to the ideal type. True, we must then supplement the ideal type theory by a statistics of the degree of approximation of members of the movement to the ideal type, very much à la Kinsey's: to what degree do people who preach chastity manage to approximate it or deviate from it? But this Max Weber would consider a part of the paraphernalia of his ideal type method.

The trouble is that every movement, when it becomes successful, as Shaw observes, becomes a bandwagon and accrues bandwagoners who conform to a different ideal type. The ideal type of the bandwagoner is described in the Bible as the rabble and the mixed multitude, and by Kuhn as the normal scientists. Yet the situational logic of both members and bandwagoners is fairly obvious from the very start. We have, likewise, the rebel turned a conservative leader. But I shall leave this.

Similarly with the functionalists who simply ignore the expressions of members of a tribe in favor of what they think is true. They cannot ask, what is the intellectual point of a mistaken self-appraisal since this will short-circuit the functionalists' mock-relativism. They therefore give a blanket explanation of all false-self-appraisal as symbolic expressions. I cannot go into further detail here. One need not criticize functionalism here. Suffice it to argue the impossibility of a functionalist explanation or even allowance for the fact that scientists often preach positivism and practice metaphysical speculation. I have explained elsewhere the situational logic of scientists who behave so. The erroneous views of scientists, in particular, may be poor expressions of proper sentiments – such as over-strong reactions to bad metaphysics, fear that young intellects will be devoted to speculation and neglect experiment, etc. Further, situational logic views function, as a rule, as the consequence of human action, usually unintended but at times intended (especially in cases of successful social reform). The technological function of science can serve as an excellent example, as it is often unintended but at times intended.

Situational logic is not the last word in sociological method, but it goes much further than most studies in the sociology of science. Even the most popular and hard-nosed current sociological study of science, the so-called citation-index studies, prompted and sustained by Derek J. de Solla Price and Thomas S. Kuhn, are extremely poor sociology. They mix ritual citations with scholarly ones, for example –an error easily corrected by functionalism and by situational logic. Also they mix favorable and unfavorable citations: this is correctable by the ideal type theory and by situational logic. It agrees with the Hollywood maxim that all citations are good, in which names are properly spelled. And, we know, the highest citation index in sociology a few years ago was given to a misspelt or perhaps fictitious name. Situational logic is good enough to disentangle such matters with great ease: the non-existent author was cited merely ritually.

There are, however, problems that go beyond situational logic. The most important one is, what serves as a means of coordination between different thinkers concerned with the same problem? My answer is that in fact metaphysical systems do so, much in the way later described by Lakatos, though he does not address the problem of coordination. Rather, trying to improve upon existing views about metaphysics, he ignored both the existence of competing metaphysical systems and other means of coordination, and other sources of interest, such as technology or even the love of music which led Helmholtz to his acoustic researches in his effort to defend old fashioned pre-Bachian musical tastes.

To conclude, some men of science are autonomous, some heteronomous. Subsequent to the latter, science itself is a fairly autonomous and powerful guild, much as described by Fred Hoyle's *Black Cloud* which therefore won love and admiration among scientists. Do we want science to be or remain autonomous? Do we want the different scientific branches to maintain their autonomy? Shall we encourage normal science and government grants for it? How, in brief, do we want to see science develop? This is today's agenda. Certainly we cannot answer these questions without using our normative or prescriptive views; but not only these. We have to say what we like about science, what are the aspects we wish to play down. And here I freely follow Popper and say, give more autonomy to the individual, and critical discussion is the

the instrument of individual autonomy. So much I go with Popper. But discussion can degenerate into scholasticism or into small topics within normal science. And so Popper, at best, is not good enough. To prevent normal science we may wish to institute preferences for large doses of general problems and of disagreements – metaphysical, artistic, or political: we can assault the autonomy of science.

CHAPTER 7

THE LEGITIMATION OF SCIENCE

1. SCIENCE AS AUTONOMOUS

The problem of the legitimation of science may have a different significance in one time and place than in another. Thus, at the dawn of modern science, when science became increasingly the intellectual competitor to religion, both science and religion, i.e. the spokesmen of both, each questioned the legitimacy of the other system. Once science won the hegemony, the problem arose afresh, but on a smaller scale and within science, every time a controversy raged. Soon science threatened the existing political order, and so representatives of the Establishment and of science questioned the legitimacy of each other's framework, and thus the whole system of science could be questioned afresh. But this did not happen: The representative of the Establishment only questioned the possibility of a social and political science – of a science of man. In the nineteenth century within the scientific community some saw the possibility of the science of man and remained radicalists, others raised doubts about the very possibility of the science of man, thereby giving legitimation to the Reaction.

Today the picture is different, and more subtle. The importance of the natural sciences for the conduct of human affairs is so obvious that everybody gladly declares admission and endorsement and acceptance of it. By emphasizing and enhancing the social and political import of the natural sciences we have managed to take off the agenda the question of the legitimation of the human sciences. Each political party which is worth its salt has its own battery of social and political scientists to legitimize its positions, more or less, and the question of the legitimacy of the human sciences is played down by the very insistence on the legitimacy of the natural sciences and on the importance of this legitimacy in all walks of life.

Yet, even when we confine our discourse to the natural sciences alone, and even if we refuse to question the system of natural science as such, we are not done with the question of legitimacy. For, within the natural sciences themselves there still are controversies, which means that specific doctrines are declared legitimate by some and not so by others. How are such matters handled?

There are two aspects to this question. First, within the world of science, where controversy rages, how is it decided? Second, how is the outside world given the verdict? How do non-scientists know who is the true blue scientist? How does a government appoint an expert on its own committees? Does he have to belong to the right political party within science, or rather outside science, or both, or neither? These are questions outside science, but they cannot be solved adequately without a look into science.

One of the most popular theories concerning scientific legitimation is that of Michael Polanyi. Polanyi himself is not so very well known as his ideas are, partly, but only partly, because they are usually attributed to Thomas S. Kuhn; more likely they are popular because they so well express the current ethos. Polanyi or Kuhn best expresses the *Zeitgeist* as far as scientific legitimation is concerned. I will refer to their views, throughout this study, as to the current views. Of course, there are variants: Kuhn does not endorse all of Polanyi's views and each of them has developed or otherwise changed his view on this or on that matter; but mostly the differences are matters of mere nuance; as public opinion is vague, ambiguous, ambivalent, even inconsistent, on some matters, we can not be too precise anyhow, or else we miss our target.

Let me stress that the current view has priority, not on account of its excellence (though it is not as bad as one might expect, and I find it interesting in a few respects), and not on account of its vulgarity (and it is vulgar, to be sure, in both senses; it is both widespread and coarse), but on account of its being established and operative: whether we like the current theory or not (and I do not) we must notice (and I even wish to emphasize) that it works. The way things in general and theories in science in particular are legitimized is just the way people expect them to be legitimized; politically, current views always gain top priority, within the politics of science no less than elsewhere.

I do not mean to say that since the current view of legitimation is operative all government is legitimate. How one gets the helm of government is one thing, and whether and how he is legitimized is quite another. No doubt when, before World War II, the German forces invaded neighboring countries, their invasion was neither legitimate nor legitimizing. Still, Austria legitimized the invasion and by popular acclaim; Czechoslovakia, except for the Sudeten, did not. Moreover, popular acclaim legitimates only on the assumption that it is the people that legitimates: In Nazi Germany and in Soviet Russia, despite all differences between the two, it was the government which legitimized its scientists in the name of the people, not the people. But, finally, even where popular acclaim is what legitimizes, this is not to say that *vox populi* is *vox dei*: the act of legitimizing and the state of being worthy of legitimation are two quite different things.

The current view takes all this well into account. First and foremost, it demands the autonomy of the scientific community, and claims that such autonomy can be achieved and secured best (but not only) in a democracy. Second, it claims that once the scientific community is the legitimizer, it usually but definitely not always bestows legitimacy on the most worthy candidate. Third, that legitimacy is transient: we can criticize any given scientific doctrine – but not science as such, not the whole body of scientific doctrine – and replace it by a better one; and this is an act of transfer of legitimacy.

That is all that the current doctrine asserts. Everything else is either corollary or ancillary and anyway peripheral at best if not mere frills. In particular, there may be criteria for the transfer of legitimacy. These criteria may be proper descriptions of what goes on within science or not; if they are they may be the best or not. Even the best criteria, however, may be violated by the scientific community which is autonomous and is at liberty to do so. It does so, of course, only with good reasons (usually, remember; not always). These good reasons are not always articulated: when they are they become the new criteria, but there is the scientific community's verdict which is more powerful than any criterion. What legitimizes a scientific doctrine, then, is the world of learning, the scientific community. Likewise, of course, the scientific community may be divided, for good reasons, of course (usually, not always), but sooner or later the balance must be tipped. The scientific community decides. Likewise, the scientific community has a leadership, and the right one of course (usually, not always, remember), and quarrels over leadership may take place, and leadership may be overthrown and replaced, and so on. But, again, sooner or later the leadership is established, though only for the time being.

2. THE SELF-REGULATION OF SCIENCE

The contribution of Thomas S. Kuhn to this doctrine, which was largely elaborated by Polanyi and was in the air anyway, is the notion of a paradigm, which is the Greek for chief example. He uses this word ambiguously, and therefore we must find one of its uses for the purpose of studying the current view. One such use would be to describe textbooks. One can distinguish different disciplines by their use of textbooks in teaching. Some use practically none. Science within the last century at least can be distinguished by the use of standard up-to-date textbooks, a practice which has something in common with the medieval scholastic use of *texts*.

The modern science textbook, then, is the primary legitimizer of science. To enter the textbook is the ambition of many a researcher. How, then, are the authors of the textbooks to choose their texts? The textbook, to repeat, even when quite up to date, nevertheless describes out of date theories in it. For didactic purposes, these are used to introduce more modern theories. It is a sociological fact that this is the only way students are introduced to their subjects. How then are those chosen? On what grounds?

The out-of-date theories in science textbooks are usually false theories. Until a few years ago, and perhaps even to this day, scientists did not like to admit that they are false. Many of us accept the view that it is wrong (evil) to teach a falsehood, even while noticing that it is a falsehood.

Scientists often describe all theories as not false, but true with qualifications (or within limits). But it is an often attested fact that students who learn modern theories after being skilled in applying older ones, experience emotional and intellectual crises. The effort required for the switch from an older to a newer way of doing science is often not an easy one.

The question arises: Are older theories such as Newton's in physics false, or are they true within limits? This way of asking questions is unhelpful because of an ambiguity: Newton's theory unqualified is false, but qualified it is true. But it is an important fact that we cannot avoid falsehood in science. The reason is that Newton's theory qualified cannot be taught as a way of introducing Einstein's theory, for example. The reason is this: Newtonian theory qualified can only be understood as part of Einstein's theory, or as following from it. Hence to understand the qualified old theory we need to understand the new theory, whereas the old theory is a necessary introduction to the new one: we thus need the old theory unqualified at first.

Hence teaching falsehoods is unavoidable. Some may, of course, tell their students that it is false, while teaching it. But this is only done by those who accept the hypothesis that teaching falsehood is not wrong (evil). The reason is that teaching falsehood is unavoidable, as we have seen, if we are to do science. It cannot be wrong to do what is unavoidable. Honesty demands only that we admit known falsity.

This raises a most important problem. In the past, when people thought textbooks included only true theories, it was easy to see how they chose the theories for inclusion in the textbook. They chose true theories. But now we know that at least the older didactically useful theories are false. Why do textbook writers include the particular theories that they do? Why are other false theories excluded? And how do we differentiate between those to be included and those to be excluded?

Let us assume that by tradition, by trial and error, by the use of educational theory, and so on, the current textbook is geared at training the science student to acquire knowledge of current theory. Current theory, then, might count as the pinnacle of the current textbook. The textbook's author presents as the pinnacle the theory current at the time in the scientific community, or else his work will not count as a textbook. But this raises the question, how does the scientific community decide which of the extant theories to validate. How is current theory chosen? Not necessarily by any criteria, we remember; not even necessarily by a consensus, since the leadership of the scientific community may impose it; whatever explanation for the acceptance of current theory one might offer, the explanation may turn out to be false. There is no need for an explanation, anyway, since the world of science operates well enough (though far from perfectly) when left on its own and to itself.

This doctrine, known as the theory of tacit (i.e. silent) knowledge or of personal knowledge (since so labelled by Michael Polanyi), has been attacked by quite a few philosophers (e.g. Adolf Grünbaum) as irrationalistic. And rightly so, but I will not join this attack here. Rather, let me notice, it rests on a tacit political supposition. The supposition is that within a democratic society, when science is organized without interference, sooner or later the community of science is organized better than any formal rules can regulate. This way the individual scientist becomes dependent on the community of the chosen, and his dependence is explicitly defended by Kuhn; it makes the society of the chosen communal and undemocratic. This can be illustrated not only by the fact that both Polanyi and Kuhn recognize and approve of the use of means of social pressure in the name of social cohesion. It can be illustrated by the science fiction novel *The Black Cloud*, written by the leading astronomer Fred Hoyle and reputed in the world of science as an adequate expression of popular sentiments within it. The novel, as a novel, is no doubt terrific. It describes old Jehovah embodied appearing on the human scene and interacting with humanity via the community of scientists, finally leaving an uncomprehended message before disappearing, perhaps for good, perhaps temporarily. At the start, when the cataclysm of the appearance of the black cloud approaches, governments bring together the world's best scientists and they stay together in a temporary settlement for the duration of the whole novel. They have thus unintentionally created a utopia, a perfect community with no particular political problem, a democracy without the democratic machinery or institutions, without checks and balances, a true meritocracy implicitly trusted by all. Just dreadful, yet alluring to men of science today more often than is safe for democracy.

3. DEMOCRACY WITHIN SCIENCE

Sir Karl Popper has offered a totally different view of the matter than Polanyi and Kuhn. The success of the world of science rests, he says (not in the communal nature of the world of science which they describe, but) in the openness of that world and its living within a democracy. The existence of highly democratically organized scientific organizations, the democratization of all sorts of institutions of learning and the jealous protection of academic freedom wherever it is exercised, the intrusion of new ideas from outside the world of learning, the democratic competitions between different institutions of learning within one country and between national groups, all these institutional arrangements foster the critical attitude which overrides any individual scientist's possible reluctance to be criticized, as it overrides his possible readiness to exercise or submit to authority.

How shall we take Polanyi's and Popper's views? We may take them in their strict readings to say that the social structure of science is communal – authoritarian and democratic, respectively, as sheer sociological generalizations. In this case, incidentally, much of what the sociologist Robert Merton says in his earlier writings is in accord with Popper's views, and much of what he says in his later writings (particularly concerning the imaginary citation indices that now haunt the sociology of science) is in accordance with Polanyi's. Now, taking Polanyi's and Popper's views as strictly sociological generalizations, each of them can and does marshal instances supporting his views; these instances refute each of the views of his opponent. And so, taken as strict sociological generalizations, both Polanyi's and Popper's views are refuted.

What I think these theories should be taken as, however, is more sophisticated, yet still factual claims. The valuable in science, say these two opponents, is due to the scientific world being organized as a community, or as a democracy, they contend respectively. These competing views, I contend, are given to critical empirical examination.

My concern here, however, is less in the question what is the excellence of science due to, and more in the question, how does the process of scientific legitimation affect this excellence. And my thesis is, briefly, that all legitimation, within science or without, is detrimental to the growth of science – even when it is unavoidable, as, for example, in scientific education.

But nowhere is legitimation so purely detrimental to the growth of science as when science legitimizes government. That is to say, the process by which the community of science legitimates the democratic society within which it lives is not at all democratic. In particular, when science is called upon to legitimize something, it is the community of science that is approached as a community, and just then it is most easy to see the difference between excellence and renown: it is the renowned, not necessarily the excellent, that are called upon in order to legitimize. This mutual support is, I think, clearly a regressive factor detrimental to both; though, of course, by and large the interaction between the scientific community and society at large is beneficial and ought to be augmented. The detrimental matter is the undemocratic legitimation.

There is little doubt that even ideal communal life conflicts in spots with the democratic process, and both Polanyi and Kuhn are aware of it when demanding that the process of education should be that of apprenticeship. In this process the manufacture of textbooks is a major factor, and one that is both socially and intellectually very backward: most textbooks of most sciences are all sorts of intellectual hodgepodges presented in a canonical dogmatic manner. I therefore think that there is a possibility to criticize not only this or that scientific theory but also this or that aspect of the scientific enterprise as a whole – particularly in view of the fact that the commonwealth of learning has changed quite a bit and is still changing. I therefore recommend the stepping up of the process of the democratization of the commonwealth of learning, especially its educatonal aspects.

It is, of course, inevitable that the far reaching process of democratization will alter our attitude to the standard textbook, perhaps even to the point of abolishing it. Clearly, the education of scientists will still have to include the stock of common basic knowledge, but what this stock should be will be democratically decided and the teaching of that stock will not be authoritarian but critical. Democratization means not consensus, in particular not any consensus over theories, but the institutional broadening of the process of exchange of opinions and of the protection of minorities.

And clearly this will severely handicap the status of science as a legitimizer. This, of course, is the loss of the last resort of legitimation. And the world is now ready for this change.

4. A NEW APPROACH TO LEGITIMACY

For, the whole history of Western political philosophy, till World War II, is the history of theories of political legitimation. It was always assumed that two and only two kinds of authority exist, legitimate and by brute force, and the authority that rules by brute force is no better than highway robbers. Popper was the first philosopher who said, the question is not what legitimates authority, but how to overthrow it when it does too much harm. This is very far reaching and liberating, but raises hosts of new problems, such as, how much harm or what kind of harm is too much, and how easily should a government be overthrown and by whom, and so on. The rejection of the classical view must be made more radical so as to permit no return to any theory of legitimation regardless of the difficulties we may face without it. To this end let me observe the following.

The idea that government is either legitimate or a matter of sheer force is a polarization to extremes that do not exist. The very fact that there are so many kinds of theories of legitimation, religious and secular, democratic and traditionalist, radical and reactionary, only proves that we do not know what legitimates. Hence we cannot easily say with assurance which government is legitimate; but if we cannot say this with assurance, then we cannot also say with assurance which government is an illegitimate use of brute force; and such accusations must be made with reasonable assurance or not at all.

And so, all government is in part a legitimate one, in part not, even the Nazi government, much to the embarrassment of quite a few honest lawyers and political philosophers. And, of course, all governments use some illegitimate force, even the most democratic ones.

Once we see that we can ask, what kind of legitimation of a regime can science offer, and admit that science too cannot legitimize any regime to the full. Hence, there is room for democratic debate of this matter, both within and without science, that might center around questions of the principles of democracy, the limits of democracy, the ethics of democracy, the role of public opinion, and the rights of minorities.

Hence, whereas the current opinion considers the world of learning as a community, and as the nearest to the ideal, and so permits it to be a legitimizer, both of itself and of some extra-scientific political matters on which science can pronounce judgement, I propose to consider the commonwealth of learning a democracy, and one that is better when further democratized, and further democratizable by further integrating it into democratic society at large, and better integrable so when it ceases to be a legitimizer from without but seeks its legitimacy and the legitimacy of other parts of society within society at large.

CHAPTER 8

SOCIOLOGISM IN PHILOSOPHY OF SCIENCE

In a nutshell, the present chapter claims this: First, the classical problem of knowledge has recently shifted from, How do I know? to, How do we know? – from psychology to sociology. As a phenomenological matter this is a great improvement, as a solution to the problem of rationality it is erroneous and immoral. The problem, (Why) should I act, believe, etc., this or that way? is answered: You should do so on the authority of your reason. But change the problem of rationality in accord with the change in the problem of knowledge, and ask, (Why) should we – rather than I – act or believe as we do? and the answer is clear: We should act and believe as we do, because our society is as it is, and should be as it is. This is clearly the same as, we should because we should. Not very enlightening.

Sociologism appears as the authoritarian solution to the problem of rationality in works of Polanyi and Kuhn; a variant of it appears as a liberal theory in the studies of Popper and his former students who, however, do not offer any positive theory of what to believe or do; rather, they offer a negative theory of what to reject. They view this as a better solution to the problem of rationality, if not even a better formulation of it (not what and why should I/we etc., but what and why should I/we not etc.).

The major change in our view of science in the twentieth century is the shift from the psychological to the sociological. Classically, most philosophers who were engaged in studying the phenomenon of science asked the question. "How do I know?", and *via* this question they found their study shift from philosophy to psychology, ending up with the philosophers forced to attempt to found science on psychology (forced by the logic of their studies). Nowadays all this – psychologism – is often viewed with disdain; the question "How do I know?" is now often asked with the hope to avoid psychologism and it is sometimes replaced by the question, "How do we know?" Now this latter question and its parallel development up to founding science on sociology is what I wish to label as sociologism. I wish to attribute this kind of sociologism to two writers who have recently achieved much popularity – Michael Polanyi and Thomas S. Kuhn, who, I think, has been deeply influenced by Polanyi. And I wish to claim that their sociologism has all the defects of psychologism, except that it replaces the defect of radicalism with the defect of traditionalism, and consequently imports some measure of authoritarianism and irrationalism.

Yet this philosophical shift, from psychologism to some sort of sociologism or another delights me. Moreover, I would even go along with Polanyi's and Kuhn's touch of authoritarianism and irrationalism provided there is a minor omission: their theory is both descriptive and prescriptive, and I accept only the descriptive part of it: I acknowledge the authoritarian element within science which they see and which Sir Karl Popper for example is bound to dismiss as pseudo-science and as extraneous to science. But whereas they approve of this authoritarianism, I hope to see a movement, perhaps even help a little to construct a movement, which would liberate us from the yoke of science. Perhaps science has by now matured enough to allow us to be its critics without thereby becoming Luddites. I hope science can, by now, take friendly criticism even when it is severe.

I say 'I hope' because my experiences thus far are all too often quite counterindicative. When I sound off my criticisms, unfortunately, I repeatedly find myself shocking my interlocutors. I do not know how to cushion the shock. I once heard the famous Harvard historian of science, I. Bernard Cohen, talk to physicists; though his narrative was not at all in accord with their fairy tales about science, they were not shocked. This prompted me to make a brief investigation. To my regret all my informants amongst his audience told me they understood him to be merely narrating ancient piquant ancedotes illustrating no moral, saying something about the childhood diseases of physics. Of all historians and philosophers of science, only Derek J. de Solla Price speaks of the present ills of science, and of these he speaks only of a minor and technical one: too many publications for too few readers. And even this minor criticism of the celebrated Yale scholar has hardly been noticed or commented on by the multitude of writers who published too many essays on his works. I suppose they too are shocked or at least embarrassed by this unseemly open criticism. Can we learn to talk critically without shocking people? What shocks people? Why do people get shocked?

1. DIGRESSION: THE SHOCKING AND THE DISARRAYED

My reason for discussing shock is social, not philosophical. The ensuing philosophical errors concerning shock are trite and in themselves do not merit airing. Even the sociological reasons for discussing shock are not too strong, as shock is rather contingent a phenomenon. For the social function of the phenomenon of shock is merely that of preventing a student from studying the shocking phenomenon, or a viewer from observing the shocking phenomenon, etc. And, of course, on general sociological grounds there are many other reasons for not observing or studying whatever it is. As referees and friends often point out to me, these other reasons may be malice, or indifference – boredom, lack of interest, or simply dismissing something as bizarre. To this list I would add the sense of aversion or revulsion as that which is most common in human society: the ancient Egyptians (*Genesis*) and the future Incas of Bertrand Russell's fantasy ("Zahatopolk") alike display revulsion which prevents them from undertaking any anthropological observation. And it is, I think, a measure of progress that in our own society, as I observe it, in the commonwealth of learning, revulsion is fairly rare, but shock is all too common. Yet of the other alternatives mentioned above, of malice I shall not speak, and of indifference and such I will say that they seem much preferable. Indeed, the only reason which makes people prefer shock to indifference is plainly psychological, and a trite one at that.

Of the psychology of shock little needs to be said. No doubt, we will all agree, the chaste lady who is shocked by the display of nudity on stage is a bit self-righteous, and a bit of a concealed exhibitionist herself, quite unnecessarily keyed up and ambivalent about the whole matter. No doubt she could cure herself by simple exposure: shock is short-lived, since familiarity breeds indifference. In our own age one violent shock blunts the edge of its predecessors so that we shall soon become as unshockable as we can be. But, as we have seen, shockable people usually succeed in avoiding the shock. Hence, shockability is a defense mechanism, and like all defense mechanisms it acts as blinkers: it prevents the shockable person from meeting the shocking. Therefore, at best shockability justifies its bearer's indifference to certain goings on; at worst his ambivalence makes him hostile and aggressive. All this may be relinquished once the shockable person realizes that even without shock he may retain his loyalty – whether to decency on stage or to science – by merely refusing to take seriously any opposition, or by showing a lack of interest in opposition, etc. We are not obliged to justify our interests or their absence and so we need no defense mechanism to sustain it.

Setting psychology aside, then, we can come to another dimension of shock, a more intellectual one, as the following will illustrate. The psychological shocks of the early twentieth century – cultural, artistic, scientific, even regarding sheer life styles and tempos – all these are by now extinct. We need to make an effort and attain a successful imaginative reconstruction if we wish to relive the initial shock the first night audiences at George Bernard Shaw's *Pygmalion* felt when they heard the heroine utter the famous line, "not bloody likely", or even the shock which Molly's solioquy in Joyce's *Ulysses* gave to the early readers of this novel who both supported the censorship of it and read it in secret. But, though well into the second half of the century, we are still, in the intellectual sense, in a state of disarray; we have still not recovered from the transition into the twentieth century. On or about December 1910, said Virginia Woolf, human nature changed. The transition, the jolt it gives, may be momentary; yet our memory of the disarray it caused is stronger than that of the psychological shock it caused. The resultant dislocation or disarray, the sense of intellectual turn, is still alive; if Shaw no longer arouses it, surely James Joyce still does. The transition, break, shock, amounts to something odd which we still feel about his work. By now most philosophers are familiar with this sense of dislocation or disarray which has been scrutinized in great detail by various members of what might be called the linguistic school of philosophy or the Wittgensteinian school and labelled by them "queer", "odd", or "odd sounding", etc. Admittedly, many of these philosophers of the ordinary language persuasion have confused this new term "odd" with the common English adjective "odd"; yet as the American philosopher Paul Ziff has devoted more than half of his *Semantic Analysis* to clearing up this confusion, I need not dwell on it. Let me, instead, describe some instances of the intellectual oddity or sense of dislocation or disarray, leaving the psychology of it alone.

Let me first distinguish feeling odd from incomprehension. One can easily see that in order to feel the oddity one has to be knowledgeable. The oddest things – Picasso and Jackson Pollock, Schoenberg and John Cage, Ionesco and Beckett, Einstein and Bohr, Keynes and even Freud – do not so much strike the uninitiated as odd but rather as utterly or almost utterly incomprehensible. When Picasso's bird was first displayed in Chicago, bystanders expressed perplexity and indifference rather than a sense of dislocation or oddity; frustration was their strongest negative response. Picasso's "Demoiselles d'Avignon" looked odd, of all things, to his ardent admirers: they were greatly disturbed before his new style won acclaim.

Let me, in the second place, distinguish feeling odd or feeling a dislocation from the disturbing sensation which often accompanies it. We all experience this when we acknowledge greatness even though it disturbs us. We fooled no one but ourselves, to be sure: while disturbed we prefer not to acknowledge greatness. We might have thought we would never ignore today's Schubert, but only yesterday we ignored Charles Ives throughout his relatively long life. Einstein was not ignored, but his greatness was accepted only after the great wave of disturbance subsided; similarly Picasso and Stravinsky. Planck, the doyen of physics, Nobel prize winner and all, died deeply disappointed with the poor reception his colleagues had given him. Niels Bohr, absolutely everybody's darling, was met with a frank sense of bewilderment everywhere; his admirers seldom understood even his diction, let alone his ideas. Evidently they preferred it that way. I wish I had counted the many occasions when physicists have told me to read Niels Bohr, though they had not and I had. It is even hard to say, in many cases, whether the innovation which is odd is accepted or not, since the odd is neither the unfamiliar, nor the familiar, but the familiar in an unfamiliar perspective, the familiar dislocated or disarrayed, put into the wrong location or the wrong array. This is why it feels so unusual, and this is why those who wish to stick to their perspective are disturbed by the oddity (and are sometimes even shocked by it – but let us ignore that). But let me give examples.

The simplest example is the standard hostile response to Picasso. "I do not see what he paints! This is not how a guitar looks!" In other words, the observer can easily identify a guitar in the painting but the guitar is odd and thus disturbing. This is why one can reject Picasso and love Mondrian. But let us move to more cognitive cases.

My first intellectual dislocation came in my first proper college course. I took advanced calculus, and it was taught in accord with Landau's exposition. After a poor introduction, the lecturer plunged in, declared the course open, cleared his throat ever so slightly, and pronounced the first sentence: Peano's axioms. Axiom No. 1: One is a number. I was dumbfounded. The second axiom – every number has a successor – was not much better. I well remember that the axiom of induction seemed both slightly problematic and less odd. Obviously these two qualities were linked. Paul Ziff says, rightly, it is for different reasons that it sounds odd to say, "I have an elephant in my bathroom", or to say, while walking down the alley, "Pass me the salt, please", or to say, "Two plus two equals four, you know". All these cases are odd, and even in the ordinary sense of the word; yet for different reasons. Platitudes of a certain kind, he observes, even tautologies, are not odd at a cocktail party, but are very out of place – odd – in a serious conference. What Ziff tried to do is to criticize philosophers who ignore the fact that oddity depends not only on unusual use of words in a sentence but also on the unusual use of a sentence in some context or another. No doubt, he has shown that some philosophers at times implicitly assumed that odd sounding sentences have something wrong in their structure rather than in their location. But I am sorry that his criticism is so scathing. For, the same holds for some axiomatized systems: Poincaré's outburst at Russell for taking so much space in order to come to the profound conclusion that the sum of two and two indeed does equal four are so understandable. Poincaré was mistaken, but he was no fool. Proving that twice two equals four is quite out of place in mathematics, but not in the study of its foundations. Axiomatics is odd, no doubt, even though some of us have become familiar with it and have had the sense of oddity rubbed off; it is no longer odd as it is now seen in its proper context.

An even stronger sense of disarray or oddity than in my first mathematics course, I experienced in my first philosophy course – on the history of modern philosophy. I shall not dwell on it, since most philosophers will be able to provide similar anecdotes about their own perplexity upon their first sudden encounter with some metaphysical system of another.

2. DIGRESSION CONTINUED: ATTITUDES CONCERNING THE SHOCKING

Let me comment briefly on what seems to be a widespread method of adjusting to the shocking, or at least a social phenomenon which is gathering strength: it is the increased disposition to mix the old with the new. The new, admittedly, is not fully understood prior to the full possession of knowledge and understanding of the old, but it may be partly understood, or bastard versions of it may be fully understood, so that the finished product will be a person well versed in both the old and the new, except that he does not know what is a shock and a dislocation – he is spared the pain of transition. Thus, for example, while developing our student's ear we can give him not only Mozart and Schubert, but also a dose of Duke Ellington and even an occasional piece of Alban Berg, not to mention Bartok for children, neo-classic Picasso, etc. In science we are now developing Einstein for children; in the meantime children in Boston suburbia where I live study the new science textbooks and say DNA and RNA and even carbon 12 or carbon 14 as easily as their great-grandparents said the catechism; and as uncomprehendingly. And in philosophy we even now do not know whether to give our students a chronology of philosophy, from the Greeks to date, or a mixture of all periods in the salad of an introductory course. Indeed, philosophy is particularly problematic in this respect: to understand Greek philosophy one needs a certain degree of historical knowledge and of sophistication, yet to understand Descartes one needs some measure of Greek philosophy, if not of scholasticism proper. And these difficulties are even prior to the introduction of twentieth century philosophy.

I shall leave these difficulties to those concerned. As to myself, I find this approach of mixing the new and the old with the aim of avoiding the shock of transition too cowardly and too conducive to dogmatism. But I do not side with Sir Karl Popper or with The Living Theater, either: I do not advocate intentionally shocking people. To pound shocks one after the other may be coarsening; but equally to circumvent shock is to make schooling more of a debilitating hot-house technique than was ever found in the history of Western education. What we need is to eradicate the very problem, to avoid the very development of intellectual defence mechanisms and neuroses and blinkers; what we need is to learn to view the dislocating as at times delightful and challenging and perhaps even as opening new vistas.

3. HUSSERL'S CRITIQUE OF DESCARTES

After all this delay it is time I got around to something recognizable as philosophy. *Cogito ergo sum*, writes Descartes. What is Descartes talking about? What is a thinking substance whose existence is thereby established? No, I shall not go back to my sophomore days; I shall plunge even deeper. Edmund Husserl, *Cartesian Meditations*, paragraph 10: "Digression: Descartes' failure to make the transcendental turn". Alas, poor Descartes! I shall make a digression too; a meta-digression to be precise. Not only is Husserl's digression excellent, but in many works I find the digressions the best parts, and I adore genuinely digressive authors like Cervantes, Galileo, Lawrence Sterne, Samuel Butler, and Charles Saunders Peirce. I do try to emulate them – despite my limitations. I must end this digression of mine, however, and return to that of Husserl; but not before observing that all too often digressive writings are dismissed by being termed "loose" and "wordy". Were digression more respectable in our culture, and systematic exposition viewed with a little more suspicion, then the topic of shock and oddity and disarray would be less of a problem with us. We take ourselves much too seriously and that is the real trouble.

Back to Husserl, then, or to the young Husserl, to be precise. "Descartes had the serious will to free himself radically from prejudice", we read. "But we know from recent inquiries, in particular the fine and profound researches of E. Gilson and A. Koyré, how much scholasticism lies hidden, as unclarified prejudice, in Descartes' *Meditations*." Now both Gilson and Koyré denied the possibility of executing Descartes' program – which is really Bacon's – as they did not think anyone had the full ability "to free himself radically from prejudice". This is why they looked for, and found, old roots in Descartes' new thinking. Such exercises are today not uncommon; they reflect some modern sociologistic view or another in that they are criticisms of radicalism: radicalism is the recommendation to start afresh, and this recommendation is declared impossible to follow by many – including Gilson and Koyré. Husserl was not convinced by Gilson and Koyré. He declared that he had found and eliminated the cause of Descartes' failure to free himself of prejudice; Descartes had shared with the scholastics an "admiration of mathematical natural science" and a predilection for the axiomatic method. This attribution of failure is rather discomforting to any one, like myself deeply impressed by mathematics and by science, and highly appreciative of axiomatics. Yet, to some measure I agree with Husserl. Nothing is worse than those mathematical texts, especially of the Germanic tradition, which begin by hurling axioms on their readers' unsuspecting heads, and Husserl is right that Descartes is not free of this technique.

This is not to share Husserl's accusation, but merely to admit that it contains a measure of justice. Perhaps it can be more justly directed against Spinoza, the opening to whose *Ethics* is so baffling that the second to his fifth Books of it are not known to so many students who could benefit from them simply because they failed to master Book I of it. But even this, I feel, is no strong accusation; Spinoza's or Descartes' error is comprehensible, Spinoza's systematic application of it is admirable; and we may all advise our students to skip Book I and enjoy the rest of his *Ethics*.

In comparison to Spinoza, Descartes is not so baffling, as he was less ruthlessly axiomatic. Yet in another way he was more baffling than Spinoza: he is particularly baffling – I am here freely following Husserl – since he begins his expositions quite phenomenologically. We do not hear in courses on history of philosophy about John Locke's schooling, or about Schopenhauer's; but we are told in such courses, because Descartes emphasized the point, that he had been trained by Jesuits, that he had enough good sense to loathe his schooling, that he felt that philosophy should not be patchwork but of one cloth, like a town planned by one town-planner. He thus explains his desire to start afresh, and how it led him to universal doubt – which, to repeat (a point made already by Gassendi), is but an improved Baconianism – and how he found certitude in his very doubt. All this is quite phenomenological; the abstraction begins with his introduction of the abstract entity, the *res cogitans*, the thinking thing, which Husserl rightly opposes, replacing it with a more phenomenological experiencing and thinking ego. Husserl objects to the abstraction not so much as abstraction but as the highway to "transcendental realism, an absurd position, though its absurdity cannot be made apparent at this point". Husserl admires Descartes as his own predecessor and fully acknowledges his debt to him. Descartes, Husserl says, "stands on the threshold of the greatest of all discoveries – in a certain manner, has already made it – but he does not grasp its proper sense, the sense namely of transcendental subjectivity, and so he does not pass through the gateway that leads into genuine transcendental philosophy". End of Husserl's digression.

4. HUSSERL NOT SO RADICAL

There are anecdotes about Husserl's naïve belief in his own greatness. There are older anecdotes about Descartes' arrogance. Descartes looks arrogant and Husserl naïve, I suppose, only because of the difference in their circumstances; our age is in some ways more understanding. But as far as each of them was concerned, each of them viewed himself as one who has finally made a completely clean break with the past and thus "opened the gateway" and subsequently made "the greatest of all discoveries". This is no paranoia, no hubris; just the facts.

I shall not undertake here the explication of this greatest discovery, transcendental subjectivism, somewhat novel and interesting though I do think it is. For, the disciples of Husserl, divided among themselves as they are, all agree: thou shalt not put Husserl in a nutshell. No doubt, putting anyone in a nutshell; Descartes or Newton or any other thinker, has both its advantages and its disadvantages. But to disallow one to be put into a nutshell is to make the study of his works a labor of love and devotion; and thus his devotees are naturally disposed to the formation of a club if not a cult; sometimes, as against all intention and even against injunction from the Master himself. Sir Karl Popper, for example, certainly does not wish to have a cult around him. Yet, and I have it from the horse's mouth, any attempt to put him in a nutshell makes him cringe. The rest is the mere logic of that situation: effective refusal to be put in a nutshell leads to a cult or to oblivion.

But I am not done with Husserl, though I will skip his transcendental subjectivity. He observed something very significant when he saw Descartes' predilection for axiomatics of sorts, or for a system proper. As I have mentioned, it is an empirical fact that both axiomatics and Descartes sound very odd. Now perhaps Descartes sounds odd because he is metaphysical, as the followers of Wittgenstein would say, perhaps because he is axiomatic. But perhaps these two explanations are one. If we need an axiomatic or quasi-axiomatic system, if we need a viewpoint, a fundamental commitment, before we can look afresh at the world, if it is true that a new outlook can be achieved only abruptly, by a leap from one set of axioms to the next, from one metaphysical commitment to another, then, all of a sudden, we can understand to the full the phenomenon of intellectual jolt and disarray, of things all of a sudden seeming and sounding odd. When a familiar thing, or event, is looked at from a radically new angle, then the familiarity and the unfamiliarity of one and the same object of our attention wildly clash. Perhaps at the time Descartes' system was shocking in the novelty of its starting point, the *cogito erasus* and all that. Yet Spinoza was all the more surprising in his very taking as his starting point the well-known ontological proof of the existence of God. I shall go further. The old quantum theory was hardly anything like a finished product, a coherent viewpoint, an axiomatic system. Nor did it introduce anything startlingly new. Yet its appearance was startling – either as the adumbration of an utterly new set, or as enough of it to clash violently with the old without allowing people the comforting thought that the clash was localized and could be eliminated.

Assuming all this to be true, the question is, does Husserl sound odd, and if so, is it due to his unwittingly following Descartes' axiomatic or systematic tendencies? It seems to be a very clear phenomenon: Husserl is very odd. I should myself say that he too has his own axiom or system or viewpoint or commitment – whether he likes it or not, whether it is of the scientific persuasion or not. Indeed, I see no way of avoiding all this. If he is as great an innovator as he claims, how can he avoid sounding odd?

Now as to the sense of disaster accompanying intellectual disarray or oddity, I suppose it is rooted in the idea that one should not change commitments as if they were socks and shoes, that we should not effect one revolution after another in a rapid succession. This idea, I am afraid, is an axiom which Husserl still shares with Descartes, and we can criticize him too for his insufficient radicalism and for his suffering from excess abstractness.

5. HUSSERL AND SCIENCE

Husserl took great pains to separate science and philosophy as much as he could – I suspect chiefly because in his day science was in constant ferment. His doctrine of bracketing, of the *epokhe*, justified the separation between science and philosophy. Those interested may find a critical exposition of Husserl's doctrine of bracketing in a learned paper by Erazim Kohak in the *Journal of Existentialism* of 1967. "Even thinkers most directly indebted to Husserl", Kohak says in his introduction, "tend to feel uncomfortable about . . . bracketing". The chief significance of this doctrine for our discourse lies in that it permitted Husserl to be reticent about science, yet quite unjustly, and for the following reason. Descartes' transcendental realism, which Husserl declares absurd, is, at least for Descartes, deeply linked with science. Husserl replaces it with transcendental subjectivism, which, Husserl insists, has nothing to do with science. But what happens to science when deprived of its traditional axiomatic base, transcendental realism? I cannot say that I know Husserl's answer. Seemingly, Husserl decided that science is doing very well and taking good care of itself anyway; so we need not worry about it. But this decision is not the point: we may still wish to know how science manages. He also seems to be saying, realism is all right in science; when we bracket reality out we bracket science along with it. This, indeed, is how Husserl manages the divorce, or to be more exact, the mere separation. But the question remains: Bracketed or not, what is the realism of science, is it transcendental or not? Husserl would not say.

All this may account for the liberty that Husserl's disciples, such as Heidegger and Sartre, take with science. What Sartre says about Einstein and Heisenberg in his *Being and Nothingness* is sheer confusion and vulgarity. But this too is beside the point. We have a philosophical problem regarding science: Is it based on transcendental realism of the philosophic kind, or is it not? One may require from the consistent phenomenologist that he effect a transformation of the philosophy of science into a more phenomenological perspective.

Science is realistic and we want to know whether its realism is transcendental or not; whether the realism of science goes beyond the observed facts or whether it goes to the extent of observation and empirical confirmation alone. Does science use realism as a primary axiom, then, or the reliability of observation? Does science start with thought or with observation? The answer to this question is the axiomatic system or viewpoint or commitment concerning science, and all else hangs on that. Now, no doubt, transcendental realism, or a priorism, or intellectualism, is very odd, and requires a lot of explanation and getting accustomed to. Its traditional alternative, empiricism or inductivism, looks much more congenial, much more phenomenological – until one runs into the problem of induction. Then one can accept induction axiomatically. This still is much less uncomfortable or odd than a priorism. It looks more odd upon a closer inspection because the axiom of induction is both concerning nature – the principle of uniformity of nature or of limited variety or some such – and concerning the process of learning – learning theory, associationism, S-R, etc. Already Bacon declared that God has created Nature and the intellect on a par. Kant regarded this as a confused way of having things both ways; when postulating this parallel between the intellect and nature, the intellect stands above both nature and the intellect; and this can never be the case. Kant viewed this postulate of parallelism as a transcendental hypothesis, and as a particularly poor one at that; he called it lazy and other unfriendly names.

6. POLANYI'S VIEW OF SCIENCE

The truth is that both theories, "we start from observation", and "we start from theory", are both uncomfortable and abstract. The more phenomenological alternative is due to Michael Polanyi. We start science not by observing and not by thinking, he says; we start science by going to college and taking courses in science.

Now we have before us three phenomena which, in conjunction, look very odd indeed. First, Polanyi sounds commonsense to the uninitiated; second, Descartes sounds odd to the uninitiated; and third, Polanyi sounds odd to the initiated, to those already accustomed to Descartes and his classical followers and detractors. He sounds odd to them for a very strong reason: Descartes himself, we remember, went to school, just as Polanyi would advise him, and he loathed it all. He wanted to get rid of all the rubbish and start on solid grounds. Surely, what we learn today in college, inasmuch as it is rubbish it is better omitted from the curriculum, and inasmuch as it is solid it is better kept. Hence, it is its solidity which recommends our learning, not the fact that colleges happen to acknowledge its solidity! Polanyi clearly seems to argue in a circle here! This is unsatisfactory especially since, no doubt, some of the learning which is in the curriculum is, doubtless, rubbish!

Those familiar with Polanyi's work, however, will not be perturbed in the least. Polanyi often seems circular. He offers, for example, as the criterion for the novelty of an invention the very readiness of the patent-tester to allow it to be patented. Rather, it would seem, Polanyi should tell us by what criteria does our patent-tester make his decisions and by what criteria should he make them. Besides, Prometheus took no patent! Now Polanyi has no strong objection against your trying to formulate the grounds for the acceptance of an idea into the curriculum or an invention into the list of novelties. But, he says, these theories you will try to offer will all be abstract and defective, and meanwhile life must go on. If your criterion clashes with life, as it will, you may choose the one or the other; if you are wise you should choose life and eschew the criterion; hence the ordinary flow of real social courses of events is more fundamental than any criterion you may imagine. Q.E.D.

Again, Descartes had a criterion. It did clash with reality, with the school he used to frequent. He blamed school, not his own criterion. How could Descartes be so sure of himself? He tells us how: he saw the natural light; that is all. But he was in error, says Polanyi; Descartes' criterion, like any other, is in error.

I have come to the heart of the present discussion, and I wish now to present two theses, both concerning Polanyi. First, he is right in rejecting Descartes' criterion. Yet the very fact that (Bacon and) Descartes told people to forget all the nonsense they were learning in school, this fact was extremely beneficial. Hence Polanyi is in error. To put this first thesis more boldly and generally and briefly, the traditionalist critique of radicalism is valid, but the occasional success of radicalism refutes traditionalism. Second thesis. There are two readings of Polanyi, both valid yet very different. One is, there is no right and no code, so the best is to follow the leader. The other is, there is right, and there is code; but the right and the code are alterable and altered by the leader. The leader is one who is followed; today's right is today's practice. The first view is cynicism; the second is positivism. As Hegel already discovered, moving from one to the other and back is great fun. One last point: Polanyi's most distinguished disciple is Thomas S. Kuhn who abhors cynicism as much as he reveres science; and so he is an honest-to-goodness positivist; perhaps the last of a breed, perhaps the first of a new breed, of the new technocracy.

7. POLANYI VERSUS DESCARTES

To put all this more philosophically is not too hard. The classical philosophers, whether inductivists or intellectualists, did admit to the systematic or logical or axiomatic nature of their teachings, but saw no need for commitment to their axioms. Rather, they expected from their audiences nothing more nor less than commitment to reason alone. The axioms were not accepted on a blind commitment. On the contrary, the foundation of the whole edifice, the Archimedean point on which everything else hangs, was the rational point to accept. The skeptical critique of rationalism is ancient: the foundations need a foundation, the Archimedean point needs support from another Archimedean point. Kant tried to justify the Archimedean point by showing that it is, indeed, the Archimedean point. But – as Solomon Maimon complained – he lost objectivity and retained mere intersubjectivity; he thereby lost valid truth and retained mere validity. His followers even found that different axioms can serve as different Archimedean points or as different bases. They searched for a criterion of choice between these – they chose simplicity. But any such criterion is question-begging, even if it were not as vague and complex as simplicity is.

This skeptical critique which I endorse, however, is declared rather academic by Polanyi and his likes. Polanyi uses skeptical arguments to defeat radicalism, but he rejects skepticism as utterly barren in favour of tradition which is fruitful. What with all the success of science, can we allow ourselves to give it up just because of some very general and ancient and futile doubts? There is a better critique, and one which cuts deeper. The whole idea of an Archimedean point is radicalist and abstract and rests on the assumption that we can obliterate the past and start afresh. Only on such an assumption does Cartesian total doubt make good sense and recommend itself. But we cannot, in fact, start afresh. The total doubt that can help us clean the slate is a myth. We can doubt this or that item in the list of our traditional ideas and theories and methods and viewpoints; but we cannot discard it all. This is Polanyi's critique of doubt, so-called. This is how Polanyi concludes that willy-nilly tradition has authority: it imposes itself on us and we cannot doubt and reject it *in toto*.

In brief, Polanyi justifies science, as Bacon does and as Descartes. But whereas these do axiomatize in the sense in which Husserl has noticed, Polanyi does not; and, unlike Husserl, he claims no finality. The authority is not of the final, whether the true or the valid, but of the accepted by the living tradition. One does not commit oneself by an act of will, by a leap of faith. One commits oneself by a prolonged engagement, by a laborious act of integration, by living the life of apprenticeship.

Let me now explain my odd observation. Polanyi's thesis – science starts with schooling – sounds commonsense phenomenology to the uninitiated, I said, and odd to the initiated; this because to the initiate, "science starts with . . ." is the beginning of the answer to the question, what does the authority of science rest on? Here we can see that the uninitiated who does not find Polanyi odd simply fails to worry about the authority of science and thus he misunderstands Polanyi. And so we have no choice but to read Polanyi's thesis as either odd or trivial and to notice that he did not have in mind the trivial reading.

8. POLANYI'S TRADITIONALISM

Now odd as Polanyi's thesis is, the fact is that for Polanyi science cannot ever be odd: its revolutions are marginal and small, never total. But Polanyi – or is it Kuhn – softens this corollary by an amendment to his constitution of science. Science follows the tradition, but the tradition recognizes its popes and cardinals. (Or is it its heroes and great men?) Anyway, these can effect quite a sizable revolution. If they are followed, they stay at the helm; if not, this is a risk they take. On what grounds they take risks, what determines their success, these are unanswered questions; even unanswerable, says Polanyi. If I offered you a theory of this sort to explain political events I would, no doubt, be laughed out of court. In a recent book, *Public Knowledge*, by a disciple of Polanyi, the British physicist John Ziman, there is a chapter on the institutions and authorities of science. "Rereading what I have written", Ziman writes there, "I find myself wondering whether I have presented too conservative a view of Science". I like the capital *S* in Science. Somehow, Ziman manages to comfort himself. He relies on the progressivism of the scientific community, and on decentralization and competition in general. Good old capitalism. Yet, before all that he frankly concedes that conceivably science "is liable to oscillate as a whole, as fashionable techniques are propagated from field to field, but it is relatively immune from petty tyranny and charlatanism". This is all much too problematic to be left without further scrutiny. Meanwhile, the gospel is spreading. I think this is quite reasonable. The philosophy of science is in a desperate state. No psychological justification of science is possible because psychology is a part of science and so the justification would be clearly circular. The existentialist retreat to commitment, if not antagonistic to science, is at least indifferent to it. Surely Polanyi and Kuhn describe certain facts correctly and promise a fresh start towards a new justification of science. Surely commitment through apprenticeship sounds more factual and more reasonable than the mere leap of faith, than the act of sudden conversion.

But let us not be overwhelmed with Polanyi's success either. How correct is his sociology? Does it matter if we have to rectify it? Does science rest on apprenticeship proper in each period? Does it matter that apprenticeship was commoner in mediaeval Germany than in enlightened England? What exactly is apprenticeship beyond the inability to start afresh (which makes all of us apprentices)? The quietism with which more and more members of the scientific community welcome Polanyi's philosophy has put such questions in abeyance. But not for long, and I do not think his philosophy will stand the pressure of criticism, once this comes forth. Indeed, it is easy to short-circuit Polanyi's philosophy by creating some tradition of empirical sociology of science not in agreement with him. For example, one which observes some merit in seventeenth century radicalism. Such a possibility is fatal to his sociologism.

9. WHAT NON-JUSTIFICATIONISM AMOUNTS TO

I shall not elaborate all this further, but rather notice the only alternative visible. Perhaps the oddest thing to comprehend is the idea of science without a shred of justification, with no confirmation or corroboration of any kind, with no Archimedean point, no foundation, no anchor, in nature or in man, in psychology or in sociology – science as a search rather than as an achievement. Such a science does not exist yet; not as a social phenomenon; only as an idea. Perhaps this kind of idea is an error too; I only regret that too many scientists and philosophers find it too odd or bizarre to pursue, even hypothetically. Let me only conclude that one great and immediate corollary to such an attitude, is that science is free for all, that we need not be disturbed by new odd ideas, that finding ever more odd ideas is fun and perhaps also progress. I think here lies true experimentalism – not just in the laboratory where an odd experiment may be performed but in the armchair where it must first be conceived. New ways of looking at things are not easy to come by, and the suggestion seems to me remote that, unless we take care, we may experience more and more oddities until we have lost all sense of continuity and thus our sanity. Nevertheless this may be true, and the danger of excess novelty may be real; and perhaps all the avant-gardists are playing with fire when they strain their imagination to the limit in criticizing the old and imagining newer and odder things. Yet theirs is the true experimentalism, and if it is false let us experiment with it and discover that it is false. What we would do afterwards is beyond my own imagination. W. W. Bartley III is willing to concede that perhaps such a discovery will offer a justification. I confess that my imagination is not as serviceable. Anyway, we both view now all justificationism as conservative and anti-experimental, and would rather encourage bold experimentation to see where it might lead us. In brief, I propose we cease viewing any oddity as undesirable – at least for a while – even the odd possibility that one day even this may change and some authority be established.

I haven't explicitly said why I think this is a sociological rather than a psychological matter. Let me only conclude with the following observation. Boldness and experimentalism are first and foremost individual affairs: masses cannot be bold, only individuals; but societies create or destroy such individuals. We can reform our society so as to make boldness and flexibility more acceptable and experiment more rewarding.

10. WHEN IS A THEORY NON-JUSTIFICATIONIST?

Perhaps a digression on (the old) Wittgenstein is in order here. At least it was strongly recommended by a referee of the journal where this discussion first appeared. That in one sense or another Wittgenstein had a version of sociologism I think must be granted. Indeed, Peter Winch, in his *The Idea of a Social Science*, seems to suggest in the name of Wittgenstein nothing short of the view that philosophy and sociology are identical. "It would be nice", comments Ernest Gellner, the arch-loather of the Wittgenstein galaxy in a rare moment of kindness while reviewing Winch's aforementioned work for the *British Journal of Sociology,*

> It would be nice to see Wittgensteinian philosophers really following up the doctrine of the essentially social nature of thought and language, instead of merely using it as a stick with which to beat empiricism [i.e. psychologism], and to take a good look at concrete social contexts, rather than doing it in a very general way and inventing their examples *ad hoc* and without check in the course of discussion with each other. But while sympathizing with his [Winch's] conclusions here one must add that he does not stick to them consistently . . .

Gellner pays the positivists with their own coin: they are not philosophers, he says, but bad scientists, doing trivial *ad hoc* examples and no tests.

And so, thus far the sociologism of Wittgenstein has not come to much even in the hands of Winch. We can therefore declare as open the question, do Wittgenstein and his followers endorse sociologism as a new mode of justification *à la* Polanyi or as a non-justificationism *à la* Popper?

I think myself that the question not only *may* be left open, but by the canons of interpretation *should* be left open. Wittgenstein never said he worried about the problem of rationality, he never said he wished to see it solved; on the contrary, he said worrying about, or attempting to solve, a philosophical problem (or what seems to be a philosophical problem) is just plain sick. According to Wittgenstein, one has to learn to live cheerfully in spite of it, or let it dissolve, or what have you, but not ever solve it.

With this Wittgenstein's sociologism, the most popular in the English speaking philosophical community between World War II and the seventies, leaves this discussion as it started, occupying no more than a digression, full of significance or devoid of it as the case may be.

Somehow, I feel, Wittgensteinians should not take offense at all that. There are many writers, after all, sociologists or novelists or what-have-you, who did not attempt to solve the problem of rationality and whose attitudes are not sufficiently clear to enable us to say which solution to it they would favor most. There are even men of science who expressed non-justificationism, and as a philosophy, yet I would not myself present their non-justificationism as a solution to the problem of rationality. Since the Einsteinian revolution a number of scientists, such as Einstein and Schrödinger in physics or Frank Knight and J. M. Keynes in economics, even the writers E. M. Forster and W. Somerset Maugham, expressed non-justificationism clearly as their creed. Yet they did not present it as a solution and they admitted that it was problematic. In particular Frank Knight admitted that there is a measure of irrationality, and an inevitable one, in the scientist's leap to unwarranted conclusions – they can never be warranted, he stressed. He felt that something was missing here and his chief concern was to avoid giving the false impression that he allows for irrationalism, the false impression that in his view anything goes. But he could not elaborate on this.

Let me go further. If I were forced to ascribe the non-justificationist solution to the problem of rationality to a modern writer (it has ancient skeptical variants recorded by Sextus Empiricus) I do not think I could ascribe it to Solomon Maimon, the non-justificationist contemporary of Kant, and I find it hard to decide whether or not to ascribe it to George Bernard Shaw.

Shaw posed the problem of rationality, Shaw presented a non-justificationist solution to it, Shaw handled some problems ensuing from all this. But he did this in a publicist if not in an aphoristic style, and it is almost impossible for me, at least, to decide how much is there in Shaw's prefaces, or in his famous toast to Einstein, and how much I read into them.

No matter; Shaw would not have minded. But I mention this as a point of interest for those concerned with some intellectual standards. It is clear from the perusal of the scientific literature that scientists are bothered about meeting standards and that it is not easy to say if they do meet them. A man of science who has an idea presents it not only as a solution, but also as a worked out solution. How much worked out? This is a question which philosophers are tempted to view as the problem of rationality: how much we have to work out a solution before it has to be taken seriously by the solver's peers. Of course, I tend to view the question not as that of rationality but as that of the sociology of science.

A philosophy too can be worked out; or it can be implicit in some study or another without being worked out. As not-worked-out, or not-sufficiently-worked-out, no doubt, the ideas of Wittgenstein and the others mentioned here surely were non-justificationist. They found nothing imposing in reason. Whereas the young Wittgenstein found that reason imposes silence on the confused, the old Wittgenstein merely offered help for those suffering from confusion. I wish to stress this point, first made by Hayek and Popper, I think, that the liberal tradition is justificationist and so was all too often ready to impose the authority of reason. After Einstein this tended to change. But the non-justificationists did not solve all problems of rationality sufficiently well to prevent some new authoritarianism as well. It is not at all my opinion that any non-justificationist philosophy is so well-worked-out that it leaves no room for authoritarianism. And so Polanyi has his honorable place. I think the question is still open: How well-worked-out a non-justificationism has to be before we can view it as a theory of rationality?

CHAPTER 9

REVOLUTIONS IN SCIENCE, OCCASIONAL OR PERMANENT?

1. THREE VIEWS ON REVOLUTIONS

There are three views in the literature on science concerning the nature of scientific revolutions, and a few variants and combinations of them. The most important view is the radicalist view which was expressed by Sir Francis Bacon very forcefully, which was traditional since the foundation of the Royal Society and until the Einsteinian revolution in physics, and which is still believed by many philosophers and historians of science, as well as by many scientists, natural and social. It claims that science is born out of a revolution against prejudice and superstitiion but that within science itself every part is so securely founded that it cannot be shaken. The most important alternative to Bacon's view is Pierre Duhem's theory of continuity; it was born out of the crisis in physics, and it is becoming increasingly popular amongst the sophisticated. It claims that every achievement of science is capable of modification, but not of overthrow. For instance, we may believe in determinism and then overthrow it, thereby exhibiting its unscientific (metaphysical) character; but we may only modify, not overthrow, Maxwell's theory – say, by viewing its equations not as precise to the last point but as mere averages. The third view was developed by Sir Karl Popper after the Einsteinian revolution and under its impact; though Einstein and a few others have accepted it, at least in part, those who have heard of it usually consider it rather eccentric. It claims that unless a theory can be overthrown by empirical evidence it is unscientific, and *vice versa*. For example, determinism cannot be overthrown, but a scientific theory which may be deterministic or indeterministic such as Newton's theory and Heisenberg's theory respectively, can be overthrown. Bacon's theory is of one revolution, Duhem's theory is of no revolution but merely reforms, and Popper's theory is of revolution in permanence.

Let us now introduce a totally different point – the fear of losing touch with one's colleagues. This point is very closely related to the above theories of scientific revolutions in the following way. These theories share a progressivist attitude towards science. And one unpleasant consequence of progressivism for the individual is that he might one day find himself left behind.

There is little or no literature on this subject, and so one has to refer to some results of field-work, however partial and cursory (not to say impressionistic). Such results indicate that the fear is very widespread, and is based on the widespread theory that with the ever-growing growth-rate of science it is very easy to lose touch with the forefront of science. One can lose touch through neglect (even a temporary and fully justifiable one, such as brief illness), through losing one's intellectual sharpness or one's creative powers (as Freud constantly feared), or through developing a mentally rigid allegiance to theories and methods which were indeed significant in one's early days but which quickly became outdated (i.e., much less significant than they used to be). One can become old-fashioned in one's beliefs simply from being ignorant of, or even unable to understand any longer, the newest methods, ideas, and experiments in the field. In such a case it is futile to feign agreement with the more up-to-date researchers, because one cannot really believe what one cannot understand. It is clear that most old-fashioned thinkers are out of touch, since if they were in touch with the latest developments they would in all likelihood see as clearly as any up-to-date person does, what the facts of the situation are. (This is not entirely universal: a person may become so dogmatic in his attachment to fashions accepted in his youth that he would not agree with his colleagues even when he knows all the facts at hand and understands all the newest ideas.)

That such fears are deeply related to progressivism is almost too obvious to note. It may even be empirically illustrated by pointing at the similar fears which progressive artists harbor, and at some non-progressive cultures which show no trace of it. (The only way for a traditional Rabbi, for instance, to lose touch, is to become literally senile.) Now, being progressivist, the three views about science and revolutions both leave room for the fear of losing touch, and provide prescriptions as to how to avoid losing touch and becoming old-fashioned. The three prescriptions diverge, and discussing their relative merit is one mode of critically assessing the relative merit of the views which give rise to them. We shall examine views of scientific revolutions, about loss of touch, and the connection between them. If we provide clear-cut prescriptions we shall thereby design some crucial tests between the various theories.

2. RADICALISM AND TRADITIONALISM

The commonest modern view about science is that in science we can prove what we believe. This view is less common amongst men of science since the Einsteinian revolution, but not much; amongst historians of science, vulgarizers of science, and educated laymen, it is still as popular as ever. One instance may deserve mention here. The Italian philosopher of science Ludovico Geymonat published in 1957 a book by the title *Galileo Galilei* and the sub-title 'A biography and inquiry into his philosophy of science'. The book was translated by Stillman Drake and published in 1965 with a foreword by Giorgio de Santillana and notes by the translator. Professor Geymonat's thesis is that Galileo is so very important a figure in the history of science and of philosophy because he made the discovery of a great truth about scientific method, perhaps the great truth about it: it is not enough to show some thesis to be very highly probable; as long as a thesis is not utterly and completely proven, it should be viewed as unscientific. One corollary from this great truth is that one should rely only on the first hand testimony of one's own senses, not on witnesses.

This theory about the requirement of absolute demonstration in science makes a scientist "guilty of . . . (an) unpardonable mistake", to use words of Sir John Herschel, when he permits himself to say something which might be a mistake. As a scientist, better say nothing at all than say something which may turn out to be an error. This doctrine renders the life of a scientist a nightmare. After he has done his best to find the truth, and committed himself to a view after he was as satisfied with its demonstration as was possible, he may find the slightest need for readjustment the greatest burden; for the need for readjustment shows faults in one's scientific past, in one's scientific education, in one's teachers of science. The slightest need for a modification thus becomes a matter of the highest principle. The slightest criticism thus becomes identical with the most sweeping condemnation.

So Sir Francis Bacon understood the situation. If proof in science is so easy, asked penetrating Bacon, why did we live in the dark ages for so long? Because, he answered, people would rather distort every fact they observe than admit that they had erred. Hence, if you want to be a scientist make a clean slate and proceed cautiously. If you make a guess, as Copernicus did, chances are you are building a new dark age. Bacon's theory is the theory of one and only one revolution in science: the revolution of science against error. Hence, science begins with the last revolution. Physics, say Bacon's followers, begins with the seventeenth century, chemistry with late eighteenth century, and optics with the early nineteenth century. The radicalist must view the latest revolution as the starting point, as has been stressed so beautifully and forcefully by Michael Oakeshott in his *Rationalism in Politics*. Indeed, as Lakatos has pointed out, when Russell was a radicalist in mathematics, he vacillated between viewing George Boole or his own self as the father of mathematics proper. Holding to the view that there can be no revolution in science, but only science against error and prejudice, Lavoisier and his followers concluded that all pre-Lavoisierian chemistry had been superstitious, and Madame Lavoisier burnt ceremonially the books of Stahl – her husband's most distinguished predecessor. This act reflects the sentiments of many historians of science, including even such a liberal historian as Herbert Butterfield. Though Butterfield rejects the view that a scientific theory needs no modification, since he endorses the view that there was only one revolution of science against prejudice, he cannot but view the revolution in chemistry a latecomer to the scientific revolution.

Lavoisier's view is not fully endorsed by Butterfield and other historians and philosophers of science. Lavoisier followed Bacon (and Newton) in thinking that there are no revolutions in science because science proves its theories beyond doubt. Nowadays we all view Newton's theory as scientific, although we do not believe in it, rejecting the invariance of mass and action-at-a-distance. Nowadays, likewise, we do not burn Lavoisier's books even though we do not share his chemical views (to say nothing of his physics). He divided chemicals into active and inert; the active ones he divided into acids, which contain oxygen (= "acid maker") and alkalies which are without oxygen; the inert chemicals he viewed as salts combined of acids and alkalines. We think nowadays of this theory as exceptionally naive, just as we think of his view that all processes of combustion, fermentation, calcination, and acidulation, are nothing more than the combination of chemicals with oxygen (which was previously combined with caloric). Nowadays, in brief, we are somewhat more tolerant towards our predecessors, and do not call them prejudiced and superstitious as soon as we discover errors in their teachings. Should we not, likewise, cease viewing as superstitious and prejudiced Lavoisier's phlogistonist predecessors, and even perhaps Galileo's mediaeval predecessors? This is the question which engages many contemporary historians of science.

The philosopher who introduced the idea that mediaeval science is not superstitious did so by arguing that all science is always alterable. He was Pierre Duhem, a strange amalgam of a most daring and revolutionary philosopher with a most reactionary one. How revolutionary it was at his time to consider Newtonianism alterable is hard to imagine. The greatest skeptical philosopher of modern times, David Hume, believed Newtonianism would go down to the end of the ages utterly unaltered. And since the days of Hume, more and more impressive evidence in favor of Newtonianism kept coming forth. Faraday thought Newton's theory must be modified to eliminate action at a distance, but this fact had been forgotten, and rediscovered only recently. In the latest biographies of Faraday this fact is still overlooked. Poincaré considered the possibility of having to modify Newtonianism, and argued that it would be always preferable to adhere to it even at the cost of altering the meanings of its terms so as to keep it in accord with the facts. Pierre Duhem attacked this argument and declared that Newtonianism too is not sacrosanct. But for the fact that Einstein outdid Duhem at the very same time (by actually offering an alternative to Newtonianism), his position in the history of thought might have become most prominent. Duhem, however, was a reactionary; his chief purpose was to argue that mediaeval science is qualitatively no different than modern science, and that those who preached otherwise, notably Galileo, suffered from a touch of megalomania or of incredibly naive optimism about what science (and they as scientists) can achieve. Butterfield has combined the radicalist view of science as anti-mediaevalist with the reformist view of science as open to modification at any time. He has achieved this partly by failing to refer to Duhem, partly by writing so very clearly and beautifully. He simply asserted that in the Middle Ages not even the slightest modifications were allowed, whereas in the Renaissance and later, modifications were welcome. This thesis he has not discussed rationally; he has not refuted, for example, Duhem's stories of modifications in mediaeval physics. His view, however, fell on fertile ground and became very popular because it is simply so very seductively quietist. After Einstein it is ridiculous to claim that anything in science is the last word; yet this is made to sound so much less disquieting than it first sounds, after you declare that though not the last word, present day theories are not going to be radically different from tomorrow's. This gives a desired and comfortable continuity to the history of science. The continuity becomes less comfortable when it extends to the distant past, beyond Newton and Galileo to Galileo's opponents and their predecessors. So Butterfield simply cuts the line when it becomes uncomfortable.

Thomas Kuhn adopted Michael Polanyi's philosophy into a further variant on Duhem's. He endorses a continuity theory of Duhem, and with Butterfield he rejects Duhem's view of the Middle Ages as scientific. But he has a modification of Duhem's view which justifies his deviation from Duhem about the Middle Ages. Though science constantly alters, says Kuhn, it has discrete levels recognizable as the discrete standard textbooks of the different periods. The continuity is provided both in the formation and in the dissolution of each textbook. The Middle Ages, however, had no science textbooks to speak of. The astronomy textbook was ancient, and it had been dissolved to a sufficient extent before Copernicus came; the astronomy textbook, in other words, was much too dated. Other fields, chemistry for one, had no textbook at all.

The continuity theory of the history of science which views all modifications in science as small, is *prima facie* in conflict with the facts of the recent revolutions in science, whether in genetics, in relativity, or in quantum theory. Indeed, Duhem viewed the revolution in physics as utterly unscientific. However, there is some progress from the nineteenth-century theory of science according to which a genuinely scientific theory is in need of no modification whatsoever, to the continuity theory of science, which allows at least minor modifications.

One way to test a theory, it has been argued above, is to see how its applications look. Let us see how we may apply the view that genuinely scientific theories are modifiable.

3. ARE OLD MEN CONSERVATIVE?

Jonathan Swift once wrote a note to remind himself when old, what old men are prone to do which young men do not particularly enjoy, so as to prevent himself from being an old pest. It is hard to say whether we can address ourselves when old: we may by then change our minds far enough and think that we know better when old than when young, thus rejecting the advice of our young selves. Sometimes, quite correctly. For example, when we get older we may tend to become less ambitious and thus acquire a better sense of proportion, not to say become more clear-sighted. For instance, in time we may learn to feel somewhat indifferent to the question, do young people like us or not. Also, we may erroneously find it more important to improve their conduct or abilities even if they are ungrateful. In some instances it is obvious that older people deteriorate. For example, when we get older we may desperately hold to our achievements, feeling too old to have newer ones and fearing that if our past achievements are all insignificant we shall face empty lives with no ability to do anything to improve matters.

This grim possibility is the one which Max Planck saw as the common situation. Though he was one of the most distinguished scientists of the century when he wrote his scientific autobiography, this work is candidly bitter and full of a sense of disappointment at his fellow scientists. Surely, here we have a striking fact. Someone has explained this fact by reference to Planck's bitter life as a German nationalist, as a German citizen, and as the father of a victim of the Nazis. Planck's life was indeed far from enviable, yet to view his bitterness against the world of science as the mere reflection of his bitter life and thus to dismiss his complaint is, again, mere quietism.

What Planck narrates is that his teachers, Kirchhoff and Helmholtz, were unappreciative of his work. Everyone was so unappreciative of him to begin with, that he got his first academic position through family connections. Even later, when he became known, none of his ideas was accepted on his own arguments, for the reasons that he had initially advanced it. He says, in a most striking and well-known passage, that science progresses not because its old leaders change their minds but because they die, leaving the field to the young newcomers who look at the situation afresh merely because they can do nothing else. That Planck's picture is misleading is beyond doubt; even though his facts are largely true, he omits the facts which do not fit his grim view of the world of learning and of his own place in it. He probably did get his first job because of the help of a family friend, but doubtlessly he became the secretary of the Prussian physical society for different reasons. As he fails to mention the fact in his scientific autobiography one cannot know what his view of it was. He likewise fails to mention that Lord Rayleigh referred to his radiation law as soon as was possible, that his papers to the Prussian society were regularly reported, for instance, in the news column of the *Journal de Physique Pure et Appliquée*, that his *Treatise* was translated into English early in the century whereas similar Continental works are still untranslated. He mentions that all the leading scientists he met before he was a celebrity ignored him, mostly dogmatically, and in the case of Boltzmann, even somewhat viciously. Boltzmann did later become friendly but, according to Planck, only after Planck had endorsed some of his views. Planck glosses over the details of his rise to fame; all he has to say is, his ideas were accepted for reasons other than his own. Why were his ideas accepted for different reasons, and why should this have spoiled his fun? He is reticent on these points. Obviously, the points he is reticent about are such that might not gain his readers' sympathies: his readers, too, might accept his ideas for reasons other than his. But why should this be so unpleasant? Perhaps this is a symptom of a serious ambivalence which Planck suffered from when writing his own scientific autobiography: he had effected a revolution which he did not like at all: he was rejected by his elders as a rebel, and by his own followers as an old conservative. He could not accuse himself of selfish conservatism because his own ideas were accepted and for selfish reasons he would have to join the younger generation rather than keep aloof from them. He was an unselfish conservative and so he felt he was right. Which comes to show how many ways there are to be mistaken.

4. ARE THE CANONS OF SCIENCE CONSERVATIVE?

What makes a scientist conservative? Planck's answer, the overestimate of one's own contribution to science, does not apply to Planck himself, yet we judge him a conservative. The theory of selfishness which Planck implicitly proposes is thus not universally true. Priestley already refers to this theory and shows his own behaviour to be a refutation of it. Richard Kirwan's fame, says Priestley, was increased, not diminished, by his conversion from phlogistonism to antiphlogistonism. Hence, self motives should entice Priestley to convert as well. But, says Priestley, he cannot honestly endorse views so revolutionary and so poorly based on evidence, and he cannot see the complete overthrow rather than the mere modification of a view which only a generation earlier was considered by all scientists as the best established and the greatest achievement since Newton's.

Here is a very strong argument for the conservative cause against a revolution, which everyone will recognize and accept unless he is a hopeless opportunist: a revolution against science is one which we all have to oppose. But what is a revolution against science? Even the most anti-scientific revolution in modern history was not declared as anti-scientific but rather anti-Jewish. Lenard, a scientist respected before and after the Nazi era, was engaged at the time writing a book against Jewish science (Einstein), and for true science, namely for German science. Now, if the Nazi villains did not say openly that their revolution was against science, then no other anti-scientific revolution has to; yet we must find out whether the revolution is not anti-scientific, so as to oppose it, if need be. Planck was doubtlessly anti-Nazi, yet being a historicist and a German nationalist he deceived himself that what later proved to be a catastrophe of the first order was a mere aberration, a passing phase. Priestley, to take the opposite extreme, saw with horror Lavoisier's book-burning – which is surely anti-scientific – and he consequently opposed much too strongly everything related to the Lavoisierian revolution in chemistry. Similarly, Planck and Einstein exaggerated the irrational element of the revolution in quantum theory, namely the subjectivism and positivism of Heisenberg, as well as the obscurity and shiftiness of Niels Bohr. These events and a little reflection may show us that it is not so easy to avoid being a conservative: we all want to conserve something, at least our progressive philosophy etc., and whether giving up this or that and swimming with the current hither or thither is progressive or opportunistic who knows.

We are all told with horror about the way the Mozarts and the Schuberts of the past were let to die in loneliness and misery; this makes us willing to be as appreciative of and generous towards all innovators; but, in the midst of all the tolerance and willingness to appreciate, even the last generation has ignored some of the great artists of its day according to present day judgment. This is obviously much less the case today than yesterday, and much less current in science than in the arts. This may be explained by the existence of better standards of excellence in science than in the arts which permit a wider range of toleration and a clearer view of what is impossible. But the standards are neither perfect nor absolutely universal, and this accounts for the errors of serious men of science concerning their attitudes towards scientific or allegedly scientific innovation.

There is little doubt that the standards of science cannot be perfect: the disagreements concerning them and changes of them through the ages are sufficient testimony even for those who would not accept the general view of the imperfection of man. Yet somehow we fail to see that such standards may lead to conservatism on the one hand and to opportunism on the other. So many people, especially historians of science, accept as scientific and thus as ever-lasting, any idea on which the scientists are agreed. Even philosophers of science often say so almost explicitly. Herbert Feigl, in his essay in honor of Karl Popper, says Popper must agree that the law of conservation of energy is well founded since for over a century no scientist has doubted it – with the single and very ephemeral exception of the famous paper by Bohr, Kramers, and Slater. This remark is particularly amazing since Poincaré has shown in *Science and Hypothesis* that the law of conservation of energy cannot be supported by experience just as no conceivable evidence can lead to its rejection. Now, if Feigl can today commit such an error after it has been shown to be an error, why could not men of science have committed the same error long ago?

There is more to it. Whatever the canons of science are, it has always been agreed since Galileo, Bacon, and Boyle has insisted on it and since it became the standard of the Royal Society three centuries ago, that clarity is the hallmark of science. Obscurity is condemned as one of the greatest violations of the canons of science. That Bohr was obscure, however, no one ever denied, least of all Bohr himself. Yet, whereas Bohr was merely worried about his obscurity and merely tried to do his desperate best to clarify his view, certain physicists reacted much more radically than Bohr. Paul Ehrenfest was doubtlessly much disturbed by the problem whether his opposition to Bohr was not as old-fashioned as the run-of-the-mill opposition to Einstein's relativity was. Niels Bohr, in his classical report on his debates with Einstein, refers to Ehrenfest's remarks to this effect as teasings of friends; Einstein, in his (much earlier) obituary notice on Ehrenfest, describes him as a depressive self-doubter who could commit suicide because of such a doubt. Einstein says clearly that indeed the primary cause for Ehrenfest's suicide was his doubt whether his opposition to Bohr was not old-fashioned. The variance between Bohr's story and Einstein's is formidable. This should make us all see how serious and how involved the problem is: even the problem whether the problem at hand is relevant to the suicide of Ehrenfest is too difficult to solve without, at least, much study and deliberation, including the study of the testimonies of Einstein and Bohr.

Yet, hard as the problem is, it is obvious that certain past solutions to it were mistaken, and the mistakes need not be repeated. For instance, the idea is erroneous that Einstein was against science because he proposed to modify Newtonianism though Newtonianism had been so strongly verified by experience. Even the most strongly supported view may be in need of improvement. Joseph Priestley, we saw, was willing to consider a well-verified theory modified one way or another, and he himself studied a number of modifications, some of which were of his own invention, before he settled for Cavendish's modification. But he could not settle for an overthrow of an established theory. Those who agree with Priestley in principle, must deny either that phlogistonism had been well established or that Lavoisier's theory was a break from phlogistonism. Indeed, already Hélène Metzger, Duhem's chief disciple, opted for the second alternative. James B. Conant, another disciple of Duhem, and Kuhn's teacher, settles for a compromise between the two. Some of phlogistonism looks to him not too scientific, some of it looks to him a close approximation to Lavoisier's theory.

The case of the Einsteinian revolution as seen by the continuity theorists is not different. Duhem allowed modifications of Newtonianism, but not as drastic as those Einstein proposed. He dismissed Einstein as anti-scientific. Whittaker, on the other hand, invested much effort in presenting relativity as a natural development in small steps from some nineteenth-century studies.

Exercises like these are very legitimate and partly even interesting, yet the cost of taking them seriously is the readiness to give up hope of rendering the continuity theory applicable to practical problems such as Priestley's, let alone Ehrenfest's. Though the continuity theory (in all its versions) applies against Einstein's opponents who forbade any modification of Newtonianism, the continuity theory permits some modifications but it must forbid large ones, of course. For, if we cannot know from sheer appearances whether a doctrine was scientific to begin with, and whether a modification to a scientific theory is small enough to be acceptable, then we may just give up hope of providing workable criteria. Popper's theory, conversely, does not oblige us to defend any theory against any modification, no matter how well-supported the theory was or how radical the modification proposed. Is this approach not too radicalist?

5. IS CONSERVATISM TEMPERAMENTAL OR INTELLECTUAL?

It is no doubt the case that whether one is progressive or old-fashioned much depends on one's beliefs; yet though most people think so, it is a mistake to identify being old-fashioned with believing out-dated theories or being progressive with believing every day the theories of that day (or the next). This popular error is particularly hard to eradicate because it leads to distorted history, and distorted history provides ample evidence in its favor. Thus, when someone was progressive but held old-fashioned beliefs our historians gloss over his beliefs, etc.

Newtonian physics ousted Cartesian physics, and those who advocated Cartesian physics after the publication of Newton's *Principia* are condemned in many history of science textbooks as old-fashioned; naturally, you will not expect these textbooks to contain the information that Newton himself was a Cartesian, as Euler was, and that even Laplace had a strong Cartesian tendency; yet this information is true. To say that Euler was not progressive because he held old-fashioned beliefs is preposterous.

Equally preposterous is it to praise scientists who jumped on the band-wagon of a new school without understanding it sufficiently to have left the old school or even while consciously trying to compromise between the two.

Helmholtz is praised for having held the theory of conservation of energy. Actually, he first advocated the view of the conservation of force, and not as a pioneer but as a compromiser between the old and the new. He said that Newton's third law assures us that the sum of all forces at any time is zero so that the law of conservation of force is quite legitimate. When he realized that this idea led to the construction of fields of force in empty space he first rejected it as mad and then accepted it either on a model of the ether or as a pure mathematical construction devoid of all physical meaning. It is clear that Helmholtz was old-fashioned in physics (though not in physiology and psychology) yet he joined the right band-wagon and even made contributions to the field.

Who cares much about the fact that H. A. Lorentz could never believe in relativity? He was one of the best relativists of his day, his own beliefs notwithstanding. Conversely, who cares that Kelvin joined the thermodynamicist school in the nick of time, just before it won? His contributions to the field until then bore little or no significance to the dispute.

All this comes to show that the problem, whom should we believe, is a misplaced problem or a misstatement of a genuine problem. Let us go to the arts again. The problem there is not of truth but of beauty. Now beauty has to be enjoyed, and so the problem who is today's Mozart, or Schubert, can be translated into, whose work should I enjoy? But the real question is not as subjective; it is, whom should I appreciate? Appreciation is both more objective than enjoyment and of a wider compass: we can explain our appreciation and discuss it critically, we can appreciate without enjoyment, we can appreciate even without seeing beauty: think of all the influential artists – painters, composers, and authors – who were in their own days artists' artists and then sank into oblivion; think of the geniuses who influenced posterity and whose works are devoid of all beauty, such as Wagner; think of Dadaism, whose immense impact did not save it from oblivion as it has nothing interesting to offer us any more; not a single exciting poem, not a single interesting canvas. And now back to science.

The analogue is clear: it does not matter what one believes is true but what one considers important or interesting – what one appreciates. Make the following experiment: look for an old-fashioned thinker who gets along well with the young, and look for the old fogey who only follows the young. You will easily observe that usually, the old-fashioned person whom the young appreciate is one who understands them, rather than agrees with them; who can expound their interests. The old fogey tries hard to agree with the young, yet they view him as merely ridiculous. This paragraph contains enough material for a few suggestions of experiments which the interested may perform. (See Bertrand Russell's reflections on his 80th birthday.)

6. THE ADVANTAGE OF SENSITIVITY TO PROBLEMS

The idea suggested here is that we avoid being old-fashioned, no matter what we believe, by being able to understand the interests of the young; but to make it fit the phenomena closely or not at all so as to make it applicable we must specify who is familiar with the interests of the young and how such a familiarity can be acquired.

To this the answer offered here is this. He who is familiar with your problems, and can to some extent explain their significance to you, can claim that he knows what your interests are. There are some striking instances of older people who were able to understand the problems which beset the younger generation and thus be active in the progress of learning even though their own major preoccupations lay elsewhere. The case of Niels Bohr is perhaps a famous contemporary case. Another case, more impressive but virtually unknown, is that of Joseph Priestley, the arch-conservative in the whole history of modern science. The facility with which he could move from one theoretical system to another, compare and contrast them, and examine their limitations, is a source of immense pleasure to all his readers (few as these are). He understood the problems of his opponents all too well, even though he was a bit too dogmatic in his conviction that these were insurmountable. Because of his religious and political heresies, the mob of Birmingham was provoked into burning his house. He fled to London, but because of his philosophical heresies, he found no friend there. He went to Pennsylvania and died there in almost total desolation. Almost; for he made friends with Humphry Davy, a daring young upstart who rose to relative fame from a rather humble walk of life. Priestley understood Davy very well, encouraged him and advised him, helped him in preparing the overthrow of Lavoisier's doctrines. In his *Elements of Chemical Philosophy* Davy speaks of Priestley with exceptional warmth, commending him particularly for his openmindedness and readiness to alter his view at a drop of a test-tube.

Davy was a revolutionary scientist, a rebel, an aspirant. When he refuted the doctrine of Lavoisier by extracting oxygen from alkalis, his success in finding a publisher for his discoveries led to threats (by Poisson, no less!) to call the police. Even on his triumphant tour to the continent of Europe he continued to destroy accepted views, including his own! (He thought that only oxygen and chlorine can oxydize, and so suspected iodine of being a chlorine-compound but soon destroyed his own suspicion.) He never accepted Dalton's views, but this did not in the least disturb his researches: he understood Dalton well enough to be able to use his ideas and he even improved upon Dalton's experiments of weighing gases. Yet his unwillingness to believe Dalton was a source of vexation to his and Dalton's mutual friends, who therefore decided to have it out with him. The story is told by Thomas Thomson and it is very human and very funny, but it has little or nothing to do with the problem of atomism which these people studied.

Davy had no difficulty in understanding Faraday's opposition to Dalton: in this respect Faraday was a close follower of Davy. But Davy could not understand Faraday's interest in Oersted's circular forces, and soon he lost touch with his closest friend and disciple. He opposed his candidacy to the Royal Society allegedly on personal grounds (Faraday was suspected of plagiarism), but really from a loss of touch. Faraday's problems meant nothing to him from 1821 to his death in 1829 because in that period Faraday was struggling with new problems which most scientists could not share with him.

7. REVOLUTION IN PERMANENCE

It may be doubted that the view offered here is specific enough. Suppose it happens that he who shares the problems of the young avoids being a reactionary regardless of his own beliefs. Can we say that anyway he who shares the young one's interests also shares their beliefs, so that finally the view offered here amounts pretty much to the received opinion?

Without much disquisition, one can push the difference by discussing a further stage in the practical problem: suppose you do not know how to make yourself believe what you don't believe, nor how to be interested in what you find so utterly uninteresting. To declare that you do agree with the young ones, or that you find their work so very interesting, merely in order to be on the right side, is opportunism and folly – quite apart from the fact that even all the young scientists together may be barking up the wrong tree. What you can do, is try to find out why the young ones are interested in whatever it is. It may turn out that they do bark up the wrong tree, or that they do have a genuinely important interest which they somehow fail to state clearly and correctly! If this kind of a discovery will be of value, then surely this will show the superiority of interest over belief.

But how does one go about interests? Interests are presentable in terms of problems and of the assessment of their relative significance. We have to explain this and provide an instance.

When all the scientists around begin to show concern with models of the ether, one may ignore this interest in an old-fashioned way or in a hypermodern way; how do we know which is which? The answer is simple: there is a problem behind the interest in the ether; those who ignore the interest in the ether and the problem as well, may be losing touch; not so those who can address the problem while declaring that the ether does not exist – as Faraday did. Nobody can call Faraday old-fashioned because he did not join the new fashion of looking for models of the ether, because he knew the reason for the search and found an alternative way of taking account of it.

This story shows that the chief aspect of the current interest need not connect at all with current beliefs but may connect with problems.

Hence, according to the present proposal, if one concerns oneself with current problems one does not lose touch even when one is very old-fashioned. For another example we may take Priestley who was well aware of the problems of his opponents and thus was always in the frontiers of science, quoted by the best students of chemistry until his very death.

But what if the problems of the young ones look to you so very trite and uninteresting? The answer to this may be, try to solve the problem, why do all the young members of my profession concern themselves with a dull problem? Doing this you may either find out where your mistake lay and thus save your skin, or where the error of your profession lies and save your profession. These things are not too likely but they do happen on occasion, and interests of very few individuals do sometimes turn out to become the interests of the whole profession in the matter of one generation or less.

To conclude, Popper's theory of science as a critical debate with empirical criticisms enables us to offer clear-cut recommendations for keeping abreast and so it can be further examined by observations and experiments. The continuity theories of science as always permitting reforms but never revolutions, either offer no clear-cut recommendations, or clearcut recommendations which ought clearly to be rejected. The radicalist theory of science as the utter overthrow of all that is unstable and thus becoming utterly stable, offers a clear-cut recommendation which evidently ought to be rejected. As to the problem itself, as to the desire to keep abreast, it concerns so many scientists possibly because their view of science and its progress is rather nebulous, and their anxiety is merely an expression of their bewilderment. Because it concerns many, it has been discussed here; whether it should concern anybody is a different matter altogether. Perhaps it is preferable to concern oneself with interesting scientific problems than with one's own place in science. As long as one is interested in problems and is intrigued by them, one need not bother so much about the judgment of posterity. But perhaps this is merely an alternative formulation of the above proposal to keep abreast by studying contemporary problems: if we bother about an interesting problem, either it is a current problem, or we may render it a current problem by our studies. It was Faraday already who considered, amongst other kinds of contributions to science, the announcement of problems to be solved.

CHAPTER 10

CULTURAL LAG IN SCIENCE

The phenomena to be discussed below are two. First, the existence and extent of cultural lag within the scientific community. Second, the denial of this fact by the scientific establishment. The theoretical proposal that will be made below is to replace Kuhn's idea of distinguishing the scientific from the non-scientific fields of inquiry. He uses the qualitative distinctions between those which have a paradigm and those that do not. It is preferable to use a more quantitative idea: of two fields of inquiry, that one is more progressive which has a smaller mean time lag between the appearance of an innovation and its public recognition. The proposal that will be made is of some institutional reforms within the commonwealth of learning, particularly such as to reduce the incentives for the Salieri effect and increase the incentive for talent scouting.

Let me make an obvious observation: cultural lags are synchronic, and time lags are diachronic and pertain to cases of closing cultural lags. That is to say, a society or an individual concerned with the achievement of some cultural item but devoid of it, may be said to suffer a cultural lag with respect to that item; the time lag of its acquisition makes sense only if the cultural lag is closed.

1. CULTURAL LAGS

The term "cultural lag", says the *Dictionary of the Social Sciences*, was coined by W. F. Ogburn, and appears in his *Social Change* of 1922. He spoke of it on no more than two pages, yet these stood out. Ogburn assumed that a society is divided into smaller units, and that these interact. The interaction, he observed, is not always so strong as to have all of them progress equally, yet not always so weak as to allow one of them progress ahead of the others without some repercussion. As an example Ogburn mentions industry and education. And not by accident. For, it is the paradigm: it was just at that junction in world history, after World War I, that the United States discovered its industrial superiority over Europe, while its intelligentsia could not boast an equal superiority. Cultural lags, said Ogburn, create tensions and call for adjustment. In this he was expressing the sentiment of many a hero of a Henry James novel, not to mention F. Scott Fitzgerald. Ogburn further proposed the hypothesis that though cultural lags create tensions and thus incentives for improvements, the delay in implementing these improvements is inevitably considerable: large time lags are unavoidable.

So much for Ogburn. It is hard to say what might have come of his prediction: very soon after he wrote Europe descended to a cultural level of unpredicted barbarism, and its intelligentsia migrated to the United States whenever it could. The cultural gap was closed fast.

Yet cultural lag, as a phenomenon, gained import: poor nations learned of the economic gap between them and the industrialized world, and so from the state of sheer poverty they jumped to the state of backwardness: tensions due to cultural lags set in. Backward nations were soon called by the euphemisms emerging nations and developing nations and their likes, and the pious hope was expressed that the time lag was minimal. It was not. This change in the paradigm case of cultural lag, from a materially rich society with poor education to a country poor in both material and educational ability, also changed the very concept of cultural lag: it was incumbent on Ogburn to argue that the economically rich but educationally poor is under strain, but not on students of the backward societies of the present: as Bernard Shaw noticed (*Major Barbara*), it may very well be the case that the uneducated rich prefer to stay uneducated, but the poor want wealth for sure. Marcel Proust illustrated (in his *Recherche*) the situation of the parvenu in a cultural society; living among the educated, his lack of education stands out, and detrimentally, and so he learns to catch up or at least look as if he does. With the rich American this need not be so. Ogburn merely claimed that the industrial and educational sectors of American society are sufficiently related to force the rich to seek education and pay well for it. The case of the backward nations is still too obvious. The most primitive people in Oceania learned about the material wealth of the industrialized world and at once coveted it. More than that: they expect it to come very soon, and as a result of mere worship, a ritual or a religion that is known as Cargo Cult. The tension of the cultural lag here is maximal and concerns mere material culture. It is not a tension between two segments of a society (industry and education in Ogburn's original example) but between the extant and the expected. And the tension expressed in, or caused by, cargo cults, is between individual people's real poverty and expected wealth. But then, tensions are quite generally a matter of expectations, of fears and hopes: we have more tensions within families than without, and only because we expect families to live harmoniously, and tensions between enemies are due to fears of defeat and expectations of victory. And the more there is at stake, for example, the more one tends to lose in a moment of the outburst of a violent conflict, the more one tends to be tense before the moment of decision and have the tension released afterwards. Cargo cults only keep tension high and enable people to live in high tension.

And so the need to redefine cultural lag seems very real and very obvious. Indeed, even the worst case of cultural lag, that of cargo cults, is something to do not only with material wealth, but also with knowledge: it is the very learning about the riches of the rich that makes the poor poorer, said Karl Marx. Hence, we can make cultural lags somewhat more specific and speak of those differences between realities and expectations which are rooted in the growth of knowledge and the resultant growth in expectations which create the tensions in question. One of the profoundest observations made by Karl Marx is that the poor may very well stay content in their state of abject poverty, that only the workers of the large industrial plants have the incentive to fight for the improvement of the conditions of their lives, that through such a fight they develop a solidarity that enables them to make the revolution from which a better world will emerge, chiefly because the fight will be frustrated and the tensions grow until explosion becomes inevitable.

My admiration for Marx is not meant to cover up the fact that all his forecasts misfired, nor to blame him for the vast inhumanities performed in his name. Nor do I wish to have Ogburn's theory absorbed in Marx's, since it is not of the same ilk: only when his concept was altered and loosely applied did it become more like that of Marx, and applied, ironically contrary to Marx's express statement, to poor people who are not industrial workers. Moreover, Marx's specific concern was not cultural; rather, he tried to explain how poverty, which usually is well tolerated, at times becomes intolerable. This, however, is linked with Ogburn's idea: uneven progress between interconnected segments of a population leads to tension through education and the rise of expectations.

2. DELAYS IN SCIENTIFIC PROGRESS

There is the obvious fact that diverse fields of learning develop at different paces in different times. There is little doubt that there were golden ages of classical physics and of modern physics, of evolutionist biology and of molecular biology, and so on. These do not synchronize. Is there, perhaps, tension due to all this? Or are the different fields so loosely connected that there is no tension? Or is Ogburn's hypothesis false?

I think tensions exist. The prestige of scholars in prestigious fields is relatively high. Often, however, their power and pay are not higher. This does lead to tensions. The intellectual standards of one field are much higher than those of another, and this is reflected in recruiting: the more ambitious tend to study the more challenging fields and eschew, say, theology. This is very regrettable when theology matters to all of us, for example when theologians and religious leaders are by-and-large too uneducated to see the risk of overpopulation. Anyone sensitive to this who attacks theology from the outside only raises the tension, and this might boomerang. Yet sending the best to theological schools is likewise time consuming. (Thus Ogburn's idea is more subtle than it looks.) But then theology is not a science, perhaps, and so we can return to our question: what happens when there is tension between scientific fields, subcultures, or segments?

Examples abound. Theoretical clashes exist not only between science and theology, but also within science. It was the clash between geology and biology, after all, that has set biologists, Darwin in particular, athinking afresh about evolution and that has raised the theory of mutations in opposition to Darwin. Social statistics is another obvious example, and its effect is more intriguing. When a new technique is introduced into a field, often many of its practitioners are unable and unwilling to use it. They may be mortal enemies of the innovation or they may divide the field into two and try to stay equal but separate as much as possible. Even in physics, where new techniques may spread like wild fire, they need not become pervasive. When new mathematical tools were introduced to elementary particle theory, many introductory courses were opened for professional physicists under the guise of workshops; those who did not go or who went but failed to benefit, simply dropped out of nuclear physics. Yet, whereas a physicist not able to master the new techniques found his field of study regularly contracting, some social scientists have managed to ignore the results of studies with the aid of new techniques and even justify these on ideological grounds. What is the scientific status of this kind of irrationalism?

Michael Polanyi defines the sciences as the outcomes of the activities of the scientific communities. He insists that we should not put constraints on science and force it into our own preconceived notions, but allow the experts of the profession to act as they understand. This understanding, he adds, is often tacit, and so can hardly be criticized, since criticism presupposes articulation. In a famous essay Thomas S. Kuhn identified this tacit knowledge with what he calls the paradigm of the field to which the expert with the tacit knowledge belongs. Now, Kuhn explicitly allows for a biparadigm case; both the progressive and the backward part of a field, then, may enjoy scientific status if each is sufficiently well defined and separate from the other.

To make matters clearer, let us consider the case of dissent within a given field.

Polanyi does not deny the existence of dissent, and he is not insensible to the tensions it causes; he expresses regrets at its presence, which he pronounces inevitable. For, when Polanyi defends the practice of the scientific community as the best, and pleads non-interference, he is still not a utopian and he does not claim that the best is good enough. He only thinks that the best way for progress to come is from within. And at times, he admits, this causes tensions within the scientific community.

Polanyi was personally very impressed with a case in which a very surprising empirical discovery was reported by quite a reputed physicist, causing no response. Polanyi himself tried to find out why. He found that other physicists had heard of the report and remained suspicious. Rightly so, it turned out. This is not to say that all suspected discoveries turn out to be non-discoveries. Polanyi himself made an important discovery in physical chemistry which was overlooked until it was rediscovered a generation later and then accepted. Polanyi sympathized with those who at the time overlooked his discovery: it did not fit the paradigm and simply had to be overlooked. But not all oversights of valid discoveries are valid: science is no utopia: it is only the best we have, and can and should be improved – but from within.

This view seems to be fully endorsed by Thomas S. Kuhn: paradigm switch, in particular, is a *Gestalt* switch, he says, and a *Gestalt* switch is, psychologically speaking, instantaneous. Admittedly, some preparation for the *Gestalt* switch is necessary, and some individuals are incapable of it. Yet he seems to suggest that the switch is instantaneous once the need for it is comprehended. It is important, however, to notice, as Kuhn observes, that some members of the scientific community will simply not switch. He cites Planck's scientific autobiography to say, old scientists are not open-minded, only young ones are. What happens, then, to the stubborn old ones? They lose their position at the helm, say Polanyi and Kuhn. Both those who move too fast, then, and those who move too slowly, lose their positions of leadership. Moreover, there can be no theory of the proper speed: The proper speed is a matter of feel. Time-lag is, thus, impossible.

Of course, tensions still exist. Stephen Toulmin discusses Priestley's tenacious adherence to phlogistonism by the claim that the need for a paradigm-switch was not evident as long as Priestley could – as he did – produce valid arguments against it. Whether Toulmin's historical claim is correct or not, his line of argument says, again, there is no time lag for science; a scientist suffering from it is outside the community of science.

One last point. If the progress of a science has an internal clock, there may be incentives to retard the clock as a whole – even to grind it to a halt. Examples for non-progress in learned communities are, indeed, more common than examples of progress. It was Robert K. Merton who has claimed that much tension is created by competition for rewards for innovators: priority and patent may be lost in a very short time. This, however, well accords with Polanyi and Kuhn: competition presumes the possibility of being recognized by the community, learned in the case of priority and industrial in the case of patent. And when a discovery can be more quickly sold as a patent, as was the case of Edison's patenting his discovery of thermionics, or when a patent can more quickly be recognized as a discovery – *vide* Faraday – the cultural lag is exactly between industry and learning, exactly as envisaged by Ogburn. After World War II, however, the close collaboration between science and technology makes it reasonable to assume that cultural lags of this kind are closed within the shortest time lag possible.

Hence, again, we find no time lag to speak of.

3. THE HIDDEN INJURIES OF SCIENCE

There is a sociological study, described in a volume by Richard Sennett and Jonathan Cobb, *The Hidden Injuries of Class*, of 1972. It records interviews the authors had conducted with urban laborers. These workers, we are told, invest enormous efforts towards the goal of upward mobility; they make sacrifices for these goals even though their chances for success are very slim; indeed, the sacrifice may be the ritual test to decide whether the individual in question deserves to remain in his place as it is, rather than progress; they may submit to the test not so much from desire as from succumbing to outer pressure; they may make the sacrifice and then be rejected anyway:

"The terrible thing about class in our society is that it sets up a contest for dignity" (p. 147). "How, then, is the believability of the reward system to be maintained? Badges of ability are good . . . because distinctive ability seems to belong only to the few. But . . . even ability criteria produce more eligibles than can possibly be rewarded" (pp. 154–5),

they observe, meaning, most contestants, even most able contestants, must be gypped. Why do they allow this? Because "sacrifice turns a man towards the future" (p. 164), because "class makes people conceive of themselves as spectators" (p. 165) of the hardly possible future for which they sacrifice the present, because they succumb to pressure (p. 166) and deception (p. 169). Above all, people's sense of personal worth is repeatedly questioned, and by humiliation, by appeal to their inadequate education, by appeal to high ideals. Consequently,

> a system of unequal classes is actually reinforced by the ideas of equality and charity formulated in the past. The idea of potential equality of power has been given a form peculiarly fitted to a competitive society where inequality of power is the rule and expectation ... In other words, social differences can now appear as questions of character, of moral resolve, will, and competence (p. 256).

The more we accept the system, the more we endorse its verdict of our individual selves as failures.

So much for the book in question – except for one observation: the authors describe mainly members of the lower classes, blue-collar workers, but only because they intend their description to hold *a fortiori* for class society in general.

To return to our own topic, the above description should, *a priori*, fit the commonwealth of learning second to none but the free professions. And, indeed, there is nothing there which Polanyi and Kuhn would not endorse as both true and proper of the commonwealth of learning. This may be surprising; indeed, some people have discovered with some measure of surprise that the picture of the community of science as described by Polanyi and by Kuhn is not liberal and not even democratic. Yet, this fact is no secret and in response to the challenge, in his famous reply to his critics, Kuhn repeats that he is quite outspoken about pressures in the world of research, a pressure which he perceives and commends, since it is unavoidable. The only comfort is Polanyi's observation that one need not view the community of science as ideal, but only as the best: once we realize this, we can admit that some injustices are indeed committed in the learned world, and we can take these as misfortunes and try to reduce them. This, undoubtedly, reduces the injuries of class somewhat.

I have thus far presented as favorable a case for the Polanyite image of the learned world as possible. It is as benign a defense of any society as at all possible, and it is applied to the admittedly best segment of society we have – the commonwealth of learning. Undoubtedly, the Polanyite image of it is more realistic than the traditional image: it is not a republic, it has its leadership and its followers; it does not have free entry, but workshops where apprentices get training the harsh way; and so on. And yet, the fact that there is no free entry, that it is an elitist group, may very well be an ameliorating factor: one is not forced to become a scientist; and, of course, this is why exclusive clubs need not be as democratic as city and state governments have to be. Let me add that Polanyi's philosophy is immensely popular among scientists interested in the social and/or philosophical aspects of science, and its popularity there is on the increase. At times Polanyi's philosophy gains currency in circles not familiar with his writings or even with his name. At times it is ascribed to different writers, at times its source is identified as the science fiction novel *The Black Cloud* by the astronomer Fred Hoyle. This way Polanyi's view gets the validity of a paradigm and the potency of a myth. In brief, the establishment of science now endorses the view that the establishment ought to adjudicate. This is hardly surprising, and is bound to minimize conflicts as well as all cultural lag, as I wish to argue now.

Let me take as my authority Lewis A. Coser, the former president of the American Sociological Association and the leading living authority on conflicts. I refer to his 'The Termination of Conflict', in the *Journal of Conflict Resolution* of 1961, where the author explains why so often battles fall short of the final act of complete victory and failure, why "most conflicts end in compromises": "both parties agree upon norms for the termination of conflicts", he observes. This is so because, obviously, only when opponents have absolutely nothing in common is there incentive to fight to the very end; usually conflicts are means to finding relative strength, and once this has been done incentives are stronger for conflict termination. This can be done more efficiently, or less.

> To the extent that contenders share a common system of symbols allowing them to arrive at a common assessment, to that extent they will be able to negotiate. Symbols of defeat and victory thus turn out to be of relevance in order to stop short of either.

More than that, compromise, "the chance of attaining peace without victory depends on the possibility of achieving consensus as to relative strength and on the ability to make the new definition "stick" within each camp". The essay ends with the recommendation for a change in research policy: "research directed toward an understanding of those symbols which move men to accept compromise or even defeat might be as valuable as research to uncover symbols which incite to war".

All this applies best to the community of science. Old fogies, as Planck and Kuhn observe, may be as prejudiced as you wish, and yet they neither lose their jobs nor their positions of leadership. Young Turks often take over the actual leadership, and they are usually given places at the top, but to conclude that old fogies are demoted is to conclude hastily and falsely: they are given the usual homage, to the very end if possible. Of course, infirmity of body, perhaps also of mind, may lead them to confinement; stubbornness and insistence may even lead to a rare case of open ostracism, as was the case of Einstein and the case of Alfred Landé in the community of physicists. But these exceptions were of people who simply refused to compromise, not of a leadership determined to oust them: their stubbornness was regretted and even their ostracism which was made amply clear to members of the profession was not made public by the leadership or by the rank-and-file.

It seems clear that there are powerful means of conflict resolution within science – facilitated by the supply of the proper symbols, by philosophers and historians of science, preferably amateurs who are professional scientists, preferably leading scientists. It seems clear that this reinforces the power of the establishment and entrenches the established order by convincing all outsiders that they deserve to stay outside and all victims of all injuries of class that they receive their just desert. Why, then, should there ever be any revolution in science? Why, then, should not research be limited to ever decreasing fields of specialization and ever smaller minutiae? Planck had the answer: new recruits are not burdened with the prejudices of the old and they give the situation a fresh look.

I think this cult of youth requires examination.

4. THE SALIERI EFFECT AND THE WORKSHOP MENTALITY

Are young people less prejudiced than old ones? Planck said, yes; Buber said, no ("The Prejudices of Youth"). Buber said, young people are as opinionated as old ones, and have the benefit of less life experience; hence their judgments are bound to be more preconceived ones than those of their elders. Yet, it is obvious, there are matters on which young people have no judgement for want of any familiarity with the questions involved. A young person may encounter a question together with the old and the new answer to it, and he may then see the advantage of the new one. Whereas the older person had encountered the question before the new answer to it existed. He, then, is under a cultural lag: a new and better knowledge is available and he is expected to have the very best but is prejudiced, is too old to learn, etc.

The discussion recorded in the previous paragraph seems rather naive. For one thing, it overlooks other options. Planck's maxim was criticized intelligently long before he published it, and by Gertrude Stein ('How Writing is Wrote'; the maxim was reported in Florian Cajori's *History of Physics*, in the name of S. P. Langley): the new generation advocates the new ideas in order to stave off still newer ones. By the time, that is, that the establishment advocates one theory, it does it in order to resist another, and they only switch from the oldest theory to the intermediate one because the oldest is not strong enough to resist the newest. No doubt, there are examples to support Gertrude Stein's view. All sorts of views that were deemed heretic by the establishment of physics early in the nineteenth century were used by them as means of preventing the electromagnetic field theory from gaining currency later in the century. (See my *Faraday as a Natural Philosopher* and W. K. Berkson's *Fields of Force*.) Nevertheless, it is very unclear why the newest theory should at all be offered, and why the establishment must fend it off. Why should there be a cultural gap in the first place? Why should the young bring new ideas and the old resist them? This question is particularly pressing when we notice that in most cultures there is no pressure of new ideas, that some new revolutionary ideas are offered by established middle age people (like Max Planck, 42 year old secretary of the Prussian Physical Society), and that some elders of the community welcome new ideas of young rebels (like Charles Lyell, who encouraged Darwin even though he was critical of his views). Where lies the diversity of these activities?

The starting point is the system of incentives. Not that these are either necessary or sufficient, but I will not discuss this point but endorse the standard approach to incentives now. The paradigm for resistance of the new through a selfish motive is the once famous and now almost forgotten composer, Antonio Salieri, whom Beethoven called his master, who taught Schubert as well, and whom Gluck considered the best operatic composer at the time. His greatest claim for fame is the allegation that he had Mozart poisoned, which allegation he denied on his death-bed, yet it stuck because he openly sabotaged the career of Mozart whom he frankly admired. Indeed, his very last encounter with Mozart was friendly on account of his admiration of Mozart's latest work, on which Mozart commented in one of his last letters, saying he hoped that Salieri's friendliness may bespeak an end to his financial troubles. It is doubtful that Salieri could ostracize Mozart so effectively, since money came from pockets of aristocrats, many of whom admired Mozart; more likely it was the Archbishop of Salzburg who caused the trouble, since he considered Mozart his rightful property. Yet the story is neither here nor there; the only impressive point in it is that Salieri opposed Mozart because he felt Mozart put all other composers in the shade and thus threatened their livelihood. Not quite: the remarkable fact is that Salieri was both aware of the reason for his hostility and candid about it. This is not to say that

Salieri was right. No doubt, posterity ignores Salieri's operas, and even Haydn's, because anything worth noticing that they did, Mozart did better. But Salieri could not prevent this except by murder; and murder, to repeat, was not in his repertoire. What he was concerned about was his immediate future. And it is hard to deny that he was a bit over-anxious. Yet over-anxious all established people are, and so they fear competition from their better juniors. This is all there is to it, I think.

Not that all such fears lead to such dramatic conflicts. Usually, elders try to suppress young upstarts, either as potential opponents or as protégé-disciples who are a bit too pushy. Of course, this is largely a matter of temperaments, as evidenced from the fact that while Salieri was sabotaging Mozart, Haydn tried to help him. Perhaps it is also a matter of concern for the matter at hand, since Haydn also tried to learn from Mozart and acknowledge his debt, whereas Salieri had no trust in his ability to do likewise. Yet it is Salieri who concerns us here, since he is the one exhibiting the kind of conflict which Polanyi acknowledges regrettably exists also in science and declares inevitable. What seems to me obvious is that the Salieri effect is very common, that Polanyi's philosophy both minimizes it and calls us to accept it as inevitable – thus closing the gap between expectation and reality, namely the cultural gap, thus making it less and less normal to expect new ideas to be accepted fast, thus making incentives for innovation ever smaller.

Why, then, are conflicts in science inevitable? Because science is autonomous and organized in workshops. It is conflicts between and within workshops that Polanyi notices, and he thinks science is impossible without workshops: to be a master scientist, as to be a master artist, is to grow in another master's workshop, to accept authority. The authority of science is not a matter of plain rules which make it open to all to observe. Plain rules, we remember Louis Coser's observation, are conducive to smooth conflict resolutions; but there are no such rules; in particular, there is no principle of induction. One major asset of a master is that he possesses, personally, tacit knowledge on such matters. The knowledge is conducive to conflict resolution, for sure, but its tacit dimension permits conflicts to arise.

Polanyi's authoritarianism, thus, is rooted in his irrationalism which is rooted, in its turn, in the defeat of classical rationalism. One may tackle it, then, either by attempts to resurrect rationalism in a better way, to create a new rationalism, that is; or else one may criticize its positive assertions and proposals. Any attempt to create a new rationalism is met by a blanket criticism of Polanyi: if the new rationalist theory is false, it is dangerous to follow it, and if it rests on a proof, then it is classical rationalism and must therefore be rejected. Classical rationalism, he adds, is not only erroneous on account of its failure to deliver the proof – after all, this defect may be remedied by some miraculous discovery – but also because proof dispenses with tradition altogether, workshops and all.

Therefore it is advisable to criticize Polanyi – and Kuhn too if one cares to – by reference to his view of the workshop. In workshops the incentive system may be one of diverse sets which history offers us. Some workshop systems or sets are all too clearly opposed to innovations. They do offer innovations, nonetheless, of course; nobody is utterly consistent. Yet we reject them. What makes the scientific workshop system different? Polanyi says, the autonomy of the science workshop. He is here doubly in error: The medieval guild system was more independent than physics after the big bang. The Renaissance workshop system put high premium on innovation and eliminated conflicts within workshops. As a result, when Leonardo's master noticed that he made angels best, he let him make them, yet they were his, not Leonardo's, until the apprentice left the workshop and decided to strike out on his own. This is not very different from the widespread tradition in physics where a doctoral dissertation's abstract is published in the learned press by the director and the successful candidate as the senior and junior authors. This does not solve all problems, but it is a tension reducer. It does not help in the case in which director and directed disagree, in case a discovery is made by a young post-doctoral researcher, and more.

The reason why Einstein had such a wonderful change of luck, and turned from a reject to a leader within less than a decade, was his great ideas, of course. This gives the impression that only talent counts, no workshops, no social organization, minimal intervention from the background. This impression is false on two counts. First, we are not all Einsteins, and few of us have both the intellectual ability and the moral courage that went into his work. Second, he did belong to a workshop, but not of the authoritarian kind Polanyi describes. It is the one described by Lewis Feuer in his *Einstein and the Generation of Science*. I reviewed this book in *Philosophy of the Social Sciences* and had a sharp debate with the author in the pages of that journal. Yet, I think he has enough evidence to support the view that Einstein belonged to an egalitarian workshop of young and powerless enthusiasts. Moreover, like others, Einstein needed access to public platforms, and these were, of course, centers of power very much as Polanyi and Kuhn describe them. He was lucky to send his papers to Planck who, though not young and quite powerful, was impressed at once, and indeed took drastic steps to put Einstein on the map.

Talent scouting is an art. It needs developing, it needs incentives.

We should open a workshop for talent scouts.

To conclude, Polanyi's philosophy is very interesting and very important in a few respects, not the least of which is that it is self-verifying: it will be true if and only if we accept it. And I propose we reject it because it views too many injuries as unavoidable without further test and because it makes scientific progress more sluggish by inviting us to be too patient with the scientific establishment and its self-serving bureaucratic built-in sluggishness. Polanyi's view seems eminently reasonable – indeed, it has only its sweet reasonable stance, its immense sense of proportion to back it. Yet it is most unreasonable to admit the existence of the tensions which, according to Ogburn, cause and reinforce cultural lags, yet to deny the existence of the cultural lag. Even the claim that both the tensions and the cultural lag are minimal is questionable: by what mechanism are they kept at the lowest level? Merton may wish to explain this by reference to high incentives and competition. Yet these only raise tensions and create incentives for causing time-lags of all sorts. We need, to repeat, new incentives to encourage the building of workshops for talent scouts – not scouts who woo Nobel laureates to their departments, but ones who spot autonomous scholars at a glance and who tender them help and moral support.

CHAPTER 11

STORAGE AND COMMUNICATION OF KNOWLEDGE

Due to current myths and to publication pressure, science is often pictured as a reservoir of knowledge, where all published material is accumulated and processed. This picture is sheer myth. Scientific literature is chiefly a means of communication, not a reservoir but a channel-system; and the channels are now clogged.

This chapter has four sections. I first (1) describe some of the most obvious facts of the matter; then (2) the view in defense of the status quo; then (3) the limits to improvements due to the ambiguity of the importance or otherwise of new items and the reasons the scientific leadership uses for enhancing the ambiguity and slowing down progress in the name of playing it safe. Finally (4) I make some simple proposals for radical reform that may easily overcome the worst obstacles. In particular, I recommend the establishment of university and learned society clearing-houses which will publish lists containing only *quality* items *to be deemed as published*, copies of which items should be available upon demand. Such an arrangement will be cheaper, more efficient, and less harmful, than pseudo-publications of unread journals.

That scientific knowledge is both stored and transmitted for all sorts of use is an accepted fact. The popular view of science as a storage or deposit of knowledge leads one to the problem of retrieval and transmission of stored knowledge – i.e., to the problem of scientific communication. The view of science as a communication network, on the contrary, raises the routine problem of storage, i.e., of temporary storage of scientific pooling of information to be transmitted at will. The view advocated here is of science as a communication system, and so it raises the question how should knowledge be stored so as to become most readily available at will. Indeed, the claim from which I begin is that publishing books in the traditional manner is a poor mode of publication just because it relies on old modes of storage.

1. SCIENCE AS A SYSTEM OF COMMUNICATIONS

The problem of communication each person faces is technical: he is a producer or consumer of information in search of both channel and counterpart, where counterpart to producer is consumer. As usual, the search for a channel and the counterpart are interwined since consumers are all too often identified with a channel, e.g. readers of *Science* and readers of *Scientific American* and readers of *Nuclear Physics* and readers of the *Bulletin of Atomic Physicists*, readers of government daily papers and of the official organ of this or that party, of the leading independent paper in the country and so on and so forth.

Traditionally, control of the channels was political power, but not an unlimited one since when the demand for a channel is strong enough yet ignored by officialdom, new channels are created. The new channels may be illegal or semi-legal, despised or not respectable. But this is the struggle of a channel for survival.

A struggle indicates a conflict – of interest or of opinions about them.

The fact that authors have limited access to respectable scientific literature should be obvious from the very consideration of the employment benefit accrued to any author who gets published and from the fact that all established intellectual leaders, except Ludwig Wittgenstein and Don Juan and a few other gurus, are published first and established later and become leaders still later on.

Some material, however, comprising much scientific literature, has readership which may be more easily accessible through personal communications. That is to say, much material published in the scientific press that may be science proper, and which gains the status of science proper from being published in the scientific press, is very cumbersome as communication. What should be done about it is to find a new way of according it scientific status, while employing better means of communication than printing it: status and printing need not go together.

Also, some material which is genuine communication is barred from the scientific press; both rightly and wrongly; to wit both on correct and incorrect judgment of lack of scientific merit. The question here is, who judges the judges. But a deeper question is here, how can we improve the criteria, the rules of judgment, so as to improve matters.

There is, also, the trade press: no press is an island. The interaction of the trade press with the academic world is complex. At times a publisher voluntarily submits his material to the profession's censorship and thus he raises again the problem of critic and of criteria. Also, at times the trade press publishes material of obviously low quality with high pretense. At times, the academics try to resist this, but they cannot possibly win. At times, the most trashy trade literature gains recognition from the learned press, again at times rightly, at times not.

But on the whole it is much harder to get published in trade channels because their readerships are less definite and so, for purely commercial reasons, they can only cater to mammoth readership – with the exception of the vanity presses, including most university presses, of course, which print books not for any distribution, i.e. not in order to communicate, and except for trade series that are deemed a must for any self-respecting library, whose publishers print them only after they have presold enough copies to cover all costs and mark-up to boot.

It seems that the open market has its own rules, and that planners of improvements of learned communications should let them be. But this ignores the interaction between the trade and the learned press, or, more precisely, the institutionalized side of it. This includes, to begin with, standard textbooks, standard monographs that have been essential for any given profession, and the complex process of book reviewing in the learned press. This is far from being all. Anyone who has tried to publish a kudos volume, whether a collection of essays of one author or a collection by different authors in honor of one, or a proceeding of a humble conference, must have noted that there are unwritten, unarticulated taboos. The first question put to a referee is, is the book up to standard and he judges this not by content but by context: is the kudos that publication would endow warranted? The second question concerns marketability; and referees make it dependent on the first – as is at times true, but at times patently false, e.g., in cases of popular pseudo-science.

The question here, then, is how can the institutional attitude to trade publishing improve the service of the free market to the scholarly community. There are other obvious means like subsidizing and like institutionalized guidelines for controlling purchase in university libraries and even university bookstores. But these are refinements.

Thus far, I have approached the problems with minimal initial information and repeated, in their context, the standard socio-political questions about standards, modes of applying them, of checking the application, of testing for efficiency and for quality control. This hardly scratches the surface.

2. SCIENCE AS THE BEST OF ALL POSSIBLE WORLDS

The idea of the world of science as a utopia here and now is very much in the air. Again and again one hears that many scientists consider Fred Hoyle's charming *Black Cloud* the way a Catholic may consider *Lives of the Saints* or a member of the counter-culture Hermann Hesse's *Siddhartha*. It describes a colony of scientists that quite of necessity becomes a utopia when it had to govern itself. Of course, other science fiction stories describe scientific dictatorships far from perfect and the stories are neither here nor there; what matters is the widespread readiness to endorse a simplistic utopian view of the sociology and politics of science.

To apply this to communication, I have tried a small experiment. I have expressed a certain highly naive view to a number of natural scientists. Some accepted it, some confessed their inability to reject it despite grave doubts. The view is this: science is a body of knowledge and publication of worthy material is the addition of that material to that body. As long as this is taken as a myth it seems unanalyzable. But the current Lévi-Strauss theory that is so much in vogue these days is just the means for analyzing a myth from the outside in terms of communications.

In terms of communications, the myth is that science is the best of all systems of communication. Best may mean most excellent conceivable and so better than very good and it may mean much inferior to very good but can hardly be improved upon under present circumstances. The first reading permits complacency, the second permits and even encourages reform from the inside: we can try to alter circumstances so as to permit improvement and the Establishment should be enlisted to effect the reform; true enough, the Establishment will not incite any revolutionary alterations, but these are not called for anyway. Thus, the two views are the quietist and the moderate traditionalist. The quietist view vis-à-vis scientific communication is openly and vociferously advocated by Derek J. De Solla Price and by Joseph Ben David. The moderate conservative view is advocated by Michael Polanyi and by Sir Karl Popper, though the latter is a bit more reform conscious. Thomas S. Kuhn holds a position somewhere between the quietist and the moderate traditionalist; perhaps he holds both, but this, to employ Claude Lévi-Strauss' idea, means a middle position all the same.

Let me state my own position. I take science to be a *conduit*, not a reservoir, of contributions of any import. Science is a social institution, a tradition, a culture; the declared canons of science, of the requirements for explanation, empirical evidence, probability, etc. These declared canons are, first and foremost, not necessarily the real canons, as noticed by Duhem, Freud, and Malinowski. Secondly, any canon is not one universally observed. Hence, whatever is the true canon, characteristic trademark, or touchstone, of science, the scientific tradition also embraces, perhaps only marginally, a sub-tradition that violates it.

I take the communication of scientific material to be the prime factor of the integration and rapid growth of modern science. I take it that some scientific communications channels are more institutionalized than others, that some scientific items of information are communicated more as an institutional factor – e.g. scientific textbook material – than other items – e.g. scientific gossip.

The communication system is, as most systems, partly a natural growth, partly a matter of design and legislation. For example, the idea of a scientific essay, of a standard style of one, belongs to Francis Bacon and Robert Boyle. The idea of a book review belongs to the Royal Society, perhaps to others. The idea of surveys belongs to Francis Bacon and Bishop Spratt. Yet Spratt's execution, as well as Joseph Glanvill's, were so poor that the idea had to be resurrected in the 18th century by various historians of science, such as Joseph Priestley. The 1812 *History of the Royal Society* of Dr. Thomas Thomson is a survey and has already an old-fashioned ring. Surveys of recent scientific studies, such as Faraday's anonymous survey of the electro-magnetic literature of 1821 and J. J. Thomson's of dielectricity of the 1880's, stand out, but the diverse encyclopedias of the early nineteenth century made surveys a fixture, as did Thomas Young's medical dictionary and similar works, not to mention the rise of the review articles in the learned press, review journals, and monographs like de la Rive's *Treatise on Electricity* of the mid-century.

I cannot here discuss the canons of scientific publication as implemented by the Royal Society, and further evolved through the ages. Let me only say that Robert Boyle insisted that scientific work is chiefly a communication – and he put all else as secondary to this single consideration. He recommended the essay form as a means for quick communication and his book of essays, *Certain Physiological Essays* of 1661 was a model for the scientific journals that started appearing at that time and it remained a model for generations of writers, including Joseph Priestley, whose style was praised by Goethe who endorsed verbatim the praise lavished on him by Lavoisier: his writing, they said, was facts uninterrupted by thought. For what we wish communicated most, says Boyle in his Proëmial Essay to his *Certain Physiological Essays*, are facts, not conjectures.

So much for science as the best system of communication to date. But the picture is more complicated. Some facts are deemed too uninteresting to publish and even the most inductivist editor will reject papers reporting them – at times on the pretext that the papers in question are too long – and some conjectures are very seminal and should be published, as even Boyle and Priestley noted. What is the guideline? Is there one at all and can there be one?

But, no doubt, the chief end Boyle had in mind is now severely frustrated. Papers are refereed for months and years, and then shamefacedly rejected. Even when finally accepted, they are on a waiting list. (A paper of mine was four years on a waiting list before I protested and got it published after a few more months.) The reason is that people now publish less in order to communicate and more as a result of publication-pressure. What is to be done?

3. THE INHERENT AMBIGUITY IN THE SYSTEM

The chief argument that traditionalists can offer against all radical reform proposals is that radical reform is based on standards which are untested and may be erroneous and harmful. This argument was forcefully presented by Michael Polanyi, in his classic *Personal Knowledge* and elsewhere. Yet he is evidently in error: the scientific tradition is not one organically evolved (if there ever exists any that is), but one developed by the rational growth of rules – rules that were often found erroneous and in need of mild or of radical modification.

Nevertheless, clearly, we must have lax rules, and we must have rules to take care of the fallibility of our rules as a system: constitutional rules concerning constitutional amendments.

Not only do we need constitutional amendments; we also need much leeway and the benefit of doubt. Admittedly, the case of great ideas overlooked by the scientific establishment is less damaging than the same case in other quarters, such as politics, the arts, and religion. These days even religious Establishments are not totally blind to new ideas. The scientific press is only one small part of the Establishment, and it has its share in the game played by the Establishment, the game of increasing ambiguity. Let me explain.

The well-known fact is that some contributions – to science, art, politics, religion – are direct answers to prayers and seem so at the very first blush. They may be theoretical – like quantum statistics – or empirical – like Eddington's eclipse observation – but they fit certain specifications that are well accepted by all students of the relevant questions. Other contributions, though unexpected, are noticed as important the moment they appear. Oersted's discovery of currents' deflection of a needle is the paradigm here. It took Einstein to put his authority behind the electron wave of Davison and Germer, but once the idea clicked its instant success was assured, Einstein or no Einstein. But some contributors are discouraged from the start – Mössbauer is a paradigm. And, at times, they are subject to mini-witch-hunts for decades before they make it – Oersted is a paradigm again – or even after they should be seen as important discoveries – Brownian motion, for example.

The citation index that Derek Price has devised cannot but prove success and ignore neglect of what should count as success: the ideas that are neglected are simply not cited. Citation is traditionally – since Bacon and Boyle made it so laudatory – a means of stimulating interest in the game. But the citation index can also be used to show that the Establishment's game is hardly ever right: they neither back the right horse, nor do they back no horse while letting the better one win; the citation index shows that the game is, only if you can't beat them, then you must join them. The result is a constant sluggishness of progress that is necessary for members of the Establishment to readjust: unlike the public speakers in Orwell's 1984, they cannot change sides in mid-speech, and unlike the scientists in the science fiction literature of Asimov, Merton, Popper and Bunge, they are unable to say, what I thought yesterday to be true, turns out to be false. When forced to say so, at gun point, they say of a refuted theory that it was true yesterday, or that times have changed but the theory – especially Newton's theory – is still all right. Here it is that the worst and most reactionary philosophical rubbish, relativism and/or instrumentalism, comes handy. But only at gunpoint. In less trying times even the scientific establishment is not as reactionary as Polanyi and Kuhn.

The rationale for the sluggishness, I say, is the maintenance of the scientific leadership. When it is not a sufficient means for maintaining control, even while crowning a young Turk the king of the day, then specialism splits to sub-specialisms so that the leaders have less and less to learn. The effect is that any study of a whole sub-specialism – one which transcends the area of subspecialism – is handled with extra care and caution, i.e. with extra sluggishness.

There is a limit to sluggishness, though, imposed by the non-academic trade press. The paradigm still is Helmholtz' *Erhaltung der Kraft* of 1847, rejected by a journal editor but well sold in the open market. It does not always work. John Newland used the trade press to publish his Law of Octaves, yet was beaten by Mendeleev. Some will say, Newland did not get the table of elements right; the answer is that Mendeleev did not either; that if anyone did get it right, then it was Niels Bohr and no one before him.

Nor do we know whether to honor a by-passed innovator in retrospect or not; Newland was not honored, but the celebrated Mendel was. It was touch-and-go in his case too, but this is a different story. It was for generations quite customary to blame Mendel's contemporaries for blindness and prejudice. This attitude is now passé: we now fully recognize the difficulty of recognizing the import of a novelty. And reactionary philosophers of science such as Thomas S. Kuhn and Imre Lakatos declared as a general law that all recognition must come too late. Lakatos even cited on this the Great Reactionary Philosopher, Georg Wilhelm Friedrich Hegel: The owl of Minerva, he said, flies in the dusk. All this means is, clear sight is hind-sight and in the twilight of the day comes the recognition of glory and achievement!

Let me concede that ambiguity is always there. Even when we all know what we want and recognize it at a glance, this is so because we all accept a certain intellectual framework that may soon be overthrown and this overthrow may totally alter the picture. Remember that in the mid-19th century action-at-a-distance won over field theory, and when later field theory won textbooks and histories were rewritten. But obviously we should minimize ambiguity. The system of scientific communication, to recapitulate, is the best we have, but it is not perfect. Partly it's defective because its responses to innovations, such as publication pressures and research grants, are more spontaneous than deliberate, partly because the deliberate part of it is out of date. Yet there are clearly defects that can be overcome, especially the evil current mythology of science as perfect; and those that cannot be overcome, especially the problematic nature of many possibly important, possibly worthless contributions. Yet the sluggishness caused by the ambiguity need not be enhanced as it nowadays is. Hence, radical reform is possible and may be tried out.

4. SOME OBVIOUS PROPOSALS

First and foremost, comes the desire to keep the scientific press as a means of communication plain and simple. Of course, it is an ideal, but one worth struggling for. Since publication pressure cannot for the time being disappear – until someone has a bright suggestion – we may designate publication for kudos purposes one way and publication for communication another way. I observe here an existing practice: all U.S. doctoral dissertations are now already deposited in a certain clearing-house in the U.S. and are thereby deemed publications and photo-copies of them are made on demand. Two things are wrong with this, fundamentally excellent idea. First, the clearing-house receives dissertations that can be exposed as pure charlantanry at a glance, as I can testify in court of law. Second, it is confined to doctoral dissertations. We should have more such clearing-houses, and they should make money by publishing lists of available items and providing efficiently photocopies or computer reprints of these on demand, and send to publishers and/or distributors items for which demand is sufficiently high.

Also, clearing-houses may publish their policies, debate them publicly, etc. This would be a factor highly democratizing, diversifying and encouraging. And if the clearing-houses will belong to learned societies, then debating their policies will increase participatory democracy in them yet, on the whole, this is a long term project. The short term project is to make scientific publication more frankly a reservoir so as to facilitate possible flow to any conduit as desired.

Not only that. Publication pressure leads to much repetition and to minute studies of all sorts of insignificant detail for which there is almost no demand. It would be much better if instead we encourage the publication of book reviews, surveys, proposals for institutional reform of the organization of the sciences and their institutions, etc. Rather than see an ex-scientist desperately try to pretend to be keeping up-to-date, I would suggest he write essays on his experience as a scientific administrator.

There is more that can be instituted, but I am more anxious to discuss implementation now. When I put this proposal before the public, it may pass unnoticed, but it may draw attention since it deals with a problem painfully felt by many. If it be noticed it will surely draw criticism, and to the extent that criticism may be just, it may have to be modified, or be replaced by better alternative proposals. Anyway, sooner or later a reform will have to be tested in the field.

For this, what now seems to be needed is a number of leading members of the commonwealth of learning who will be ready to lend the project their full moral support. This means that the venture will have some limited currency. If it succeeds – and in modifications prescribed by new experiences – then it will both increase its limits of validity and be emulated to varying degrees. This will lead to the proliferation of competing standards, competing intellectual frameworks and more.

It will be a step towards making science more democratic, more open, more encouraging, more offering of options to all sorts of oddballs.

But the major objection to the proposal made here is the objection to all radical reform: science functions very well and much better than any other aspect of human society. As Bacon already stated, the conservative says, things could be much worse and the reformer says, and they can be much better. The troubles people who use the academic press for communication encounter are real. Some of my younger colleagues are complaining and with justice. I plead with those who reject my proposals to remedy the situation to come up with alternatives, not to defend the status quo, not to overlook just complaints. I admit, leaving the status quo could mean deterioration; but it can also be progress.

One final technical point. Vanity publication – an author publishing at his own expense – is an abortive form of publication geared to authors' demand. As in movies, nowadays, production is a minor venture as compared with marketing and distribution. The reason private vanity publishing (and movie making) fails is the absence of large scale clearing-houses. University vanity publications channel material through clearing-houses or have clearance sales just like remainder clearing-houses of the tradesmen. The trade presses and the trade clearing-houses are the middle men to be cut: all we need is the private or vanity publisher who is thus far inefficient because of lack of distributors, plus the distributors in forms of influential advertising, clearing-houses which can cater *ad hoc* to increased demand. Whereas, today a trade clearing-house will sell university press remainders but not vanity press remainders, thus cashing in on the reputation of the university, it is better that the university not have a university press but a university clearing-house. In this way it will be able to transcend the primitive printing press and use more modern means of duplication, better employed to fit demand.

CHAPTER 12

THE ECONOMICS OF SCIENTIFIC PUBLICATIONS

The market of all published materials in the broadest sense, including books, magazines, movies, is very unsatisfactory. The market of scientific publications fares particularly badly, and this despite heavy subsidies. The fact that scientific publications serve the market so poorly just at the time when reproduction on a small scale is available to all scholars at small expense and when the system of subsidizing publication is largely diminished and partly collapsing – these suggest that now is the time for a decisive move for a reformer who wishes to see a radical change and for the entrepreneur who wants to make a kill. What is wanted, I shall argue, is a simple plan that I shall briefly outline, and the blessing of a reputable learned body, which I think is also coming, and even in a big way.

My own concern in the matter is very simple. In these days of excessive publication I know of many good people who have dropped out of the academic world and the community of scholars simply because they could not get published. So, rather than fight the flood of publications or its causes, especially publication pressure, I wish to rationalize the system so as to make room for more deserving young people.

1. THE QUALITY MARKET AND THE MASS-MARKET

I am anxious to begin with a strong repudiation. I will have nothing to do with the trend, fashionable or notorious, of condemning the mass-market. The mass-market simply offers at a low price what other markets can offer at higher prices; and availability of goods and services is all to the common and private good. Economically speaking, there is no trash-market – every commodity is sold at a fair price on the open market. I shall soon discuss the question, how open the market really is. Sociologically, there is no élite that can direct the taste of the market better than the consumer; politically, an open market is essential for democracy; educationally, the available trash on the open market only drives out of the market worse trash; banning it will bring back not good culture but the older and worse trash; legally, the matter is unnecessarily complicated and so I shall here discuss only legal material and ignore all illegal matters, inflammatory, obscene, or samizdat.

There is still some complaint that has to be examined. The complaint has to do with the allegation that the mass-market drives out the quality market, and in two ways. First, competition is tough; second, tastes deteriorate or at least do not improve. Let us examine each. I will try to argue against the first – that poor stuff drives out good stuff – and defend the second – that mass-markets foster conservative tastes.

It is a fact that whereas the latest trash best-seller is on the rack at every drug-store, the quality book is often to be had only by a direct application to the publisher; that the better movies can only be seen in the movie-club or cinematèque; that good reproductions are seldom available and then they can be bought almost only in museums; etc., etc. How can we judge whether this fact bespeaks the mass-produced commodity ousting the quality commodity? It may well be argued that prior to the advent of the mass-market high quality and low quality commodities, all produced in relatively low quantities, were less accessible than now. Last century, after all, it was so common to write to a publisher, a bookseller, and even a wine-merchant. The famous philosopher Hegel wrote quite a few letters to his wine-merchant, for example, yet I know of no living philosopher who writes to his wine-merchant even if he drinks and entertains much more than Hegel: he either buys his drinks from the mass market around the corner or 'phones his specialist! The mass-market seems not to reduce but increase accessibility of old products and of new. Is this so? Empirical evidence is too complex to be relevant by itself and theoretical considerations all go round the question of openness: if a market is open, economists say, accessibility is optimal. Is the market open?

No.

In a classical essay on the method of economics, in a paper that did more harm to the economic profession than any other single paper, Milton Friedman said, the fact that the market operates as if it is open suffices. Moreover, he explains why the market is pseudo-open: shares of different economic concerns are on the open market. But whether the market is or is not pseudo-open is an open question, and the fact that the tobacco companies, to use Friedman's examples, sell shares in the open market, makes shares and dividends subject to the laws of supply and domand, not the cigarettes: anyone who knows something about tobacco knows that American cigarettes are all very similar to each other as compared with, say, Gallois. The Friedmanian economist will have an answer of course: there is never a short supply of answers and of excuses in the Friedman stock. The answer is that one can buy Gallois cigarettes in the United States too, either at special places or by a special order with the aid of an import license. Anyone who will protest that import licenses are hard to get will only have Friedman's sympathy and agreement and accord: all such interferences with the free market, of course, should be abolished *tout court*.

And so you start attacking Friedman and you end simply playing into his hand. Only on the way home, on the stairway, as the French say, it occurs to you that since an average American customer cannot import French cigarettes since it is too complicated to obtain an import license, and since there is no free entry to the market, his choice of cigarettes is limited to American cigarettes for all intents and purposes; unless he happens to live near a cosmopolitan center or near a university where foreign cigarettes have snob appeal. To this, too, Friedman will readily agree: the market mechanism optimizes but works no wonder.

The question that seems most interesting here, economically, is so obvious it is surprising one has to mention it, as my experience in debates with economists shows: Why is the market not open? As Friedman and others notice, there is no free entry because initial investments are prohibitive. This is true, but ridiculously unsatisfactory all the same. It only is the claim that there is a stable mass-market – say in cigarettes; and that to enter the market often means to enter the mass-market. For, only the mass-market has a very high threshold of initial investment. After all, everybody knows that small quality markets exist, in cigars though, hardly in cigarettes, or in hand-made shoes and tailor-made suits and chef-prepared soups, etc., etc., and these markets are not that closed. Perhaps the prevalence of bakeries all over Europe is the best example.

Does this mean that Friedman does not have to worry about there being no free entry? Can he answer the objection by saying, but you can start a small cigarette factory any time you want? Not at all: the quality good is hardly a substitute for the mass-produced good.

If so, how come they compete? Why do we all say that the Beatles drive out Beethoven? Is it perhaps not fair that they do? There is a confusion here, since most of Beethoven's better known works are on the mass-market. But do Beatles and Beethoven drive out the music with less mass-appeal, the music that is not so much in demand as to be mass producible? I do not know. But the producers and dealers with mass entertainment concede the point. When in the United States some entrepreneur or another tries his hand with pay-TV – and let me say right now that for my money pay-TV is the wave of the future, since it allows more voice to the consumer, and the model for the future of scientific publications, since these should be most consumer-oriented – whenever someone wants to introduce pay-TV, the established TV networks and their distributors as well as the movie world, movie distributors, and owners of local movie-houses, they all begin a witch-hunt. They admit that pay-TV will broaden the range of choice and raise the quality of some viewing; but they claim that this improvement will be temporary and very soon after the introduction of pay-TV the older standards will return except that the customer will have to pay.

Why do these entrepreneurs care whether the public will pay or not? The answer is that they fear that pay-TV might oust commercial TV. This answer is, of course, an exaggeration. The more reasonable expectation is that the market will be divided between consumers who pay for TV shows and those who watch the commercials. Is this fair? Perhaps it is, since after all the cost of commercials is covered by the market as a whole, so that the pay-TV watcher seems to pay for the commercials as well as for not watching them. Now, this is supported by the facts: the governments of a few countries subsidize the top, cultural television channels not showing commercial advertisements much more than they support those popular channels that do show them: the production costs of channels with no commercials are relatively higher and they bring no income. The subsidy is thus made in two ways. It also helps siphon the discriminating public from the general public in two ways. In the US, too, there is a subsidised national network, yet, there clearly seems to be also a market for pay-TV.

I hope all this indicates how complex the situation is, and rather than carefully analyze it, let us take a short-cut: how is pay-TV prevented or impeded in the United States? First and foremost, of course, the question obtains because initial investment for pay-TV does not begin to compare with initial investment for a broadcasting system. Hence these are other mechanisms which impede its growth, friction being only one of them. The competitive model is here not very useful as a tool for explaining the facts.

2. NEW PRODUCTS AND THEIR ENTRY TO THE MARKET

We can generalize our result. The major problem for commodities not mass-produced is of prohibitive costs, and the excess costs are less and less of production and more and more of distribution. It is always expensive to make a film, but experience shows that a quality movie can be made for much less than the ordinary studio film budget. The cost is not at all prohibitive, but a film privately made will almost certainly not be distributed and so the costs will constitute an almost total loss. Some of the loss may slowly be recovered by playing the movie in quality movie houses, as happened, for example, to

Hallelujah the Hills, but all the same, it is non-distribution that makes for the loss. And distribution costs are indeed prohibitive; distribution in the mass-market, that is!

This is why large producers of mass-products, whether Twentieth Century Fox, McGraw Hill, or Chrysler Corporation, are willing to make the initial investment in designing and testing a commodity with much less discrimination than to make the real investment of mass-producing and distributing it. This is very hard on the artists that make the aborted movie, the author of the aborted book, or the engineers and designers who perfected the turbine car. This is particularly true for the company which is sensitive about its reputation and invests in prestige goods. An author may feel complimented when his publisher calls his book a prestige book, and then he will be bitterly disappointed to learn it means they do not intend to advertise it heavily if at all and they will make no effort whatsoever to distribute it. This happened to all my books thus far. And some of them, I feel, are potential bestsellers which have not made it for want of recognition.

For, here is the fatal error of the economic theory of competition. It is all right to view all the deviations from the competitive model mere friction and say it takes time till the demand for a certain commodity builds up and till its availability comes up to standard and so on. There are many ways to defeat Friedman's argument that friction is of no concern to the economist any more than to Galileo who started developing science by ignoring the air friction of falling bodies. The analogy, for example, can be pushed further: Robert Boyle tested the excuse that Galileo had for the falling feather: by creating a vacuum he reduced air friction and found the feather's performance much improved. Moreover, soon Newton took over and developed his law of friction, as later did Stokes and many other physicists. And so, if Friedman's analogy is to be pursued, we need both to check the claim that reality and the model diverge only because of friction, and we also need a model for friction. Both. But another way to argue against Friedman is to show what the mass-market does to a new commodity. For, if there is any rationale in economic theory for free markets it is that of Adam Smith: the allocation of resources, he argued, is optimal if done by the actual vote of the consumer – a vote performed by the use of the market mechanism. Is the allocation of resources in the free market optimal even though there is a high threshold for entry? Is the mass-market optimal? To repeat, I do not mean, will the mass-market be better if we had less Beatles and more Beethovens, since, as I have said, the mass-market handles both adequately. I mean, how does the mass-market decide whether to cut a record of a new *avant-garde* composer or of a new band? Except for very few famous composers, the *avant-garde* seldom enter the record mass-market. When they do they often do so by fiat. For example Elliott Carter's double concerto had no chances to enter the open market since all critics condemned it as too difficult. But through a subsidized quality market it entered the mass-market all the same. Bands have no subsidy, and so when one critic voices doubts as to a new band, its fate may be negatively sealed; but then there are those entrepreneurs who gamble on new bands, just as publishers gamble on any textbook they can, since the gamble is very favorable, the initial investment is small and any hit is a real jack-pot.

Thus we have come to the argument of taste: the mass-market works on enormous investments of each peculiar good; prior to investment as many consultants are asked to express opinion both on quality and on saleability; and consultants are established if at all possible; and established consultants enhance established tastes. The success of the conservative expert is at rare times upset by avalanches or floods of new taste – which are the very exceptions that prove the rule. This, then, is not only friction; this is ousting almost all innovation of taste. It is bad enough for economists to declare tastes exogenous; they cannot possibly openly support a model whose strong propensity to keep tastes constant is so successful. Hence the argument from friction to explain away the absence of free entry to the market does not work: the market does not behave in a seemingly free manner: tastes in it tend to be bound by assessment of large entrepreneurs. Expectations are self-validating that changes of taste are very improbable!

To repeat, the example from Elliott Carter shows, as does the example of the Beatles, that at times a change of taste occurs in the mass-market contrary to the specific expectations of the mass-producer but in accord with his expectation of some such rare events, and on the occasion of this rare event the mass-producer at once follows the maxim, if you can't beat them, join them. But thus the producers in the mass-market curb changes of taste even though they adjust to them with the optimal speed. Nowhere is this truth more obvious and deadly than in the field of scientific publications, where it is almost self-understood that the newcomer must either serve as apprentice for a long period or beat the expert. This self-understood truth has been introduced to the philosophy of science by Michael Polanyi and Thomas S. Kuhn.

Before coming to this, let me sum up my simple observation. We have the mass-market, and we take as optimal the allocation of any resources that are already represented there. We have the quality market that does not compete with the mass-market even for goods that look quite substitutive. What is missing is the middle range between: there is a trend which I wish to combat, a tendency to concentrate the market into two and only two clusters: the large mass-market where entry for a single product is a matter of a gamble and subject to constraints of conservative tastes, and the small quality markets.

Though my concern is with scientific publications, my description is very comprehensive. Let me mention, as an example, architecture. As long as the production of a single house has to do with the mass-market, it is constrained by conservative tastes. For example, it is hard to get a loan or a mortgage on a house that is not built to a conservative taste. Quality architecture does exist, but is notoriously scarce as any budding architect will tell you.

How, then, can we fill the gap between the mass-market and the quality market?

3. FREE MARKETS AND CLEARING-HOUSES

The classical model of the free market has not been touched by any further change and development in its image of the mechanism of supply and demand. This is, I think, quite a remarkable fact and is explicable by the popularity of the stupid positivistic theory of revealed preferences. Classically, the producer is supposed to either produce at the equilibrium point or quickly move towards it, and he moves towards it by producing and marketing somewhere near it. In other words, tastes are only expressed by consumers when they actually purchase! Thus, the producer first produces and sells; if things go well he increases production and if they go badly, he reduces production. In both Marx and Keynes, it is a crucial fact that a catastrophe ensues when production is forced to be reduced further by each reduction until a standstill is reached, only because prices fail to fall at a sufficiently fast rate to stop the recession. In other words, both Marx and Keynes assume that the supply and demand models operate in the open market even though they deny the classical view that it is always all to the best. As to the marginalists, including Friedman – the main point of the marginalist revolution was to explain how a margin of actual transactions is what determines the whole market's behavior.

Now, clearly, the situation is in fact quite different, and not only due to quasi-monopoly or other sorts of friction. There is a qualitative difference between finding the equilibrium price of a commodity by playing in the market proper and of finding by working out models and testing them, of making it by market manipulation, etc.

Playing in the market proper has its costs, whether of flooding the market with goods that find no customer or of advertising, etc. The initial cost can be prohibitive, and thus raise the point of our previous discussion, namely it narrows the range of tastes to cater for. There may be a smaller initial cost for research, or no cost at all, for example, if the research is conducted by a professor who has a grant or who trains his students. Alternatively, there may be no initial costs when a book is printed only after enough standing orders from libraries are in to cover all initial production costs. There is a qualitative difference between the following two: on the one hand the classical behavior of a publisher who cannot possibly publish less than ten thousand copies of a paperback book per year, and who thus goes on publishing as long as the going is good, and a little more; on the other hand the behavior of the portrait painter who only works to order, or on demand. The market mechanism for portraits does exist, but there is no supply seeking its demand here, but supplier seeking customers.

For, the extent to which one works on demand depends on demand, and it may be suggested that there is less friction in this case; but this is so only if there is free access for all who may wish to place and order: if demand is not expressed by actual purchase, then it needs another channel through which to be expressed. But even this is true only for some variation on the competitive model.

To take an example, consider the publisher who publishes on demand and sells on the free market the remainder that he rightly considers windfall. Windfall on a regular basis however is only possible in an oligopolistic model; and indeed, the publishing market, not being as free as the classical model demands, since there is a high threshold for entry into it, is in part competitive, in part oligopolistic; that is to say, established houses that have reputable series can have books published in these series with no risk – with orders making them profitable before they go to press. Why then can others not compete with these publishers on small scales until profits go down to the equilibrium level?

The answer is, you need to acquire reputation for quality markets; and this is not yet done openly enough. And so, though mass producers can enter quality markets by purchasing reputations first (as initial investment), they have to do so gingerly. As a result one and the same firm will be able to cater for two different markets and cut the costs both in economies to size and in siphoning commodities back and forth between the two markets in accord with the standard model for oligopolistic competition. Hence, our question is answerable by discussing the market for reputation. The more classical this market, the more competitive the whole market. And, from the viewpoint of one interested in medium size markets, since reputation may come high, medium, or low, this might very well regulate and create intermediate sizes of markets, between the mass-market and the quality market. What makes mail-order sales so shabby is their low reputation; otherwise there can be, and there starts developing a new industry, of medium reputation mail-order clearing-houses; particularly for books, but also of men's clothing. The advantage of such clearing-houses is that they truly reduce friction; they compete with the reputed quality producers on a basis akin to that of the mass producer: they can produce on demand to varying sizes of markets, where demand depends partly on actual order but also partly on estimates; except that usually clearing-houses do not themselves produce, strictly speaking; usually they place wholesale orders or retail orders or anything in between, and then retail with the aid of the postal service, telephone service, and whatever else they find handy. But as an exception a producer may have his own clearing-house, as Harvard University Press and Cornell University Press are now doing, and as many European manufacturers are now learning to do.

It therefore may be useful, not to say profitable, to look at the clearing-houses market as depending on the market in reputation. For, clearly, once we can enter the cost of reputation into our books we may find whether it is worth our while opening a clearing-house. And clearing-houses, I think, are the only way to make publishers more competitive and more consumer-oriented: instead of writing to a publisher it may be easier to receive a letter from a publisher – provided it is not discredited as trash-mail!

4. THE MARKET IN REPUTATION

Once upon a time reputation could only grow slowly by the lengthy exhibition of the reputed quality, whatever that may be: to be reputed for promptness or for quality one had to exhibit promptness or quality year in year out, and under trying circumstances too. Nowadays this is no longer the case. (Michael Polanyi's theory of science as a guild is old-fashioned since the way to acquire scientific reputation is now very different from what it used to be.) It is customary to put the reason for the change at the doorstep of the public relations officer and the advertiser and such. For my part, I beg to differ. I think the reason is not the advertiser but the cause for demand for his services; it is the newly acquired social stability within social unrest and change that constantly demands the change of sets of tastes and of related expectations. But if a social group should be declared the cause of the newly acquired attitude towards reputation, then I would place it at the doors of the universities, not on Madison Avenue; in particular, it was the professors succumbing to the administration's administration of publication pressure so-called. (And this, of course, is quite in accord with Polanyi's theory and with his demand to restore the autonomy of the guild which the administrations of universities do destroy.)

Universities trade in reputation; they must both own and confer it. The rating of the degree of reputation acquired by a given university in Europe is still to a large extent an estimate that remains stable for decades if not centuries – and can be decades or even centuries out-of-date, of course. In the U.S.A. the rating is partly done by counting annually the number of Ph.D.'s in the faculty of different universities. The administration of Harvard University consequently forbids the hiring of even a temporary instructor, unless he has a Ph.D. degree, no matter where from. They actually prefer a Frankfurt Ph.D. to an Oxford B. Phil., since the statistician does not recognize the uncontested fact that the latter takes more years of harder work to acquire. This, like publication-pressure, is but manipulating the meter.

Manipulating a meter is nothing new. Any government controlling prices is tempted to control particularly harshly the prices of those commodities that were originally sampled almost at random to be used as the meters of inflation. It is hardly necessary to notice that this is but a form of lying.

Publication pressure came as a result of the choice of publications as a meter, and the meter was needed because the older ways of assessing the quality of a scholar proved not so much inadequate as inapplicable in the world of quick expansion and change. Once publication was a major meter of quality it became automotically the criterion of quality: the administration of a university may want high quality staff and so they may wish to hire a reputed person, particularly one who has published much research; and if there is no new place on the faculty, the administration may request a member of the faculty with no research published to leave and vacate his place for one with a lot of research published – unless he publishes on his own, and pretty quickly at that.

Except that thus publication ceased to serve the market and began to serve the author in an indirect way. And so it became worth his while to subsidize his own publication. This has to be done gently at times since vanity publication, i.e. publication at a loss whose purpose is to serve the vanity of the loser (who is usually the author), has a low repute and so usually a researcher's publication at a loss must be done obliquely. He may have his department publish a journal at a loss to the university; he may get the subsidy from a reputed organization as a part of a grant; he may invest the subsidy in the form of high membership dues to a small super-professional learned organization that does hardly anything except subsidize a journal devoted to its own sub-sub-speciality. These days a few sociologists in the U.S. are grouping to organize a new sort of learned journal, one in which contributors cover production costs and all will be done openly on a cooperative basis. Of course, there is no telling how such a journal may fare. It may be ill-reputed as variity publication and so both fail to be distributed and fail to be recognized as a learned journal for the purpose of promotion and tenure. But if it will manage to be successful and reputed because it would serve a public frustrated by the conservative tastes of the recognized learned competitors, then, who knows, it may beat the expert and thus be a trail-blazer.

For it is in the field of innovation, where a larger portion of the general public have a taste for novelty, and whose ideology officially keeps all doors open to novelty, that the treshold may be more easily lowered.

Let me speak of the threshold first. Since it is only reputable scientists that deserve to publish a treatise or a collection of essays, no academic of rather low repute will manage to enter the market with such commodities. Competition in the pecking order of scholars interacts with the competition in the open market very strongly.

An author not reputed enough to have a collected essays to his name will have his peers sabotage his effort to publish such a collection on the open market. Yet he also may become reputable overnight by having published a collected essays in a reputable series on the open market. The reasons an editor of a reputed series may accept his collection for publication is in part that he, the editor, possibly likes the content of the essays, in part he approves of of the position the essays advocate, perhaps, in part he expects the essays to do well, not to be torn to pieces by experts. Were the competition more open, with less friction – friction in so many senses – then an editor would only think of marketability and customers will purchase only if they thought the product worth the price! No other market can be so open yet is so closed, as the academic market in scientific publications!

In brief, that the market for reputation exists and that entrance threshold to it is high is a fact: examples galore illustrate it: as to the mass-market, the fact that entry to it is barred is illustrated by the fact that once the barrier is overcome there is overcompensation. When I was barred from the reputation I may have coveted, my descriptions could be called sour grapes; now that I suffer excess reputation, I hope my wish to let younger people have better chances will be taken at face value.

The fact that the market in academic reputation is so much more complex and friction-ridden than the open commodity market need not be surprising, since reputation, like many other privileges, is not a commodity but a privileged access to commodities and so not just the cause of friction but friction par excellence, friction that the advocates for the free market combatted from the very start. There is thus a great difference between the reputation of an artist on the mass-market, which he needs in order to gain entry to that market and the reputation of a scholar or an expert who helps control it. We can only know the most reputed artist, and so it is lack of information which makes his reputation important. And lack of information is by definition a form of friction. But the reputation of a scholar or an expert is not what makes us know of his product; it makes us value his judgment of other people's products in the first place; there is the world of difference between an expert producer and an expert judge of products. And, of course, the ideal is that everyone has the courage of his own judgment as to his own preferences. Let me conclude, then. We have producers regulating productions by sales in the market, and we have those who study it. The latter consult experts. The experts are not necessarily economists and other students of markets – they can be experts in tastes. They are the real enemy.

5. THE QUESTION OF TIME-LAG AGAIN

It is not that I am trying to preach here the autonomy of consumers and the desirability of allowing young producers better chances. It is that I think there are ways to do so that are sufficiently profitable to sufficiently many so as to be operative. I intend to beat not the expert but the expert's control of the market. What is really needed is public discussion of the matter. Let me explain briefly, first, why it is easy to beat the system: I have in mind a brief social analysis.

Consider now the factors peculiar to our changing yet stable world. Experts are a stabilizing factor, yet on condition that their expertise is not checked too often. Also on condition that it is checked from time to time; a modern rapidly moving system cannot afford too much dead-wood, particularly not in its control centers. Now the question is, how often can we check expertise. The expert wants to prolong the period between checks; so as to prolong the period between revolution. He is normally revolutionary when young and relatively unrecognized and the more conservative he becomes the

the higher the threshold he passes. But he cannot have his way for long; times move. So it is a question of time-lag. And the more we learn to move with the times in relative security the less we need artificial stabilizers and hence the less chances the expert has to keep the time-lag long. Once the time-lag is near zero I think the expert will be a friend rather than a foe. Indeed, the increase in life expectation already does that, and so calls for increasing delays in publication of good stuff.

Publication pressure plays into the hands of experts. They are editors and referees, and so in charge of learned journals; they are consultants of publishing houses, especially university publishing houses; they control grants that likewise lead to publication, both in offering a scholar time for research and write-up and in offering subsidy for publication. The result is that young upstarts have to wait many years before they may be heard; they need a longer and longer breath, and some fall by the way. So many good scholars could not make it because a paper was rejected by one journal after it kept it for one year and another journal after another year and similarly with a book that a publisher's reader kept a year in his drawer. A young economist I know was told by a leading publisher his textbook was good but could not be published since the author is not reputed enough; clearly this was a veiled invitation to co-opt a reputed parasitic co-author. In the First International Conference of Scientific Editors new winds blew. An editor and mathematician, for example, reported a case where an editor's referee kept a mathematical paper in his drawer too long only in order to have his own student finish a similar project. Such desperate measures must lead to protest and protest was well heard in that conference.

What was common to most people who had some complaint, to repeat, is a keen sense for innovation. In other words, here we have a case where the attempts to stifle changes of taste must fail. But the attempts may cause a time-lag and the time-lag may be sufficiently long to exclude from the game those who do not have the time to invest to beat the time-lag. Again, we see, the allocation of resources suffers. The way I would like things to move is that of killing all reputation, to saying to anyone who wants to trade on his credentials, *hic Rhodos, hic salte*! But this is practically impossible for many reasons. We can only improve the system of credentials by making it more fluid. The main instrument I wish to offer now is that of creating reputations with as small a time-lag as possible. The result will be a wide range of reputations that might create middle range markets for scientific publications.

To begin with, there is no need to print papers accepted for publication that statistics shows are not read by anyone save perhaps author, referee, and publisher. The American Institute of Physics reports a high rate of scientific papers published not read by more. If a reputed clearing-house publishes a catalog of its publications and declares all or almost all its publications to be on demand publications, i.e. printed more or less to order, then the reputation due to publication can be had at a smaller cost than that of running a journal. The current artificiality of journals having to print so and so many pages per so and so many months will thus also be given up. The publisher will have to employ scientific editors with status – i.e. reputation – comparable to that of editors of scientific journals; but no journals.

We now have university publishing houses that are large scale vanity presses on the verge of collapse despite heavy subsidies that at best operate as their own clearing-houses and, often they use commercial, ill-reputed, clearing-houses. We can have a reputed university clearing-house that will cooperate with a reputed commercial publishing firm, not as MIT Press cooperates with Wiley, but on a rational division of labor. The organizations of scientific editors – regional or the global one, of all or of special groups of professions – can offer both blessings and controls. The chief concern should be better deployment of resources, a wider range of choice for the consumers, a broader spectrum of means of production, from the xerox and the multilith to the video-tape, and a more competitive market.

Once we consider entry to the market as entrepreneurs, it is clear that there is no free entry and that it matters not much more in publishing than elsewhere in terms of inefficiency and the difficulty of entry of individual authors or individual works. Once we consider entry not of entrepreneurs but of authors, and seek the economic organization most suitable for such entry, things brighten up. The idea still is this. Start with experimental marketing, achieve a small quality market, try to broaden it, so as to permit the work a chance to enter an even larger market. This is prevented by the gap, the enormous gulf, between the quality market and the mass-market. The gap can be bridged in different ways. One way I have outlined here: replace publishing houses by distributors. Rather than print in a book 'published by *X*', print in it, 'distributed by *Y*'. And so on along the line. But there are other ways and means, and the only criterion should be, does a certain reform reduce friction or increase it. We need a better theory of friction and experiments in the field. And scientific publications are the best place to start since the taste for innovation is strongest in science. A suggestion to editors of learned periodicals: advise your contributors to replace name and place of publisher with ISBN whenever applicable, or else omit them altogether.

CHAPTER 13

REVISING THE REFEREE SYSTEM

The referee system reflects standards accepted by the profession, yet economic and social factors are no less decisive in determining the success or failure of an attempt to publish a (seeming) contribution to learning. If we want to improve the referee system, we may wish to propose the improvement of both standards of publication and the economic and social means of their implementation.

1. THE ECONOMICS OF PUBLICATION AGAIN

Publishing a paper or a book is helping author and reader communicate, and is thus the position of a middle man. Commercial publishers proper would then accept for publication a book they think will sell well on the open market and they often even commission authors to write them. On the supposition that publication is no more than that, all one needs is the open market. This, of course, is not the case. Publishers assess the marketability of commodities which they do not put on the market for trial runs, and there is no market research for them. Initial investments in books, and particularly in textbooks, are prohibitive. Learned journals and university presses are subsidized. All this makes the market mechanism fairly useless. Publishing learned papers and books is controlled by more complex mechanisms.

The control of the book open market is, nevertheless, more satisfactory than the control of university presses and the learned periodicals. Publishers may accept for publication books that seem not too promising, and as mere gamble. They may be induced to publish when a spirited volume by an unknown author is introduced by a leading authority, or at least if they can issue brochures with recommendations from known people in the field, i.e. from people whose opinions are *a priori* respected by prospective customers. Even there, however, the problem remains: individual publishers do not try to outdo their competitors by far, and the authorities in the field may combat a young aspirant sufficiently uniformly to discourage any publisher from attempting to publish his book. Publishers ask their referees both whether they value the book considered for publication and whether they deem it sellable. But both referee and publisher conflate the answers. Hence, young aspirants are pushed to the academic publishing systems – the university press and the learned journals.

University presses were devised as subsidized, non-competitive, publishing houses. They are considered, by and large, to be failures. There are financial and social reasons for their failures. Financially, subsidies required for a proper operation of a press that cannot compete in the open market are enormous and must continue on a regular basis. Moreover, expenses pile up and this means that unless drastic measures are taken to remainder old books, destroy stocks, repay debts promptly, etc., subsidies must increase. Therefore many university presses have closed down. The social cause is really shameful: university presses emulate the commercial presses in all aspects except that they are choosy. This is disastrous.

What is happening these days is that more and more university presses all over the Western world act as mail order book-sales institutions. Their only asset here is that they appeal to a narrow market which they can easily reach through universities and learned societies membership lists. But here, too, they compete with the commercial press in learned publications, and despite subsidies they often do less well. Nevertheless, some of them go on functioning, as prestige publishing houses that are more vanity publishing houses in reality (vanity publishing is publishing at a loss covered by the author), banking on subsidies for both the press and the author or journal editor, not to mention the fact that publication pressure may bring a desperate author to seek ways to finance his own work out of his own pocket.

The financing of learned periodicals is done in two ways. First, many research grants include money to cover publication costs. Second, membership in a learned society is often a must and membership fees usually include obligatory subscription fees. This makes the editors of learned periodicals public servants in more than one sense: they are both acting in the interest of the public and the public pays for their services. Often enough, editors will deny this on the ground that editors of learned periodicals seldom receive money for their services. This will not do. Editors acquire enormous powers in these days of publication pressure and hard competition for every job opening, and they often obtain fringe benefits in the form of extra secretarial help, travel expenses, and more. It is a fact that editorship is coveted. If an editor is not careful, a journal, even a journal he has founded, will be snatched from his hands. (This, I am happy to report, has happened to me too. I only regret the journal I have founded is not as interesting as I wanted it to be.)

Economically speaking, to conclude, the publishing market is not a free market because of the absence of free entry to it, and thus what really significantly differs in publishing is the absence of competition which leads to stagnation and the impact of reputation that may fight stagnation. The question that matters, then, is how reputations signify. This, of course, leads to the question, why are people reputed. Thus, to take the two extremes, if an academic is reputed because he has published a textbook that makes a lot of money, then, clearly, his reputation acts as an added contribution to the over-all stagnation. If, on the other extreme, the reputed academic has gained his reputation because he has effected a scientific revolution, he may be trusted to use his reputation to push the revolution throughout the literature and so be a very vitalizing force.

Of course, experience is not that extreme. Very often one who was revolutionary when young, successful at middle age, and is now an elder statesman, will be assigned the task of writing a text. He may assign the task to his young lieutenants and then try to give it the finishing touch and mention the helpers in his acknowledgements with generous compliments that might help them advance their careers. And the young lieutenants may even have a spark of the revolutionary in them, though hardly enough, of course. More often, however, a publisher would commission a textbook from a leading authority and hire ghost-writers to rewrite it. Then real author and ghost writers quarrel and may reach a compromise. Textbooks are bad not by accident but by design. The design takes care of the interests of as many powerful people in the field as possible, so as to insure sales.

2. CRITERIA OF EXCELLENCE

All this realistic description may sound impressionistic and cynical. It is not. It is the result of much experience and it would make no sense were there no agreed standards of excellence. Anyone familiar with the economics of mass-produced commodities where there are no standards (but, by definition, mere fashions) knows the difference between mere fashions and fashions related to standards. Standards, however, need not be correct, and dealers with quality art products know that there are conflicting standards of quality that govern differing markets, yet the differing markets are not as fickle and unstable as the merely fashionable ones.

It is therefore very interesting to examine the standards of scholarly publications and see whether there are competing standards, whether standards can be improved upon, and how.

Publication standards were introduced, as far as I know, by Sir Francis Bacon and by Robert Boyle. Boyle's standards became the official standards of the Royal Society of London and, through it, of the learned world at large. Any new fact – an observation or an experiment – described sufficiently clearly to enable the reader to repeat it, should be published – with or without comments, explanation, or even speculations. Similarly for mathematical innovation. Of course, this is problematic. What fact is new? How much space should an author of a page or a new fact be allowed? The "Reports" section of *Science, Physical Review Letters*, and other periodicals, have solved this problem only recently. They do publish every observation or experiment which is at least mildly new, and as briefly as possible. But in many fields such solutions do not exist. Moreover, new experiments cover a small part of the scientific endeavour. Furthermore, it is not uncommon that an editor returns a report on a new fact, either requiring supplementary facts or rejecting the new fact as not really new or as not really interesting.

Undoubtedly, some new facts are not really new and not really interesting. The question, which new fact is interesting, which not, may be interesting, but an editor need not concern himself with it, and he need not even be in possession of a tentative answer. As long as he has a good trustworthy judgment that helps him sift the interesting from the uninteresting facts amongst those presented to him, he will be doing well and serve the public well.

This simple fact is of supreme importance. It has become the cornerstone of the philosophy of science of Michael Polanyi, who has christened the ability of the editor described in the previous paragraph as "personal knowledge"; this philosophy is becoming increasingly popular amongst diverse groups of scholars, especially in the formulation it has received from the pen of Thomas S. Kuhn, who has invented the term "paradigm" to denote this personal knowledge, and who speaks of a paradigm-shift as a change in fashion regarding what is interesting to the scholarly profession or group served by the journal in question.

Personal knowledge goes a long way. It covers all research activities, not only the finding and reporting on a new fact. On the contrary, both Polanyi and Kuhn reject the (inductivist) emphasis on new facts that is so characteristic of Bacon, Boyle and many other participants in the scientific revolution. For my part, I have much sympathy both with their de-emphasis on facts and with their accent on editors' good sense and good judgment. My dissent from them concerns their demand to trust editors. There is no need to show that trust leads to corruption, that not trust but checking and criticism breed the trustworthy. This is a paradox that all students of democratic institutions get used to: we can personally trust while instituting means of control, checks, criticism; and we need training in complaints.

I wish to stress the fact. Polanyi and Kuhn are right to observe that so many scientific editors use good sense. The paradigm still is Max Planck who was excited by reading papers from an unknown young patent-tester – Albert Einstein – and who published them rapidly. Yet they are mistaken to conclude that all is well. A new Einstein would not have his paper published nowadays without waiting for years in line. There are waiting lists everywhere.

Of course, some editors follow the rule, first come first published; others let important papers, or papers from prestigious people, have priority. Some editors prefer not to publish two papers from one author within a given period of one to five years – particularly if he is controversial – and some editors give priority to young newcomers. Most editors of collections of essays or of series treat the young, the established, and the famous, in totally different ways. Is this right? Does it serve the reader best? Does it serve the interests of the profession best? Or does it serve best the interests of the commonwealth of learning? Can we assume that all these interests coalesce? If not, which takes precedence?

Personal knowledge does operate here. And this is how a conservative editor finds uninteresting what a radical editor finds exciting and *vice versa*. There are, no doubt, different tastes and different criteria. Polanyi and Kuhn simply overlook the fact that increasingly many periodicals, especially organs of learned societies, make public their criteria for publication. They overlook it, since in their view the criteria are seldom clearcut and must be viewed, anyway, as mere guidelines, hardly binding anybody. Yet, were this view true, complaints would be impossible. Yet there are complaints all the time. Most of these complaints reflect power-struggles. Yet power-struggles would be empty politics were they unrelated to objective standards. My own complaint about these complaints is that they are seldom made public and so may be, and often are, petty. The reason they are not usually made public is self-reinforcing: the very absence of customs and institutions to that effect makes a public complaint a very strong form of protest, acceptable only in very extreme cases. The result is that one who makes a public complaint may do harm to the ones he complains about, and even excessive harm; and he is therefore pressured not to complain publicly, or else attempts are made to minimize and isolate the episode, and the plaintiff may well find himself rebuked. As I was; and I understand those who try to avoid the rebuke, yet I think they should be encouraged to complain, not rebuked. In other words, we can improve and democratize the system of publication of learned writings, and since the very employment of a scholar may depend on his publications, this may improve the commonwealth of learning at large.

3. REVISING THE REFEREE SYSTEM

First and foremost, it must be noted that it is almost never right to complain about a referee. His task is thankless and hard. He works for free with indefinite criteria, at times on material whose value is hardly assessable; what is more, he cannot say who should have the benefit of doubt. Not that he must create his criteria: a journal can have a policy or its opposite and he merely takes it as given or refuses to act as its referee. There is nothing wrong, I think, in one journal's policy to reject any controversial paper or any paper on a controversial topic; and there is nothing wrong, however, and indeed much to praise, in another journal's opposite policy. That many journals have no expressed policy, or no revealed policy, regarding this matter, is thus understandable and so not in the slightest condemnable, and it may make the conscientious referee's work harder but also the callous referee's work easier. So much for the criteria; as to their application, it is harder to say. We may imagine that when an ambitious paper is submitted, only top people can assess them, and they may refuse to act as referees, or be too much imposed upon by a flood of ambitious papers. An unassuming paper may be of the limited importance its author claims to it; and this limited importance, as described or (slightly) misdescribed by its author is hard to assess, etc. Let me say, in parentheses, that I have a few papers I like very much for which I claim no status of innovation, and which are unpublished because I am tired of reading referees' reports noticing that even I admit unoriginality and of wondering if my referees believe that each printed paper is novel. So much for the application of the criteria. There remains the benefit of doubt, which editors may allow themselves in deciding matters. Some journals are very respected and so very coveted and so flooded by many contributions and so can pick and choose and so can afford to claim the benefit of doubt and so at times they do, closing the door to risk and to opportunity at one and the same go. Others do not.

Nevertheless, at times referees do feel obliged – though it is above and beyond their line of duty – to spot daring papers and give them the benefit of doubt.

It is clear that since there is no control over the referee's work and since editors are not obliged to accept them – they can either publish on their own judgment or look for other referees – it is always the editor's responsibility to decide whether a referee's report is not too unjust. And since in declaring a contribution sub-standard a referee declares its poverty to be obvious, this declaration must be either backed by very powerful evidence or invite the editor to glance at the paper he rejects. If he does not do so, then he may be committing an injustice, and of the kind easily remedied upon a reasonably good challenge.

Let me stress. I am here making the most out of my claim that the sub-standard by definition stands out, that a regular member of an audience to whom an intended publication is addressed should be able to see with ease the flaws of that work for it to be sub-standard. Otherwise, even if the work is flawed, and even if it is so flawed as to count as no contribution at all, we cannot justly deem it sub-standard. Were this not so, then we would be unable to escape the conclusion that almost all papers published in the last decade in any given field are sub-standard. In retrospect it is all too easy to see how rare any lasting contribution to any field is. We should not condemn editors for that, and we should insist on the benefit of the doubt. Yet as long as there is some filtering of materials for publication, it is clear that the first to be filtered out is the (obviously) substandard. To use Thomas Kuhn's jargon, the papers filtered out first must violate the (current and well-known) paradigm.

The importance of this discussion is only this: all the problems of refereeing, soluble or insoluble, may vanish if a referee is asked for an opinion on a sub-standard paper. Perhaps even a thoroughly bad paper may have a pearl hidden in it somewhere; so the referee may feel obliged to scan it for a pearl. But beyond that even the most conscientious referee need not try to go.

It is therefore particularly useful to complain, if complaint is called for, when a referee calls a paper sub-standard and it is not. Such cases are not only the most clear-cut, but also the most basic. For, if we do not even agree about a sub-standard paper, then we are all at sea and we better admit this fact and take it from there. The greatest abuse of the referees' duty, then, is to call a paper sub-standard when it is not.

That is to say, what is sub-standard must be obvious and conceded by the members of the intended audience. No matter who should benefit from doubt and in what circumstances, one benefit of doubt must generally be conceded: what is not obviously recognized as sub-standard may be poor, but is not quite sub-standard. This fact is well recognized by people who move between a high-standard crowd and a low-standard crowd, and it may be instantiated in diverse ways. (Kuhn tells us that he developed his view when he moved from the high standard company of physicists to the low standard company of some leading social scientists.)

Suppose a referee calls a paper sub-standard and it is obviously not. Even then it is hard to complain. The referee cannot possibly have done wrong, except in the sense that when he misrepresents his own true opinion he violates the rules of ethics; for, when he is requested to referee and he volunteers, or even if he receives from a commercial publisher a nominal honorarium, he is a public servant of one sort or another. Moreover, the anonymity of the referee makes it impossible to judge whether his lapse of judgment, when he calls a passable paper sub-standard, is not a mere slip, or even an occasional exaggeration tagged to a well reasoned and just proposal to reject a poor piece of work.

The complaint, then, must be launched against the editor. The editor, too, may plead a lapse of judgment. But if his peers are serious people, they would want to hear details: how come he has not read the referee's report and how come he did not see its poverty or did and failed to discount it. This may be unpleasant. To counter it, however, many journals have instituted revision procedures – resubmitting, rebutting the referee, requesting a new refereeing, a grievance committee, and more.

The institution of such bodies is not sufficiently widespread, not sufficiently known, and not always helpful since people are reluctant to complain. (The Consumers' Union reports this fact regularly and repeatedly explains the importance of complaints.) And, of course, the study of all this has hardly begun. The nearest to it is the fact that in a few professions there are organizations of editors for the exchange of information, and an international organization of scientific editors has already been formed which, doubtless, will encourage such studies.

Let me conclude by recommending that members of scientific organizations be encouraged to complain – and to make their complaints impersonal. Also that as the best complaints we should view complaints against judgments of papers as substandard. This will enable us to fix the minimum standard well, and then to raise it, perhaps. Also, it will help us formulate the standards of scholarly publications – as tentative hypotheses to be tested by application.

Polanyi and Kuhn are right, thus, to say we do not quite know what our standards are, and that they are binding nonetheless. I still find their views objectionable since there is always a possibility to try to formulate our standards and to try to improve them. Moreover, for a rational process we need the attempted formulation in order to attempt an improvement.

CHAPTER 14

SCIENTIFIC SCHOOLS AND THEIR SUCCESS

A view of the social structure of science is now popular, according to which each discipline of science is autonomous and its membership is divided into the few leaders and the rank-and-file, also known as normal scientists. The leaders declare what is the model scientific theory in the field, also known as a paradigm. A scientific revolution is a paradigm change, and is effected by the leadership, old or new. Normal scientists, finally, solve routine problems, also known as puzzles, in the light of the paradigm. This terminology is of Thomas S. Kuhn. The view it reflects is now popular, its advocate is Kuhn and its originator is Michael Polanyi.

One obvious oversight in the above picture of science is the omission of the existence of the supernormal scientist and of his role. It is, however, the supernormal scientist who entrenches the current paradigm. The reason he is needed is that there are always competing schools in science, and the cases where one paradigm reigns supreme in one field are cases where one or more supernormal scientists give one school its prominence. Hence the Polanyi-Kuhn school of philosophy cannot overlook the supernormal scientist without loss. Yet once this role of the supernormal scientist is clearly recognized this philosophy can easily be challenged. More generally, the view of science as being the exclusive monopoly of one school of thought in the face of the existence of competing schools is rooted in a theory of rationality as obligatory which makes it obligatory to belong to one school and to denounce its competitors in the field. It is a tacit assumption of all writers on science that the dominant school in any scientific field is the one possessing rationality. A field in which this does not hold is simply declared unscientific or irrational. All this deserves better critical examination.

1. GENERAL

Science is traditionally viewed as the paradigm of rationality. Rationality was always defined as a clear characteristic of views, opinions, theories, or beliefs; presumably there was always one most rational view in any set of competing views; that one view was thus rationally obligatory. Consequently, there can hardly ever be a rational disagreement. My own view of rationality, expressed elsewhere (*Science in Flux*), is that it must apply to disagreements, since some but not all disagreements are eminently rational. The external or sociological expression of this requirement to observe rational disagreement is the claim that there are schools of thought, some of which contribute to our rationality. Whereas Popper considers Freud and Adler, for example, unscientific and so hardly rational, I consider their initial contributions, when they were highly controversial, very rational indeed: they have influenced our ways of thinking, and for the better – since they opened to us new ways of thinking.

I will not discuss here the rationality of any specific scientific schools or their contributions, however; rather, I shall discuss the "official" denial of their very existence or diversity, and the ugly consequences this denial has. No doubt, schools are problematic, and the very admission of their existence or diversity brings a political aspect to the theory of science; this is, no doubt, quite troublesome. It is no accident that the "official" view of science banned politics and to that end it also banned all scientific schools, open scientific controversy, and even the publication of controversial material in prestigious scientific journals. Yet the "official" view is an error, and ignoring politics does not do away with it, especially at times when theorists of science are trying to bring in the import of the social dimension of science. Indeed, most of them have described science both internally, by reference to its theories, and externally, by reference to its social setting. And the two definitions of any one theoretician may, and often do, clash violently. We may either try to resolve the clash, or exploit the different views of science for political purposes by a simple technique of cheating: we can switch from one to another according to our convenience. The refusal to allow for scientific politics makes it hard to begin any scientific political party, but also to stop ones which exist, and so cheating is made all the easier.

Sir Francis Bacon, the father of modern theory of science, whose views were for long the "official" views of the Royal Society of London and all its daughter societies, called schools kingdoms of the mind, and their leaders tyrants and conquerors. Science, he said, is true, and truth is one. He did, however, also offer an external, social demarcation of science: it is the source of abundant invention: invention is the "mark" of true science. He was even aware – in the immense acuity that is so characteristic of all his writings – of the fact that the two demarcations do not quite merge: invention can be made by sheer accident. This is why he stressed that the scientific venture, once it got going, will produce not occasional or accidental inventions, but inventions "in streams and buckets."

The very first philosopher to be elected fellow to the Royal Society of

London as a theoretician of science is Sir Karl Popper. In his view the demarcation of science is not truth but the possibility of exhibiting falsity if it is there: a theory is scientific if it is testable, i.e. refutable in principle. The external or social demarcation of science he gave is that of a society which institutes means of encouraging and fostering the freedom of speech, especially criticism, including "empirical criticism," to use the expression coined by Michael Faraday over a century ago. Briefly, Popper's is a democratic theory of science, yet it is a theory of direct democracy, so-called. For, he permits no parties within science, or at best he tolerates their existence as a mere accretion. For, he has a criterion by which to choose a theory amongst a set of competitors: the most highly testable.

It may be interesting to make a comparison here. For Bacon the internal criterion of science is inductive proof based on accidental discovery and his external criterion is predicted invention in steady streams – both criteria are apolitical. The Royal Society of London tried to abide by both criteria. Its internal criterion was deemed a great success, especially in view of the success of Newtonian physics and the claim that this success is due to Newton's having followed the prescriptions of Bacon's inductive philosophy. The external criterion Bacon offered was dropped silently – presumably as a failure. Inductivist philosophers, particularly David Hume, John Stuart Mill, and the American pragmatists, but also others, replaced the Baconian external criterion by the democratic criterion now usually ascribed to Popper. The reason for this injustice to Hume and Mill is, perhaps, due to the fact that inductivism is populist in politics rather than democratic in the critical sense of democracy. (Thomas Jefferson, for example, was inductivist and also populist in matters of science; yet he was a critical democrat in politics proper.)

Though critical democracy is an external demarcation of science which obviously fits Popper's internal demarcation of science, it is simply unacceptable: science is not direct democracy, critical or otherwise, but a party democracy. The Newtonians, for example, were a party who denied that they were a party and so they denied the very existence of a competing school in physics (see my *Faraday as a Natural Philosopher*). Yet when the field school won it too denied that it was a school and it invited historians to rewrite the history of the rise of field theory as if all the time the Establishment showed it no unkindness and as much sympathy as it deserved.

Looking at the history of science as presented in accord with the "official" view, then, one can see a one party system, but one can also discern, if one is sensitive, different party-lines at different historical periods, bounded by explicable sudden switches.

This historical image is central to the philosophy of Michael Polanyi, Popper's greatest opponent. With slight reservations, it is also advocated by Polanyi's famous follower, Thomas S. Kuhn, who is these days (as Robert Merton reports) the most influential living theoretician of science among men of science next to Popper. Polanyi declared science not internally demarcable, only externally; there is the community of scientists, whose party line may change; they have leaders who recommend changes, thereby taking the risk that by not being followed they lose their place at the helm. Imre Lakatos refined this view by declaring that the leadership can be justified, but always only in retrospect. The owl of Minerva, Lakatos quoted Hegel's famous aphorism, flies only at dusk.

The situation is confused, especially since Lakatos's requirement to use hindsight is his license to tamper with the evidence. The history of science, said Lakatos, is a caricature of its rational reconstruction. With this he doomed to failure the attempt to distinguish the slow rise of one school through a prolonged struggle, from the quick rise of another school through a land-slide. In both cases the rational reconstruction presents a steady growth but hardly a struggle. In the first case the reconstruction omits the allegedly irrational struggle; in the second case the reconstruction blows up thin gossip into history. The school which is small and poorly organized yet accedes to much social and political power due to an unexpected landslide, finds itself pretty much in the same position as a small political party after a landslide victory. Some of its members or even new recruits, may swiftly build up an organization out of thin air, or out of gossip and romance; they may pretend that what has happened is no landslide but the slow steady rise to power of the new group which in truth is just coming into being. Or they may claim to have no ancestry but a radical novelty, and either conceal the new organization they are building or marvel at its appearance ex nihilo. Yet, whereas a new political party has to be described as the growing opposition, a new scientific school has to be described as growing either in a vacuum or in the lap of the Old Establishment.

Since the history of scientific schools is no-man's land, the leadership – usually not the top but the second-from-the-top leadership – is free to describe matters as they wish, hardly noticing that they use no critical machinery to check their stories, and feel justified in that their dedication to the cause of science is unquestionable. This situation is pregnant with much bitterness and intrigue; but these are burried within one generation since they tend to remain unrecorded or records get destroyed.

We can approach such matters more rationally and with a bit more sociological sophistication than is exhibited in run-of-the-mill histories and philosophies of science. Even oral history of science records, a relatively new innovation, are still not sufficiently indicative of the readiness to notice that bitterness is unavoidable in science as in any other human activity. Even Michael Polanyi's strong emphasis on this inevitable unpleasant aspect of science is not yet taken as license to record the seamy side of science, though we have exceptions such as Gerald Holton's intriguing essay on Paul Ehrenhaft whose honesty drove him, through a strange form of ostracism by the scientific community, to a kind of crankiness. Yet recording controversy, bitterness, dissent, and ostracism is often dismissed as not very enlightening. The stories thus presented may easily degenerate into gossip, biography, stories of rare strange events with or without moral. The sociological dimension of dissent is what gives it significance and depth: today's dissenter may be gambling for tomorrow's position of leadership.

2. SOCIOLOGICAL

There are two kinds of sociology of science, popular and systematic. Popular sociology is exhibited in the kinds of reasons that are given in discussing specific cases. In the learned world, for example, we are supposed to be impressed when told that the object of our gossip is heir to an important predecessor's chair. This is not applicable in the United States, or even in cases of personal chairs in Britain (such as Popper's). Yet the fact that heirs of most great lights in the past are hardly household names signifies that heirs to great men are often small conformists. This, I suppose, is why Gilbert Ryle denied in print that he was influenced by R. G. Collingwood, whose chair he had inherited.

Systematic sociology of science is recent. It began when sociologists noticed the social phenomenon of science. They did so in passing first, looking at science as an occupational stratum, and then, more thoroughly, as a subculture. Yet with the growth of science as a major industry, noticed by Derek J. de Solla Price in his classic *Big Science, Little Science,* time was ripe for a new fashion, which is to science what the fan magazine is to the popular arts in general, and to the cinema and the spectator sports in particular, yet decorum requires that the science fan literature put on scientific airs. Derek J. de Solla Price invented then the citation index as an objective measure of a scientist's popularity. This was particularly propitious since most large universities offer computer services, often at no charge, and mock-statistical studies can easily be done with the aid of computers. Any criticism of these mock-studies is welcome, of course, since the sociologists of science can use them to present improved versions of their studies. The sociological fact that these citation indices are measures of popularity rather than of import is also no impediment on the supposition that some correlation between importance and popularity may be found. That is to say, the fundamental assumption here is not statistical but an assessment concerning the source of the reputation of men of science: somehow it is scientific and hence well deserved.

This is how descriptive sociology is harnessed to serve the interest of the scientific establishment. And one immediate corollary to this is that the growth of a scientific school in reputation and power is in measure with its objective significance and its contribution to human knowledge.

How can this be questioned? To this end we may consider the internal and the external definition of science and see how well the two correlate. And in each case we may take first paradigm cases, i.e. undisputed examples, of science and then competing definitions of science. The paradigms for the internal definition should be theories and factual discoveries, of course, perhaps also inventions. The paradigms for the external definition can be social institutions, social characteristics (such as intellectual freedom), or individual men of science. As paradigm scientific theory and facts we may choose Newton's theory of gravity, but we also have, for example, Archimedes' law and Boyle's law, Franklin's kite and Priestley's oxygen, and Mendel's genetic theory, and Jenner's vaccine. As a paradigm scientific institution we may take the Royal Society of London or the Pasteur Institute. Scientific societies, research institutes, and modern universities will have to do for the time being, even though ancient science existed without them. As for scientists, things tend to get out of hand with professionals and free-lance amateurs, whether full-time or part-time, inventors and engineers and mechanics and lab-assistants, not to name great cranks, eccentrics or charlatans like Paracelsus and Mesmer. There are, admittedly, leading sociologists of science who define science externally as run by professional men of science. This is preposterous in view of the fact that science is ancient and yet professional science, sociologically speaking, does not exist prior to the twentieth century. Let us try, then, to ignore the individual man of science for a while.

We have, then, two components of science, external or institutional or sociological, and internal or the characterization of a body of theories and of factual information. Yet, ever since it was noticed that theory may change, a more stable factor entered sociological theory of any social entity that has to do with ideas, and this more stable factor is known as ideology. And so, in addition to scientific theories and societies, there is one regular component:

ideology. What is scientific ideology? Here is a tremendous occasion to get off target and instead of being descriptive become prescriptive, or, worse, confuse the prescriptive and the descriptive (as when (allegedly) espousing official ideology, when pouring hell-fire and brimstone on deviants, etc.). The "official" ideology is a product almost entirely of one man: Sir Francis Bacon, who viewed his ideas as prescriptive, not as descriptive. He saw no theory which he could call scientific and no institution which he thought was conducive to research: universities and monasteries were all the institutions of learning that he knew; and he loathed them most sincerely. His prescriptions became the "official" ideology of the Royal Society of London. And so, for Baconians that society is the paradigm of a scientific institution.

This way, then, the "official" view is both descriptive and prescriptive. It is true of all writers on science, from its early days to this day, including even the most descriptive sociologists of science such as Robert Merton and Joseph Ben David, that there is a strong tinge of prescriptivism in their descriptions – and perhaps quite legitimately so. Thus, when Merton discusses science, he has no room for dysfunctions and he finds even duplicate researches and the severest competition in science quite wholesome. A stronger example, perhaps the strongest, is the way Benjamin Rush is treated by his biographers and editors. Most conspicuously, the editor of his *Correspondence*, L. Butterfield, declares unabashed that Rush's bloodletting was something he performed not as a man of science or of medicine but in his capacity as a private citizen. This calls for comments. Today we do not approve of lobotomy, at least not of the butcher-job that went under that name when the practice was introduced after World War II. If we declare that such operations were performed privately rather than medically, we thereby make a few extant medical institutions, as well as a few individuals who are still alive, and the estates of others, liable – criminally and/or financially. The same, of course, held at the time for Benjamin Rush, whose excessive blood-letting was declared unjudicial. He took to court the party which decried his practice unjudicial and won a libel suit. And yet L. Butterfield declares his excessive bloodletting private because unscientific! The explanation for this is, obviously, that Butterfield is using the "official" view of science: following Bacon, he identified science with objective and uncontestable truth. But since Einstein has contested Newton, this cannot stand.

Now anybody is free to prescribe as he finds fit, of course, whatever he thinks should be prescribed; and presumably everybody will agree that sound policy is more likely to emerge out of debates about the many prescriptions thus offered than any other way. But whereas mixing the descriptive and the prescriptive idealizes and thus entrenches the establishment of science, the separation of the two can best show that the described state of affairs falls short of certain reasonable prescriptions which we may wish to implement. This is not to deny the greatness of the achievements of science, or of its traditions, to date. But what is good in science can hardly be credited to its current organization – from elementary to higher schooling to national and international conventions. The scandal that is the Nobel Prize is our open secret, but we still speak of its attainment as the highest proof of greatness. It is time, then, to give science and its organization a hard look rather than smugly sing the praise of praiseworthy achievements while attributing all success in science to its Establishment.

On second thought, there is a diffficulty here. Anyone is free to prescribe as he finds fit. I say, and in a clear-cut sense this is trivially true: we all – meaning author and intended readers alike – admit this liberty both in morality and in politics. The moral freedom, in addition, is limited by the rule that we do so responsibly and honestly, and, in our case, praise and condemn this or that aspect of the social institutions of science as we find fit. Yet there is a sense, equally obvious, that makes things look quite different, and has to do with the conditions under which one prescribes. It is legitimate, though in my opinion misguided, to prescribe hard work as morally good. Yet one who does so is quite at fault if he approaches a hard working person and exhorts him to work hard: it is a kind of affront. Also, a philosopher of science who prescribes devotion to science or to experimenting may anger a working scientist. And since philosophy of science inevitably includes prescription and since a philosopher who incurs the anger of a scientist is in bad shape, philosophy of science becomes an unpleasant occupation. This anger need not be limited to scientists. In a public meeting Sir Alfred Ayer, the leading British philosopher, expressed anger at Sir Karl Popper's exhortation to be critical, saying, we do not need to be told that by Popper. But Popper at least could and did publicly express his views. Young philosophers of science find it difficult to publish their views because, trying to be descriptive they fail to notice that they are also willy-nilly prescriptive and thus anger old pompous scientists.

3. THE MYTH OF UNANIMITY

The internal demarcations Bacon gave between science and pseudo-science are very interesting. Science is pure of metaphysics, religion, politics, etc. It is purely accidentally observed facts and empirically demonstrated theory;

nothing else. The accent is on purity. Pseudo-science mixes science proper with religion, metaphysics, conjectures, and politics. Indeed, the reason pseudo-science exists is simply political: there are tyrannies of the flesh and of the mind; tyrants of the flesh and of the mind. Schools are dominated by intellectual leaderships and perpetuate dogmas. Science cannot have any leadership: the leadership that allegedly comes to promulgate science is only self-serving.

Bacon conceived of new institutions of science; of a new social basis of science, we would say today. He offered two. One was the secular university with its research laboratory and research organization. The other was the amateur researcher who is a gentleman of means. Both ideas are incredibly bold and admirable. The first disciple of Bacon was Sir Henry Wotton whose (worthless) Baconian researches were published posthumously. He was the founder and first provost of Eton, and we know from John Beale's correspondence with Robert Boyle that when they were his students Wotton infected them with Baconianism. In the very days of turmoil and civil war England was infected with the Baconian vision of a secular college and filthy rich Robert Boyle was fingered as the man to finance the venture. Instead, Boyle implemented Bacon's and Wotton's vision of the disinterested amateur scientist who is a gentleman of means. The rest is history.

(It is clear that Boyle's preference for an amateur society was not a matter of expediency but of ideology. He recommended amateur science in his first publication, *Seraphick Love*, whose appearance was instrumental in calling the meeting in which the Royal Society was founded. He explained his view in the opening to his first and most important book on the elasticity of air: a philosopher needs a purse as well as a brain. And when some of the confiscated Irish lands were registered in his name by his friends to make him instrumental in founding a secular college he used the money thus accrued to him for purposes of charity and religious mission, much to the annoyance of biographers and historians of science. The paper in which I discuss all this is still unpublished, perhaps because editors and their advisers insist on viewing this as a prescription on my part to restore amateur science.)

Bacon's amateurism was noted with unjust derision by professional Justus von Liebig a century ago. Unjust though Liebig was, he noted a significant truth: the Baconian methodology is the methodology of the amateur gentleman, assured by Bacon he will be a serious contributor to science with a little time off, some money, and good will. The good will is important since the first step one must take in order to qualify is to renounce – to give up all his past views and thus all his intellectual affiliation. Thus, the first step in becoming a student of nature is an act of conversion. Next the Baconian philosopher can put the little time and money he has dedicated to science to simple use, to perform simple and easy experiments and observations. For example, does alcohol burn faster or slower when table salt is added to it? Liebig showed that even this simple experiment was poorly designed by inept Bacon. But we do not mind this fact these days.

Bacon, to repeat, had a touchstone or a hallmark of science: invention. Genuine science, that is, must serve technology. Whereas pseudo-science, he said, may at times yield some invention, if at all, only by some accident, the discovery of the truth leads to the true mastery of nature, to discovery and invention "not in treacles but in streams and buckets."

There is a matter of great confusion here to be cleared. Bacon uses the word "accidental" in two different senses. (1) A discovery does not follow from any theory – it is a surprise, he says; it is accidental in the sense of being logically independent of theory, and a discovery made by one who has renounced all theory and approaches nature with no preconception and who thus cannot depend on any theory. (2) The convert's mind is turned to nature, fit to make discoveries systematically; the one whose mind is blinkered by his preconceptions cannot make discoveries systematically but accidentally, meaning only on occasion. Here we see that one who makes independent discoveries makes them systematically and one who does not make them independently makes them rarely, yet the word "accident" at times denotes independence from theory, at times a rare occasion.

This slight confusion is widespread: not that Bacon was confused about it; but his unfortunate choice of words did confuse many writers, both concerning scientific method and concerning invention. Yet, confusion did not obstruct in view of the fact that Bacon saw science and technology as two aspects of one and the same venture.

This is a central point of Bacon's philosophy of science, both prescriptively and descriptively. Prescriptively it was the invitation of philanthropists to join the cause of science. Descriptively it links, perhaps unifies, the internal and external – methodological and social – characteristics of science: his secular college has a high standing in society because of its services to technology. Historically, the Royal Society took this philanthropic point much to heart and was determined to assist humanity by mastering the various crafts, from husbandry and forestry to shipbuilding and gun-powder making. It supported the spread of potatoes in Ireland and the implementation of Newcomen's steam engines in mines. Not much came out of it all at first, but soon the learned world enthusiastically supported first agricultural experimentation and later industrialization. Nevertheless, Bacon's "mark" of science, his externalist characterization of it as technologically fruitful, was dropped. Perhaps it was dropped because the traditional dispute in philosophy of science, from antiquity to date, is between the realists who view truth as the aim of science and the instrumentalists who view improved prediction as its aim. And, no doubt, Bacon was a realist; yet, of those who tended to notice his accent on technology, many mistook him for an instrumentalist – even in our own age. But perhaps the reason lies deeper. There is no *a priori* reason to assume that the discovery of truth and of usefulness must go hand in hand, and, obviously, at times falsehoods are useful, at times truths are not. And so, in a realist mood, Bacon's external characterization of science as invention was dropped and his internal characterization of science as demonstration or proof was a near universal "official" view until the famous crisis in physics.

Contrary to Bacon's view, however, there is no complete demonstration in science. For many Baconian philosophers this point is but a minor irritation. When they are squarely confronted with the absence of certainty, they hope that high probability will do. Bacon himself already did this, perhaps also Newton and myriads of others to date. They are quite right to insist that high probability, even not-so-high probability, is no different from certainty: the very methods of rational thinking which accord one view certainty, may accord another view probability, high or not so high. But certainty and probability would make rational men abide by the verdict of science. For, Bacon insisted, science spells unanimity and so permits no schools.

Amazingly, science is obviously myth-ridden, yet this is seldom noticed. George Bernard Shaw noted this, yet he made little of it: where myth is operative, e.g. in biology and medicine, he saw no science but mere pseudo-science; where proper science rules, e.g. in physics, he said myth was marginal. He never attacked the myth of unanimity in science, even though he required that we argue the pro and the con of any case, and criticize each thesis as strongly as possible – Darwinism in particular. Unfortunately, however, he tended to view Darwinism, and more so Pavlovism, as pseudo-scientific proper. This puts him in the tradition of the early nineteenth century Cambridge philosopher William Whewell and the late nineteenth century philosophers John Venn and Stanley Jevons. This tradition's hypothetico-deductivist or Kantian variants of inductivism or of Baconianism, mingled well with the "official" view of the commonwealth of science. Among philosophers the ideas of Whewell fared less well: John Stuart Mill successfully questioned the rationality of Whewell's view on account of Whewell's claim that we need luck in order to invent a good hypothesis.

The latest variant on Whewell's doctrine is that of Popper. He too sees the rational method as the arguing of the pro and the con of a theory, and in particular the subjecting of a theory to the severest criticism possible. Yet he too thinks there is one rational method, which singles out the theory which best explains the phenomena, which is best testable, which is thus far best corroborated (i.e. best stood up to severe tests). And so, for Popper, too, there is no room for scientific schools of thought. His later writings, however, inconsistently with his earlier ones, are pluralistic, but only in a permissive mood. He says almost nothing about scientific schools of thought, and he has not retracted any of his old anti-pluralistic views of science.

4. RATIONALITY: UNANIMITY VERSUS PROLIFERATION

Popper has offered both an internal and an external characterization of science: internally, the theories of science are empirically testable; externally, the institutions of science are those conducive to empirical tests. This external point was made – at the same time or thereabout – by Robert Merton who was then, at least, quite an orthodox (Whewellian) inductivist.

The merit and import of Merton's contribution to the sociology of science is in its inductivism, being an extension of the sociology of knowledge to scientific knowledge proper. He has himself obscured this fact by generously ascribing this extension to Karl Mannheim. The sociology of knowledge is the culmination of the application of Bacon's theory of prejudice; Bacon simply dismissed all school doctrines as mere prejudices: truth is one. This led to the neglect of all history of ideas. When histories of ideas finally turned up, they were histories of science, and the same up-to-date science textbooks with a bit of dates and historical anecdotes thrown in – as I have described in detail in my *Towards an Historiography of Science*, 1963; (facsimile reprint Wesleyan University Press, 1967). What was not science was condemned as superstitious. The Romantic Reaction to Enlightenment went for the collection of folk wisdom, which later Baconians, especially Wilhelm Wundt, were ready to endorse as the natural history of superstitions. Marx meanwhile explained all prejudice as class-prejudice (and Freud as neuroses). Karl Mannheim's sociology of knowledge is a semi-Marxist analysis of all sorts of prejudices. Merton said we can offer also such analyses of the discovery of scientific truth. He is thus the father of the sociology of science proper.

The institutions of science are seldom conducive to criticism, Popper and

Merton notwithstanding. A drastic counter-example is the discovery of electricity in metals, of metallic conductivity, etc., which is the beginning of 18th century electric research. The man who made this discovery was one Stephan Gray, of whom little was known until recently. Historical detective work discovered that he was banned from the Royal Society because Newton disliked him – as he did many a critic, potential and actual – and so he had his papers published in the Society's *Transactions* only after Newton's death. Another is the case which I have published first, about the attempt of the Lavoisierian establishment to block the publication of Davy's refutation of Lavoisier's doctrines in France with the aid of threats to call the police. The police, of course, had more important business at the time.

Bacon's harsh strictures on schools are often just, and Popper's strictures on them – which are updated Baconian – are quite accurate. Yet Bacon's view of all schools as unscientific is false. Popper's paradigms of pseudo-science, Marx's, Freud's and Adler's schools, will be judged to have contributed quite substantially to the body of scientific theory, and more so to the progress of the sciences; Popper himself notices this in passing: see his *Open Society*, Chapter 13. School activity positively relates to science, as was admitted recently within philosophy, by Stephen Toulmin in passing, and by Thomas S. Kuhn unwittingly. In his study of the controversy between Priestley and Lavoisier, and in an attempt to vindicate Priestley, Toulmin suggested that at times two doctrines may compete with no clear-cut decision to say which way the trend will go. Though decision may become clear-cut and so, in retrospect, look like condemning one of the two schools, this is a hindsight and so unhistorical. Transition may legitimize dissent between the old and the new, then, and the transition periods are of indefinite length, yet, he seems to suggest, they are rare on the whole. Kuhn endorses Toulmin's point. He calls a school doctrine a paradigm. Following Michael Polanyi he declares the paradigm obligatory and he observes approvingly the existence of repressive measures against all opposition. Yet common and justifiable as he finds repression, he permits pre-paradigm periods, leading at times even to bi-paradigms, before one paradigm system becomes dominant as it should.

Kuhn's pre-paradigms and bi-paradigms do exist. Absent is the case which he calls normal, namely the paradigm-case for a Kuhnian paradigm. The nearest history comes to exhibit a paradigm is Newtonianism. In my *Faraday as a Natural Philosopher* (1971) I outlined briefly the ancestry of Faraday's heresies which went back in an unbroken line to Leibniz, the greatest of Newton's contemporary opponents. The social history of the opposition is not very clear beyond the fact (noted first by Owsei Temkin only in recent decades) that scientific opposition was largely underground.

The existence of contending schools of thought, overt or underground, is a challenge to the very traditional idea of single rationality, according to which there exists a single method by which all reasonable people can and should settle all disputes within a relatively short time. Immanuel Kant noticed that philosophical opinions concerning science, split into two schools which go back to antiquity. He declared all views subject to traditional disputes to be beyond the limits of reason. Modern positivists have endorsed this: philosophy lies beyond reason, and hence, they added, beyond the limits of language. ("The mystical" is Wittgenstein's neologism.) The next step is an irrationalist theory of rationality also endorsed by Polanyi and Kuhn: rationality cannot be articulated; it is a way of life, akin to art and religion; a member of the commonwealth of learning knows what's rational and should rationally be listened to, even when he cannot verbally justify his judgment. Such claims were often made in romantic circles about artists and art critics. Consequently, romanticism prescribes forcing budding and deviant artists to submit to authority; only geniuses who manage to overcome the pressures of the existing authority may become the new authority. This was applied to the sciences most systematically by Michael Polanyi, and later, with some variations, by Thomas S. Kuhn and by Imre Lakatos.

Polanyi and Kuhn, then, accept the status quo as the only standard of justice, and justify the repression of the rebel, yet also the rights of the rebel once he acceded to power. It is a theory of the Establishment as such; conservatism. It is applicable to politics, the arts, and the sciences alike; perhaps also to religion and philosophy.

Is there any reason to think the conservative philosophy applies better to science than elsewhere? Pierre Duhem said, yes. He used for this Bacon's myth of unanimity of science, though without Bacon's argument that truth is one: he did not think science aims at the truth about nature. The official philosophy of science, we remember, is Baconian; official deviations from Bacon are grafts of hypothetico-deductivism. (That the graft is inconsistent with the stem need not bother us here: mythology is seldom consistent anyway.) The non-Baconian part of the Establishment is a 20th century innovation – pragmatist, conventionalist, instrumentalist, which endorses Duhem's claim that the unanimity in science is its hallmark. Hence all disagreements, it is now as "official" as before, even between scientists, belong to metaphysics. Hence interpretations of scientific theories, said Eddington, are not scientific – simply because interpretations need not be unanimous.

This forces one to declare dissident scientists pariah, or the topic of dissent non-science, or both. Thus Einstein's debate on determinism was metaphysical and he was "officially" viewed as senile.

The superiority of Duhem's view over Polanyi's is not on matters of fact: they both see dissent not in science but in metaphysics, religion, politics, etc. Both view the rationality of science as rooted in its internal logic; but Duhem denies any rationality to religion and metaphysics, where commitment alone reigns supreme. Only when Polanyi makes science a matter of commitment as well, only then does it become possible for him to apply the same view of rationality both to science and to religion (see his *The Logic of Liberty*). But then it is baffling that Polanyi allows diversity in metaphysics, in religion, in politics, and in the arts, but not in science.

5. CENTERS OF LEARNING AS SCHOOLS OF THOUGHT

The topic of centers of learning is now somewhat fashionable in sociology of science. Two historians, John Herman Randall and Alistair C. Crombie, started the fashion when they discovered the important role that the university of Padua played in the late Middle Ages and the high Renaissance. Great and/or influential scholars studied there, or at least hung about there. Robert Merton and Joseph Ben David, the most prominent active sociologists of science today, have sought to generalize the idea.

Randall and Crombie hardly specify what were the influential Padua ideas, yet they insist that it was specific ideas about scientific method which put Padua on the map. Crombie expanded on these ideas in his book on Robert Grosseteste, Bishop of Lincoln, England, in the 13th century, whom Crombie singles out as his paradigm. Crombie says the Padua methodology was hidden in Aristotle's methodological books, was made explicit in stages by various scholars, and came to its full expression first in the 13th century writings of Grosseteste and later in writings of Bacon and Galileo: the true method of science is the combination of the inductive and the deductive method. No Aristotle student has ever ascribed this idea to Aristotle; very few philosophers (Justus von Liebig and Friedrich Engels) have explicitly endorsed it. Galileo and Bacon were certainly against it: the one wrote scathingly against induction, the other wrote bitterly against deduction. Also, Bacon and Galileo both openly opposed Aristotle; though for contrary reasons, of course. That the revolution of science is the result of the rise of scientific method is, of course, a classical view; in its classical version, however, the view was that the method was the inductive method, fully expressed for the first time by Sir Francis

Bacon. Pierre Duhem strongly opposed both the idea of a scientific revolution and inductivism. Crombie's view of the scientific revolution is idiosyncratic, and is partly a compromise between Duhem and the classical view. It will not stand even a superficial examination, either philosophically or historiographically, and its detailed exposition suppressed and distorts many unpleasant details, not to mention the flagrant misrepresentation of the views of Aristotle, Bacon and Galileo.

Merton and Ben David are quite different, and accord more with Merton's ideas of the institutions of science: they explain the significance of centers of learning by reference to no ideas, scientific or methodological, but to institutional characteristics which facilitate the growth of science.

There is some leeway here. It is usual to ascribe to the scientific revolution a methodology, as well as a center: inductivism and London. This does not preclude the claim that Galileo's methodology was different, or that after him Paris became a center with yet another methodology. All it means, really, is that we need not – indeed, we cannot consistently – accept everyone's views on scientific method, as Duhem noted with much (justified) emphasis. How then was a positive revolution aided by a false methodology? Bacon said it cannot be done: we cannot succeed unless we know enough methodology. But he was in error on this point too, since his false methodology was useful; perhaps it was useful in its providing a system that facilitated scientific discourse, as Merton and Ben David claim.

Centers move, at times because of political oppression of ideas – from Italy to France and England after the trial of Galileo, from Germany to Britain and the United States after the rise of Nazism. The emigrations accord with Merton's sociology, but not the immigrations: why from Italy to France and only then to England? why England rather than other Protestant countries? why France of the Ancient Regime and of Napoleon? There are a few studies on these details, primarily by Merton and Ben David. They do not tackle well Lewis Feuer's critique. His view of science as hedonistic and liberal is better than Merton's view of science as austere and Protestant. Feuer also better accords with Randall and Crombie and goes even further than they. Centers of learning are not only institutions which express the formal qualities of science and of scientific research, but also some more definite institutionalized contents, i.e. school doctrines. Not only liberal hedonism but any stimulating views which one goes to learn at the center so as to see how one gets stimulated by them. Randall and Crombie ascribe to centers of learning some rather silly methodology. Yet in principle they are right: schools are seminal; people go there to learn and to join active research. There is a serious difficulty here: what and how do they learn this? Polanyi says, it is impossible to articulate what makes science great any more than regarding the arts: to become an active artist one becomes an apprentice to a reputable studio; and the same goes for science. Yet art critics and historians do try to articulate. We have less understanding of both the scientific character and the sociology of, say, antiphlogistonism, than of impressionism. At most Ben David draws attention to the Paris of antiphlogistonism as a phenomenon; he does not even attempt to treat it as art historians and sociologists treat the Paris of the Impressionists.

Why is this so? Because impressionism is a known school, but not antiphlogistonism. Indeed, since we recognize the Copenhagen school of quantum theory as a school proper, we have no more trouble about it. We can even tell about the encounters opponents of the school had with it. But when historians speak of opponents to antiphlogistonism, whether Priestley or Davy, they still denounce Priestley and conceal the evidence regarding Davy!

6. PHILOSOPHY OF SCIENCE AS EMPIRICAL SOCIOLOGY

Wittgenstein's disciples have claimed that every claim is either a part of logic or empirically verifiable. They were asked whether this very claim was empirically verifiable. They struggled with the question and failed. Popper has claimed, alternatively, that the theories of science are empirically testable, i.e. empirically refutable, and the Wittgensteinians asked him the same question. Unlike them he can view his claim as a verbal definition, or as a conjecture. The conjecture may, but need not, be empirically testable, i.e. refutable, and hence scientific. As a verbal definition his claim is, of course, both legitimate and uninteresting. It becomes a conjecture once we have examples of scientific theories, of which it says that they are refutable; or examples of non-scientific ones, of which it says they are not refutable. Is the Book of Genesis science or non-science? Most of us will say, non-science. Was its first chapter refuted by geological observations if taken literally? Many members of the scientific Establishment say it was. Hence it is both unscientific and refutable. Will it then count as a refutation of Popper's claim taken as a conjecture? How can we adjudicate this question?

Of course, the statements of the Book of Genesis should be taken literally, and in their straightforward meaning, in order to clash with statements made by geologists on empirical grounds. And, as is well known, many fundamentalist scientists reject the geological evidence, many geologists reject the Book of Genesis, and many deny that the literal straightforward meaning is the acceptable one. It is no doubt in accord both with the common view and with Popper's that the reinterpretation of the Bible makes it unscientific, and that the fundamentalist scientists who would rather interpret geology than the Bible, deprive both the Bible and their geological theories of scientific status. Yet those who reject the Bible on empirical grounds usually deny that it is scientific, when by Popper's criterion it thereby seems to be scientific. How should we assess this situation? Shall we say that Popper's criterion is thus refuted?

It is hard to say, yet what the discussion in the previous paragraph shows is that, as Popper has noticed early in his career, theories are not what Wittgenstein and his followers thought; they are not sentences, but claims which come with certain attitudes as to their meanings, and meanings are neither absolute, nor capriciously decided by individuals.

Thus, before answering the question, is Popper's claim refuted, we have already the need to approach matters in a rather complex sociological fashion. Hence, we may decide to reject the question, how should we characterize science as a body of theories, whether proven, provable, probable, or refutable, or refuted. It is better to approach science sociologically: science is a specific tradition, including some traditional doctrines and practices associated with its institutions, exhibiting continuity but also deep changes: when rudimentary market research first appeared in the scientific press around 1820, for example, it was ridiculed; today things look different. My *Towards an Historiography of Science* describes how diverse opinions of the history of science are strongly colored by a few central doctrines in the philosophy of science. I shall not do the same here regarding the sociology of science. First, because there is much less sociology of science literature than history of science literature to survey. Second, it is a bore to repeat oneself. Third, in my survey of the history of science I could not discuss at length my own view of science and its possible impact on the way I write its history. It took a few years before I found an occasion to publish a history of science to my own liking. But I would like to discuss my own sociology of science straight away. This can be done now more easily: Popper's philosophy, which I was advocating in the early sixties, is now recognized as legitimate: he is nowadays recognized as the head of an established philosophical school. And so, I can now discuss his opinions on a par with the opinions of others, such as Bacon or Duhem or Polanyi. This will enable me to advance my disagreements with Popper especially regarding (metaphysical) research programs and their impact on scientific research which have also, I am very pleased to observe, been popularized by some established writers, such as Watkins, Feyerabend, and Lakatos. Luckily also the climate of opinion has become congenial to the concern I advocate: a while before I published my own views Thomas S. Kuhn published his views which are by and large Polanyite, yet which include the characterization of groups of scientific researchers by reference to research programs, and Sylvain Bromberger made similar suggestions, so that today quite a number of detailed philosophical doctrines whose sociological implications are very obvious are popular enough to merit examination. This is a very exciting situation which they miss: we have here philosophical doctrines which have sociological implications and which are therefore empirically testable!

The doctrines in question are all descriptive rather than prescriptive. This is so in view of the fact that there exist diverse phenomena which are viewed as significant instances of scientific theories and of scientific institutions – and uncontestably so. In particular, Archimedes' Law, Newton's theory of gravity, Lavoisier's chemical theory, the wave theory of light, electromagnetics, the germ theory of disease, Mendelism, the Weber-Fechner law of psychophysics, all count as indisputably scientific doctrines; the Royal Society and its derivatives and emulators among the older universities, the national and international scientific societies, ad hoc institutions such as the *International Geophysical Year*, are all undisputed scientific social entities. Whether the Ford Foundation, the Rockefeller Foundation, NASA, and such, are scientific institutions or not, are still open questions though not yet publicly discussed. The question, who is a scientist is often read to mean, who is a professional scientist. Ben David and others have recently repeatedly claimed that psychology, for example, is a young science because professional psychology is. This idea is backed strongly by Michael Polanyi, who defines science as the activity and body of doctrine of the professional society of scientists, and whose views are now popularized by Kuhn.

The way to test Polanyi's view, perhaps, and that of Ben David, is to ask whether the originators of the scientific theories listed above were professional scientists. This is not clear as yet, because thus far ideas and institutions were discussed, not professions. Science is in part a free profession even today, and it was a hobby with such great minds as Archimedes, Boyle, Coulomb, Priestley, Franklin, Lavoisier, Dalton, and many others. Also, for centuries universities disfavoured scientific research, so that for many academics engaged in research, this activity was a mere hobby with them. And some of them, particularly Newton, were rather embarrassed about their academic status. In 1830 David Brewster, Newton's biographer, strongly sympathized; not much later, this amused Augustus de Morgan, who was a professor at the modern University of London and a commentator on Brewster's work. Could we, then, say that as a Cambridge professor Newton was a professional scientist?

The way to circumvent this problem is to claim that institutions are scientific to the extent that research goes on around them and in them, regardless of profession. This claim will take the Ford Foundation out of our discussion; the Center for Advanced Study in the Behavioral Sciences which it finances, and which finances its scholars, will count as scientific, of course; so will count the library as the British Museum which does not support its scholars beyond offering them its services free of any charge. Universities will then likewise be made scientific institutions by this claim if and to the extent that scientific research takes place under their wings, and the scientific societies like the American Sociological Association will be research institutions for about three days a year, at most. Even the institutions which that association sponsors and which have research conducted under their wings the whole year round, will not make it more scientific by this claim than Ford Foundation does to Ford Motor Company. But even so, this claim does not solve all problems. For example, is research in industrial institutions scientific? Or in the Consumers' Union's Laboratories?

In order to make out theories empirically testable we want more detailed characterizations of scientific theories, of scientific institutions, and of the contribution to science which an institution has to make. For the time being, the very fact that we have undisputed instances of both scientific theories and scientific institutions will make our discussion descriptive throughout and so also hopefully open to empirical examination. This, in particular, with no reference at all to professions, but merely to the activity of scientific research.

Thus, to conclude, I propose, very much in opposition to Ben David, to ignore all professional manifestations of science. My proposal would be refuted by a system, such as envisaged by the science fiction writer Isaac Asimov, where one cannot publish a scientific paper unless it reports the results of research supported by a grant. Such systems, I say, cannot produce science. Also, I propose that scientific institutions be related to scientific research, regardless of scientific doctrine, professional scientists or any sort of sub-culture. This is not refutable as yet, since I have not yet characterized scientific research. I do propose that what counts as scientific research, however, varies in time. Thus, to repeat my examples, though the writing of the Bible will never count, market research at times does, at times does not. This calls for further discussion.

It is time to relate all this, however briefly, to scientific schools. The correlation is simple. Institutions lend respectability: if one person who is rather unknown is a member of a respectable institution, then he is respectable. People who dabble in university administrations and busybodies around universities and professors when discussing other professors from fields on which they have no judgment, they all judge a professor by his rank and affiliation. And research institutions are the fortresses of schools: if you wish to conquer a paradigm, apply for a respectable grant, convert members of respectable institutions, or, still better, get a respectable grant to collect members of respectable institutions in an enjoyable conference where the new paradigm is explained. "Paradigm" here is a euphemism for a school's doctrine. This, however, is easier said than done. Nevertheless, it is an empirical fact that not so long ago it was not possible for disciples of Popper to get such grants and enter most respectable institutions, and that things are rapidly changing. Why?

7. HOW METHODOLOGY AND SOCIOLOGY FUSE

It is a strange sociological fact that we have two characterizations of science, one internal, relating to theories, one external, relating to institutions. In university classes, when discussion elicits the realization that the two are not coextensive, a strong sense of confusion ensues. Suppose we declare science to be empirical certainty and admit that no empirical certainty is possible. Hence, science does not exist. This conclusion looks incredible, because we have an external definition of science, with members of our science faculty as men of science who do produce real science.

Bacon and Descartes did not share this quandary, since they rejected all that they found as fake. In other words, as long as the definition of science is only normative, our difficulty does not arise. Does any modern writer define science merely normatively? No; though possibly Popper says he does. But if he says so, then he is palpably in error, since he admits the existence of paradigms of science, such as Galileo's and Newton's and Einstein's theories of gravity. It was one of his first disciples, J. O. Wisdom, who stressed the importance of the existence of paradigms for any stipulation of what science is, since *a priori* with no paradigms each stipulation is quite unproblematic. But what makes a paradigm case and why? Popper begins his discourse by observing that Einstein's theory is a paradigm for science, astrology for non-science and problematic cases such as Freud; and he defines science so as to include Einstein, and he (gladly) lets his definition decide the problematic case (in the negative). Is this fair? What would happen if we had two stipulations, both admitting all admitted cases, excluding all excluded cases, but differing on the problematic cases?

Popper would have the right to dismiss this question if his theory were the only one that admits all the admitted cases, such as Einstein, and excludes all the excluded cases, such as astrology. Yet even then the question would be intriguing enough to pursue.

One may ask, what makes a paradigm of science or of pseudo-science just that? Of course, we can say, the proposed stipulation, for example, Popper's, explains this. Now in a sense this is eminently true: Popper explains the scientific character of Einstein's theory by the claim that it is empirically refutable, and the pseudo-scientific character of astrology by the claim that it is empirically irrefutable. There are many facts that Popper's theory throws light on. For example, Eddington's crucial test between Newton and Einstein was found impressive just because experience could go this way or that way and nobody could know in advance which of the two predictions could come true. Likewise, one impressed with the evidence an astrologer marshals in favor of astrology is bound to be disenchanted once one pays close attention to the disregard with which the astrologer handles the facts. But what about the refutations of all astrology presented in *Recent Advances in Astrology* of 1975?

There is a rather uncontestable sociological observation that I (but not Popper) deem of crucial importance: most of the people concerned with the problem of demarcation of science, whether as laymen, as philosophers, or even as scientists, do not examine closely the theoretical situation, but they usually consider matters much more sociologically: Einstein is credited by the world of science, astrologers are not; relativity is a subject in the university curriculum, astrology is a parlor game. When one mentions to physicists an uncredited physicist, to biologists an uncredited biologist, etc., they do dismiss them as uncredited.

We tend to be highly critical of this conduct, since public opinion is no substitute for one's own. Nor is public opinion any guarantee, especially since it usually is behind history: it takes time for the public to learn about what the avant-garde is doing. Yet, though public opinion should not be a substitute for one's own opinion, and though it is not reliable, still, the more advanced a society the more dependable its institutionalized public opinion is! And when one is quite ignorant one takes public opinion as an initial guide. This is how science integrates in our culture, and is embedded in it. This is important, since even those who diverge from public opinion are affected by it.

Hence we cannot ignore public opinion when discussing science. Yet how is public opinion formed? how come people choose to view Einstein's theory as scientific due to a criterion of scientific character that they do not know? Even if Einstein's theory is scientific because it conforms to Popper's new criterion, how come people act in accord with a criterion which they do not know, which indeed conflicts with the "official" criterion they openly profess to accept?

This difficulty is quite general. The most obvious hypothesis is that people are familiar best with what they themselves feel and think, with their own criteria, etc. When Freud found out that people do not know how they feel, he intuitively found it necessary to explain this fact. And so he did – by the censor hypothesis, i.e. by the hypothesis that certain self-knowledge is avoided since it is painful and anxiety-generating. This does not explain the sociological fact that one is institutionally geared to act in accord with one criterion, while professing another. Malinowski, who should be credited with the discovery of this fact, was not inclined to explain it. Indeed, he viewed this as the hallmark of a social science proper as it blocks reducibility to psychology: the social rationale of one's behavior need not be known to one, and when it is not, then one's psychology will not replace one's sociology since it is deficient. Yet the logic of this argument is questionable, since there is agreement here between Malinowski's opinion and Freud's, and they both consider the acceptance of social norms as the result of socialization. Malinowski, but not Freud, wanted to explain the survival of the norm itself; and he explained it as the survival of the fittest. That people do not actually know the norms of their society Freud explained by the fact that their socialization was painful and they prefer to forget pain. Malinowski explains the same ignorance by people's want of scientific training. Freud thought it takes no sophistication to discover one's norms; Malinowski disagreed. He saw himself as the first social anthropologist with a title to science, and Durkheim as the first social scientist proper. And so, whereas Freud expects knowledge and explains the ignorance he finds instead, Malinowski expects ignorance and only from scientists he expects better. Popper shares neither of these expectations: we all have views that are subject to test and improvement, he says. And, indeed, the prevalent fact is that people do have views about their norms, criteria, and institutional settings; and that their views are oftener false than true. Why, then, do people hold erroneous views of their own norms? To this Malinowski gave an evasive answer: their views do not matter, only their social behavior. Indeed, insofar as he at all noticed views these were institutional – for he was a sociologist, not a psychologist – he declared institutionalized views to be not endorsed views but merely ritual utterances.

This is where Malinowski is hard to swallow. Ernest Gellner and I. C. Jarvie have criticized him and take people to mean what they say, even when expressions shared by all members of a social sub-culture also constitute an instituted view which also has diverse social implications. Still, we do not know all of their implications, especially since others draw new implications from them in new conditions. This holds for all sorts of social groups, be they peasants of a given primitive society or men of science in the Western world; their view is itself an institution, and one that binds, and their view entails varieties of social actions. And the same holds for Malinowski's views which are, of course, institutionalized in his school: the functionalist school.

And when a member of a social sub-culture stops believing the instituted view a new situation arises: It turns out, then, that one reason Malinowski could prefer to ignore his subjects' own institutionalized views of their own institutions is that these often were in conflict, and he assumed that all societies are functional and so really devoid of all conflict.

But functionalism is false and conflicts abound. And when in a conflict a person may be unable to act, be paralyzed, and even break down. For example, the psychologist D. O. Hebb describes the breakdowns which psychology graduate students had when they failed to confirm their hypotheses in a manner they were supposed to. Had they been less critically minded they could find ways to escape their predicaments, but they were trained to be critical: their training forced them to try the impossible: as scientists they were expected to act in accord with the internal criteria of science – confirm their hypotheses – and as scientists they were expected to act in accord with external criteria, and submit successful results within specified time limits.

Thus, we have institutionalized criteria to demarcate science, which are vague and conflicting, offering one person leeway – to use the view more comfortable under the circumstances – and another person a double-bind: if he is exempt by one criterion he can still be judged guilty by the other. The methodology and sociology of science, in brief, do not always live comfortably side by side.

8. LIVING UP TO ONE'S STANDARDS

When we look at science we may look at it internally, and so notice first and foremost what looks to us the best by the given internal standards – whether of precision, prediction, utility, verification or testability. But when we look externally at a sub-culture whose task is to live by the standard of science, we see another picture. There may be no "official" standard to the sub-culture, or a few conflicting ones, or one they preach but not practice, or one they preach and try to practice but do not live up to.

For example, the expansion of affluence and literacy permits whole new social sub-cultures to contribute new members to the scientific sub-culture. And so we find, for the first time, many scientists coming from homes with no traditional ties with the world of learning. The result is that they learn the "official" mythology of science, say Baconian inductivism, for the first time in class in college, mingled with their science proper. Not being immunized to myth, they take the myth for a fact and so try to live by its light. And so it happens that, as D. O. Hebb narrates, they try to conduct their research in the light of the theory of research which they were taught. Hebb reports sufferings of casualties of high standards, and notices that these may and should be prevented, yet it does not occur to him to question the standards themselves. Merton unwittingly justifies the standards, as did already Max Weber who, Feuer notices, was a willing victim of high standards.

Much as I object to Polanyi's theory of science as a profession with apprenticeships and mystiques, allegedly like the arts, it has a great merit, first because even though science and mystique do not mix, many professors of science profess mystique instead, and second because it enables us to observe the fact that not all those who pay an entrance fee are lucky enough to enter. But Polanyi's analysis is most inadequate and has undergone almost no empirical examination. There is a great need, as Paul Feyerabend has observed, to apply standard anthropological techniques to science. It can be observed, and with little difficulty, that the same scientists apply different and even opposing standards in different cases. In particular, when confident and when different, scientists with some dexterity apply different standards. Unfortunately, they are not always at liberty to choose their own standards, e.g. if they are underlings and/or research students, and they do not always have at their disposal the divergent standards. They may be too conscientious and self-conscious to do that, or even as masochistic as Max Weber.

The standards of many cultures and sub-cultures are often too high and even quite impossible. Impossible standards can always cause damage, but one may expect them to be particularly harmful and incapacitiating in the scientific community because of its peculiar intellectualism, high moralism, and other high qualities noted by Weber and by Merton.

Hence the complex social structure of the scientific sub-culture. In particular, there are, as Polanyi and Kuhn observe, in each sub-culture, leaders and herd. But in no society, no sub-culture, no club, are there two and only two strata. At the very least we must notice, as mediators, the watchdogs, which both force the stray sheep into the fold and help them when they get lost. The watchdogs are the editors, the philosophers and historians of the sciences, old revered former shepherds who have lost all contact with the more up-to-date developments, and all sorts of rabble and mixed multitudes who find it easier to preach and offer a helping hand with a grant application than to try and do a regular scientific job.

Also, there are many dimensions of scientific excellence. A scientist may excel, in particular, as both a leader who makes a revolution and as a leader who is a terrific normal scientist. We can see him in top gear when he presents in a public meeting new data and theories, especially when his audience bombard him with requests for information, with questions, with alternate hypotheses and with avenues of possible tests for them. He is well prepared for all these, he has anticipated all of them, and has ready empirically supported answers to all the questions. It is not easy to capture the standards by which he excels. But he is a very familiar figure, even though not a very common one, and a paragon of a man of science for most scientists and philosophers of science alike. He is the pride and joy of his profession and the model his colleagues present to their apprentices. The most remarkable fact about him is his simple-minded view of science in general and of his activity in particular. He has no need for any methodologist to analyze his work, and he either sees no methodologist around or he shoos him away for fear of the evil eye. Of course, being human he is not free of problems. Truth to tell, he even clashes with colleagues and has troubles with his assistants, because he is usually very demanding. Being very demanding and yet simple he can hardly be anything else but highly self-demanding. And so, if he has a family he is in trouble with them as well. (See Michelson's life by his daughter.) But all these troubles are highly extra-scientific, of course.

Or are they?

9. SCIENTIFIC SCHOOLS AND THEIR SUCCESS

The major reason for the invention of the sociological but not the anthropological category of a sub-culture is just this: if a chief in a primitive tribe usually quarrels with his wife, anthropologists want to know about the fact, as well as about the technique that sustains estranged relations without bringing the related parties too far apart or too far together, whereas if an artist or a scientist or a business executive suffers the same, the analyst of the sub-culture will relegate study of these estranged relations to students of the modern family. This is why novelists play the role of social anthropologists of modern society: it is too hard to study the social network of a modern society as a unit the way Malinowski, Radcliffe-Brown, and E. E. Evans-Pritchard tried to do for primitive societies. Thus we must ruefully leave the social relations of normal science to the novelist who also studies the complex social relations of business and art: the social relations of men of science are equally complex. We can, instead, examine the way our supernormal scientist manages to maintain his simplistic view of science. With this I will conclude the present discussion.

Not all scientists whose views of science are simplistic are normal. Some of them may be leaders – in Polanyi's and Kuhn's sense of leaders, i.e. those who dictate the broad outline of contemporary scientific views and activities. It is hard to speak of these leaders in general since each of them may be unique in diverse ways. The supernormal scientist, however, and his simplistic view of science, are exceptional only in the sense that they fit the norm so admirably well. And the norm really comprises just the very few – surprisingly few – qualities as Kuhn has described them: normal scientists are naive, credulous, and submissive on a large scale, yet sophisticated, critical and independent on a small scale; and they are ever so industrious. What singles out the supernormal scientists is their being outstanding – a quality barely allowed by Polanyi and Kuhn, and certainly not allowed by Polanyi and Kuhn if it has many social repercussions.

And it has. For one, the supernormal scientist, not the leadership, is the trend setter. Polanyi has noted that the innovator takes a risk, as he may or may not be followed: if he is not followed he is practically ostracized. The innovator who takes a risk can take a calculated risk; he may adumbrate his innovations, privately or in small scientific meetings. But in between these two steps come the private conferences with the supernormal men of science. (Dr. Thomas Thomson describes how supernormal Wollaston convinced Davy not to oppose Dalton.)

The supernormal also benefits: he knows the trend in advance and so is less surprised and so his simplistic view of science can better be maintained. There exists an enormous contrast within his make-up between the large and the small: on the large scale he is naive, credulous, and submissive; on the small scale he is shrewd and critical. How can he prevent the mixing up of the two? By keeping them apart. But what was large-scale before a revolution may turn up as small-scale after it. Indeed, this is the kind of benefit to expect from a revolution. To buffer this the organizers of public views of scientific fashions create new myths. They do not thereby make life much easier for the little scientist, but they may at times help him. They set the fashion and relieve him of worries about it – indeed by denying the very existence of fashions in science.

But this can be changed, and to an extent has changed already. The idea of scientific fashions itself is becoming increasingly fashionable. I should mention that when Erwin Schrödinger's *Science and the Human Temperament* first appeared in 1935 it made no dent. The reason seems to be just the one he indicates: science seeks invariance and so opposes the sway of fashion. Now that Popper's philosophy is becoming popular, where objectivity is replaced by the tendency towards objectivity, fashions become permissible. Similarly Marx, and the Mannheim-Merton view, already enabled one to be both objective in answers and choose questions in accord with fashion.

The truth, however, is that fashions come in schools and schools fight for public attention and win or lose it in part – hardly ever to the full. The very existence of schools, of choices of large scale views, should worry our supernormal man of science. Indeed, if there is room for him in the Polanyi-Kuhn philosophy, he should cease being normal in times of strife and become a leader. He does not. He stays in one school all the time, perhaps changing schools once in his life, perhaps not even that. (Remember Michelson.) But he needs the mythmaker to tell him that science equals unanimity and that he always is studying problems that are in the center of things. This is not peculiar to science alone, where it originates. The growth of the Popper school in philosophy is an example of how mythmaking helped shape growth. But here I speak of science.

The myth of scientific unanimity is strong and of ancient origin. It has the intellectual rationale in a simplistic view of rationality as binding. It has a different social rationale in every epoch of science. These days its role is to buffer the simplistic view of the supernormal scientist who lends credence to the myth that good scientists are the same as competent hard-working scientists and to the related myth of technocracy. This myth is best questioned in the presence of what Kuhn calls bi-paradigm, i.e. two competing successful schools of science. This is an eye-sore for so many scientists and mythmakers alike. It can be minimized by viewing one school of science as dominant and the other as deviant. When the other can be viewed as deviant no longer, they simply switch roles and their history has to be rewritten accordingly.

In brief, transitions between doctrines in science, changes of fashion if you will, are all the more abrupt because they officially do not exist.

Not yet, this is. But the existence of scientific schools of thought cannot be kept a secret for much longer.

CHAPTER 15

GENIUS IN SCIENCE

The present chapter borrows its title from a paper by Michael Polanyi that has already been published three times (in English), and which, no doubt, will deservedly be published more. The present discussion has, indeed, developed out of comments on that paper by Polanyi – both were read in the Boston Colloquium for the Philosophy of Science, 7 April 1970. Yet it is not necessary to add the subtitle 'Comments on Polanyi's paper' because it might just as well read, 'Comments on the whole literature on the topic'.

This is an amazing situation. It is not that there are no opinions on the topic other than Polanyi's. Indeed, I shall soon quote a number of writers, mostly well-known classical ones, on the topic at hand. But there is no study of the question. The reason is not far to seek: the classical writers, i.e., those preceding the Romantic Movement, all insisted that genius was inessential to science and that one blessed with it was lucky only in that he achieved the same results as others with less effort.

Were this the prevalent view, matters would not be in such a confused state. For in the meantime a lot has happened. The Romantic philosophers found Enlightenment quite satisfactory for Great Minds, but not for the Common Man. They took it that the two were essentially different, calling Great Minds Heroes, Geniuses, World-Historical Spirits, Natural Leaders and What-Have-You. Precisely where this Romantic theory clashed most sharply with the Enlightenment, it was at its weakest. The inegalitarian condescension towards the common man repeatedly developed into contempt for him, hence lending unwitting support to racialism and fascism. We find the first exemplified in Sir Francis Galton, the second in Ortega y Gasset. This may explain the reluctance of the scientific community to relinquish the ideology of the Enlightenment, on which it is historically, traditionally and institutionally based.

But what was not officially allowed through the front door could not be entirely prevented from an occasional entry through the back door. And, at times, the entry is more than occasional. For example, among men of mathematics it is a widespread myth that he who has not made his mark in mathematics by the time he is in his mid-twenties never will. Mathematicians assume that to be a mathematician of standing you must be a precocious genius.

It is no accident, then, that it should be Michael Polanyi who addressed himself to the topic at hand, since his philosophy is, from the start, a frontal attack on the philosophy of the Enlightenment. Polanyi has not evolved a Romantic philosophy of science. Rather, as I have argued in Chapter 8 above, he has evolved a phenomenological view of science, i.e., a view of science which fits the philosophies of Husserl and Sartre better than their own views of science do. I shall not enter all this here, but merely outline briefly the Enlightenment attitude to genius, then describe a few paradigms or archetypes or images, or self-images, of the Enlightenment thinker and of the Romantic thinker respectively.

1. IS GENIUS REALLY NECESSARY?

The existence and importance of genius was a vexed issue for the classical scientific tradition. Both the intellectual leaders of the Age of Reason, Bacon and Descartes, agreed that common intelligence suffices for the pursuit of knowledge. The last great philosopher of the Age of Reason, Immanuel Kant, also endorsed this view. He considered it an insult that Fichte, the first modern philosopher to speak of genius, and a self-appointed spokesman for Kant, had suggested that common intelligence is not sufficient for the understanding of Kant's great work, the *Critique of Pure Reason.*[1] Laplace was similar. As much of his *Analytic Essay on Probability* as possible is reproduced in ordinary language in his *Philosophical Essay on Probability*. Some of its mathematics is also so transposed, but unfortunately with the reverse outcome: the historian of probability, Isaac Todhunter, observed that one of Laplace's formulae is comprehensible when put mathematically, but, ironically enough, not when put in ordinary language!

Admittedly, modern science had its 'hall of fame' from the start. Already in Bacon's *New Atlantis*, which names a scientific Utopia, the community erects statues in honour of great scientists. There are statues made of precious metals, of stone, or of wood, depending on the status of the heroes they represent. But the hero is not supposed to be a genius. The paradigm of a scientific hero for Bacon was Christopher Columbus, and the reason is all too obvious: his greatness was not scientific. Yet, as it happened, the heroes of classical science turned out to be Newton, Galileo and Bacon himself. Bacon was in the 'hall of fame' of science, not as a scientist, but as the father of empirical scientific method. (There is an amusing story that proves his dedication to science above and beyond the call of duty, namely his having died from a cold caught while collecting snow to freeze a chicken [in order to study refrigeration: the story was endorsed by Macaulay in his famous essay on Bacon, 1837]). Galileo was in the hall of fame of science as a genius; but his real claim to fame was that he was a pioneer of and a martyr to science. Newton was clearly a man of genius and the greatest hero of all classical science – not only of physics. There was no way around that. Laplace, in a memorable passage at the end of the *Système du monde*, speaks of him as both most gifted and most fortunate – fortunate to have been born just at the time when enough factual knowledge had been amassed to enable one to come up with the grand generalization. This uneasy attempt to compromise between Enlightenment egalitarianism and the theory of genius did not succeed. The question persisted: was Newton's unique talent essential for great success? Sir John Herschel remarked that both Boyle and Hooke were capable of doing the job, though they missed the opportunity.

All this is a far cry from Bacon's claim that all reasonable men can be equally successful in science. Bacon stressed that scientific method is like a compass and a ruler; that is to say, it renders talent superfluous by making research accessible to ordinary intelligence. In the mid-nineteenth century the question was pressing. Robert Leslie Ellis quoted Hooke's reiteration of Bacon's egalitarianism and added that it had been of great importance throughout the history of science. This is very unsatisfactory. It has been claimed that genius is inessential for science. We ask, Could Newton's genius be replaced? Herschel says, Yes, by the genius of Boyle or of Hooke. This is not to the point: could Newton's genius be replaced by any non-genius? If so, how? The claim that genius is inessential is declared very important, yet the crucial question is left hanging.

The solution is sought in the nature of scientific method, which ensures that everyone can be a scientist whether or not he is a genius. No doubt, genius cannot be acquired at any cost, whereas the 'compass and ruler' that replaces genius can be acquired. But this does not mean that they can be cheaply or easily acquired. Briefly, what is required is not only the techniques of building science but also moral virtue! One must be humble enough to wish to employ scientific method! Thus, throughout the eighteenth century, and even in the early nineteenth century, failure was repeatedly explained by moral failure, viz. lack of humility!

If scientific method is hard to acquire, and if in the seventeenth century there were no more than three people capable of achieving what Newton did, then this does make it possible to claim, however dubiously, that genius was not essential. Indeed, Herschel interpreted the work of Boyle and Hooke as failed attempts to achieve Newton's results. In the same year Sir David

Brewster published a biography of Newton where he presents him as a morally impressive man – which suggests that Newton's genius was not essential even to Newton's scientific success.

Later on, Augustus de Morgan forcefully ridiculed Sir David Brewster's idealization of Newton in his *Essay on Newton*. DeMorgan makes it quite clear that, but for Newton's scientific genius, we would barely pay attention to him as an individual or as a person, except perhaps to censure his prima-donna mannerisms and his great dogmatism, perhaps even to ridicule his morality as terribly conformist and rather cowardly.

Today we can take all this and more quite calmly. We find Frank E. Manuel's recent *Portrait of Newton* intriguing, but without world-shaking philosophical corollaries: we no longer think moral rectitude an essential ingredient in scientific method, nor do we declare scientific method an essential ingredient in the making of science. Both Frank E. Manuel and his readers willingly grant Newton the status of genius that he richly deserves.

What this change of attitude indicates is a change in our philosophical background, which enables us to accommodate it. The pressing question, whether Newton's genius was really necessary, has not yet been adequately dealt with. Presumably, the fact that we grant genius to Newton so easily indicates our willingness to give up the old egalitarian point of view. But there has been little explicit debate on this point. Even scientists who were influenced by romanticism, for example John Tyndall, hardly touched upon this question, except to say that what is described as scientific method, though essentially correct, is so much stuff-and-nonsense because of an omission: the omission is, of course, of imagination – of the spark of genius. The debate dissipated before it began.

Such is the literature on the question of genius in science. There is, however, another side to it, to which I now turn.

Let me briefly describe the classic paradigm or archetype of a philosopher and contrast it with the romantic paradigm or archetype of a hero, to provide the background for the first fully-fledged theory of scientific genius, namely that of Michael Polanyi.

2. PARADIGMS OF GENIUS IN THE LITERATURE

The *classical paradigm of a philosopher* is well depicted in the 1930s by the famous bibliophile John F. Fulton, in his well-known essay on Robert Boyle. Boyle's greatest significance, says Fulton, is that he invented (with the assistance of Sir Francis Bacon and Sir Henry Wotton, I might add, not to mention

Socrates, Plato and Aristotle) the paradigm of *the gentleman philosopher*. The philosopher had to have 'a brain as well as a purse', said Boyle. (Joseph Priestley repeated this remark verbatim over a century later.) He wears no gown – he is no clergyman and no professor – and he experiments in an attic, in a cellar, or in a potting shed.[2] Boyle stipulated, for example, that disputes amongst scientists should be rare, and dignified; Watts stipulated that it should not go beyond two rounds or so, and that one should not be too eager to win a debate. What was so very important in that tradition, however, is not its trappings, of course, but the idea of self-education. That idea made Boyle view the natural home of science as not the university and not any profession, but the voluntary association of disinterested 'curiosi', as he called them. Their purses were at first pretty fat. But whereas the majority of 'curiosi' in the seventeenth century were upper-class, in the eighteenth century they were middle-class and in the nineteenth century they were working-class. The ideology – of Boyle, Watts, and Sam Smiles – was the same, but with Smiles it became too demanding, and so the working-class man of science tended to be professional. He could be a high-class technician, for whom promotion-avenues could be created; or he could be a university man, a public lecturer, a literary man of one sort or another. He could not be a research scientist, until the advent of the present age of post-World War II. In the meantime the idea of the old-style gentleman philosopher has almost disappeared. Just recently Michael Polanyi has revived it, and also amalgamated it with the idea of the expert professional scientist, to create a new image of the scientist. But we should leave this now, and move briefly to the romantic paradigms of the hero, who must be a man of genius, of course, and often enough is.

The most famous *romantic paradigm of a hero* is the young Keats or Schubert. Young and lonely and freezing in his attic, the hero is busy forging human destiny while racing with the angel of death who is gnawing our hero's lungs: Heinrich Heine long ago poked good-natured fun at the prevalence of consumption amongst romantic heroines, in his *The Romantic School*. Let us not make too much of this frivolity. No doubt, the young and lonely romantic hero was graciously allowed on occasion to be middle-aged, as Paul Gauguin was, and even have a number of good friends, as Schubert had, provided he was rejected by the multitude while forging the future of mankind. Here even Faraday fits the bill. (We are fortunate, says Maxwell, that Faraday was neglected by his contemporaries, as he was thus left alone in his basement to develop his mighty fields of force.)

The romantic philosopher considers the hardship and loneliness of our young hero an essential part of his apprenticeship: it is both the forging of his character and his trial by an ordeal. Indeed, the ordeal is one and serves two functions and more: one function of test, one of education. (The paradigm here is Elijah or Jesus going into the wilderness.)

Romanticism is a reaction to radicalism. Whereas the radicalists of the Enlightenment or Age of Reason took self-education to be the ideal, the romantics took tradition to embody the wisdom of the ages and declared it beyond ordinary mortals to break away from tradition with impunity: when ordinary mortals do try it, they bring about calamities like the French Terror and the Napoleonic Wars. But extraordinary immortals, any immortals in fact, are another matter. Hence, when someone breaks from the pattern we should all come down heavily against him. If we break him and send him back to the old pattern, that is good for him; and if we fail to break him, that, too, is good. Here, then, our cruelty is a great merit one way or another.

The cruel 'we' in the previous sentence is the leadership: political, intellectual, artistic, etc. Here, the *middle-aged romantic paradigm* is the leader. The leader is not the young genius neglected by the angel of death for awhile; this, we remember, is like Gauguin, the exception. Rather, the middle-aged romantic hero is a dependable father-figure. He appears at a crucial moment out of thin air to hold the sky while the earth is shaking violently, and he returns to thin air as soon as the very worst is over. He is Jean Valjean and Shane; he is Bambi's father in Bambi's youth, and Bambi himself in middle age; like the figure of Elijah or of Jesus, both appearing mysteriously for awhile among the common folk when things get too bad to bear any longer.

So much for the young and the middle-aged romantic hero. It is old age which causes trouble to the romantic lore, and it is old age which plays a remarkable role in our own post-romantic lore, as we shall soon see. Traditionally, before Enlightenment and Romanticism raised so many difficulties, the old man was a paradigm of *a sage*; and a sage has to be listened to. This point has been stressed by Gilbert Murray, who noted that the ancient Greeks consulted their elders and when they failed they went to consult their ancient, long deceased sages who were in Greek called 'heroes'.

That is to say, traditionally the old man is the sage and consequently is the leader. Now the romantic hero cannot be a leader except if and when he actually delivers the nation, in which case he is the middle-aged paradigm. The romantic old man must be sage, and the romantic sage – old or not so old – is problematic: he embodies tradition, and so he cannot deliver us from it to something new. Therefore, *there is no romantic paradigm of a sage*. One must here come to the rescue of Romanticism: the difficulty it faces is quite genuine, and has to do with the incongruity inherent in the very concept of the romantic sage. We can see this from unsuccessful examples which purport to present us with a romantic sage. Thus, Jesus of the synoptic gospel is awkward as a sage, and so is Nietzsche's Zarathustra, when either is considered from the purely dramatic point of view. We have a good and romantic sage in old Moses in the *Book of Deuteronomy*; we have an echo of Moses in Seldon the founder and guide of Isaac Asimov's recent famous science fiction trilogy, *Foundation*. We have almost nothing in between in world literature worthy of mention as a romantic sage. The real difficulty is not dramatic but philosophical. The romantic lore's difficulties can be overcome by ingenious artistic devices within the romantic philosophy. But the limitation of the romantic philosophy cannot be transcended unless we transcend romanticism itself. Let me explain.

Romanticism as a philosophy presents a dramatic tension between the individual and society. Ordinary conservative philosophy plainly sides with society and against the individual every time the two are in conflict. The conservative paradigms are, aptly, the tragic young heretic (Absalom), the loyal youth on his way to success and glory (David, Solomon), the father figure (David, Solomon), and last but not least the sage who embodies the tradition. The radicalist movement of the eighteenth century, of the Enlightenment or Age of Reason, tries to do away with tradition as a whole, and the *radicalist paradigm* has *Prometheus* as a model. The romantic movement of the nineteenth century has inherited this much and the idea of progress from the radicalist movement. Nineteenth-century romanticism is not only conservative but also backward-looking and reactionary – reacting to radicalism, that is – yet it is also progressivist. And so it allows a *romantic paradigm of a rebel*. It allows the young hero to break away from tradition and to create a new one – or rather a new variant on the old one (since radicalism, or starting afresh, is denied). The young hero is the only one permitted to stand up to the old sage – and even that is only in retrospect. He has to be made of pure gold, he has to be kind and dedicated, to suffer a trial by ordeal – loneliness and all that – and he must, absolutely must, be a man of genius. There are two strong reasons why the young rebel must be so outstanding. First, his excellence resolves the conflict between the conservatism and the progressivism within the romantic philosophy: the rebel is a progressive force, but he must be idolized, in order that there will be no widespread attempt to emulate him: the greatest majority are followers, and only a few can be leaders. Second, only the most excellent can transcend the tradition, yet thereby enrich it rather than destroy it.

Here is the reason, finally, that young rebellion is only permitted in retrospect. As a traditionalist, the romantic can only judge with the aid of traditional criteria. Hence, all rebels are judged as culpable, and should be so judged. But as a progressivist the romantic permits some breakthroughs, some rebellions which lead to innovation, to the rise of new traditions. And by the new criterion of the new tradition the rebel who has founded the tradition is a hero. Hence, both the oppressor of the young rebel, and his later worshippers, are absolutely right. Hence truth is relative – relative to the tradition, that is.

Can we rectify this and recognize a rebel-hero when he is still a rebel? Can we, in other words, recognize the one rebel, among the many rebels, who is destined for success? Surely that will both reduce the injustice to him and the pains of the revolution. The idea that successful revolutions are predictable amounts to the idea that there may be predictions in history, that we can write the broad outline of the history of the future of a tradition. This embraces the doctrine of historical inevitability, but the converse is not true: Hegel accepted the claim that historical processes are inevitable, but declared historical prediction impossible. Indeed, we can see his point: historical prediction will help us deromanticize the young rebel since it makes it no longer essential to have him pass a trial by ordeal. Through science, thus, the romantic hero becomes the *deromanticized rebel in the Marxist paradigm of a hero, a scientific revolutionary*.

Since we live in an era which has little sympathy with romanticism, I shall not criticize any of the doctrines I have just outlined. Rather, I wish to indicate that the advent of romanticism and its offshoots in the nineteenth century could not influence the scientific tradition and the lore of science – except marginally and stealthily. The romantic hero could be an artist or a soldier, a social reformer or a political rebel. But he could not be a scientist. When a medical man, for example, became a hero, it was not in his medical activity, in his advancing science as such, that he showed heroism. It was in his fight with the politicians, in his attempt to implement his innovation, that he met all the obstacles and his (Semmelweis') martyrdom. The split between 'the two cultures', the scientific and the artistic, is rooted in this contrast, as Michael Polanyi had observed: the tradition of natural science remained radicalist and so stood as an odd man out. The social scientists both demanded the reliance on some (irrational) traditions and the break-away from them on occasion. Science has use for neither. However strong a social or an artistic tradition was, there was always the possibility of improving it by a rare and most excellent rebellion – indeed, the very excellence of a tradition made people follow it closely and so made it stagnate and so made room for a rebellion. Not so in the house of science, where the sage was absolutely right every time by the very definition of science as demonstrable knowledge: science had one revolution to end all revolutions; it created a tradition free of all arbitrary traditions; hence it needed no rebel, not even a rebel of genius. And genius was anyway quite inessential: any intelligent person was capable of joining the house of science, where all is serenity and where no conflict is allowed. Men of science, like Semmelweis, even Pasteur and Lister, could be heroes who fought superstition, ignorance, ill will; but their fight is in the public at large: within the house of science there was no rebellion possible.

3. THE ROMANTIC THEORY OF SCIENCE

This, then, is an historical sketch within which I wish to place Polanyi's theory of science, of the hero of science, of the genius of science. The first thing to notice is that Polanyi is anti-radicalist, yet he maintains much of the radicalist image of science: where order and harmony reign almost supreme even though claims for finality of every item of scientific doctrine are no longer maintained, and even an occasional revolution is allowed. Let me stress, further, that Polanyi distinctly rejects the image of the young hero, even though his own case history would have served as an example for it had he chosen a more romantic philosophy.

In 1914, Polanyi himself, a not so very young physical chemist – he had abandoned a medical career beforehand – published a theory which was rejected politely. 'I survived the occasion' he tells us (*Knowing and Being*, p. 89), 'only by the skin of my teeth.' Strangely the theory has now been accepted for a decade or more, by the community of physical chemists: the obstacles to its acceptance had been removed for reasons that had nothing to do with Polanyi – perhaps because in the 1940s Polanyi himself left the battlefield and changed to a third profession, namely sociology.

What is remarkable about all this is Polanyi's philosophical attitude. He does not feel triumphant, he does not shout 'I told you so!', perhaps because he had left the battlefield before the day of victory. Nor does he complain. The theory he was offering ran contrary to current central views in physics, views on which the main investigations of the time rested. All of which, as he disarmingly tells us, he was plainly ignorant of. In the light of those investigations Polanyi's results seemed implausible and were therefore overlooked – erroneously but understandably. Let me quote Polanyi's chief point.

The dangers of suppressing or disregarding evidence that runs counter to orthodox views about the nature of things are, of course, notorious, and they have often proved disastrous. Science allows some measure of dissent from its orthodoxy. But scientific opinion has to consider and decide, at its own ultimate risk, how far it can allow such tolerance to go, if it is not to admit for publication so much nonsense that scientific journals are rendered worthless thereby.

These are strong words, supporting enlightened censorship not in the community at large, heaven forbid, but in the scientific community. The community at large, I understand, must permit the publication of such nonsense; the scientific community cannot. Who is better off? Is it possible that a serious young rebel will be rejected by the scientific establishment, defy the sages, publish privately, and then win the day? Polanyi, in the above quote, concedes the possibility. But not the likelihood. Interestingly, he is quite right: we do have odd cases of rebels winning from without, such as John Herapath, Michael Ventris, or even Pierre Duhem. And we likewise have odd rebels who failed to win only because they could publish only outside the professional literature which was overlooked – such as John Newland. But more often than not even ostracized rebels like the young Thomas Young, or Humphry Davy, or the old Faraday, published in the community's periodicals, not outside them, and won their battles from within, not from without.

'Discipline *must* remain severe and *is* in fact severe', continues Polanyi's passage. He explains that he got his 1914 theory published and accepted as a Ph.D. thesis by luck, that as a professor of physical chemistry he could not teach his theory to his students since external examiners demanded knowledge of current views. 'The authority of current scientific opinion is indispensable to the discipline of scientific institutions . . . its functions are invaluable, even though its dangers are an unceasing menace to scientific progress.' Indeed, Polanyi ends his paper by saying that today there are examples of 'the most dangerous application of scientific authority' and he cites one.

All this is reminiscent of the recent controversy concerning the fact that in his academic capacity Galileo taught not Copernicanism but Ptolemaism. Can we subsume Galileo's case under the severe academic discipline of the year 1600 or will Polanyi deny that it was properly academic? Will he defend it, in short? I think he must defend it. Should he therefore turn against the rebel Galileo who broke out of Academia? Not necessarily. And indeed, he will not: *Galileo can be exempted on the ground of being a genius.*

Though anti-romantic, Polanyi uses genius like the romantics, in the same way and to the same end. He wishes to have some limited rebellion against the authoritarian system; and so he limits the rebellion quite *ad hoc* to those whom we can *post hoc* worship as men of genius.

Recently Thomas S. Kuhn has enriched Polanyi's system with the idea of a scientific paradigm. Though there is no universal scientific method, no compass and ruler for routine scientific work valid for all times and places, there are partial methods, valid for limited periods, and enabling the progress of dull and uninspired, or 'normal', scientific unquiry. From time to time the paradigm changes, new compasses and rulers are handed out, new scientific opinions are institutionalized. The authority of the establishment of science is maintained, but scientific opinion does change. How is it changed? Kuhn is vague on this point. He rightly speaks of difficulties which the old paradigm handles only inelegantly and in a cumbersone way. He rightly tells us of sleepless nights. And he rightly tells us of the recognition which the scientific leadership accords the new paradigm. The last point hints at a significant truth: the sleepless nights may be shared by sage and rebel alike; but the sage approves of the new paradigm, *ergo* the new paradigm comes from the rebel. How, Kuhn does not say; but the hint is that the rebel is quite a genius. The hint comes in the distinction between 'normal' science, i.e. dull routine science – which slavishly, almost mechanically, follows the paradigm – and the science of crisis which is pregnant with a new paradigm. And so, finally, we have arrived at the obvious and genius has been accorded some place in the fabric of science, even though the genius needs both talent and luck – the luck to be born in a moment of crisis

4. THE ROMANTIC SCIENTIFIC LEADERSHIP

There was a reason as to why it took so long for the philosophy of science to recognize genius: genius was introduced in the romantic manner, the manner that irrationally condemns all rebels in their day yet praises some rebels after they succeed. The irrationality of this is something that runs deeply against the scientific tradition. *The romantic has acceptance or popular success as the criterion of correctness, whereas the philosopher of science wishes to have correctness as the criterion for acceptance*. It may take time to prove the correctness of an idea in science but, the traditional philosopher of science insists, as long as the correctness of an item of scientific theory or information has not proven correct we all must suspend judgement, and when it has proven correct or incorrect we all must accept or reject it as the case may be.

As we shall soon see, the rejection of the traditional philosophy of science expounded in the previous paragraph enables us to develop a new view of genius. But before coming to that let us notice that objective criteria of acceptance militate against the romantic view of genius. Let us further notice that it is very hard to admit that genius – romantic genius or any other kind – is so useful in science that at times it may be even indispensable. As we say, science relies on objective criteria; and that implies, also, that we do not want science to depend too much on this individual or that. I propose that the root of this refusal is the desire to view science as devoid of all possibility whatsoever of establishing any authority of any kind, however benign. Suppose you reject the romantic view of genius and allow genius all the same. How are you going to recognize genius? We have the radicalist criteria that refuse to recognize genius, and the romantic criteria that clash with the radicalist ones. As long as we hold objective criteria for scientific acceptability, the question 'Does the accepted idea come from a genius?' becomes less important than 'Is the idea acceptable?'. And so, if you have a new idea, genius or not, you may propose it, anonymously if possible, and the scientific community will judge it on its merit, with no connection to who you are and what kind of person you are.

This kind of utter impersonality is linked – rightly, I think – in everybody's mind with radicalism, absolute democracy and the absence of all authority. It is traditionally also linked with science. Polanyi rejects this very link. Indeed, in the autobiographic essay I have already quoted, Polanyi disapprovingly quotes Bertrand Russell to say that there is no authority in matter scientific. Instead, he presents a view of a community of scientists, an elite, of people invested with 'personal knowledge'. With utter democracy gone, with the authority of institutionalized scientific opinion firmly established, with only a limited measure of dissent to check this authority, one may ask Polanyi, how does he think science progresses so well? – Or does it? One wonders why men of science prefer to publish in the highly – though not fully – censored scientific literature than in the much more open and relatively free book market?

Polanyi believes that every community exhibits both a certain degree of freedom and a certain degree of authority. And so, the question 'Why does a scientist remain a scientist?' is for him no different from a question regarding any member of any community. What he has added to his theory of science when he has developed, quite late in the day, his theory of genius, is the answer to the question, 'Who is the scientific leader?' The answer he gives is, 'The genius, the one who is better equipped to discover recondite scientific truths.'

Polanyi has not given up the idea of scientific freedom and democracy. He admits that the leader may be in error. He admits that the community is suspicious of its leadership and tests its results – thereby making them more accessible to more men of science. But the accessibility is a matter of degree, and so is the authority of the leading scientist – both *vis-à-vis* other, less keen-eyed scientists, and *vis-à-vis* the common man who can barely see a thing.

It is here, again, that Kuhn's contribution clarifies Polanyi's philosophy. There is little in the writings of either Polanyi or Kuhn about the layman – educated or otherwise – except for the claim that scientists know more than they can explain to laymen. The role of the common man, however, is given by Kuhn to the common, rank-and-file, run-of-the-mill, average, 'normal' scientist. He accepts the leader's authority and brings the leader's vision closer to the rank-and-file and its fruits to society at large. He also thereby exhausts that vision, thus opening the opportunity for the next vision of the next genius.

What, then, is the difference between the romantic and the new kind of hero?

First, the romantic hero appears to the society at large, whereas the new hero appears to his guild. Second, the romantic hero has to struggle and preferably die young. The new hero does not struggle but grows in the community of his colleagues, within his guild. And when time is ripe he takes over – like Bambi – and proves his mettle by leading the membership in a time of crisis – indeed, he is only in evidence in such times.

5. AGAINST SCIENTIFIC AUTHORITY

The question 'What role does or can genius play in science?' clearly hinges much on our views of both genius and science. Let me start, then, by rejecting the Polanyi-Kuhn authoritarian view of science, not so much because it is authoritarian, as it is contrary to what is valuable in science. Let me confess that in my opinion almost all 'normal' or routine science is of very little scientific value. (Some of what is called 'normal' science, incidentally, is dull science but exciting technology, or even exciting technology with no science to it at all.) The authority of scientific opinion, which Polanyi views as dangerous but highly beneficial – indeed quite indispensable – is, in my opinion, simply obnoxious. I think myself that genius will be more commonly recognized if we destroy some taboos and encourage everyone to develop his own way, *pro* or *contra* the established opinion as he may wish. We may recognize new kinds of genius all over the place, then. This will not re-establish equality, since not everyone is a genius of sorts; but it will go a long way towards it. It will go further if we agree to leave more open the question of who is a genius and who is not: almost everyone, after all, is possibly a genius of some sort, perhaps of an unheard-of sort. We can all *try* to rebel even though we cannot all be successful. This possibility should suffice to establish all freedom of dissent except that which is clearly conducive to violence. Polanyi supports censorship in order to protect the scientific literature from being flooded by worthless writings. I find this unacceptable. First, the flood of valueless writings is harmless and censorship is harmful. Second, censorship has failed to stop the flood but succeeded in lowering the standard since the censors do not dare to decide which unusual paper is outstanding and which is scandalous and they easily decide that an average paper is not scandalous. Third, scientists do not read what is published but what they choose to read or what they are told to read and so the flood of publications is immaterial. Fourth, and naively, men of science may want to have the chance to decide for themselves between competing views; or at least they may want to examine the criteria employed by their censors to make such momentous decisions. Which means that they do not necessarily have to wait for a crisis or a revolution to hear what is going on. Polanyi and Kuhn's system makes for the prevalence of inner circles, who have their information through private channels, and the scientific public which gets information as published after screening by worthy censors. For my part I think only this inner circle is scientific; when it grows it needs its own publications which cannot be secret in democratic countries. And so Polanyi's system of scientific censorship within democracy is an unstable phenomenon. Let us hope it is transitory.

So much for my rejection of Polanyi's image of science: it is too authoritarian in its description, since it ignores scientific defiance, and it overestimates the authoritarianism that does exist in science, since spirited scientists defy it, and it underestimates the democratic society in which science flourishes, since defiance of the establishment of science may take place outside the scientific establishment.

These points are all to be found in Popper's theory of science and in his theory of democracy. Let me briefly comment on them.

6. TOWARDS A RATIONAL THEORY OF GENIUS

My own theory of genius is a corollary of Popper's theory of science. It arises out of a very new theory of the relation of the scientist to his own tradition. In 'Towards a Rational Theory of Tradition', Popper summarizes his theory that the scientist is not trapped in his own intellectual tradition, because science is an institution which is constantly critical of itself and its intellectual presuppositions. He has extended this doctrine to provide an account of the Periclean theory of democracy: only a few may come forward with proposals, but we are all able to judge them. How does this solve the problem of genius?

The Romantics believed that genius was necessary for the overthrow of tradition. If science is self-critical, however, we do not need the romantic genius among us for that purpose. The Enlightenment made genius inessential because the scientific method produced results by (inductive) algorithm. If science is conjectures and refutations, as Popper maintains, then genius is not only not an embarrassment, it is essential. Instead of an egalitarian theory of mediocrity, or an elitist theory of genius, I propose an egalitarian theory of genius.

According to Popper, each single step in the advancement of science is a matter of genius, big or small. Popper himself says, at times, the democracy of science is assured by the fact that even a simple man can criticize a man of genius. This claim contains two sub-claims, an objectionable one and a true one. The true claim is that of impersonality, which has been criticized by Polanyi. The objectionable claim is that criticism is easier to generate than a conjecture; in truth things go sometimes this way sometimes the other. Arthur Koestler makes much the same point in his *The Lotus and the Robot*, where he says that even a successful choice of word in his writing is a matter of a minor inspiration. In brief, according to Popper's theory we are all inspirable, and in different directions and to different degrees.

Popper's theory of democracy comprises, I think, two insufficient theses. First, that of equality before the law. Second, of the possibility of the legal overthrow of government by peaceful means. I think even the fact that we try to equalize educational and medical services proves this hardly matters to the discussion of the quality of members of the scientific community. It does, however, pertain to the question 'Can a scientific community function within, say, a savage society kept savage by a small tyrannical cast?'. Functionalism tells us *a priori* that this is impossible; but functionalism is only a coarse approximation to the truth. The questions of social relations of science are still hardly explored. It is clear, however, that they exist, that they signify, even to the point of preventing science from totally stagnating. Contrary to Polanyi, I should say, unless scientific results get regularly vulgarized, the mediaeval picture he presents of science – with masters, apprentices and unarticulated personal knowledge and mystique – might come true and plunge us into a new Middle Ages.

As to the possible overthrow of the leadership, it is insufficient: the leadership must provide viable programmes to prevent frustration with democracy as such. In the scientific community, however, in spite of what passes today as the scientific establishment, there is no establishment and no leadership: the most the so-called establishment can do is, first to prevent the curious from finding some pioneering results, and second, to prevent the public from knowing what generally goes on in science – both mere delay actions. What the scientific establishment so-called really does is to keep the education of the general and the scientific public at its present abysmal level and to allocate public resources to the wrong hands. Here the Polanyi-Kuhn theory of science is descriptively quite correct. Anyway, this has to do with the social relations of science, not with the scientific community.

What the scientific establishment so-called does achieve, to be more explicit, is a trial by an educational ordeal which should bring only the best to the scientific community. In fact, however, the ordeal breaks the back of all those described by Kuhn as 'normal' scientists. The more 'normal' one is, the more one depends on the guidance of others, preferably the men of genius.

Once autonomy, both moral and intellectual, is achieved, we can return to the classic view and admire genius, and even admit almost everybody as a genius, big or small – i.e. inspired – and admit that certain summits can only be conquered by men of big genius, but appreciated by others as well.

The question of how we can achieve a moral autonomy (even while trained as scientists) is educational. The question of how we can achieve an intellectual autonomy is central to our discourse. It can only be solved by denying the scientific community any intellectual privileges, by admitting that no one is fully intellectually autonomous, that intellectual tutonomy is a matter of degree. That is to say, intellectual autonomy depends not only on one's readiness to be responsible for one's own choices – which is an all-or-nothing affair – but also on one's readiness to take the trouble to critically study the object of intellectual attention – which is a matter of degree and of circumstances. And here, I think, there is much truth in Polanyi's view that we take much of our intellectual make-up on trust; but his trusting of the expert is, I think, quite undemocratic and anti-autonomous. It is here that the genius of the vulgarizer does a lot to keep up the intellectual autonomy of the community at large – including the scientists. Here Popper's view that people are best when constantly under surveillance can be wedded to Polanyi's theory of the genius, namely the genius vulgarizer of science.

It is a social fact that democratic science vulgarizes, that mediaeval science is that of master and apprentice. Polanyi's theory of science as a mediaeval guild is empirically refutable. His theory of genius, then, has to be expounded against the background of a more democratic theory of science to retain its initial great appeal.

NOTES

[1] In his *Critique of Judgment* Kant describes genius as discrimination or connoisseurship plus imagination. It is clear that he views this a matter of degree, whereas romanticism views genius as a new quality.

[2] In his *Chemical Manipulation* of 1827 Michael Faraday quotes Dr. Marcet to say, Since I spend so much time in my laboratory, it should be housed in the best room in my house, not in the worst one. Nevertheless, Faraday's own laboratory was in the dark basement of the Royal Institution: Faraday belonged to the earliest generation of professional scientists, but he consciously and conscientiously emulated the old paradigm of the philosopher, of the gentleman scientist. He even refused to draw patent applications or to be paid for his commissions as consultant. His intellectual sources were Boyle's works, as well as those of Dr. Isaac Watts.

BIBLIOGRAPHY

Agassi, J., 'Methodological Individualism', *British Journal of Sociology* **11** (1960), 244–70.

Asassi, J., 'Institutional Individualism', *British Journal of Sociology* **26** (1975), 144–155.

Agassi, J., *Towards a Rational Philosophical Anthropology*, Martinus Nijhoff, The Hague, 1977.

Agassi J., 'Sociologism in Philosophy of Science', *Metaphilosophy* **3** (1972), reprinted here as Chapter 8.

Asimov, I., *Foundation*, 1st edition, New York, 1951.

Brewster, Sir D., *Memoirs of the Life, Writings, and Discoveries of Sir Isaac Newton*, Edinburgh, 1885.

Chandler, R., 'The Simple Art of Murder', in *The Second Chandler Omnibus*, London, 1968, c. 1962.

de Morgan, A., *Newton: His Friend: and His Niece*, London, 1885, pp. 140–8.

Ellis, R. L., General Preface to *The Works of Francis Bacon*, Vol. I, London, 1857, p. 25.

Fulton, J. F., 'Robert Boyle and His Influence on Thought in the Seventeenth Century', *Isis* 18 (1932), 77–102.

Hammett, D., 'Tulip', in *The Continental Operator*, New York, 1974.

Herschel, J., *A Preliminary Discourse on the Study of Natural Philosophy*, London, 1830.

Koestler, A., *The Lotus and the Robot*, New York, 1961.

Kuhn, T. S., *Structure of Scientific Revolutions*, International Encyclopedia of Unified Science, Vol. II, no. 2, Chicago, Second edn. 1970.

Lakatos, I. and Musgrave, A., *Criticism and the Growth of Knowledge, Proceedings of the International Colloquium in the Philosophy of Science*, London, 1965, Vol. IV, Cambridge, 1970.

Macaulay, T., 'Lord Bacon', in *Critical and Historical Essays*, Vol. II, London, 1865.
Manuel, F. E., *A Portrait of Isaac Newton*, Cambridge, 1968.
Maxwell, J. C., *Scientific Papers*, Cambridge, 1890, esp. 'On Action at a Distance'.
Murray, G., *Five Stages of Greek Religion*, New York, 1925.
Ortega y Gasset, J., *The Revolt of the Masses*, New York, 1960.
Plekanov, G. V., *The Role of the Individual in History*, New York, 1967.
Polanyi, M., 'Genius in Science', in *Encounter*, January 1972, pp. 43–50; also in R. S. Cohen and M. W. Wartofsky, eds., *Method and History: Essays in the Natural and Social Sciences, Boston Studies in the Philosophy of Science*, Vol. XIV, 1974, pp. 57–72, and *Archives de l'Institut International des Sciences Théorètiques*, Tom. 18, *De La Méthode, Méthodologies Particulières et Méthodologie en Général*, Bruxelles, 1972, pp. 11–25.
Polanyi, M., *Knowing and Being*; essays, edited by Marjorie Grene, Chicago, 1969.
Polanyi, M., *Personal Knowledge, towards a Post-critical Philosophy*, New York, 1964.
Polanyi, M., *The Study of Man*, Chicago, 1963.
Popper, K. R., *Conjectures and Refutations*, London and New York, 1962, 1963. Reprint, Harper Torchbook Edition.
Popper, K. R., *The Open Society and Its Enemies*, in 2 volumes, London; Princeton, one-volume edition, 1952. 4th revised edition, 1960. Reprint, Harper Torchbook Edition.
Salten, F., *Bambi*, 1923.
Smiles, S., *Self-Help*, London, 1862.
Todhunter, I., *History of the Mathematical Theory of Probability from the Time of Pascal to That of Laplace*, Cambridge, 1865.
Tyndall, J., 'Scientific Use of the Imagination', in *Fragments of Science; a Series of Detached Essays, Addresses, and Reviews*, New York, 1897.
Watts, Dr. I., *The Improvement of the Mind*, London, 1809.

P.S. I had intended to include a movie list which pertains to the present study, including names of such movies on artists as *Beethoven, Rembrandt*, and *Lust for Life* (Van Gogh); on scientists, such as *Life of Louis Pasteur, Madame Curie, Dr. Ehrlich's Magic Bullet*, and most impressive, *Freud*; and others such as *Scott of the Antarctic* and *Edison the Man*, or even *The Amazing Dr Clitterhouse*. Each of these movies offers an insight to the influential views of genius, and some of them even offer an insight to the working of the genius' mind. I simply cannot say more here.

CHAPTER 16

SCIENTISTS AS SLEEPWALKERS

1. BETWEEN LUCK AND WIT

Does the research worker know what he is doing, and to what extent? I shall not answer this problem but merely present and explain it.

The above title averts to Arthur Koestler's *The Sleepwalkers*, whose subtitle is: *A history of man's changing image of the universe*. According to Koestler, it seems, scientists are like sleepwalkers: ignorant of what they do, they do it with perfect assurance and with success. But I am not quite clear about that. Only in one passage does Koestler speak openly on this issue. He likens Kepler to a sleepwalker there, and in that manner: with the sleepwalker's assurance, etc. That passage, by the way, concerns two errors of computation which Kepler made and which cancelled each other out. This startling fact is explicable, and hence it is much less startling than it seems. Kepler calculated to a very high degree of precision, far beyond factors which might influence the accuracy of his results. Indeed, by sheer probabilities, he could have made a few more computational errors without being much the worse for it. It seems, then, that Koestler's likening of scientists to sleepwalkers fizzles out at once.

Let me stress that it is not at all clear whether Koestler likens Kepler elsewhere to a sleepwalker, and whether he likens other men of science to sleepwalkers. But it looks that way. Koestler takes seriously Kepler's speculation of the *Mysterium*, according to which the planetary orbits are captured in perfect polygons, and Kepler's hope that this would explain the number of the planets and the ratios of the distances of the planets from the sun. Koestler says that this was Kepler's *leitmotif*, the erroneous but fruitful idea, the prejudice which miraculously led him to his destination – rather than to blind alleys – much like a sleepwalker, I suppose.

I shall go further. I will say that Koestler's metaphor is part illusion part truism, yet a truism well-worth stressing. The illusion is that a sleepwalker who has survived all the hair-raising obstacles and arrived has arrived because he was predestined to arrive. Though Koestler does not assert the doctrine of predestination, he repeatedly harps on it – and harping on a popular prejudice is worse than explicitly asserting it. So much for the illusion. There is the truism that sleepwalking or awake we walk on a tightrope and that looking hard at the abyss is dangerous. It looks as if this amounts to saying, ignorance is bliss. Of course it is not, since we may more easily avoid risks when we know them; but we need some optimism as a better method than looking hard at danger all day long.

Consider the fact that Kepler's speculation is false. Had he known it was false he would not have undertaken his researches. Hence ignorance is bliss. So suggests Koestler. I say "suggests", as I am not clear about Koestler's thesis. Anyway, the suggestion is not convincing. Had Kepler known his speculation to be false he might have tried a better one, or he might not. We do not know. Moreover, as his speculation is false, though it spurred him on to do research it also blocked his progress. Indeed, Kepler was so prolific in producing his ideas and publishing them, he was so careless at times with his deductions, that it is all too obvious that he repeatedly found himself in blind alleys and said so out loud. Koestler quotes him occasionally admitting that he was in a blind alley, and at one time, as I said, he speaks of him going to his destination with the assurance of a sleepwalker. Koestler, it seems, cannot lose. Kepler's great progress proves he had to progress and Kepler's retardation proves he had to progress in spite of all obstacles. As I say, the only hard evidence Koestler has is that Kepler did make great discoveries, that he also goofed many times, that his main idea which was the motivation for his actions, successful and failed ones, was a failure.

Well, we all know of tactical success in the midst of strategic failure. What of it? So much for Kepler. But Koestler is in error about sleepwalkers too. Admittedly they are assured; yet successful they are not: they too meet with blind alleys and other mishaps. One does not have to be a sleepwalker to find oneself in a blind alley or to meet with an accident; but it helps. And so, the very image of Koestler of a scientist moving towards his unknown goal quite successfully yet closed-eyed is not really the truth even about sleepwalkers as such, only about an amazing feat of an amazing sleepwalker, about a quaint fact which has caught people's fancy.

Let us forget the lucky sleepwalker and ask, how does the scientist arrive? Is it by luck? Is it by predestination? Is it in any other way? Is it perhaps the case that quite unlike the sleepwalker the scientist knows what he is doing?

There is a theory of science which declares the scientist to be the one who has the sixth sense to guess right more often than others – and to guess, not to deduce. If science were merely mechanical deduction, so the argument goes, it would better be performed by computers; and if it were to make in a series all the errors there are to make, then common men would be better at it than men of science. No, says this theory; there are too many deductions to make and too many guesses to explore; the scientist senses, smells the avenues worth exploring, the paths worth taking. Science, thus, is an amazing feat, and it justly catches people's fancy.

This theory says, we are all blind, we are all sleepwalkers; most of us naturally end up in blind alleys, and some end up at the right place – and thus earn the title of scientists. It is quite possible that this is what Koestler has in mind. It has been adumbrated already by many writers, including Knoblauch, Tyndall, Sir Oliver Lodge, and others. It was incorporated, I suppose, in the philosophy of Michael Polanyi and echoed by Thomas S. Kuhn. I suppose that to the extent that Koestler propagates a consistent doctrine he may have this very suggestion in mind.

For my part I find this suggestion not very interesting, as it is permeated with materialism, namely the worship of success: We all fumble and some of us nonetheless arrive; hurray for them! I shall not stoop so low as to attack such a doctrine. Were Koestler advocating nothing but materialism, nothing but success worship, his work would not be so important. Admittedly, his approval of Kepler made commentators dislike his chapters on Kepler and his disapproval of Galileo made the same commentators dislike his chapters on Galileo. This, however, only makes Koestler's commentators more materialistic, more success-worshippers, than he is. But I still wonder, what does Koestler's *Sleepwalkers* tell us about science in general? It tells us that all astronomers up to Kepler's contemporaries loved circles but he broke the magic and invented the elliptical orbit. It tells us that science is a mixture of the rational and the irrational, as practiced by the Pythagoreans and their successors up to and including Kepler; but excluding Galileo and his followers who attacked religion as irrational and who wished to see science as purely rational.

I do not know how these two theses relate; how can it be both that circles are all bad and yet that they illustrate partial rationality which is good; how was Galileo alone too bad because he held the circle; is this *because* he advocated a purer rationality; how the pure rationality of modern science is all bad even though science developed so much. Koestler himself notes the great modern development – but only when he compares mankind to a psychopath – morally backward and physically strong enough to be self-destructive. It seems that Koestler only notices a fact when it helps him moralize. But I must leave Koestler now and center on pure rationality versus partial rationality, as it may hopefully offer us the key to his appealing metaphor of the sleepwalker.

The appeal of the metaphor is partly in its paradox: the sleepwalker is ever so blind, so poor in rationality, and the scientist is ever so rational. Partly the appeal is in a fresh idea which the metaphor contains. We are so used to contrasting rationality or reason with irrationality or unreason, that we forget the more commonsense idea of reason, the contrasting of reason with stupidity and the fruit of reason with sheer luck. In Yiddish the successful is normally expected to express humility by applying the expression, "with more luck than wit." Here the fruit of reason, the wages of hard thinking, is contrasted with grace: success may be a reward for thinking or not: – the Lord, says the Psalmist, protecteth fools. Within such a universe of discourse, there is no room for irrationalism; even the irrationality of the hothead is here viewed as but one example of a case where not all possible foresight was employed: irrationality, from the viewpoint of common sense, is but a rationalization of foolishness, and foolishness is but an affliction.

Let me elaborate a bit on this common sense idea and show the difficulties inherent in any attempt to apply it to scientific research proper. What I like about this idea is its immense common sense. From the common sense viewpoint rationality is so straightforward and desirable that irrationality is automatically placed not as a rival but as an affliction – as a matter of course, of course. But the rationality which can be taken so much for granted, itself takes too much for granted; it is too naive to be of use for science. For, when common sense contrasts wit with luck, it implies that wit – being foresight – leads to success without luck; that wit, in other words, must deliver the goods repeatedly and at will.

Let us take a simple example. If a patient recovers from a fatal disease, for example, we immediately employ common sense and say, either it is by luck, or the doctor used his foresight – he has a new method of cure. We do not distinguish between a recovery due to a lucky event in the domain of the patient or in that of the doctor's; we distinguish between the case of an accident which has caused the cure – by luck – and the foreseeable case, the one based on a routine which causes the cure, which surely is more a matter of wit than of luck. Of course, if the doctor's routine increases the chance of recovery from very low to very high, then, surely, we will all say it is not luck but wit. Even if the doctor's routine merely increases the chance of recovery, say from 30 to 60 percent, *if it does so regularly*, if it *is* routine, then the recovery is in part by wit, indeed, 30 percent wit. We can mix wit and luck, then; but they do not overlap. Or do they?

2. SCIENCE AND COMMON SENSE

For now comes, indeed, the sixty-four dollar question. Given the method of cure, we say, a given cure can be foreseen as the effect is repeatable, and it thereby makes the cure more a matter of wit and less a matter of luck. But is the discovery of the method of cure itself a matter of wit, or is it a matter of luck? When I am in a cautious mood, I tend to say, common sense is limited in its application to common phenomena, and the discovery of a new cure is anything but common; and so, my question does not obtain. In a bold mood one may repeat the Einstein-Born hypothesis: science is but an extension of common sense. This hypothesis is very bold, and perhaps even false. It is in any case fascinating and worth pursuing for a while.

In our own case, pursuing the Einstein-Born hypothesis will force us to say, if a discoverer can discover at will, he does so more by wit than by luck; and if it is by luck then it is not producible at will. This fits extremely well with Einstein's claim that his successes were more due to luck than to wit: he had little control, he said, over his results. I do not think many people would accuse Einstein of false humility, even of the false humility which is accepted in polite society as *comme il faut*; it is well known that Einstein's humility was both considerable and sincere – it was the humility becoming to people as members of the species. We do not have a scientific method, and so all discovery is a matter of luck.

Before debating this point, let us acknowledge that we owe it to Sir Francis Bacon, one of the most brilliant and ingenious philosophers of all times. He was a man in many ways the opposite of Einstein: cruel, dishonest, quarrelsome, vain; a strange mixture of conceit and arrogance, and unlike Einstein he believed in scientific method. Yet, like Einstein he was brilliant, and like Einstein, he contrasted wit with luck; even like Einstein, he put ingenuity on the side of luck rather than of wit.

Discovery, says Bacon, can be compared with the drawing of a straight line or of a perfect circle. With a ruler or a compass you can perform this task repeatedly and with ease. The knowledgeable draftsman is usually the one who uses these tools and with their aid performs the task repeatedly and at will. Without the tools, one needs immense luck to perform them, to draw reasonably straight lines or round circles. Bacon, unlike Einstein, thought that scientific method was possible. He also thought that it is given only to the one who is humble enough yet ambitious, nimble enough though cautious, etc. And, he humbly confessed, he was lucky enough to fit this description very well and so he intended to become a great discoverer – if he only had a bit of time and a few assistants. These he unfortunately never had, much to the lament of his disciples, including the great Amos Comenius.

The drawing of lines and circles according to wit is an art given to one schooled in the trade of draftsmanship. Also, by sheer luck, on occasion, the same art is given to any Tom, Dick, or Harry; perhaps for a moment or two in his life. Also, on a rare occasion, we do meet a gifted person who, without schooling at all, has the same art for his whole lifetime: he can deliver the goods at will. His gift or his knack *is* his luck: though he can reproduce the job at will, he himself is a lucky streak; his knack, to be precise, is. We cannot, that is, produce at will people who, without schooling, have the ability to discover. And this, by our very criterion, makes the existence of people with a knack a matter of luck, even though a knack is something which enables its possessor to repeat his performance at will.

To show how common sense this view of Bacon's is, let us take a simple problem. Take a trade school – say, Harvard University Architecture Department. Suppose it produces streams of able draftsmen every year. Is this a matter of wit or luck? In a sense our question is already answered: since the able draftsmen are produced regularly, clearly Harvard University Architecture Department exhibits a certain wit, which we call educational proficiency or such. But we may question this. We may say, one in a thousand is, by sheer luck, a born draftsman, and he naturally gravitates towards Harvard, the Fair Haven of the born draftsmen. If this were so, – personally I will not enter this debate – clearly, we would all agree, Harvard only can draw the best, not produce the best. It does have wit, we would still admit; namely, the wit of repeatedly drawing the best, of repeatedly creating the illusion that it repeatedly creates the best (rather than attract them), etc. Right or wrong, this discussion as I have outlined it is very common sense, regardless of the question of the truth or falsity of specific claims it contains. Also it is in complete agreement with both Bacon and Einstein about the contrast between wit and luck. If so, then we can take what the two thinkers agree about and approach their central disagreement. As it happens, we all today agree with Einstein and disagree with Bacon. There is no scientific method, no science-making algorithm; and therefore, we must conclude with the philosopher Einstein, that the scientist Einstein – or any other scientist – is successful more due to luck than to wit.

There is an obvious objection to this, which I shall briefly note before winding up my present commentary on a Yiddish expression (regarding luck and wit). The idea that there is no science-making algorithm, one may say, overlooks the fact that Einstein made quite a few discoveries, whereas most of us make none. Though there is no science-making algorithm, Einstein had a higher than average chance of making a discovery. Admittedly, Einstein had no guarantee – the like of which Bacon had vainly promised – that effort over a given span of time would lead to discovery. Yet, he had a higher than average likelihood, and this is better than mere luck. Just as a doctor would prove his use of his wit by a high success ratio even if he could not give perfect cure procedure, so the very existence of an Einstein does prove that something like a science-making algorithm does exist – perhaps some partial algorithms.

So much for the objection. It has been answered here already. I shall repeat the answer. Never mind the question "do partial algorithms exist?" Had we been able to produce an Einstein every generation we would, thereby, create something akin to a science-making machine – whether because Einsteins do use partial algorithms or for any other reason. But we cannot produce an Einstein. Let us assume, from now on, that an Einstein does and will appear in every generation. It is still doubtful that the objection holds. For, it is one thing to have by luck one Einstein in every generation of billions of offspring and another thing to say that we produce him. Of course, there are people who say that Einsteins are produced in the best intellectual hothouses. The facts speak differently. And so, Einsteins, to conclude my answer, are not produced: they just happen.

3. *APRIORISM* IS ALIVE AND WELL

My answer to the objection is not complete. We may nevertheless insist that we do produce an Einstein every now and then, a big one or a small one, at long intervals or short ones, simply by what is euphemistically known as replenishing the earth. Since this sounds so very silly an idea, let me draw attention to the fact that this is what a well-known philosophy, namely apriorism, amounts to. If knowledge is inborn, then the only way to bring about knowledge is to bring about birth, the only creation of knowledge is procreation. Since apriorism is so commonly held to be a rather silly and old-fashioned doctrine, I shall now argue that it is far from being extinct or even unpopular – it has simply gone underground.

Descartes lived at the peak of the Renaissance, when the sense of awakening was the strongest, when it was easiest for a thinker to say, my teachers were great ignoramuses, and I am a great scholar – my knowledge is truly mine. And yet, this very stance raises a brow: what is so special about Descartes that he should be the first learned man? How come he, of all people, knew more than all his teachers put together? Descartes must have been asked this question. He said, seemingly in reply, that being so special is not so surprising; for example, that a town planned by one man is better than one planned by many different people, which is to say an unplanned town. Descartes could say, perhaps, too many cooks spoil the broth. Yet, no doubt, all this is but an aside. For a town or a broth reflects one man's taste, whereas knowledge reflects not tastes but the truth. What made Descartes so special, said Descartes, is the fact that he was methodical. Here we come again to Bacon's ruler and compass idea. There is a difference between Bacon's detailed views of method and Descartes'. To the insider the details are of supreme importance; to the outsider what matters more is the fact that they both recommended a conscious and methodical application of a fixed and simple set of rules. Indeed, for Gassendi, not even much of an outsider, Descartes was an immediate disciple of Bacon; for Mersenne the difference was more troublesome, and Descartes tried to pacify him.

The point to stress is that both Bacon and Descartes explained the allegedly sharp transition from the depth of the Dark Ages to the enlightenment of today: We have an algorithm, we know that we have an algorithm, and so we can start cranking the computer. Bacon and Descartes both said explicitly that the knowledge of the method must be conscious, and so must be its rigorous use. In a moving passage, in his posthumous fragment *Valerius Terminus*, Bacon says, unless one knows the seats and pores and passages of the mind, one is not qualified to do research. Clearly, this raises serious doubts about the starting point: how do we do research to find out the algorithm if we need the algorithm to do research?

My intention here is not so much to raise doubts about Bacon and Descartes, however, nor to question their apriorism. My intention is to notice their stress on the need for an explicit knowledge, as well as for conscious and extremely methodical application of set rules; this stress is too much common sense to require comment. But let me give you a simile. We do have computers, and at times we do use them intelligently, whereas our forefathers had no computers and many of our own contemporaries just fiddle with their computers until they run out of public funds. Why? What is so special about us? How come only we use computers properly? Obviously, we know the rules, and we use them properly to build and employ computers. This much, one must notice, is our seventeenth century heritage. We view our privileged ability to use properly our computers the way the seventeenth century pioneers viewed their privileged ability to use properly their minds. Here I must speak briefly of partial algorithms – a topic on which I have already given a few papers. That these algorithms exist is a fact, although they are not always as precise as to be called algorithms proper. But analyzing and synthesizing chemicals is a good example for them; some people devote their scientific careers to jobs which yield results with regularity.

Polanyi and Kuhn view this as irrational as there is no guarantee for the success of the algorithm. I find this hypersophisticated since by common sense following an algorithm is rational, repeatable etc. And if the algorithm is exhausted, then the one who applies it and fails for the first time is extremely lucky as he found – by serendipity – a much greater find than the one he was looking for, to wit, the limit of the partial algorithm.

I do not wish to endorse the view of Polanyi and Kuhn about normal science. Contrary to their view, I assert that many ordinary scientists do little of any significance, and that many ordinary scientists perform small scientific revolutions, i.e. refute small accepted scientific hypotheses. But I do agree with them that some scientists follow partial algorithms – a fact which I consider rational and they view as irrational.

I have mentioned all this not so much in order to criticize Kuhn as to illustrate a common sense view of rationality which, when over-generalized in an over-optimistic mood leads us straight to Bacon's and to Descartes' view of science as the epitome of rationality. There is, then, an element in these people's view which exceeds common sense by over-stretching it. Yet they had a strong common sense aspect to their philosophy of science which we may profitably keep in mind.

Then came Newton, and what method he used God only knows. What he said about method is either puzzling or not worthy of his great intellect – as E. A. Burtt has noticed. To take one example, he said that we have to adopt a generalization which is not violated by facts until it is, and then cling to it with proper qualifications. This would make a European say all swans are white except those in Australia, and it would make the Australian say, all swans are black except, etc.

Perhaps I am unfair to Newton; perhaps his idea of generalization has nothing to do with the color of swans. It was a bold statement that Max Jammer and Stephen Toulmin made on a different occasion, to the effect that "all swans are white" is unscientific. I do tend to agree, but will not say whether Jammer or Toulmin himself would stick to this extremist and highly Kantian remark.

Kant said that only a generalization cast in the conceptual scheme of science is worthy of the name of a scientific experience. Before we can generalize about Mars's orbit, for example, we must have developed our mathematical framework so as to have the concept of an ellipse. Kant's theory, I suggest, fits Newton's *Mathematical Principles of Natural Philosophy* best: we build a framework, as in Newton's first two books, and then generalize within it, as he does in his third book. Yet Kant's theory is also a disaster: there is no doubt that Newton himself would not have accepted Kant's analysis of Newton's intellect any more than he would accept a Freudian analysis of his psyche.

It is well known that Kant was weak about the application of his theory of knowledge to the history of science. He mumbled some traditional apriorist remarks about observations awakening knowledge which is, however, *a priori* valid. His chief contemporary critic, Solomon Maimon, flatly declared inapplicability to history to be the chief weakness of Kant's theory of knowledge. This, like most of Maimon's contributions, was overlooked.

The root of the trouble is, of course, in the difficulty we have seeing a man systematically planning to do one thing and doing another. Freud has meanwhile shown us that we can act this way and make fairly good sense. But this is all on a rather primitive level. On a more sophisticated level the subconscious and the id are just too dumb. We can imagine, thanks to Freud, a man seemingly concerned with problems of other people, yet with personal interest at heart. But we cannot, in any like manner, imagine a man thinking he plays a piano while he effectively programs a computer – or vice versa. It is too much even for a sleepwalker to follow Kant while thinking that he follows Bacon. Things get so bad, that the best advice is often to ignore the question altogether. It came to me like a brick on the head that nowhere in I. B. Cohen's *Franklin and Newton* is there a discussion of the fact that the method these gentlemen had allegedly employed is different from what they said they had employed, that indeed they staunchly opposed the method Cohen ascribes to them. The problem is just overlooked by almost no reference at all to this puzzling fact, and by engagement in polemics with those who today share the view of Newton and Franklin on science.

It was Pierre Duhem, I think, who started it all, but at least he was very explicit about what he was doing. And Sir Karl Popper gave Duhem's idea the rubber stamp of approval, while using Einstein's famous joke. Do not listen to what a scientist says he does, said Einstein, look at what in fact he does. This was merely the application of the inductive method against Newton's claim that he had employed the inductive method – or rather against the claim that Newton knew best what Newton did. Clearly this is untrue, and Einstein's joke is the reduction of the claim to an absurdity. Einstein himself, of course, meant no more than a criticism, as he makes clear in his comment on his joke: he says, this should silence me too, but I shall not be silenced. The question remains, if Newton thought he was employing one method, how could he employ another, and systematically so?

As I said, psychology cannot offer an answer, as subconscious tricks are not sophisticated enough. But social anthropology can offer something more sophisticated since when we follow a tradition we may be following something more sophisticated than we can produce ourselves. As Malinowski showed empirically in the case of the Kula in the Trobindian Islands, tribal rituals may be a sophisticated method of keeping the economy going, yet without the knowledge of the practitioners who ascribe to the ritual a mere magical symbolic meaning. To apply this functionalism, as it is called, to our case is not difficult. The model here is not the scientist operating a computer while thinking he plays Bach, but the apprentice who has fairly mystical ideas about the computer, yet who operates, to the extent that he does, impeccably. I have mentioned the apprentice in allusion to Michael Polanyi. The leaders of the scientific community, unlike other chieftains, have to innovate. Thereby, Polanyi adds, they take the chance: if the herd does not follow them they lose their positions as leaders. This sounds so trite, perhaps even tautologous, that one can understand Kuhn's failure to mention it. Of course, if you are not followed you lose your leadership. Yet looking at it thus, Kuhn fails to mention, and his readers thus lose, the incredible hypothesis concealed in the discussion: when the leader makes a good move he is followed by the herd, but his bad move is not. Hence, the mob of scientists, the normal scientists so-called, have the sleepwalker's assurance: ignorant and dogmatic as they admittedly are, nonetheless they follow the right leader to their destination. This is part and parcel of the Polanyi-Kuhn lore.

And so, the introduction of social anthropology into the picture only replaces the individual sleepwalker by a crowd of sleepwalkers. This is true not by virtue of our subject matter but by virtue of the subject matter of anthropology: not only Duhem created a problem, but Malinowski created its analogue. Indeed, he even felt the problem, for he tried to solve it – by invoking the name of Darwin. I do feel, by the way, it is high time we coin the phrase "Darwin ex machina." Natural selection, says Malinowski, eliminates the unfunctional cultures, so that all existing cultures are, *eo ipso*, well functioning, or functional. Applying this final touch to the Polanyi-Kuhn sociology of knowledge will make the crowd of scientists not sleepwalkers but lemmings, some of which are selected or elected, some of which are doomed. It makes me shudder. I prefer sleepwalkers to lemmings.

I do not share the Polanyi-Kuhn hypothesis about leaders and rank-and-file.

I think the sociology of science is a complex and variable phenomenon. I think, with Dr. William Whewell and Sir Karl Popper, that genuine scientific discovery, small or large, is in part a matter of genius, creative genius, creative imagination. Partly it is also a matter of luck, and, of course, partly a matter of wit or schooling or established scientific practices. I do not doubt, myself, that the established scientific practices, though not quite algorithmic, are rational par excellence, that common sense deviation from them according to the context and its requirement are equally rational (as they occur simultaneously to various investigators), that utilizing any chance insight is similarly rational. I assume, with some misgiving, that the chance insight is not rational and we cannot produce it at will though we can produce it by worrying about problems. So much Whewell and Popper. Query: does this amount to a science-making sausage machine in any sense of the word, however weak? Whewell evaded this question. Popper says emphatically, no. I am at a loss to find the answer or even the way to approach it. Are insights matters of pure luck?

The fact remains: most great thinkers were imbued with a profound sense of their being very lucky, with a strange sense of gratitude even. We find it in Kepler, in Galileo, in Einstein. When a man like Newton avoids such expressions, it is obvious that he is tremendously ambivalent. At times he thought nothing of his works, viewing himself as a mere child standing at the shore of the ocean of truth merely picking up a shiny pebble; at times he valued his work some, and then he said he was a dwarf standing on the shoulders of giants. It is no surprise that when he was forced – as he falsely felt he was – to defend his originality, his own significance, he was venomous.

I say this not in order to reduce the greatest man of science to a psychological case, I say this as an admission that psychology as well as social anthropology do enter the picture. Yet the fact remains that Duhem, Popper, I. B. Cohen, all say Newton only thought he was employing his method, but in fact he consistently employed theirs. This is *apriorism* the like of which even Descartes never dared pose. How lucky of Newton to be an unknowing disciple of Duhem, Popper, or I. B. Cohen!

Is the sleepwalker lucky each time? Is there a consistency here, a lucky streak? A lucky streak, as opposed to a case of luck, is what we are talking about. Repeatedly science advances, and nobody knows why. Perhaps the Darwin ex machina is right. Perhaps failure is left behind yet success accumulates. If so, then, as I have said before, the method of creation is the method of procreation. In a sense we all think so, Darwin or no Darwin. We all think, if we only survive as a species, and if we avoid stupid disasters like atomic warfare and population explosion, then all will be well. Wealth is bound, then, to accumulate; material, technological, intellectual. Of all the science-fiction novels and novellettes and stories which came my way, I found only one which contemplates a different possibility, and it is E. M. Forster's lovely short story, 'The Machine Stops'.

Can the machine stop? Popper keeps saying, science is not a sausage making machine, it is not guaranteed to advance ever. True; we can all imagine all sorts of stoppage. But we do not for a moment doubt that given enough occasions, and chances however small, we must win. By Bernoulli's law, we say. True, Bernoulli's law plus the hypothesis that knowledge accumulates will guarantee success without a science-making machine, by sheer random efforts. Random effort then *is* a weak version of a science-making machine, even.

Moreover, though we do not even know if Bernoulli's law meaningfully applies here, yet we think anyway we stand a better chance than mere randomness. Is this luck or is it wit? Where do we stand now anyway? A glance at a projection, whether of scientific advisers to governments or of the Hudson Institute, shows the facts: we are optimistic without knowing what about, never mind why. Do we trust our luck or our wit? What is genius – in science, in the arts, in devising a new chess gambit; is genius luck or is it wit? Is genius at all describable with the help of these distinct categories? Is genius the ability to take a chance at a lucky moment or is it inspiration? Can genius offer us the way to break away from the dichotomy between luck and wit? Can systematic luck? Or can a lucky streak? Is sleepwalking unharmed a kind of a knack or of genius? Is there a method to develop the knack, say of a gypsy violinist? Is a gypsy violinist so very different from a Yehudi Menuhin?

The idea of scientists as sleepwalkers looks absurd at first. Only at first. Science, we say, is the peak of rationality, and sleepwalking is as blind as can be. But if scientific discovery is inspired and if sleepwalking is a knack, can any of us here say whether a scientist was inspired or had a knack? A thinker or an artist walk in a terra incognita, like a blind man, yet he does what he has set out to do, and as best he knows how. This, it seems to me, is the epitome of rationality. Calling it sleepwalking is but a poignant way of stressing that reason is limited, that the best vision is merely partial. The problem is, can we delimit and describe this partial rationality?

CHAPTER 17

THE LOGIC OF SCIENTIFIC INQUIRY

Is methodological theory *a priori* or *a posteriori* knowledge? It is perhaps *a posteriori* improvable, somehow. For example, Duhem discovered that since scientists disagree on methods, they do not always know what they are doing.

How is methodological innovation possible? If it is inapplicable in retrospect, then it is not universal and so seems defective; if it is, then there is a miracle here. Even so, the new explicit awareness of rules previously implicitly known is in itself beneficial. And so, improved methodology may make for improved methods. Hence, methodology is in part descriptive, in part prescriptive. Knowing this, a methodologist might improve his own studies. For example, Popper would then not hasten to conclude from the fact that past scientists depended on positive evidence that they had better do so in future as well; perhaps a lesser concern with confirmation may increase the productivity of scientific inquiry.

1. IS METHODOLOGY INNATE?

The question I wish to present now is strikingly simple: Do we learn scientific method, and if so, how? Historically, in the best classical tradition of epistemology, the question universally recognized as central, was, do we know *a priori* or *a posteriori*? Now we agree that we do not know *a priori*; perhaps not even *a posteriori*, though something *a posteriori* does happen: we do learn from experience. This, however, does not settle the issue at all. The question to which the present chapter is devoted is just the new version of the classical question, and looking at it makes it much easier to comprehend the present state of affairs in the field of the philosophy of science. The question is, do we possess knowledge of the logic of inquiry *a priori*, or do we learn it? If we possess it *a priori*, how can our studies in methodology progress, as they unquestionably do? If we learn it, if we learn about learning, do we possess the ability to do so *a priori*? Or can we say, perhaps, that we are methodologically ignorant?

When I say I wish to present a question, I mean a question, rather than alternative answers for it. A question well-put, they say, is half the answer. Well, then, my target is only half the answer, and I may fall far short of that too. It is a strange fact that philosophers barely pose problems, and seldom discuss questions, being so much in a hurry to answer them. But we can take our time now, and center on a question, for a change.

The question seems abstract. Let me try to concretize it a bit. It used to be taken for granted that scientists know all about induction and use what they know. This was criticized by Duhem who said, it is an empirical fact that different scientists have different views about scientific method. Duhem tried to discredit the view that scientists know all about methodology because traditionally scientists were inductivists: the paradigm was Newton who preached induction and allegedly practiced it too. And Duhem was an anti-inductivist. Einstein gave the idea which Duhem criticized the coup de grace in a mock-inductive fashion: do not listen to what a scientist says he does, look at what in fact he does. Now, as Newton did not practice the method he preached, what method did he practice? Evidently, the right method. Whose theory describes the right method? Mine, of course: if I did not think a theory was true, I would not be advocating it. Hence, says Duhem, says Popper, say many others, Newton preached inductivism, but – quite unknowingly – he practiced the newest methodology ever, my methodology.

This is a remarkable fact, well worth exploring. It came home to me slowly, and two events helped to bring it about. One was during my substituting in a class in the London School of Economics when I said, as a matter of course, that Newton was an arch-inductivist. One student, who had listened to Popper's lectures for quite some time and was well-versed in Popper's doctrines, was quite shocked. This, considering that for two centuries Newton was the inductivist paradigm, seemed to me quaint enough. More recently, in the Boston area, I was invited in a series of guest lectures to follow a lecture by *the* Newton expert, I. B. Cohen. The students there were also shocked to hear that Newton was an inductivist. A week earlier Cohen told them what method Cohen thought Newton had followed; and they all assumed as a matter of course that Newton knew what he was doing; and hence, they all ascribed Cohen's views on method to Newton. Clearly, Cohen had no time to explain in one evening how Newton could practice the method of hypothesis while preaching against it; he could not even mention in that one evening that though Newton speculated systematically, he thought experimental philosophy had no place for speculation. For my part, I have now the whole chapter to explain it, if I only knew the explanation – but at least I have already mentioned the problem. And, as I say, I took a long time to arrive at it, in spite of these helpful events. Now the simple fact is clear to me: it is almost providential that Newton should have practiced the right method even while he preached the wrong one. Well, perhaps he did not. The claim that he did is a very strong version of apriorism which does not demand, as Descartes did, any degree of awareness.

The reason that this doubt, which I have discovered, has not been noticed before, even by critics of Duhem and of Popper, including myself, is quite obvious. It is not the matter of respect for Newton who somehow must have done the right thing. Heaven knows, and Augustus DeMorgan has graphically enough illustrated, how often Newton could do just about the wrongest thing possible. The matter is not of history, but of the philosophy of the social sciences. The historical school in the philosophy of the social sciences, notably Hegel and Marx, declared that all societal laws are confined to historical domains, are bound to space-time regions. The opposite or generalizing school of Rickert, Windelband, and Weber, has won the upper hand: the qualifications of the specific space-time conditions of historical personages can be described in universal terms and so the societal laws applicable to them may be stated as strictly universal. Now scientific research is subject to certain laws of human activity and so describable in universal terms. Hence, my methodology, if true, *eo ipso* applies to Newton; hence, we can test it by applying it to Newton's case.

Let me report briefly how I have arrived at the discovery that something is fishy in what I have just described as the central argument supporting the claim that Newton (unknowingly) practiced my methodology. My story will render my problem somewhat more concrete, I hope, and will also provide a case study to which we can apply any part of our inquiry. Moreover, it concerns the empirical support of a scientific hypothesis, variably known as verification, confirmation, corroboration, agreement or accord between fact and theory, and by other names. I am tired of empirical support; I have argued to my own satisfaction that science is not after empirical support; technology is, and even science can use empirical support to other ends, such as increasing the degree of explanation of a hypothesis; and other kinds of support may come handy in science and more so in metaphysics. But science is not after empirical support as such. My arguments have not been attacked by anyone except Popper, – Wisdom has later reiterated it – and even he has offered only a measly small and poor attack, and only as a token of general esteem, not as an expression of esteem of my views on empirical support. Since it is all I have to sharpen my wits on, I centered on it much more than I otherwise would. And I do not regret it.

Briefly, I have criticized Popper for an inconsistency. He must decide whether science is a process of but conjecture and refutation, or also of corroboration. This Popper has not taken up. Also, I have said, I prefer conjecture and refutation to conjecture, corroboration and refutation. I even do not like conjecture and refutation too much since I qualify it so much as to make it look quite different from the way Popper has intended it to look. But here this is not the place. So, let me repeat, however inaccurately or misleadingly, I prefer conjectures and refutations to conjectures, corroborations, and refutations. To this Popper has retorted by saying, if not for corroboration history might have developed differently, since scientists were importantly encouraged by corroboration or empirical support.

I do endorse this historical statement. I take it for granted that science in the 17th century gained much moral support from empirical support – whether genuine, like that for Boyle's Law, or spurious, like the empirical support Newton's theory of gravity gained during Newton's own lifetime. This, of course, takes us to the following question. Assume the (alleged or true) historical fact that scientists did seek empirical support, and for a strong reason. Does this refute my claim that science does not seek support as such? I do not know. Let us take it slowly.

Let us suppose, for a moment, that what encouraged scientific research was an external event so-called, an event unrelated to what is now called the internal history of science. Suppose, say, that it arose as an act of indignation against the humiliation which Galileo suffered at the hand of the Inquisition. I say "suppose" for lack of information, but surely it could have happened; we have such an instance in the case of the attack on Darwin from religious quarters, in a period when such attacks were far less frightening; but the 19th century is not the one in which support for science as such was needed to keep the enterprise of science going. Supposing, then, that the Inquisition caused a back-lash which helped science develop. Will this be a criticism of the view that, internally, science is conjectures and refutations? No. Why is the claim that support plays a positive historical role in the rise of the sciences more damaging to that view than the analogous claim about the Inquisition?

The answer is obvious: whereas the act of the Inquisition is clearly external to the history of science, it is not at all clear whether support, the search for it and the use of it, is equally external. How, then, do we judge externality or internality, and how does this make the observation that without empirical support science would have died a criticism of my contention that science is not after empirical support, that support plays no independent role in science?

In the second biennial meeting of the Philosophy of Science Association, Imre Lakatos drew attention to the fact that, insofar as the distinction between internal and external history of science signifies, it must be viewed as theory laden. The theory in question is, of course, a historiography of science, and the major ingredient in any historiography of science – I think it is by now agreed – is a view of scientific method. It should therefore not be surprising if I were to conclude that I view the historical role of support as external, that Popper views it as internal, that we are both consistent, but that he has no criticism of my view thus far.

There is, however, a snag here. The more I stick to my logic and insist on a subtle distinction between internal and external history of science, the more subtle this distinction becomes, the more it loses its significance. This, I think, was Kuhn's comment on Lakatos. As long as the distinction is applied to such things as the Inquisition's humiliation of Galileo on the one hand and on the other to the fact that Galileo could measure time to the precision of the order of magnitude of one second, the distinction is powerful. Once we cut it finer, diminishing returns set in fast.

This fact seems to me very interesting. I have observed elsewhere that all methodologists who are worth noticing were reformers of methodology, yet most of them systematically refuse to say whether they legislate how scientists should act or observe how they do. Now the testing ground should, of course, be history: a reformer cannot, but an observer should, apply his theory to history. It turns out that things are not so simple, and for the following, rather obvious reason. The history of science is largely the history of what *we* value in science, and so the reformer can and indeed should rewrite history. As Borges says, every man creates his own predecessors. Nevertheless, the reformer, in the very act of recreating the past in a new way, may perhaps show himself as a reformer. This is not clear-cut since theoretical insights are enriching too. In particular, the insight into existing scientific method also helps us recreate the past. And so we need to know the difference between the discoverer's reconstruction and the legislator's reconstruction – which leaves us exactly where we were before. How can the reformer apply his ideas, say, to Newton, if Newton knew nothing about these ideas? Borges answers this question ingeniously for the history of the arts: we select what we value and we value events which reflect our ideas, events which were not especially noticed beforehand but are stressed now. The same may go for other aspects of events, of course. But can we apply all this to the history of science?

In a way this has happened already: the new or renewed ideas about the metaphysical foundations of science made Burtt, Koyré, and I. B. Cohen, rewrite its history. Now, of course, what they did is to link science very closely to some factors which are, strictly speaking, external to it. The very externality of the metaphysical factors may, indeed, explain a lot. When Newton, in the highly speculative *scholium generale* of his *Principia* says, I feign no hypothesis, he may be puzzling; even if he means a Cartesian hypothesis, as Sabra argues, it is still somewhat puzzling. But when Newton says in the highly speculative Last Query in his *Opticks*, hypotheses have no place in experimental philosophy, he does not sound so puzzling; he says, it seems, this Query is not illegitimate as long as it does not masquerade as science proper.

Thus, the restriction on our conception of science imposes a severe restriction on our internal historiography of science, which consequently prevents us from recreating the past of science as freely as we recreate the past of literature. And yet, this very restriction offers us a new way of overcoming it – by adding as a significant component in our history of science the external history of science, where not only politics and religion, but even ideas which men of science held are considered strictly external, such as scientists' religions and/or metaphysical views of the world. Now the question is, should we consider the history of methodology internal or external to the history of science? In a way, this is but a reformulation of our first question: do we know methodology *a priori* or do we learn it? Do we know what is science after? Do we know what we want? This, of course, is treated in Plato's *Meno*. You will forgive me, I am sure, if I repeat it in brief.

2. THE SOCRATIC PARADOX OF LEARNING

The paradox begins with these assumptions. First, whatever it is we wish to acquire, we do not already posses it. Second, we do, however, already possess knowledge of it. (We may defend these assumptions, but I shall not do so here.) Now, when we substitute 'knowledge of something' for 'whatever we wish to acquire' we get the result. 'When we wish to acquire knowledge of something we both do not possess knowledge of it and we do possess knowledge of it'. We can present the paradox a bit more formally. We assume,

$$\vdash Wish\,(x, P\,(x, y)) \supset Kn\,(x, y).\sim P\,(x,y).$$

Substituting '*Kn*' (for know) for *P*, we get,

$$\vdash Wish\,(x, Kn\,(x, y)) \supset Kn\,(x, y).\sim Kn\,(x, y).$$

The reason I use this clumsy notation is to show that the paradox is a paradox in the original Greek sense – it is no antinomy but merely an unintuitive result. For, the conclusion is equivalent to the assertion

$$\vdash \sim Wish\,(x, Kn\,(x, y))$$

that necessarily we can never wish to know anything. It is easy to see that the result is hardly even counter-intuitive, since the knowledge wished for in our discussion is detailed and certain knowledge, and, no doubt, even the detailed knowledge that Sherlock Holmes wishes to have he may regret once he has it; when wishing for it he has at best a vague general knowledge he *thinks* he wishes to fill out. Socrates shows that the result is less counter-intuitive than it sounds when he says, even if we finally do achieve knowledge, somehow there will be no knowledge that the knowledge wished for is the knowledge acquired. In other words, the root of the paradox is not in the concept of wishing but in the – rather unusually strict – concept of knowledge.

There is very little literature, to my knowledge, regarding the Socratic Paradox of Learning, and it hardly ever goes beyond Plato's *Meno*. There is, I have the feeling, an allusion or two to it in the writings of Sir Francis Bacon, who offers a unique solution to it, when he took all knowledge to be his province. For Sir Francis Bacon, any attempt to acquire any specific piece of knowledge is defective, directed towards an arbitrary goal. For Bacon, the whole enterprise of science, of learning and of knowledge, is a matter of all-or-nothing. He spoke of two kinds of knowledge, of little knowledge and of all knowledge, namely all the knowledge we can ever acquire. He declared all the criticisms of knowledge we know to be attacks on little knowledge only. All knowledge can be attained once the mind is utterly free and empty; the empty mind's capacity to learn guarantees full success; but when the mind is partially filled, it is prejudiced. It is prejudiced because the very partiality is something which is prior to its own empirical justification, and hence prejudicial. The very choice of a question must be arbitrary or based on a hypothesis – not on knowledge which is of the answer to the question. Of course, we always lack the empirical foundation to that which we as yet seek; hence, it is arbitrary to seek any item of knowledge in particular; but not all knowledge.

Bacon's fundamental theory of learning took off from the mystic tradition of his day, from the cabalist and from the alchemist, from Natalis Conti who assimilated these traditions into a Renaissance pseudo-Greek mythology. The fundamental division of minds was into the willful or arbitrary or *ad hoc* or evil on the one hand and the submissive or accepting or simple or good on the other – oh, sacred simplicity! Willful minds, said Bacon, put Nature in chains. Submissive minds collect simple facts as an act of worship of attending to the whims of Nature. She is allowed all caprice, of course, and as a reward She will reveal her secrets to man; as we all know, submission is the best road to mastery.

The Baconian folklore has been accepted to an unspecifiable degree, the sifting of myth from reason in it has never been done, though obviously the mythical part of inductivism is less conspicious now than it was then. This is obviously a symptom of increased confidence in science – a fact of no mean significance for my debate with Popper. Even Bacon himself felt that he was too much of a mythmaker. He said in his *Novum Organum* that it matters less how we start than that we start; that when we get going who knows what the outcome will be; he said that we may even learn, by induction, something more about induction. This was a passing remark, and I never made much of it because it was said in passing, because it seemed to me to be an admission of the criticism of induction from infinite regress, and because it occurs in an extravagant passage – extravagant even for Bacon – which says that even logic may be learned by induction.

On a closer look we have here the second order of the Socratic Paradox of Learning. Let me expand this slowly.

It is well known that some games, dances, rituals, etc., were consciously created and their rules were specified in writing by their originators. There are others which have evolved and which have never been recorded – say among pre-literate people, which are met for the first time by the anthropologist, say Malinowski.

When a Malinowski goes to a new place to observe a new tribe he operates scientifically, following strictly the rules of the book. He observes facts, hard and fast facts, such as people playing diverse games. But of course from the facts he derives the rules of the game: he is interested not in the specific but in the universal, not in the dance he observes but in the custom it exhibits or is an instance of. Moreover, he wants to achieve a higher-level induction, of course; he wants to generalize from ethnic generalizations to social anthropology in general, and again he does it by the rules of induction, by using his ethnic generalizations in a comparative study or in comparative sociology, creating what Bacon called Tables of Similarity and of Differences, etc., etc.

Now, of course, one of the tribes in question is the social anthropologists; and Malinowski here has to use scientific method in order to discover the customs of his own tribe, which include scientific method. But if he is already in possession of the rules he need not study them; whereas if he needs to study them he has no tools by which to pursue a study – of anything; of the rules or of anything else. This is a complete impasse – not the infinite regress but the zero regress, the Socratic Paradox of Learning itself.

I have the sneaking suspicion that this discovery – the discovery that scientific method is hit by the paradox – is not original with me. It was already Hegel who said in Kant's philosophy there is learning to swim before plunging into water. Following up on this Lakatos says that fish do not know hydrodynamics yet they can swim. Newton's learning, like the fish's swimming, was quite natural, then, and not the application of known rules. This is just an appealing manner of putting an idea which is more often put in the academic manner given it by Gilbert Ryle and adopted by Michael Polanyi: the distinction between knowing how and knowing that. Right or wrong, these views are solutions to the second-level paradox. Though they are all witticisms, they must count as expressions of their authors having noticed this second-level paradox of learning, or, as I shall call it, the paradox of swimming.

Now, if you were to think that with these witticisms in scholarly garb the matter is settled, then you would be making the same mistake that I made for quite some years. Now I know that they settle nothing. The one person who saved me from my error was Nicolas Malebranche, the occasionalist. If you were the one who lifts your arm when you wish to lift your arm, said that learned clergyman, you would know how you do it, wouldn't you? Funny, the very expression, I know how to raise my arm, which lends credibility to the Hegel-Ryle-Polanyi-Lakatos resolution of the paradox of swimming, by a slight alteration turns the table: I know how to raise my arm: and I do now raise it so; yet do I know how I raise my arm? If I did I would be a leading physiologist.

We all reject Malebranche's occasionalism; we all do think that we do raise our arms even though we do not know how we do so; we know that we raise our arms, but we do not know how we do so. Ryle says we know how, but not that; I have just reversed the order. Where is the rub? What is the significant corollary to all that?

Living in the post-Hegelian and post-Darwinian era, and in the day of artificial brains, we do not nowadays accept any static solution. We see as common to both the inductive and the apriorist views a claim which we vehemently reject (*pace* Piaget, Chomsky *et al.*) namely, the claim that we are born with a logic of scientific inquiry. Inductivists and apriorists disagree about the nature of methodology. But they agree that it is inborn. It follows from our rejection of that claim that we cannot discover the logic of scientific inquiry by applying it. We neither know how to inquire, nor that inquiring is conducted this or that way. The book, the set of rules of the game, is not imprinted on us from birth, we think, but we can knowingly or unknowingly apply it. Therefore, we must conclude that if we unknowingly apply it we will never notice the fact. Nor is it written anywhere so that we can read it and try to learn to apply it. The book, we feel, is a mystery. We have here Kafka redivivus. Lakatos' fish is paralyzed! For, the fish is either able to swim, or able to learn to swim, or will never swim. Put learn for swim, and you get the fish is either able to learn, or able to learn to learn, or forever ignorant. This, I contend is an impasse. But I shall take it slowly to pinpoint matters.

We can say, *a priori*, that swimming is given for the fish *a priori* or *a posteriori*, or not at all. Symbolically, the trichotomy may be written thus:

$$\vdash Asw(f) \vee Lsw(f) \vee \sim sw(f)$$

When Kant accepted Hume's refutation of empiricism as logically watertight he employed the refutation and deduced from the above trichotomy the dichotomy:

$$\vdash Asw(f) \vee \sim sw(f),$$

and then proceeded with what he called the transcendental argument, which begins with the allegedly empirical facts

$$sw(f)$$

and concluding

$$\therefore Asw(f)$$

which, he suggested, is equivalent to

$$\vdash sw(f).$$

This final step of viewing the opinion as demonstrable was contested by Solomon Maimon who took Kant's apriorism as a mere hypothesis much to Kant's annoyance. C. I. Lewis reinvented Maimon's position and arguments.

But we must hurry along. We had the alleged trichotomy as a tautology:

$$\vdash Asw(f) \vee Lsw(f) \vee \sim sw(f).$$

In our post-Hegelian post-Darwinian evolutionist manner we reject this as too static. We have here some *a priori* knowledge improven genetically, we say, and it enables us to learn more through experience. This is *the* modern solution. It does not work. Take any given fish, and put 'Learn' instead of 'swim' in our formula. We obtain:

$$\vdash AL(f) \vee LL(f) \vee \sim L(f).$$

Assuming, as before, $\sim L(f)$ to be false, we get

$$\therefore AL(f) \vee LL(f).$$

And so, the static view of knowledge which we now reject reappears as a static view of learning! To call one disjunct knowing how and the other knowing that is neither true nor intellectually satisfying, methinks. Moreover, without much deep thinking we feel that perhaps

$$\vdash LL(f) \supset L(f)$$

i.e. if a fish can learn to learn, then it can learn, and even perhaps that

$$\vdash L(f) \supset AL(f),$$

that some learning must be inborn or else no learning will ever occur – compare a human with an insect, for example, or with a computer. Now, if we accept these two dubious claims, we get at once

$$\therefore AL(f)$$

The fish is born with some *a priori* knowledge of scientific method. Whether this *a priori* knowledge is analytic can, again, be contested *à la* Maimon and Lewis.

3. COMMON SENSE VERSUS METHODOLOGY

Can we then do methodology before finding out the inborn methodology and the way it grows? Clearly, somehow, all this sounds most incredible. The present discussion, such as it is, concerns scientific method; that is, I am already doing a social anthropological, an ethnographic study, of my own tribe – the only case, perhaps, of perfect participant observation. How am I doing it? Like Malebranche's wise guy who raises his hand, I just do it, and I don't know how.

Common sense has taken over, and it can do so *de facto* or also *de jure*. Moses has said already, for the teaching which I am giving you today is not in the sky that you should say, who should go there in order to bring it to us, nor is it overseas that you should say who should go there to bring it to us, for the teaching I am giving you today is in your heart, etc. I have argued already that the strong point of inductivism is exactly this sentiment, expressed by Bacon in different ways in his *The Advancement of Learning* and in his *Novum Organum*: it matters less how you start and more that you do start. Once you get going, the ball will roll, one thing will lead another, and great deeds are just beyond the horizon.

The peculiar fact about all this is that when common sense is turned from *de facto* to *de jure* it receives a religious fervor. Some of us are still uneasy about the idea of the religious fervor of science; some of us oppose it with genuine religious fervor. The fact is, the strongly common sense fact which we so often refuse to observe with the religious fervor of the dogmatist, is that science is, in fact, a religion. Further, that the religion of science, as any religion, is not common sense – that even the religion of common sense is not common sense. Historically, the religion of common sense was a version of positivism, of the religion of science. Philosophically, *de jure* common sense is a contradiction in terms; and *de facto* common sense is as tautologous as humid water.

The reason is quite common sense. Common sense is skeptical, placid, pliable, unauthoritative; its present position is inherently non-inherent, *de facto* in principle. This makes the Socratic Paradox of Learning not only less counter-intuitive; it makes it plain common sense: whatever we wish, whatever we achieve, whatever we have, our appraisal of it, is all fluid. What we achieve, we achieve partly as a result of our wish to achieve it, no doubt, but only partly. Which establishes the Socratic Paradox as a fact of life.

Common sense philosophers, whether of the Scottish school or the Oxford school, tried to combat skepticism with common sense, since they viewed skepticism as wierd. This was a misconstrual of skepticism and the making a dogma out of common sense. Now, no doubt, there is a dogmatic aspect or part to common sense, and there is the exaggerated or inflated or misconstrued skepticism to go with it. Thus, sometimes, in an appeal to common sense, people suggest to give up a search because it has not thus far led to a final and solid result – i.e. to a dogma. Thus, in the name of common sense, many so-called philosophers have suggested that we give up scientific method as a bad job. Should we say, all scientific method is but theory, is but a hypostatization of a living, pulsating, common sense? I must say, this sounds – I am speaking empirically – most convincing, most common sense. When I am in a missionary mood I feel, common sense is changeable, sometimes for the better, and so I ought to try to improve it. It will greatly improve if and when all proposals to give up searches which do not lead to finality will cease sounding so appealing and common sense. I am not in a missionary mood and so I shall address only those interested in my reason, to say, for what it is worth, why I think we would be better off when it will not sound so plausible to dismiss attempts to capture the spirits of common sense because it is, no doubt, *eo ipso* a hypostatization.

Let us go back to social anthropology. The social anthropologist captures the common sense of the natives. Indeed, this is the pinnacle of his ethnological search. There is little reason to doubt that the natives' common sense is changeable too – it may be less changeable internally – for internal reasons – than the Western common sense, but it is therefore all the more changeable externally – as the outcome of contact with the West.

The standard complaint against traditional Malinowskian social anthropology, as presented, e.g. by Ernest Gellner in his classic 'Time and Theory in Social Anthropology' in *Mind* **67** (1958), 182–202 (reprinted in his *Cause and Meaning in the Social Sciences*, Routledge, London and Boston, 1973, pp. 88–106) is that it is static. It hypostatizes the local custom and fixes it. Now, no doubt, there is no need to fix it forever, but I cannot see how one can avoid hypostatizing it as long as one describes a time-slice. The only possible way to give it some movement is to do it historically. But when we come to pre-literate tribes the Baconian injunction against speculation turns into the Radcliffe-Brownian injunction against conjectural history (the term seems to be Sir Edward Tylor's), and this dooms the project to failure. Well, not quite. Thor Heyerdahl, also a Baconian, says in his *Aku Aku* that history can be revealed by the spade. But I shall leave the pre-literate tribes now, and move to the most literate one of them all, the scientists, for example the anthropologists.

Malinowski, we remember, made two hypostatizations, perhaps both necessary. One was to fix today's custom, one was to banish history. When we study physics, for example, we must do the one hypostatization, we must fix today's physics – at least long enough to give it a good look, to teach it to our young rascals, etc. But we need not banish history. Of course, inductivist historians of physics put the history of physics in the margin of today's textbook of physics. This hypostatization, which I have christened, for short, as up-to-date-science-textbook-worship, is quite redundant. Though we hypostatize today's physics, we can acquire a sense of change by looking at its history. We can treat theories of scientific method likewise. But when we treat the living actions of scientific inquiry we hypostatize not only today's method, which is inevitable, but also yesterday's, which is unnecessary. Newton, we say, preached inductivism but practiced my methodology.

In the post-Hegelian and post-Darwinian era, we have allowed knowledge to evolve; after Einstein we see little difficulty to admit a modification of our knowledge here and there; this makes knowledge much less universal than was envisaged during the Age of Reason. But at least we want reason, when it is equated with the ability to learn or with the logic of inquiry – we wish reason to be universal to all men. If not, then the categorical imperative itself will collapse, the ideological base of Western democracy will need a drastic revision!

As I have already shown, this argument is invalid. It may well be the case that Newton's logic of inquiry and Einstein's are identical only at birth, but that they evolved very differently. Let us assume for a brief moment that ontogeny recapitulates phylogeny, that one's own development is a capsule version of the development of the race. It follows that sometime before Einstein sprang into full maturity he was a Newton of sorts; that perhaps he believed in Newtonian mechanics or even in Newtonian methodology – which was a variant of inductivism, we remember. We may be tempted to say, perhaps as far as opinions are concerned, ontogeny does not recapitulate phylogeny, but in the logic of inquiry it does. Even a superficial knowledge of the history of Einstein's career, however, will show the opposite to be the case, that Einstein never was half as much an inductivist as he was, in some sense, a Newtonian. Yet the more important fact is that ontogeny never recapitulates phylogeny, that even Einstein's Newtonianism, such as it was, was much removed from Newton's.

There is no reason to assume that in any sense the logic of inquiry endorsed or employed by Newton was the same as Einstein's. Suffice it to assume that at birth Newton's and Einstein's inborn abilities to learn were identical. Even this assumption seems to me highly dubious. Even at my birth I do not think I was as good as Einstein was at his. Nor do I think this destroys the thesis of the unity of mankind.

On the contrary. I see the unity quite in the opposite place: even I can learn, or acquire, a methodological insight from Einstein! Without going into the venerable ancestry of this position, ancient and mediaeval, let me say: it is this position, or this variant, for which I am most grateful to my teacher, Sir Karl Popper.

One puzzling implication of this is really so obvious I can mention it right away. The obvious line of thinking is as follows. If scientific method is described truly, and if Newton was a scientist, then the description applies to his method. Hence, when I assert my view of scientific method, I already assert by implication that Newton practiced my method. Denying this obvious line of thinking seems to be putting a severe limitation on social anthropology; yet I do deny that and I do wish to avoid putting any limitation on social anthropology. Can I do so?

But perhaps I am moving too fast. Before asking about the peculiarity of the theory of scientific method, let me ask, does anyone have a theory of scientific method? Bacon and Descartes surely did. Yet we ignore them today. Did Duhem, did Popper, offer a theory of scientific method? Clearly, both Bacon and Descartes said, a scientist must be aware of his method. Must. Duhem and Popper say, he need not be, and is, in fact, seldom aware of what he is doing. Query: according to Duhem and Popper, is his awareness a factor significant to his research in any way whatsoever?

Suppose we say, no. Then, first and foremost, all arguments from the way people happen to have looked at science become strictly external. In particular, the fact that they were encouraged by empirical support may hinge on nothing more than their ignorance of the very method they were following. Suppose we say yes, then even if the new view of method, be it Duhem's or Popper's, applies in some sense to all science, clearly it applies one way to those ignorant of the new view and another to those familiar with it. In this respect, even the very discoverer of a point in methodology is *eo ipso*, an innovator: the very increased awareness is beneficial.

This is generally true of all human affairs. Take Freud's discoveries and see how their discovery makes us better off. But I must rush to my conclusion.

There is here a very obvious lacuna in views of Duhem and Popper as compared with Bacon and Descartes. Clearly the old methodologists offered complete theories of method, something akin to algorisms or, as Popper says, science sausage-making machines. One important and oft ignored part of the algorism, incidentally, is the awareness of the scientist. Duhem and Popper both stress that the making of science is subject to no algorism. But they do not tell us how much of the phenomenon is explained by their theories. Their writings consist partly in the exposition of historical counter-examples to preceding doctrines. This is almost always fascinating stuff, but not yet explicanda. In science, counter-examples to the old theory are the explicanda of the new. Is it so in methodology? Do Duhem and Popper explain the counter-examples? In particular, they indicate the absence of awareness. Can this be explained? Surely not fully, because of the Socratic paradox of learning. I feel that the very idea of the growth of science – central to all methodologists – is here crucial. Inductivists have explained – quite satisfactorily –

the surge of science in the 17th century: at that time the true scientific method came into awareness, etc. Duhem and Popper rightly reject this explanation. Duhem has also rejected the claim that there ever occurred a scientific revolution. We nowadays agree that, as a matter of fact, there was one. Can we explain this? I think we can: the false methodology gave rise to false hopes. True, the false methodology required empirical support. This was whipped up. Pascal's brother-in-law's excursion to the mountain was deemed an empirical support. Newton's crude calculation of the perturbation of Saturn and Jupiter and his fiddling with Flamsteed's data, were empirical support. But now we can ask: was all this really necessary?

CHAPTER 18

THE CHOICE OF SCIENTIFIC PROBLEMS

1. ACTIVISM AND PASSIVISM IN PHILOSOPHY OF SCIENCE

The student accustomed to classical philosophy is bewildered at the contemporary scene, no less the student of classical art and science. What I think helps reorient oneself when bewildered, is to try and see the shift in concern and try to understand it and its roots in the older situation. To take one very small example, classical astronomy was increasingly concerned with increased precision of observation and calculation of planetary orbits. This concern has not vanished but become fairly marginal, and, of course, its purpose has changed with the monumental change of status that Newtonian mechanics has undergone since Einstein appeared on the scene. Nowadays the concern with ages of stars and distribution of objects, whether in galaxies, clusters, or space, becomes of great import, which fact could scarcely make sense to classical astronomers.

The greatest shift in the philosophy of science, indeed the most important one, is that from the problem of choice of scientific theories to the problem of choice of scientific problems. This shift was gradual, to be sure, and it is interesting to see in retrospect, what could not be seen in advance, how the concern shifted. For, at least to begin with, but actually until now, it was not clear enough that the two concerns are extremely different. But this much can be said, and should be said outright. Roughly, the problem of choice of theories points at a passivist view of science, as I shall soon explain, whereas the problem of choice of problems points at an activist view of science.

The passivism of the classical theories about science can easily be overlooked. When Galileo said in his theological works that it is a mistake for religion to prescribe to a man of science what to believe, it certainly looked active, not passive: I want to be the judge of what I say is true; I do not want the Church to tell me what to believe. Similarly, there is an activist accent, an emphasis on search, in our reading of what Bacon said: the great harm that a false belief does is that it makes its holder live in a fool's paradise, and instead of searching for the truth he is put into the false satisfaction that he already is in possession of the truth. For, according to Bacon's theory of the idols, i.e. of accepted errors, error stands in the way of the search for the truth and so it is very important not to accept error, not to give one's assent to a false theory.

It was Bacon's theory that made the problem of choice of theory so very central. We might ask the philosophers who sought, as they did for centuries, as many of them still do, why do you seek for criteria of choice of a theory, why does it matter which theory to choose? The philosopher may have sunk to the depth of scholasticism and then he would say, the task of philosophers of science is traditionally this and do not question it. Alternatively, he will give two answers, both Baconian, and both seemingly activist. First, he might say, the wrong choice of theory impedes the proper search for the truth. Second, the better theory is technologically much more useful. This is Bacon's aphorism: knowledge is power.

Yet, despite this, the theory that dominated classical philosophy was passivism. Whether apriorist or aposteriorist, philosophers had one simple problem that they found utterly insoluble: how does the making of a hypothesis advance science? Not knowing the answer they gave up and advocated passivity.

The question is much more tricky than it looks, and I wish to stress that activists, too, could not solve it. Indeed, the philosopher who first tackled it seriously was an activist, the great nineteenth century philosopher, Dr. William Whewell; and he failed. His failure made passivist philosophers, though not active scientists, dismiss him. This is how the wonderful philosopher that he was drew no attention from philosophers and historians of philosophy until after World War II, and even then he did not, and still does not, gain half the attention we may profitably give him.

Whewell said, scientists make conjectures and refute them. Since in principle for every problem there may be infinitely many solutions – since every set of facts, to be more precise, may be explained by infinitely many hypotheses – a scientist may sit endlessly on one given task, try to explain his facts by as many hypotheses as he wishes, and refute them one after another. In other words, says Whewell, there is no guarantee for scientific progress, for scientific success. But, says Whewell, and here is the crux, from time to time, with luck and ingenuity, the man of science may hit on the true hypothesis, confirm predictions based on it, and thus verify it. He thus brings to science some progress and enables researchers to move to the next problem – to the next explanatory task.

The expression, with luck and ingenuity, or with ingenuity and luck, is Whewell's; it was the center of John Stuart Mill's attack on him. This attack, the biographer of Whewell tells us in the *Dictionary of National Biography*, has destroyed Whewell's position amongst the philosophers of the succeeding generation. As we know, Whewell was very popular among scientists, in Britain and on the Continent, among physicists, biologists, and especially geologists.

Yet the ancient claim is uncontested: for every set of facts there is, in principle, an infinity of hypotheses that may explain them. This makes the problem of choice of hypothesis so hopeless that some philosophers felt it should be denied. The greatest commentator on Bacon's philosophy was Whewell's contemporary Robert Leslie Ellis, the renowned editor of Bacon's *Works*. Ellis said explicitly that the whole of Bacon's philosophy of science rests on the implicit or tacit or unnoticed presupposition that the number of possible theories about the world is *a priori* small. Lord Keynes, in his famous *Treatise on Probability* of over half a century ago, says the same of John Stuart Mill. Mill's principle of simplicity of nature is just that. And to make the assumption explicit and stand out, Keynes gave it a new name: it is the principle of limited variety, or the principle of limited variation: *a priori* there is a small number of possible explanations of the known facts already given to us.

The principle of limited variety does not work. A man lost in the forest knows *a priori* that there is a limited number of preferred directions for him to choose in order to get out of the forest. But he does not know which; and this is why he is lost in the forest. Until Keynes gives us some help in determining which hypotheses are of the limited variety and so *a priori* are open to choice, we do not know what to do. I shall come to this most important point later: the *a priori* choice is so vast that we want some constraints put on it, *a priori*, before we come to make any definite choice. Keynes gave us no aid in this: he offered no constraint; he did not say, of all the possible hypotheses, we eliminate *a priori* those which have this and that quality, and consider only the hypotheses that have the other qualities in order to choose from.

I do not mean to say that Keynes was the only one to tackle the problem. At about the same time Keynes wrote, also Sir Harold Jeffreys wrote, and both these thinkers can be considered the giants who led the rear guard, the backward-looking sector, of the philosophers of science. Jeffreys said, even if there are infinitely many hypotheses, we can order them and assign to them *a priori* probabilities in proportion to elements of an infinite series of a converging sum. We can then say that the hypotheses at the tail end of the series are of so low a probability that they can be ignored. In other words, the solution Jeffreys offers is in principle different but in practice the same as that of Keynes. It suffers, in practice, from the same shortcoming.

And so, given a set of factual statements, there are infinitely many competing possible hypotheses to explain them, all but one false; and there is no reason to expect the scientist to hit upon the true one by sheer luck.

Bacon was the philosopher who used systematically the claim I just made: there is little chance to hit upon the truth by the use of a guess. The active approach of making a guess, said Bacon, the Method of Anticipation, as he rightly called it, is very gratifying. It gives one a good feeling to participate actively and to devise short-cuts that are, admittedly, at times quite useful. But the cost is high. The Method of Anticipation leads to disagreements, to schools, to school dogmas, to dogmatism, to stagnation, to the end of all research.

Bacon recommends the passive method. The Method of Interpretation of Nature he called it, the method which, he said, gives the intellect not wings but weights. Do not make a hypothesis; do not jump to conclusions; wait; be slow and patient; let Nature force conclusions upon your intellect!

This is powerful stuff. I have heard it since childhood from teachers, colleagues, friends, passing men of science. The idea that learning is better when boring and dangerous when exciting is Bacon's leading idea, as Justus von Liebig insisted in his famous scurrilous unjust attack on him, over one century ago. Liebig, of course, was one of the pioneers of the activist school.

I wish to stress, as I leave passivism, that the apriorists or intellectualists were generally not less passivists than the aposteriorists or the inductivists or empiricists. On his death-bed Réne Descartes, the father of modern apriorism, proved himself a worthy disciple of Bacon. He regretted, he said, that at times he showed impatience and curiosity in his research. Activity of the mind is suspect; of course; why should Mother Nature be obliged to accept the whim of an impatient guesser?

All this may be theoretically gratifying. The answer to skepticism may indeed be this. If I have a criterion by which to choose a theory it may raise the skeptic's question of the criterion by which to choose the criterion; but perhaps I do not have a criterion and the truth simply impresses itself upon my mind. As Spinoza put it, truth is its own criterion; when it appears, one simply cannot doubt it! Let us accept this theoretically. What does it mean in practice? Suppose revelations are the truth. When, however, do we have a revelation? And so the problem of choice of theories remained and gnawed. It is still considered by the many philosophers of the rear-guard as the cardinal problem of the philosophy of science.

Yet times have changed, and the mere change of scenery plays tricks on what we focus our attention on. The problem of induction, of the choice of the objects to fix our beliefs upon, the problem of choice of a hypothesis, the problem by any other name, has altered as the result of the growth of modern science, in particular of modern physics, as a formidable specialization. The enlightened intellectual of two centuries ago was called by the problem of induction to justify his own beliefs in the theories that he had learned; with effort and some self-deception the same may be said of the last century; today it is the philosopher who definitely feels called to justify the choice of hypotheses, but now it is the scientist's choice, not his own. For the philosopher himself there is usually no choice but simply the acceptance of the choice of the man of science.

2. THE ROOTS OF THE AUTHORITY OF SCIENCE

A glance back to Galileo, if I may, Galileo preached against the authority of the Church and for the autonomy of the individual. Today we replace the authority of the Church with the authority of Science. It is in this sense that Science does now play the role of a surrogate religion.

I am not fighting this. When I was young I tried to fight the idea of science as a surrogate religion, but I gave up. Rather, let me consider the situation in a more serious vein and say, science as a surrogate religion exists and is not a serious matter because those who accept it simply ask the local scientist what they should believe. Hence, the problem of choice of beliefs for the individual may matter a lot, and it may matter very little or not at all: as Bacon said, the choice is of great import for research and for technology; and, as I have shown right now, for the layman it may be of no import whatsoever, being treated as a mere religious courtesy.

And so we have as important only the authority of science for the individual man of science engaged in scientific research. Does research physicist X believe physical theory Y, does research biologist U believe biological theory V? If yes, why yes, and if not, why not?

The questions, we remember, impinge on both research and technology. But these days of specialization separate research from technology, and even scientific research from technological research. And, clearly, what the technologist believes matters little as long as he uses in technology the proper scientific theory. And so the problem of induction does not arise in technology, it seems. It is not surprising, then, that many philosophers who thought the problem of induction insoluble hoped to get around the problem in science as well, by making science declare itself merely the theoretical part of technology. That would make science never what we call pure science, but rather always what we call applied science: not descriptive of any metaphysical import and not informative of the nature of things but equations of mere instrumental significance. Not surprisingly, the philosophers who make this move are called instrumentalists. And the instrumentalists say, even what a research scientist believes does not matter to science. Hence no problem of choice of a hypothesis to believe in. But inductivists and instrumentalists alike have to admit that this still leaves the layman in the cold. Of course, the layman's beliefs do not matter much, but what about his daily decisions? He has to follow science, and in physics it means listening to experts who listen to Einstein and in molecular biology it means listening to experts who listen to Crick and Watson; but in psychology does it mean Freud or Skinner, in economics does it mean Marshall or Keynes, in political science does it mean Adorno or Albert, etc., etc. Laymen are engaged in practical affairs, too, not only technicians. How are they to act?

For the authority of science can be examined both on the individual level and on the institutional level, and on both levels the problem of choice of hypotheses is tied up with both thought and action. Either way, we face the problem of demarcation of science. We want to find the right justification for believing and/or implementing the right theory. Now we do not find either, and so we are lost; but at least we want science and justification to go together. Thus, we may accept a demarcation of Freud's theory as scientific and a justification of it; and we may go the other way. We wish to exclude both cases of theories – e.g. Freud's – demarcated as scientific yet unjustified and of theories demarcated as not scientific yet justified.

I should state right now that this is exactly what I intend to do: I intend to demarcate Freud's theory as being non-scientific and yet justify it as scientific. I know it sounds absurd, but only because until now we spoke of the choice of hypotheses or of theories, whether for belief or as bases for actions. It will look different when we arrive at the choice of problems to devote time to study.

But this is anticipation. Back to my discussion of the problem of choice of theories. The authority of science, I say, is individual or institutional. And once we separate the two, the problem of induction, the most difficult and annoying and insoluble question in the field, becomes very easy to solve. At a high price, as we shall see, so the solution is no solace. But this is a different matter.

For the individual layman, we remember, the problem of induction, as the problem of choice of objects of belief, is not a pressing problem; as long as he knows that he believes the most authoritative scientist, he can rest assured. Even the Roman Catholic Church was willing to accept this, and St. Robert, Cardinal Bellarmino, only wanted from Galileo assurance of his authority: proof. This is not the whole story, as we can have the assurance of authority with no proof. Indeed, Bellarmino himself accepted the authority of the Church Fathers and was not willing to change authority lightly. Changing authorities too often, I presume, is quite detrimental and so must be avoided, if authoritarian philosophy is to be upheld.

Unlike the individual layman, who needs to be told what to believe, the individual research scientist does not have to believe; on the contrary, as Bacon knew and explained, the less the researcher believes the more he seeks. If he seeks, and if he consequently finds interesting or useful facts and hypotheses, even then he need not believe them. No one ever complained about the fact that Einstein did not believe in Einstein's theory, that Schrödinger did not believe in his equation and his theory of perturbation, that Dirac came to hope that the whole of quantum theory, including Dirac's equation, would soon be overthrown and superseded.

That takes care of the individual – layman or researcher – on the condition that we can take care of the institutional or the social side of science. The institutions of science get their authority by social mechanisms – laws, conventions, rules, court decrees, ministerial and administrative regulations, and, for our purpose of particular importance, research grants committees. But we can speak in general of the social aspect of science, of the scientific tradition and customs. This is a very important fact that was overlooked by all philosophers of science before Popper, Polanyi, and Robert Merton put much stress on it, perhaps as a reaction to the crazy Soviet idea of guided and planned research, which in the thirties and early forties was advocated by communists and fellow travellers in the West. According to this idea the Party can and should tell the scientist what to research and how to conduct his research – to the alleged benefit of both science and society.

The difference between the communists whom Polanyi attacked so forcefully and Polanyi himself, however, is only technical. For the communists never thought that Stalin, the Sun to the Nations and the leading light though he surely was, that Stalin himself could tell a scientist what to do: only fools could believe that Stalin knew enough physics and biology and linguistics to tell all researchers what to do. Rather, the idea was that the Party organize the scientific community to work in accord with rules coordinating research that originate at the top in accord with the wise precepts of Democratic Centralism as Lenin prescribed it. Democratic Centralism is benevolent tyranny of the tyrant who looks for much feedback from his people, and gets it not by the occasional mingling with the people as Haroon al Rashid did if the Arabian Nights stories are to be believed, but by having the same members of his organization who dish out and execute his orders also bring back the responses they elicit while transmitting his orders to the people.

Now Polanyi, and after him Thomas S. Kuhn, advocate democratic centralism too, but well within the scientific community. They want no Party and no politicians to help the scientific leadership impose its will. This, they say, is dangerous, since it will block out scientific revolutions even when they are scientifically necessary.

Neither Polanyi nor Kuhn speak of the means by which the leadership of the scientific community impose their decisions. They observe – rightly, of course – that science does wield power and authority, and that scientific policies are unavoidable. But they shift the problem of induction, or reduce it, to the problem of political power within the scientific community, and without further ado. This is a big mistake.

When we have political power, we must look for its source. If it is guns, then the revolution brings the revolutionaries to seizing armories; if it is armies, then the revolution leads the revolutionaries to the colonels; if it is the media, it leads the revolutionaries to the broadcasting station. What is the source of power in the scientific community?

Polanyi and Kuhn offer mere hints. Scientists do perform research. Research centers on problems. Problems are dictated. They stem from a paradigm, says Kuhn. Hence, he rightly concludes, revolutions are changes of paradigm. He who wishes to effect a scientific revolution then must attack the paradigm and impose an alternative to it: he should offer researchers new problems.

I have thus explained, I think, the fact that the critics of Kuhn, from the day his book appeared till now, center on the problem, what is a paradigm. For the revolutionaries may well want to seize the paradigm. But Kuhn hedges. He says, in brief, whatever the paradigm is, it is first and foremost what the leadership has. This is terrible, since he does not tell us the mechanics of power in the scientific community and even blocks the way for our finding it for ourselves.

Now I must stop and look around, because a terrible thing has happened to me. I tried to tell the story of the problem of induction, the rear-guard problem of choice of theories in science, and all of a sudden I now find myself concerned with the problem of choice of problems for scientific research!

This is no accident. Every time I try to outline the transition, I find that

I have made it before I described it. Let me backtrack and see what has happened to me and why.

It is easy to correct my mistake. Let us focus on Bacon's view that belief directs research and forget all about problems. Indeed, Bacon, who knew that problems are powerful tools for guiding research, none the less opposed them, and very logically so. Assume that one can never choose a problem without previously endorsing a belief. Assume, further, that we want to start our scientific search with a problem. It follows that the belief guiding the choice of the problem is quite unscientific, and so quite probably an error, and so quite probably misleading. We should therefore ignore problems: they are either bad things based on wrong beliefs or else they are good things based on good scientific beliefs, which will take care of themselves once we take care of beliefs only. Let us repeat Kuhn's view while, likewise, trying to concentrate on beliefs and not on problems:

Beliefs of individuals outside science do not matter; beliefs in science are institutional, and based on beliefs by scientists; and beliefs by scientists are dictated from above and only change in revolutions. This view is pseudo-Kuhn.

This pseudo-Kuhnian view makes no sense. The whole thrust of Kuhn's view is that research is on problems, the choice of problems is dictated by paradigms, and paradigms are dictated by the scientific élite.

Is a paradigm a belief, a theory? Is there the problem of choice of paradigm? This, then, may be the Kuhnian variant of the classical problem of induction.

If so, Polanyi and Kuhn solve the classical problem of induction in two ways: psychologically and methodologically. They do not answer it by reference to truth, since this move is question-begging. Psychologically, says Polanyi, everyone must believe something. He can change his mind, doubt this or the other part of his belief, but he cannot doubt all his beliefs, reject them all, and start afresh as a *tabula rasa*. The importance of our system of belief, however, is just that it is problematic, that we do try to improve it.

Here Kuhn comes and adds the following criterion: the better system of belief is one that dictates more and better problems, mostly small and manageable ones, that the thousands of men of science can work on, solve, and thereby bring progress. But how do Polanyi and Kuhn know what progress is? Polanyi says, I know but cannot tell a layman. This is mystification and irrationalism that understandably has aroused the anger of Adolf Grünbaum. Kuhn, however, remains reticent. Hempel says that Kuhn thinks there are the normal inductive criteria of progress. This makes Kuhn's philosophy an elaborate joke, of course; which is precisely Hempel's point.

To escape this criticism, Kuhn might take an instrumentalist stance and say, the success of science can be judged by extra-scientific criteria accepted by society at large. But then we have three, not two approaches: (1) the individual, whether layman or scientist, (2) society at large and the place of the diverse scientific and pseudo-scientific institutions in it, and (3) the internal institutional structure of the world of science. And it is here that problems reign supreme. I do not believe in scientific leadership, but allow me to play the Polanyi-Kuhn game for a little while longer before I drop it and move on. Leadership is constantly replaced. The way it is done is smooth or rough or in between. In a smooth transition the leaders or the public address any young person aspiring to a position of leadership and offer him some small tasks which may lead, if well executed, to bigger tasks. Luck and competition and all sorts of contingencies may intervene, and success in a proper series of tasks does not always lead to a proper reward, etc. But smooth transition, as much as it exists, is just here. It has been observed in the biggest political machines, and it has been observed among chimpanzees and baboons. Take this seriously for a moment and ask, what are the tasks of scientific research? Say solving difficult problems, and there you are. But say conducting experiments, say teaching classes, say any other thing, and you have a totally different picture of the situation.

In other words, the merit of Kuhn's philosophy is that he speaks of scientists as of research scientists, and thereby as of solvers of problems, usually small (within a paradigm) and occasionally big (paradigm change – a scientific revolution). But this is a loose part in his philosophy. If you do not like it, you can replace it and have pseudo-Kuhn, i.e. Kuhn's philosophy without problems; or, if you like problems and still dislike Kuhn's philosophy, you can take the problems part and forget the rest.

Once we see that, we may ask, how come Kuhn speaks of problems rather than, as he would if he were more traditional, talk of experiments and of calculations, and of confirming hypotheses, etc.?

There are two parts to the answer to this, one bad, one good. First, the bad part. As long as scientific method is a set of simple clear-cut rules, we do not need the authority of the leadership of science. It is of the essence of Polanyi's philosophy, and which Kuhn has explicitly endorsed, that were the rules clear enough there would be no need for authority, for leadership, for paradigm. Paradigm, says Kuhn quite early in his career, is Polanyi's personal knowledge – knowledge not given to clear articulation. Irrationalism lies at the center of their system.

Problem orientation, however, is the good part, the saving grace.

And the good part is so good that we can keep it without the bad part. We cannot solve the problem of induction, but we keep looking, says the rear-guard philosopher, being quite sure that scientists regularly choose the right hypothesis – by a rule regulating his choice, i.e. by induction. But just as the inductive philosophers say, we do not know yet but we keep looking, so men of science can: they can say, we do not have a principle of induction, we do not know how to choose, but we keep trying to choose properly.

This presents science as problem oriented. It does away with the problem of choice of hypotheses, but, as we saw, leaves us with the problem of choice of problems in science.

3. THE CHOICE OF THEORIES VERSUS THE CHOICE OF PROBLEMS

There is an interesting logical point here. If we can reduce the problem of choice of problems to theories, then the problem of choice of problems will again give way to the problem of choice of theories. And so, if we want to stick to the problem of choice of problems, we must try to prevent the possibility of reducing it to the problem of choice of theories. Can this be done? I think so. First of all, we remember, the problem of choice of scientific theories was important since bad choice impedes research. But good choice must be scientific, and this fact imposes on us the problem of induction. Well, one way around it is to say, with Descartes and Kant and many other apriorists, a good choice need not be scientific in the empirical sense, it may be *a priori* or metaphysical (even though it will be scientific in the sense of having certainty). This will not do. No rational philosopher today accepts any *a priori* certain metaphysics. Or one can say, we do not need certainty, any metaphysics will do if it helps research. I do not like this, as it is instrumentalist and as it says too little about what is a useful metaphysics. Indeed it makes metaphysics play the role of a Kuhnian paradigm – as Lakatos has noticed.

But we can say more. If we do not want to reduce the problem of choice of problems to the problem of choice of theory, then all we have to say is that the choice of theory is not enough for the choice of problem. It surprises me that no one ever said so. William Whewell, the great activist, said, problems concern explanation, and facts call for explanation, and so on. How the choice of any idea with which to explain is executed, he never discussed: he ignored this point. Ernst Mach said, such choices are arbitrary and so not good, and so in order to get rid of the arbitrariness we must look at all that is done in research together. Synoptically. This, he taught, was scientific philosophy. But he was in error, since science as a whole also is rather arbitarry, since all problems that are studied today put together do not make up all the problems we know; and the arbitrariness is not eliminated but enhanced by the synoptic view: why do all physicists or all biologists of a given period center on this or that group of problems?

Duhem said, this depends on the state of our laboratories.

This is true of some cases, but as a general theory it is obviously false, since problems make laboratories no less than laboratories make problems.

The paradigm theory only says, we do center on clusters of problems; and as we saw, it says no more, and so it does not explain the fact.

I wish to stress that the coordination of a scientific community around a given cluster of problems is a fact, and one that no one has explained, as far as I know, except for myself in my 'The Nature of Scientific Problems and their Roots in Metaphysics'. I cannot testify that my explanation is true, though I think it is an explanation that deserves attention. But in line with my present discussion let me say this.

There are two different demarcations of science: scientific theories may be described as a primary given or as solutions to scientific problems. Strangely, the positivists Wittgenstein and Carnap did both, and in a tradition of calling philosophic problems pseudo-problems. For them the insoluble question, even the question not soluble with scientific success, was a pseudo-problem, a puzzle, a *Scheinproblem*.

In other words they assumed that scientific character is bestowed equally on a problem and on its solution, that a question that does not avail itself of a scientific solution is not a scientific question. And, moreover, that just as metaphysical assertions are amiss, so are metaphysical questions. In line with this they could not avoid admitting that a question answerable by a logically true answer is a logical question. But history is full of examples of metaphysical and scientific and logical questions and answers breaking such boundaries. In particular, a metaphysical theory can raise questions some of which are answerable by scientific theories. In a sense, then, a metaphysics may generate problems whose solutions are at times scientific. The way it does so is by generating a research program. For my part I have stressed the fact that the important research programs are rooted in metaphysics, and called them metaphysical research programs: sets of problems generated by a metaphysics. Lakatos, however, preferred the label of scientific research programs, since the answers to the problems generated had better be scientific and then handled empirically. There is little difference in the naming. The important question is, how do we choose questions? It seems that Lakatos has criticized me for allowing questions to come from diverse sources, metaphysics being a major but not the exclusive one; he wanted metaphysics to be exclusive. He evolved thus a philosophical research program: try to show that each significant research is rooted in a problem which is rooted in a metaphysics generative of a cluster of related problems. Needless to say, I think history does not avail itself of this treatment: Lakatos was simply mistaken in his over-enthusiasm.

And so, though metaphysics is, historically, a major source of research programs, and thus of clusters of problems, it is not so much the source of problems that matters, but the centrality of them. To the extent that Kuhn's theory captures reality, it does so because at times clusters of problems excite whole generations of researchers who drop whatever they are doing and attack the new cluster. The new cluster may be a fragmentation of one single scientific task, such as the task of testing the eight-fold way theory of elementary particles. It may be the opening of new technique to solve many outstanding problems of diverse scientific and technological characters, such as radio-active tracing in medicine and in diverse branches of scientific biology and of biological technology, especially agriculture. And it may be rooted in a new metaphysics, such as Einsteinian cosmology, or in a new combination of different metaphysical systems, such as the theory of black holes as radiating black bodies. But whatever it is, the primacy of problems in science becomes increasingly an obvious fact.

The problem of induction, then, which is the problem of choice of theories, becomes secondary. As Sylvain Bromberger says, a researcher accepts a theory not as an object of belief but as an object of study: commitment to study is bigger than that of a mere belief, he notes, since study may be an investment of years of hard work. But there is no question of justification any more: the researcher does not know that he is right in his choice of this or that road since the road is still uncharted. He has the presumption of success, at times, but this presumption is metaphysical, of course: there is no scientific presumption. The reason one chooses a problem, then, in the final resource is the curiosity one has, and the rationality which one exhibits is in the breakdown of the problem, in the plan of the attack, and so on. These are large topics, as yet hardly studied. The program of studying science as a research activity proper, as the choice of problems and the method of attacking them, has hardly begun. It is the core of the activist philosophy of science of the future.

One final question: how do we choose beliefs? We do not know. But as

Robert Boyle has noticed, as well as Spinoza and even Charles Sanders Peirce, it may not be a matter of choice to begin with: our beliefs may be given to us and not open to rational manipulation. But this is a topic for another discussion.

CHAPTER 19

BETWEEN METAPHYSICS AND METHODOLOGY

Metaphysics constitutes sets of statements or principles; methodology constitutes sets of rules. Metaphysics is constitutive; methodology is regulative. That metaphysics gives birth to rules of research – scientific or pseudo-scientific – was obvious for long; Wittgenstein and Popper, following a lead of Bertrand Russell, went further and tried to do away with the constitutive while retaining the regulative which might be necessary for scientific research. In Wittgenstein's case things obviously don't work. I shall venture to argue that Popper's case is also objectionable. (In the case of Wittgenstein, as well as in that of Popper, incidentally, I refer exclusively to the author's first and classic book.)

1. METAPHYSICS AS HYPOTHETICAL

Many essays on the impossibility of metaphysics, on the status of metaphysical utterances as pseudo-sentences, etc., litter the contemporary scene. They usually present the argument that if metaphysical statements were possible they would constitute synthetic *a priori* knowledge. This argument is really below the level of serious argument, since it ignores the possibility of proper statements which constitute neither *a priori* knowledge nor empirical science. Notice that this severe stricture holds for the said argument for the impossibility of metaphysics: in particular it need not hold for all authors who explicitly postulate that no metaphysical statements are possible; e.g., those who like the young Carnap hold that verification principle; or those who, like the old Carnap, consider it a desideratum or an adequacy criterion for a language that it allow no statement that is neither demonstrable nor scientific; for them the impossibility of metaphysics is axiomatic, and so in no need of proof. But those, like Bernard Williams, who do not accept such axioms yet prove the non-existence of metaphysical statements by claiming that these are impossible because they must be synthetic *a priori*, commit a blunder so trivial that one may well gasp in lieu of any critical comment: they ignore the fact that Kant's dual dichotomy, to *a priori* and *a posteriori* and to analytic and synthetic, applies to knowledge only: it is the dichotomies of all *known* propositions to those *known a priori* or *a posteriori*, to analytic and synthetic *knowledge*. It is, in particular, certainly not a table of all possible statements, or possible propositions. (It even excludes, we all know, contradictions which are, doubtlessly, decent enough.) If we rectify this incredible omission, we may concede that all metaphysical statements or propositions are not knowledge but forever hypothetical. To add that whatever is forever hypothetical is impossible is, of course, to add the verification-principle. Now, as I say, the verification-principle is a different kind of error than the argument that metaphysics is impossible since were it possible it would constitute synthetic *a priori* knowledge: this argument confuses the hypothetical with the hypothetical in principle or else it is not an argument but a tortured restatement of the verification-principle. And, I wish to report an impression, the argument is presented not as a restatement of the verification-principle but rather as a proof that metaphysics is impossible independent of that principle.

I hope I am forgiven this elaboration of a trite and obvious point. My aim is to be clear that all statements to be discussed here, unless otherwise explicitly specified, are hypothetical, and usually not scientific, that is to say forever hypothetical.

Traditionally, the role of metaphysical principles was to generate rules for scientific research. So proposed Descartes, so proposed Robert Boyle. Recently it was announced as the latest discovery in methodology – by Imre Lakatos and by a few other philosophers of science. Now already Robert Boyle pleaded for tolerance: scientists should not be quick to dismiss an interesting scientific hypothesis just on the ground that it fails to follow a broad research program, because it fails to conform to some metaphysical principles. He gave an example: his own hypothesis of the spring of the air (Boyle's law) is not in accord with the accepted mechanical principles of his day. Yet those who accept these principles will have to explain mechanically even the spring of a metal spring; and when this will be successful, his own hypothesis of the spring of the air will be mechanically vindicated. Lakatos and his followers go the other way. They declare as quite a general policy, which is adopted and they approve of, the principle of methodological intolerance.

But to return to history. Whereas Descartes, Boyle, and others, accepted metaphysical principles as true – the one as demonstrated, the other as plausible, but both as true – Kant took a different route. His transcendental deduction was the proof that certain principles are essential for science. His theory of ideals said that also certain other principles are essential as regulative principles for scientific research. In either case, however, he made claims for universal validity, not for objective truth. Solomon Maimon preferred to accept Kant's views as hypotheses which aspire to be true, rather than as universally valid but not pertaining to truth. Maimon said, we have choice between certainty and claims for truth; and he chose claims for truth even if as mere hopes or aspirations. Kant was irate, but did not answer. Russell, in his *History*, considered Kant flatly pre-Humean, thus implicitly strongly siding with Maimon (whom he does not mention).

2. METAPHYSICS AS HEURISTICS

Russell's own attitude to metaphysics was not very clear or stable, but certainly he was often suspicious of it as dogmatic, and at times tolerant of it as one wishing to remain undogmatic. Perhaps this is what brought him to the practice, at least once, of ignoring the constitutive and retaining the regulative: instead of the metaphysical doctrine, Nature Is Simple, he suggested – as early as in 1900 – the rule, Search For The Simplest Hypothesis Possible.

Russell's move is most interesting and most fruitful. It did not resolve his ambivalence, however: what status, he asked, did any regulative principle have? Take ethics. Should we declare it absolute and become dogmatists, or relative and condone Nazism? At times he said, let us take it as absolute but the knowledge of it doubtful. Yet he found this quite unsatisfactory because not in the least binding. All his life he was troubled by this. And so, his idea of turning the constitutive into the regulative was with him a mere adumbration that he never fully developed.

Be it as it may, this adumbration received its fullest vindication in Wittgenstein's *Tractatus Logico-Philosophicus* of 1921. The general idea of this book can be presented very easily in a manner which will expose at once both its strength and its weakness: all statements in the object-language are decidable by logical or by empirical means (verification-principle); all metalinguistic statements are meaningless, but look meaningful and significant (as rules of ethics do), since they are sound rules which can be followed but not stated. For example, the *modus ponens* is meaningless: we can infer but not assert a theory of inference. For another example, the law of causality is meaningless: we can explain causally, state causal laws in the object language, but not *assert* a theory of causality, since if we had a theory of causality it would be in the meta-language.

To attempt to criticize Wittgenstein now is really a bit late in the day. Yet a spillover from his doctrine may have lingered longer than his general idea. Or perhaps it was not his ideas at all but those of Helmholtz, of the young Einstein, of Niels Bohr, and of the young Heisenberg. In his Faraday Lecture,

Niels Bohr quotes Helmholtz's Faraday Lecture to say that by excluding action at a distance Faraday "expurgated physics from the last vestiges of metaphysics". Bohr expressed the hope that he had not brought metaphysics back to science. Einstein's operational definition of simultaneity, or rather allegedly operational, but this is another matter, was meant to "expurgate physics from the last vestiges of metaphysics" to quote Carnap, in that it made it meaningless to ask whether two events were absolutely simultaneous, or even simultaneous at all if they were within the light cone. Similarly Heisenberg declared questions of precise enough data about an electron's position in phase-space meaningless, thereby, likewise, expurgating physics from the last vestages of metaphysics. Now, in his *Logik der Forschung* Popper endorses what he calls Heisenberg's program of purging physics of all metaphysics (see last sentence of Section 13; title of Section 16, 'An Attempt to Eliminate Metaphysical Elements . . .' and, a couple of pages later, "the elimination of metaphysical elements is here achieved . . .").

My point that Popper was more influenced by the tradition hostile to metaphysics that went on within science rather than within philosophy is of no mere genealogical concern. Philosophers spoke of scientific languages, i.e., system closed under deduction (this is why Wittgenstein refused to give tautologies the status of propositions proper). Popper, by contrast, never viewed scientific systems as languages; he always maintained that under any conditions scientific systems include statements from which metaphysical statements follow, whether you like it or not.

Therefore, Popper was at the time fully aware of the impossibility of purging science of all metaphysical elements in one important sense: for Popper purely existential statements are metaphysical: "there exists an X" is for him, in isolation, metaphysical for any (descriptive) X. Yet they follow from scientific observation reports. A statement declaring the existence of certain things in certain times and places, which are scientific, entails the statement that such things do exist. Popper never tried, therefore, to purge science of its metaphysical consequences, but only of unnecessary metaphysical presupposition, of metaphysical riders or tags, such as, allegedly, Newton's postulate of the mere existence of absolute space is for Newtonianism: its elimination from Newtonianism reduces its content, but not its empirical content, i.e., not its degree of refutability or of explanatory power. Hence, concludes Popper, such elimination is beneficial. The reason why Popper views such elimination or purge beneficial is, specifically, that it increases the degree of testability of the purged theory and Popper demands utmost testability throughout his *magnum opus*.

There are still other kinds of metaphysics which Popper deals with. First, there are metaphysical principles. Examples are the principle of causality, or of simplicity, or of precision or mathematically or quantitive nature, or concerning generality and universality, not to mention principles of scientific character, or anti-metaphysics, and of the validity of scientific evidence and of the scientific theories which are supported by accepted valid evidence. All these principles are very important to Popper, but he insists on translating them from the constitutive to the regulative mood (see the index to the English edition, Art. 'Decisions or rules of method').

On the authority of thinkers so different as Kant and Wittgenstein, it is customary to consider the principle of causality and such as metaphysical. Yet Kant, at least, who was terminologically very particular, was more specific. He considered the universe of discourse of all possible metaphysical doctrines, as we should say today; there he distinguished between universal principles and particular principles, i.e., between the principles he considered accepted within any conceivable metaphysical system and the principles he considered distinctive of a particular given metaphysical system. He called the former preconditions for any metaphysics, and therefore preconditions for any scientific metaphysics, and the latter he called a metaphysics proper. (He included the preconditions only in his *Critique of Pure Reason*, and the specific metaphysics he advocated in his *Metaphysical Principles*.) Now it is preconditions only that I have mentioned thus far apropos of Russell, Wittgenstein, or Popper, such as the principle of causality. Newtonian absolute space, for example, is clearly a part of Newton's specific metaphysics, as contrasted, say, with Leibniz's, or with Kant's (as expressed not in the *Critique* but in the *Metaphysical Principles*).

Of specific metaphysical principles Popper has little to say. His *Logik der Forschung* of 1935 explicitly declares metaphysical systems of possible heuristic value for science – he chose the example of atomism and it still is the paradigm – yet, towards the end of his volume, in the last section to be precise, he dismisses the proposal to operate with a metaphysical system proper. Why not start with theories on the highest level of universality? he asks, and says, because they will not be testable. (This clashes with his view of degrees of universality as monotonously increasing with degrees of testability, but never mind that.)

The idea that a metaphysics can be presented as a rule of method has meanwhile been developed in another field – by Max Weber and Ludwig von Mises. Weber distinguishes between individualism as a faith, as a metaphysics, specific to the Protestant philosophers of the Reformation, and his own individualism of method – or methodological individualism, as Mises called it. The metaphysical version is a statement, the methodological version is a decision or a rule. As F. A. von Hayek has put it, it is the decision not to be satisfied with any sociological explanation until it will be entirely reduced to individuals and their circumstances alone. Methodological individualism was endorsed by Popper too, in his *Open Society*, a decade after his *Logik der Forschung*; but I shall not discuss this here.

A important example of a specific principle which, as a principle is somewhat of a dogma, but as a rule for research is beneficial, was given by Ernest Gellner: the functionalist school of anthropology held the view that there are no (Tylorian) survivals; that is, they asserted that every institution which exists at present has a social function at present, which is quite independent of history. Now consider the claim that every member or limb in a body has a physiological function. It doubtless is a dogma, especially in the face of the existence of the appendix; yet it generates the rule to search for a physiological function for every member we have, including the appendix, and this rule may be laudable, especially since the view that the appendix is functionless may well be in need of reexamination. Now Gellner considers functionalism, especially its denial of the existence of (Tylorian) survivals, the complete analogue to biological functionalism: it is an obvious dogma, yet it generates beneficial research rules.

3. METAPHYSICS AS BOTH HYPOTHETICAL AND HEURISTIC

All this can be generalized to a theory which may be borrowed from Émile Meyerson. The idea that simplicity is a matter of our decision was, of course, part and parcel, if not the central idea, of the Continental conventionalist-instrumentalist school of the turn of the century to which Meyerson belonged. Yet one cannot identify it with Russell's rule of simplicity quoted above without loss of the subtlety of the problem at hand: the conventionalist or the instrumentalist does not search for explanation proper, let alone true explanation; and so the task of searching for a simple theory is a technical matter, not in any way a dogmatic imposition of simplicity on Mother Nature and her complexities. Now if Mother Nature is complex, the instrumentalist may still prefer simplicity for his own convenience, whereas those who search for the simplest explanation possible and who seek the truth may then be in the soup. They may, however, go about matters in a more open-minded way; they may say, let us pretend that there are degrees of simplicity; the world is, or the true explanation of the facts is, of a given degree. We can approximate the true explanation from below or from above; Russell's rule suggest to approach it from above, i.e., start with the simplest explanation available. Why? Popper has an answer: the simplest is the most testable and so the best.

Now Meyerson acknowledged Duhem's great influence on him, but also Bergson's; he was, in other words, committed to truth in matters metaphysical but not in matters scientific. Yet he managed beautifully to show the idea of unity, for example, shared by Parmenides and Bergson, was a methodological motive: try to describe every couple of entities as one.

Meyerson's leading disciple was Alexandre Koyré. Koyré learned from Duhem and Meyerson that essentialism was false, that science cannot uncover the mystery of the universe. But Koyré also learned from Galileo and Newton that instrumentalism was false, that science is not a search for convenience, but an attempt to uncover the mystery. He developed a view not dissimilar to Popper's but not as definite, and he adopted Meyerson's idea of the role of metaphysical theories as generating programs for research. His paper on Galileo's Platonism which he wrote during World War II is a classic. Already William Whewell, a century earlier, noticed that Kepler was inspired by the program presented in Plato's *Timaeus*. But this was incidental to his critique of the Baconian view of science as based on random collections of facts, and nothing like Koyré's trail-blazing researches. Koyré was soon followed by I. Bernard Cohen who, in his *Franklin and Newton* of 1956, distinguished clearly between Newtonian mechanics, Newtonian optics, and any other Newtonian science, and Newtonian metaphysics which helped scientists carry on their researches by attempting to form hypotheses conforming to their metaphysical ideas.

It is a bit hard, after all these publicly known facts are listed, to consider seriously the complaint of both Popper and Lakatos that I did not acknowledge their priority for these ideas. Rather, let me argue in the following paragraphs, that the first works of these two authors, Popper's *Logik der Forschung* and Lakatos' *Proofs and Refutations* cannot properly accommodate for metaphysics. This is quite legitimate since neither withdrew his views that I take to task here.

As I have already mentioned, Popper expresses his doubts as to the usefulness of metaphysical speculation as a method in a book in which he acknowledges that at times such speculations were of some value – as inspirations – within science. Popper's very idea of degrees of testability is intentionally anti-metaphysical, as he bases on it his proposal to adopt the convention to always choose the utmost degree of testability, the utmost degree of simplicity, the utmost degree of precision. So is his idea that simplicity, explanatory power, mathematicity, precision, all that is good in physics, is nothing but testability. The idea that we value explanatory power because we seek the true explanation of the facts belongs to the older Popper of the mid-fifties, not the younger one of the mid-thirties. The idea that mathematicity and precision are also merely methodological demands runs contrary to Galileo's metaphysics on one hand, and to Heisenberg's and Bohr's metaphysics on the other. For, in Galileo's view mathematics is not merely the language of science; mathematicity is the very profound quality of the universe. For Heisenberg and Bohr, on the other hand, the limit of precision is inbuilt. And, indeed, Popper rejected their reading of Heisenberg's inequality from the very start (and throughout his career).

We can consider counter-examples to Popper's proposals to prefer, within science, the more testable over the less testable. The theory of the continuum, whether of elasticity or of hydrodynamics, is now universally considered not physics but applied mathematics. It is testable, or, to be more precise, it generates regularly testable theories, yet it is metaphysically untenable. Consequently, the program it generates, and its diverse executions, belong to technology, not to science. Any theory of elementary particles, even of nuclear forces, is much less testable than some simple and obvious theories of elasticity, hydrodynamics, two-phase-flow, etc., yet these are considered scientific par excellence.

Take the theory of mathematics as proofs and refutations which Lakatos presented in parallel to Popper's theory of physics as conjectures and refutations; in it there is more continuity, not less, than in Popper's; in it there is less systematic analysis, however. There is less specification, less discussion of simplicity, etc., and more discussion of new aspects of research, especially problem-shift. Often, incidentally, the idea of problem-shift helps us observe instances where high degrees of refutability are viewed as more significant than refutation, to wit, when instead of concluding a debate we shift to a more interesting one, at times to a more criticizable idea. This is in line with Popper's sentiment, with his strict and severe demands for maximal refutability. Lakatos explicitly refrains from discussing the initial mathematical problem which gives rise to a series of proofs and refutations. Nor does he discuss the question, does mathematics always accept a problem-shift if the problem is not thereby narrowed down? Of course, he cannot. For this one has to develop the theory of the role of metaphysical frameworks in mathematical research. This has been done more recently by his former student Peggy Marchi, which, I hope, will soon see the light of print.

To conclude, there is a realistic bias in metaphysics that demands more than mere technical qualities such as the Popperian ones – of content, testability, simplicity – or the Lakatosian one of progressive problem-shift. This bias has to be checked of course – by having, I have suggested, competing metaphysical systems and attempting to evolve scientific hypotheses conforming to them and test these. This theory is in the spirit of Popper's *Logik der Forschung*, and in accord with fragments of his later writings, but conflicts with much that it says, indeed with the anti-metaphysical tenor of the book as has been expounded here.

To conclude, I do endorse Russell's sentiment; it may be a dogmatism to allow metaphysics to enter, and it may be dogmatic to force it to go. It looks as if there is no other way out. But Russell himself offered a way out: making metaphysics not constitutive but regulative. This may be even more dangerous since it may mean seemingly banishing a dogma while really acting in strict accord with it. The remaining alternative I know is Kantian, having metaphysical dogma conflict with each other. Kant himself considered this result indecision and so pointless. Yet this did not stop his metaphysical researches. For the modern student, much thanks to Popper, indecision is not a vice and so need not be pointless, and it may be the generating force of conflicting series of conjectures and their attempted refutations.

Nevertheless, I conclude, it is Popper's insistance on the demand for the preference for the highest degrees of testability every time that gives his critical philosophy its anti-metaphysical flavor. I therefore suggest, first, the replacement of the idea of demand with the idea of desideratum, and second the replacement of high testability with any testability or even mere arguability.

CHAPTER 20

RESEARCH PROJECT

1. WHAT A SCIENTIST DOES AND WHAT HE SAYS HE DOES

In the opening of his celebrated Herbert Spencer Lecture, Albert Einstein said, do not listen to what scientists say, look at what they do. This, he added, should prevent him, as a scientist, from continuing his lecture, yet he went on. This witty remark is packed with serious ideas. On the face of it we have here the claim that scientific method requires observations, not hear-say, as its starting point. This claim is known as classical inductivism. Now what scientists say is inductivism, but inductivism requires that we do not listen to them. The claim that we observe and not listen to hear-say was first made against the scholastic philosophers who allegedly preferred to discuss Aristotle's reports about facts rather than observe the facts. In the present century it was the corner-stone of the revolution in anthropology, especially as advocated by Bronislaw Malinowski. Malinowski rejected reports of travellers, even reports of informants who are members of societies to be studied. Rather, said Malinowski, the society itself must be observed from within, i.e. by participant observers.

Now the demand for participant observation was translated by some, e.g. by the inductivist Max Born, to say, you must be a scientist first and a philosopher only after having accomplished some scientific achievements. Yet others said, but an accomplished scientist is an informant, not a participant observer: he has no training in philosophy or in social anthropology. But then who should train the participant observer? Philosophers who have not yet experienced participant observations? Moreover, it is hard to deny that some of the world's greatest scientists and some of the world's greatest participant observer philosopher-scientists have uttered different views of scientific method and have reported on their own methods in conflicting manners. The situation certainly is confused.

It was Pierre Duhem who said, I think that scientists cannot be relied on regarding scientific method since they contradict each other. Alternatively, of course, they can all be relied upon and the conclusion should be pluralistic: there is no scientific method and each man of science is left to his own devices! In other words, though we need not believe what informants say about general matters, perhaps we can believe their own reports! Can we? The question is complicated. Some reports are made as sheer ritual, e.g. when a scientist claims to have gained inspiration from Chairman Mao's little red book or from Stalin or Marx. Or when a scientist claims to have derived his theory from the facts: as if seeing a falling apple makes one a Newton. Some reports are distorted by a scientist's preconceived notions about scientific method, e.g. when he claims to have observed a fact by sheer accident, which, we know, is *a priori* an insufficient narrative because it omits to tell us why he noted the event and recorded it, etc.

The question is complicated, further, because only in the second approximation can we ignore a person's report and observe his conduct. In the third approximation his report interacts strongly with his conduct and so we must notice it too. Even if his report is a mere ritual, it is extraneous only to the extent that he knows it to be a ritual, and even then it interferes with his work in that it forbids reporting and discussing matters conflicting with the view he has to affirm for ritual purposes alone. Nowhere is this more obvious as when a valid critique of the view expressed in the catechism is dismissed with a ritual excuse. Let me mention two examples.

The inductivist view is the view of science as cumulative, as that of piling truths upon truths and in a hierarchic order. It obviously clashes with the critical view of science as trial and error, namely as the history of modifications of errors, i.e. of false views. By the inductivist view any scientific theory, e.g. Newtonian mechanics, is true and unchallengeable, by the critical view science may include theories known to be false, e.g. Newton's theory. When a criticalist argues with an inductivist, clearly, his claim that Newton's mechanics is false will be denied, and even vehemently, and by any inductivist. When the criticalist will try to prove his claim by the assertion that Newtonian and Einsteinian mechanics are inconsistent and so cannot possibly be both true, then the inductivist will dismiss this triumphantly by the claim that his opponent is ignorant of the facts of the matter: that Newtonianism is an excellent first approximation to Einsteinianism; that it is still employed both in technology and in research; and so on.

All these claims are true. The function of throwing them at the opponent abusively and triumphantly is that of closing the debate fast. And the closing must be fast since the next move of the criticalist is a checkmate: modification and approximation are ammunitions from the criticalist arsenal, not from the inductivist arsenal: to say that something is approximately true is to deny that it is true.

What harm, you might ask, is caused by this, admittedly rather shaky excuse? After all, the person who knows that Newtonian mechanics is both scientific and an approximation is already much better off than his orthodox predecessor of a century ago, since the predecessor refused to allow a scientific theory the status of an approximation, especially Newtonian mechanics, whereas the latter-day inductivist allows for it.

The answer is simple and very very important. It runs contrary to much that many philosophers and social anthropologists say, yet it is both simple and crucial. What is said apologetically and under pressure is not capable of being used as a regulative idea. Thus, the one who under pressure admits Newtonian mechanics to be an approximation may also under pressure admit that special relativity is an approximation to general relativity. Yet most physicists do not feel under pressure here and claim special relativity to be a special case, not an approximation, and they thereby confuse issues. Moreover, when not under pressure, say, when conducting new research rather than admitting the force of concluded research, they will not say, here is a successful theory, perhaps it is not true but an approximation to something else. Anyone familiar with the work of Lee and Yang and with the ensuing discovery of the neutrino knows that their way of thinking was just this. Confronting a difficulty, they noted, as others did before them, that it stems from a very successful law – the law of conservation of parity. And, they said, suppose this law is a mere approximation; under what condition, then, would it be a very poor approximation so that constructing these conditions would permit a crucial experiment?

An inductivist under pressure may admit that the law in question is but an approximation; but it takes more than admitting to explore the possibility. Hence, the inductivist catechism is a possible constraint on the conduct of a scientist.

Let me take another example. The classical inductivist view of research is that it is based on no conjecture, but either on older scientific findings or on nothing; the finding based on nothing is, obviously, quite accidental; the discovery based on prior knowledge is either one waiting for anybody to see it or one newly deduced from older science. On the newer inductivist theory, a scientist may start with any conjecture not clashing with old science – and the accent is on any – but then test it severely and either refute it or confirm it. These views are not quite operative in research. And so, one might ask, what does it matter that they are held? The answer is that as long as they are held research projects can hardly be intelligently written up. The damage of this fact will be discussed soon.

Supposing old inductivism be true, how can one aim at an accidental discovery? Only by serendipity, i.e. by doing something else, or by a wild guess, such as, what will happen if you mix a certain concoction never tried before and process it in the lab by using the instruments present there in any specific order. I do not think any foundation will give anyone a grant if he only says, I wish to try out this and that hitherto untried arrangement and observe what has happened. You may have heard the story that radioastronomers who tried to reduce the noise in their instrument to zero, failed and then, upon hearing about cosmic background noise decided that their failure to reduce all noise was due to cosmic background noise and, hey presto! they won the Nobel prize. That this story is incredible can be seen thus: suppose someone says, I wish to reduce the noise in my instruments to zero – and all instruments have noise! – and asks for a grant. Will he get it? Certainly not. Two objections will be made against him, one that it is impossible, and the other is that, per impossible, it would be pointless. Now I say per impossibile, since, of course, had he succeeded to perform the impossible it would be a miracle, and a repeatable miracle is a scientific revolution of sorts. The zero attempted by our Nobel prize winning astronomers, by the way, is not a real zero, but a zero by astronomical standards. And so, of course, when we say zero noise level, we may well mean, as near zero as our present scientific theory permits, namely, while taking account of Heisenberg's principle and of the cosmic background noise. And so, of course, anyone who wishes to reach as near zero as possible may be asked, what is the point? Perhaps our Nobel prize winning radioastronomers had a point, had a reason for attempting to reduce the noise, and, still better, a scientific reason to expect success. In that case their failure obviously signifies something still unarticulated.

It is this kind of reasoning that I have tried to apply to diverse cases in the history of science, especially Priestley's discovery of oxygen and Oersted's of electromagnetism. Priestley says a few times that were not a candle handy while he was making his experiment he would not have discovered oxygen. Yet I have tried to explain both how the candle came to be there and why he said it was an accident. Briefly, he expected the candle to behave in the way opposite to what it did. And so he discounted his false expectation, and so the candle remained there stripped of its rationale. As he said, by accident it just happened to be there.

2. THE HARM CAUSED BY INACCURATE REPORTS

And so, you may ask, what matters what Priestley said, if he did the right thing? The main thing for a scientist is to be undogmatic, unapologetic, and that he was. This is very similar to the case of the greatest dogma in the world of science ever: one hundred years ago all men of science swore by Newtonian mechanics. It was, they were deeply convinced, the epitome of certitude, of scientific perfection. It was not only one hundred percent true: it was not reducible to any other theory; it could not be superseded by any theory – whether one that modifies it or one that does not modify it of all. But then came Einstein and he was at once recognized. Even Schubert and Van Gogh and Samuel Beckett were not as quickly recognized as Einstein. So what does it matter that scientists affirm a dogma if they accept things opposed to their alleged dogma so fast?

The answer is, research suffers, not the recognition of its success. And research must be given some rope. It is not good to say, if you refute my theory I will concede but until then I sabotage you and recommend that you get no research grant and no job, especially since you refuse to draw conclusions from your past failures to refute me. And it is no good to admit refutations as discoveries when they happen while forgetting that they are refutations of current received opinion. It is here that my admiration for Popper is boundless – perhaps because I compensate for rejecting everything else he says, I cannot judge, but perhaps because he presents a challenge and observes the fact that great empirical discoveries are refutations of great theories that precede them.

Although I think this is an eye-opener and a tremendous relief, I think Popper is a mere trail blazer. There are many possible ways to test a theory, most of them *a priori* not very promising. Popper advises any ambitious young experimenter to seek the easiest way to tackle a citadel; but he says no more. And if the young ambitious empirical researcher is in search of a grant, he need say more about his expectations, especially since the referees and the grants committee are likely to refuse him his request just on account of his conceit and rebelliousness.

So much for empirical discoveries. As for theoretical discoveries, the picture is still more grim. Classical inductivism advises the theoretician to do the impossible: deduce theory from facts. More moderate and modern inductivism recommends to seek explanatory hypotheses, and it admits that these come out of the blue; so does the conventionalist-instrumentalist methodology. There is one word that is very exciting, that is often used in research, yet seldom in the theory of scientific method, and even then with no excuse: it is the word "problem". What is a problem? What is a good problem? How do grants committees decide whether a project is good enough we do know: they want the problem to be both good and manageable. In the whole literature on scientific method barring a few essays, mainly mine, there is no mention of these qualities, much less their characteristics. What makes a problem good, important, interesting? And what makes it manageable? When is a partially manageable problem manageable enough to warrant a grant? No answer. No mention of the question, even.

Not only is it nowadays agreed that research is conducted around problems; it is admitted, though only half-heartedly and as a mere after-thought, that even good experiments can be conducted after solving some problems. But that is not enough. On the authority of Poincaré and Einstein and Jacques Hadamard, on the authority of R. G. Collingwood and of Bertrand Russell, it is now admitted that one must immerse oneself in a problem, read all about it and study everything possibly useful for its solution, and then let go. A possible solution comes or not, and once it comes it must be examined. So little on the authority of so many giants! How come? I do not know, but perhaps it is because too few study this exciting field, perhaps because the current mythology of science concerns itself with other matters.

And here, then, is another answer to the question I repeatedly pose in the present chapter: what does it matter if scientists say ritual rubbish about what they do if they do the right thing? I have tried to argue that the ritual rubbish is a hindrance, that many scientists follow it and produce rubbish quite unnecessarily since in better accord with what they believe than the work of the scientists who produce gold. But now clearly I may also attend the philosophers and historians of science, since I am one of their crowd: what philosophers and historians of science have undertaken to do is to describe and analyze what men of science do. Yet often enough they fail to heed Einstein's advice and instead of looking at what scientists do or did they listen to what they say. One example is phlogistonism. Despite the enormous progress in chemistry in the period between Robert Boyle and Antoine Lavoisier, it is still the current opinion amongst historians of science that the ideas current between these two thinkers were silly. Some historians of science, very few of them, only defend the phlogistonist ideas as useful in their day; no one but me praises them as ingenious hypotheses that led to great discoveries. The reason is that in the Newtonian era science was deemed to be certitude and the views of the phlogistonists are, of course, false. And the idea that some ideas, though false, are interesting, ingenious, worth examining by the means of experiment, etc., all this Popperianism is news that has not yet ousted the old lore. And philosophers of science are much worse than historians of science, and much more apologetic for science. They try to fight a last ditch battle: though scientific certitude is gone, they confess, not all is lost and near-certitude, as near as possible, can be established and it is our duty to try. Right or wrong, it is a post-mortem approach to science, it leaves no norm to the study of scientific research, much less research projects, their merits and defects, and the way they are and should be assessed.

I have before me a recent and very important issue of *Daedalus*, the most established journal of the American Academy of Arts and Sciences, Spring 1978, devoted to *Limits of Scientific Inquiry* (edited by G. Holton and R. S. Morison; it is also published as a book by W. W. Norton, N.Y., 1979). The limits discussed, it turns out, are moral, social, economic, political. Not methodological. On the whole philosophy is almost totally ignored; rather than the presuppositions of science, the presumptions of science are at times discussed. I am all for this issue, and even its circumventing philosophy is a good thing, since philosophy of science is so backward and, worse, defensive and apologetic on the behalf of science and so disingenuous or at least not serious. This seems to me to be the point of the Epilogue to this volume, written by Gerald Holton, or rather of the very last page or so there. He discusses there the new concept of progress and he cites two philosophers of science, T. S. Kuhn and the late Imre Lakatos. And, I dare say, his choice is right, since they are the two influential philosophers of science (though one is a historian of science, really, and the other was a philosopher of mathematics, really, they are influential as philosophers of science) today, barring Norwood Russell Hanson whose ideas are not too clear as he died too young to clarify them. Kuhn, reports Holton, gave up the concept of truth, even the concept of the approximation to the truth. With this, concludes Holton, science must give up all moral claim for autonomy. With this, I should add, philosophy of science must abdicate, and all theory of science must come under social philosophy and of public administration (not, mind you, sociology proper, since, as a science, sociology too has no claim for either truth or verisimilitude, if Kuhn is to be followed).

Lakatos is on the other extreme. He said he had a theory of what makes a good or a progressive research program. And a research project is good, he implied, if it integrates within a good research program. And he wanted, reports Holton, to impose his ideas on research grants committees and make them conform to his ideas. For example, Holton cites Lakatos, contemporary elementary particle theory and environmental theory do not qualify and should not be supported. Holton refrains from a comment. He thinks Lakatos now had a rope long enough to hang himself. But this is true only because Lakatos is dead. Were he alive he would relent as a compromise move, grant all elementary particle physicists and some environmentalists their right to public money, and expect as a compromise counter-move from the grant committees to listen to him more attentively. Had they done so he would not have told them how to distinguish good research programs from bad ones in general – he simply could not do so; nobody can. He would, however, have given them specific advice, thus enhancing his rapidly increasing political power still further.

3. EVALUATING RESEARCH PROJECTS

What we need, instead of Kuhn's abdication of all responsibility and Lakatos's eager acceptance of it all, is to look at how research projects are evaluated, to evaluate the evaluation, and to make tentative proposals for a possible improvement. And the first thing to learn to ignore when doing this is what excites the vulgar: power politics. Let me show this in a little detail before finally coming to my point, perhaps.

The first thing conspicuous to the onlooker, especially the sociologically or socio-politically trained one, especially if he is ignorant of science, is that research grants and benefits and jobs go in clusters. In Italy things are clear-cut: one job goes to a Christian Democrat, and so another job has to go to a Communist, and vice-versa; and not only in philosophy but, when possible – we shall return to this – also in science. In the United States at times there is no politics and no perceptible academic political affiliation; rather, there are gangs and clusters of scientists. These clusters are now identified by a newly discovered litmus paper: the citation-index so-called. That is to say, members of a gang tend to cite each other more than outsiders, for obvious reasons, legitimate and not so legitimate. Had these paper sociologists of science gone out of their libraries into the real world they would have discovered the machinery by which such clusters are created. Editors, especially unscrupulous editors with no respect for the truth like Imre Lakatos, clearly make conditions: I will publish your paper if you omit these references and add others. Cowardly editors do similar things but more obliquely: if and when you corner one and demand an explanation as to acceptance or rejection you will get certain hints. Teachers, likewise, may submit former students' papers only after their bibliographies are tidied. Derek Price of Yale, the inventor of the citation index, sees in the formation and success of such scientific power centers the formation of Kuhnian paradigms, of what Kuhn talks about when he says small normal scientists follow their fearless leader. Perhaps.

What I wish to suggest is that all this is philosophically neither here nor there. One way or another, either the group selects its own works, or works that represent it, in a progressive and interesting and scientifically viable way, or it does not. If it does, then sociology apart, the methodological question is, how does it do it. If it does not, then, scientific method apart, the sociological question is, how does it do it. For the cluster or group or gang has its own values, norms, and mores, and these are either scientific in part or not. If not, the gang will remain marginal to science even if its members run academies, win Nobel prizes, and all. If yes, the question is, by what rules, criteria, convention, and custom. The sociologist may want to know how the rules are implemented. But he will sooner or later have to decide which rules. The methodologist may only want to know which rules.

Take the late Imre Lakatos again. He was power hungry and amazingly successful in attaining power. He had great ideas of his own, and these gained him the prerequisite respectability but no power. To gain power he meddled in things he had nothing to contribute to. So he pretended he had a contribution to make. His pretense forced him to pretend to be able to judge research projects. So he had to pronounce judgments on research projects of all sorts. How did he do it? By trial and error; and he clung to judgments that enhanced his power and gave up those which risked it; as to the majority of his judgment that went neither hither nor thither, he had his ego reign supreme and he judged high-handedly and arbitrarily. The result of this is that for the serious student of research projects Lakatos exists as a mere mirror of existing judgments, neither more nor less. His meteoric access to power may interest a student of academic politics or a biographer of this remarkable and fascinating person who was also a trail blazing philosopher of mathematics. To the methodologist of science his work on research programs is at best secondary.

What all this illustrates is, I think, the great handicap this field suffers from: a current myth known as the distinctness or distinction between the descriptive and the prescriptive. For, I say, the method I recommend on grounds of a social philosophy that I leave untouched – it is discussed in my book *Towards a Rational Philosophical Anthropology* – we have to study current practices, criticize them, and offer possible improvement. Yet when we focus on current practices our *a priori* preferences already come into play. When we look at scientific practices we take paradigms such as Newton and Einstein and Fermi on account of their exaggerated dimensions that make some of their characteristics visible to the naked eye. Yet Kuhn has a valuable thing to say when he notices that these days most scientists do microscopic work.

Question: what values does microscopic work have? This question pertains to science and to other fields with giants and dwarfs in them. Let me take an example that is certainly no offense to the dwarf: the string quartet on a small isolated university campus, manned by four old professors. These fellows play for fun and have enormous fun regardless of the poor artistic quality of their production. They will take no offense if I say they make no contribution whatever to the history of music or even of musical education. And certainly it is hard to compare them to, say, the Mozart family playing a trio for fun. Nevertheless, I say, these four professors are in a sense pillars of the musical tradition without which no Mozart is possible. It is very hard to say how and why, and easier to feel it, but feelings are no argument and can mislead. Anyway, the four rural professors ask for no grant from any arts council and so they need not trouble us. But they do ask for grants, modest perhaps, from the science council. How do we judge these?

It is my considered opinion that most of these professors contribute to the arts and to the sciences in the same way, that contrary to Kuhn who claims that they solve puzzles and thus contribute to the growth of science, I think they are dilettantes who make no scientific original contribution. But just as a science student must repeat some experiments, he has to have a task of research, and his professor offers at times only a taste of research and a sense of research, not the real thing. Nevertheless, dilettante science is as important for science as dilettante music is to music.

It is a fact, I contend, that humble grant requests are judged differently than ambitious ones, and for a variety of reasons, mainly political: it is a small cost for involving many more people in research than strict standards would permit. And though I dissent from the rationale of this practice, I endorse the practice itself.

Yet the practice makes sense only as long as science is viable: if all scientific research is but marginal, science would lose its point. What role does the vast grant offer? Historically, it is certainly marginal, since until World War II it did not even exist. Now that it is a generation old we can assess the good and the bad that it has caused. The bad is well known. The McCarthy era and the Oppenheimer affair, in particular, are monuments to it. The good of it is also quite obvious: once we grant a large sum of money to investigate a new theory, e.g., elementary particle theory (that Lakatos hated so), it is at once a focus of attention. How do grants committees decide that a theory is worth while? This is a general methodological problem that has been studied by Popper and his followers, myself included, but which has little to do with research projects directly, although, of course, a lot indirectly since it pertains to recent background to current research.

And so, at long last I am arriving at the question, how is a research project actually evaluated and to what extent does it insure progress, and how? The first approximation to an answer to this question is almost purely sociological in that it pertains to the sociological projection of a rather critical view of what science is. I would term it the Merton-Popper view; Merton took a critical attitude of science from the nineteenth century philosopher William Whewell; Popper applied his own views. Both agree that science is mainly conjectures and their tests. Criticism and dissent thus become fairly central to their sociology. And what they both say is, briefly, what Hume said before, namely that competition insures a certain level of criticism in science. In other words, though scientific research admittedly centers around mafias – to use Paul Feyerabend's nasty expression – competing mafias shoot down each other's ideas, thereby leaving room for better ones. Better ideas, then, are those which require better guns to be shot down. This is Popper's view. Whewell said, some ideas stand out as beyond criticism, as impervious to any criticism. Since Einstein, this idea can be deemed passé.

Now all this seems to me true, yet quite insufficient, both in that it overlooks the fact that two mafias have to be equally good to survive, which means they will unite to destroy a third potentially better one. To prevent this we have to know a bit more about research programs generative of research projects. Indeed, I think it becomes now increasingly universally agreed that research projects all too often are judged in the light of a broader research program. This, of course, invites dogmatism, of the kind Kuhn advocates, and even Poper finds some use for, or of the kind Lakatos illustrates, or of any other. How do we manage to offset this?

I cannot enter this. There are competing metaphysical systems generating competing research programs. This much can be said in general. More particularly we must take current examples. I plan to do so on some other occasion.

CHAPTER 21

THE METHODOLOGY OF RESEARCH PROJECTS: A SKETCH

There is a traditional reluctance among methodologists to study the ever increasingly important phenomenon of research projects, research project evaluations, etc. The reason for this is that projects are embedded in programs and programs in intellectual frameworks, or conceptual frameworks, or metaphysical systems. It sounds dogmatic to judge the product of research by a reference to a metaphysical system. Yet, first of all, it is not so dogmatic if judgement can go both ways, if we have competing systems at work, and if what we assess is not the outcome of a project but the existing assessments of projects prior to their implementation. Indeed, one of the most obvious things to do is to compare our assessments of projects before and after their implementations. To this end some further theorizing is required.

Here, then, is a sketch of the theoretical background to such investigations, a sketch of what has been said thus far relating to the matters (not yet) in hand.

1. THE MISSING STUDY OF RESEARCH PROJECTS

To begin with, there is an obvious lacuna. On the one hand we have a tradition of assessing research projects and programs; on the other hand we have not enough theoretical and/or historical studies of these. It is hardly necessary to mention the fact that today's budgets for single research projects and for groups of projects are ever increasing; that this makes preparation and detailed planning essential; that the assessments of the detailed plans on diverse levels of the administrative hierarchies of the research administration become inevitable since the projects are usually publicly financed; that public guidelines are developed for these assessments by public servants; and that these guidelines can be given to critical examination and theoretical explanation, perhaps also improvement.

But research projects and programs are not new. When Lavoisier put his research results in a sealed letter in the safekeeping of his learned society he wished to secure priority without letting people in on his activity – he was embarking on a new and promising research program (of the antiphlogiston school). Mathematicians are quite familiar with the Erlangen program of Felix

Klein and of Hilbert's program. The famous paper by Niels Bohr on the correspondence principle offered a research program that within a few years came to fruition in the form of the new quantum theory. The question, do we want a covariant version of quantum theory or not, was debated by physicists at a great length. Darwin set the tone of biological research for generations to come, and so did the neo-Darwinian neo-Mendelians. So, to a smaller extent, did the research with the aid of radioactive tracers to sets of given problems of adsorption and circulation and growth. In psychiatry the medical or chemical school is engaged in researches which, diverse as they are, fall intuitively to the same category, and are set apart from projects followed by members of other schools (the psychiatric and the antipsychiatric).

And yet there are no studies of the history of research projects and programs prior to the last decade or two. There is almost no study of general concerns of students of a given field at a given age. Why?

The first reason is inductivism: any project should be based on existing knowledge and so needs no further justification and analysis than that of the knowledge available at the time. If, per chance, a research project is based on a different kind of ideas, so the inductivist lore goes, then it is a preconceived notion, and so a prejudice. It is then unlikely that the project bears fruit, and if it does then its connection with the prejudice that justified it before success is tenuous and may be ignored – should preferably be ignored indeed. Any other possibility, a project based on no idea, is covered by the title of a hunch.

Philosophically, all this can be dismissed: a hunch can be articulated into an *a priori* guess which may be pursued, beneficially or not.

2. BUCKETS, SEARCHLIGHTS, AND POWERHOUSES

There is a simple oversight in the inductivists' suggestion that since a project is based either on scientific knowledge or on prejudices or on hunches, it requires no need to go into further detail. The oversight is, as it happens, of a fundamentally important point. The inductivist opposition to prejudice is that it blinds its holders to the facts, especially to the truths that conflict with it. The empiricist recommendation is, thus, to shun all prejudice, to approach experiment with no definite project. This, indeed, explains the lacuna.

Now if one with a project based on a prejudice is successful, this may be a lucky accident, but not if it is systematic. Systematically speaking there is here a fundamental philosophic difference of approach to experimentation, the empiricist unplanned one, which allows facts to come as they are and speak for themselves, and the intellectualist apriorist one which creates the pigeonholes for the facts prior to their appearance.

The metaphors Sir Karl Popper has offered are very apt. The bucket theory of the mind is the empiricist theory of indiscriminately taking in all facts as they come. Its opposite he calls the searchlight theory. Now, from the view-point of the bucket theory, there is no problem of choice of facts or of experiments – they are all welcome and of equal value. From that viewpoint there is the problem of choice of theory, of course, and it is solved with the help of all facts or of all possibly relevant facts. This is the principle of the non-selection of facts, announced by Bacon and by C. S. Peirce, and by J. M. Keynes and by R. Carnap: when assessing the probability of a theory consider all facts available, and then dismiss, if you will, only facts irrelevant to the problem at hand. The searchlight theory offers a very different approach: theory guides us in the search for new and relevant facts. The question asked about the searchlight theory is, what guides our search for theories?

For the apriorist proper, be he Descartes or Kant, the answer is, in principle, very simple: just as empiricists think certain facts impose themselves on our senses with finality, so apriorists think certain theories impose themselves on our intuitions with finality. Let us take for granted that there is no proof of finality possible in empirical science. What, then, can guide the search theory by the fallibilist holder of the searchlight theory?

The easiest answer is that the process of research is two-way: theories help search for facts, and vice versa. The best example is an experimentum crucis, an experiment helping us decide between at least two existing theories. But all in all, facts only "suggest" existing theories; they may also stimulate our imagination, but then it is the imagination, not the fact, that originates the theories. (And whether facts or coffee stimulate the imagination is of no import except for the theorist of imagination.)

Is there more to say on the topic?

It should be obvious that just as the searchlight (that a theory is) helps us find facts which as yet we do not know, but have some ideas about, so there may be something, some super-searchlight, that helps us find searchlights (or theories) which as yet we do not know, but have some ideas about. Let me call this a powerhouse. The powerhouse will not in itself light for us buckets full of facts, but make us devise searchlights that can do so. Do we have a theory of the powerhouse, then?

3. HISTORIANS FOR POWERHOUSES, POSITIVISTS AGAINST THEM

The powerhouse is, of course, the metaphysical presuppositions, the intellectual framework, within which a scientist can work. It offers both constraints and suggestions. A typical example is quantum field theory which suggests that we define Lagrangian operators and such to generate field equations, that we make these equations Lorentz invariant or such, and that we quantize them. Classical Newtonianism recommends that we postulate the existence of diverse conservative forces, etc.

Such frameworks were consciously devised. Descartes, Boyle, and Kant, called them metaphysical foundations or simply metaphysics. Even some positivist authors found these frameworks in history and acknowledged their value; they had no good word for them in general, but had to admit that in some historical juncture they played useful roles. Not so E. A. Burtt and Alexandre Koyré, whose studies were meant to show how central and substantial were the contributions of these frameworks for the advancement of science at the time of the scientific revolution. This was before World War II. After it I. Bernard Cohen extended the study to cover the eighteenth century, in his study of the Newtonian intellectual framework and its importance for the researches of Benjamin Franklin.

Rich and interesting as this historical harvest is, it will only be utilized, it seems, when the theory guiding its authors becomes explicit and critically examined. What role, then, can a conceptual framework play in the process of concept-formation and theory construction?

My starting point and point of contrast between the historians and philosophers of science in the first half of our century is C. G. Hempel's classic slim volume, *Fundamentals of Concept-Formation in Empirical Science* which sums up an era in research into the philosophy of science, that of modern logical empiricism or positivism. Some portions of that volume discuss the possibility that theoretical concepts are fully definable by empirical means, whether operational definitions or other means (the other means are known as correspondence rules). Another portion is devoted to the only theory ever offered of partial definitions of theoretical concepts with the aid of empirical concepts, namely R. Carnap's (the theory of reduction-sentences, so-called) which admittedly does not deliver the goods. Anyway, its scope was too narrow from the very start: it only came to help us define dispositional terms in terms of observational terms, and no more. In addition it is objectionable on the ground that there is no observational term that is not dispositional.

At least we have yet to meet an instance of an observational term which is not dispositional (water, for example, says Popper, entails wetness, fluidity, decomposability to two gases, etc.: fluid is viewed by Hempel as observational and yet it is clearly dispositional: a fluid is that matter which is disposed to take the shape of its container). Hempel's slim volume also contains the Carnapian theory of the adequacy criteria for rendering concepts quantitative. It contains other, even more technical points, less relevant to our topic, such as the idea of abstract axiom systems that may be empirically interpretable and so become empirical hypotheses.

Towards the end of that volume Hempel brings an example used by Moritz Schlick, the founder of the Vienna Circle. Why, he asks, should we not use the Dalai Lama's heartbeat as our universal time unit? Because, he says (p. 74), this "would preclude the possibility of establishing any laws of the simplicity, scope, and degree of confirmation exhibited by Galileo's, Kepler's and Newton's laws". There is no reference here to the intellectual frameworks of these thinkers as conflicting with that of the Dalai Lama theory. The gap is extraordinary and it wants examination.

4. THE PEDIGREE THEORY AND THE HIC RHODOS THEORY

There is here a red herring to get out of the way. There are two competing views of the judgement of the worth of a theory, the pedigree theory (as Popper calls it) and the classical hic Rhodos theory. The pedigree theory says, a theory which has been properly developed is a good theory, the hic Rhodos theory says, never mind how a theory came into being; we want to see here and now what it can do for us and we should judge it on its merit. (The label "hic Rhodos" comes from the Aesopian fable of the fellow who bragged that he once performed a marvellous jump while he was in Rhodos. Hic Rhodos, hic salte, he was told: here is Rhodes and we want to see you execute the marvellous fiat here and now.)

Sir Karl Popper ascribes the pedigree theory to Bacon and Descartes: Bacon said, a theory which emerged from experience by proper induction is a good one, the rest are not; and Descartes only substituted intuition for perception as the proper source of knowledge. Popper does not say who is the originator of the hic Rhodos theory, and I do not know who would be his candidate for it. For my part I would like to ascribe it to Bacon and to Kant, not so much because they have originated it, as much as because they gave it strong expression.

For, in my opinion, there is a confusion here between two pedigree theories, one a criterion of quality, and one an explanation of quality. The snob says, if he is an aristocrat he is good, and I shall not test his qualities; if he is not I will not be impressed by them. A reasonable conservative will say he is good, he is good in having the tested qualities of leadership, and he has these qualities because he was well-bred. The snob's pedigree theory conflicts with the hic Rhodos theory; the conservative's view is consistent and may be refuted either by an aristocrat of low quality or by a commoner with the quality of leadership. The conservative is aware of this for he has modified his theory in recent years, to speak of a high correlation only, between pedigree and quality.

Now Popper's classic 'On the Sources of Knowledge and of Ignorance' (*Conjectures and Refutations*) discusses validation only: the classical pedigree theory says, theories properly constructed along given rigid lines are all right, others not. Now no one accepts such a strict correlation today, and so the validation by pedigree cannot be good enough: there must be some stricter method of validation that selects the one rotten apple in the basket even if by some criterion there is at most one rotten apple in the basket. Yet, I should add at once, if there were no correlation of any sort between the method of research and the resultant theory, we could have no theory of research projects and programs. This, I think, is a point of cardinal importance. And so, though pedigrees are not validations, they are of interest. Notice, also, that as far as validity is concerned, both snob and conservative, if they are of the strict sort, agree on pedigree as validation, yet conservatives will agree with Popper that it is advisable to test the most aristocratic of theories whereas the snob will wince at the suggestion. Notice, likewise, that conservatives of the non-strict sort consider pedigree explanatory of high quality, not a guarantee for it.

Popper's advice to researchers is, fall in love with a problem, and let your imagination go. Very important, of course; but can we say more? What, for example, is Popper's attitude towards metaphysical frameworks? In his classic *Logik der Forschung* he says they are not meaningless, since they can become scientific; but they are too abstract for being testable, and so he advises against them. Yet in that very same volume he transformed certain metaphysical ideas, like the law of causality and the law of simplicity (every event has a cause; nature is simple) to rules of research, thus systematizing a suggestion thrown out by Bertrand Russell (*Leibniz*) and by Wittgenstein (*Tractatus*) beforehand. Now the rules of research, he said, are conventions of the scientific world, tested by their fruitfulness. In a sense the whole theory of research projects is a take-off from that; in a sense it is a criticism of that, since it proposes alternative and competing rules, based on some powerhouse ideas.

5. IS METHODOLOGY INEFFABLE OR RATIONAL?

Enter Michael Polanyi, and on his coat-tail Thomas S. Kuhn. Polanyi begins by recognizing induction as the only plausible rational procedure of producing scientific theories. And he accepts the traditional arguments against inductivism, from Hume's to Popper's, as well as the fact that men of science undertake research projects with certain discrimination. There is, then, a powerhouse, there are searchlights, there are discriminations everywhere. But there can be no theory of the discrimination, no theory of the powerhouse.

There is an important subtlety here that is regularly obscured in the literature. Neither Polanyi nor Kuhn reject entirely the traditional or modern criteria of excellence of scientific theory, such as testability, confirmation, simplicity. These criteria, however, are far from enough in their views. They are irrelevant to the task of theory construction, and when theory construction runs against accepted criteria it may override them, like the traffic policeman who can overrule traffic-lights.

Polanyi insists that there is something ineffable operating here, a tact or an intuition of the scientist, or rather of the leading scientist (who acts as a policeman); miraculously, scientists function in accord with fellow-scientists: They agree on leaderships and on proper conduct. The accord between men of science is neither explicable, nor dispensable with; it is the outcome of a scientist's apprenticeship in the tradition of science. A scientist may be critical of an aspect of science but not of science as a whole. Hence, Hume's critique of inductivism, though valid, cannot hold against Polanyi.

Kuhn has added to Polanyi's theory the idea of the paradigm. A paradigm can be a scientific theory which serves, *pro tem*, as the model of a scientific theory; it can be a fragment, and it can be a whole intellectual framework. It is that idea round which scientists rally, it is the core of their unity. For Kuhn, as for Polanyi, the ineffability of the scientific procedure is important. But he stresses that this does not make the man of science live in a precarious situation akin to the mystic whose ineffable doctrine is so notorious. On the contrary, there is normal science, which is the routine practice of the normal scientist, and it comprises a small task, a small problem (Kuhn calls this a puzzle, in accord with ordinary use and in conflict with that of Russell and that of Wittgenstein). The peculiar thing about these small problems is that somehow they integrate, by virtue of the existing paradigm. When their degree of integration is much lower than expected it is time for a revolution, for a change of paradigm.

How a revolution occurs should be discussed perhaps in Kuhn's best selling *The Structure of Scientific Revolutions*. It is not. The leadership, or some Young Turks if the leadership fails, spend some sleepless nights, and by a feat of daring intuition, *à la* Popper, they develop a new paradigm. This is endorsed by the rank-and-file if and when they can apply it, with the use of their personal knowledge, *à la* Polanyi, to many routine problems.

All this is terribly vague and unsatisfactory, but at least it recognizes the fact that scientists work on fairly long-term research projects.

6. THE RATIONAL WAY TO ASSESS METAPHYSICAL FRAMEWORKS

The problem we started with is that of discrimination. In fact some projects are judged to be good, some not good enough; and the judges may be good or not good enough. Is there anything that can be done about this last fact?

Says Polanyi (and concurs Kuhn), only experts can judge projects, only experts can judge experts. Moreover, neither judge of project nor judge of expert can fully articulate his standards. Moreover, criticism is always partial, never total, so criticized expert usually stays in and improves his practice. And if not, he is simply left out of the game; sad, but at times it happens.

Polanyi and Kuhn demand that we leave things exactly as they are. Especially, our theory should not go beyond the present situation. But once a sociological study of all this proceeds well enough, Polanyi and Kuhn will have to endorse it by their own tokens, and so find themselves inconsistent in accepting and rejecting it at once.

The Polanyi-Kuhn theory is based on observed uniformities in science. But we observe uniformities, deep and superficial, over groups large and small, and regarding doctrines and rules of method articulable to this or that degree. And there were important loners, of course, throughout the history of science. Also, the plain uniformity they postulate leads to a plain view of a change in the climate of research in a given field, whereas in fact the change may be rapid or slow, drastic and far-reaching or not, sweeping the whole profession or a marginal group in it, and the marginal group may or may not lead to victory, either to achieve supremacy or to see at once a new kind of dissent. All these phenomena are reflected in the researches in which participants in a field of research are performing during the time in question. How do research practices change?

In order to get a new kind of research going on a large-scale – of the character dubbed a paradigm by Kuhn – some rational debate is necessary, at least among the leaders of the field in question. Indeed, plausible arguments for and against the change in the kind of research are heard, and so they can be recorded and assessed. I recommend that this be done, that the arguments be improved upon by the attempts to notice and eliminate irrelevant factors, such as university and national power-politics considerations. Such researches have already started, and Lord Snow and Daniel Bell provide examples of researches cluttered with irrelevancies. Such arguments can be further improved by replacing whenever possible poor inductivist and conventionalist arguments by realistic and critical ones. Also, I recommend that the good inductivist and conventionalist arguments, when presented, be noted and translated to some more modern parlance.

The articulation of paradigms, I conjecture ('The Nature of Scientific Problems and Their Roots in Metaphysics', in my *Science in Flux*), is a metaphysical system, a powerhouse. The interaction of a metaphysical theory and a scientific one can go both ways: each may require a revision of the other. Also, a scientific theory and an observation report of empirical fact can go both ways – especially when facts confirm a hypothesis which conforms to an opposite metaphysics. For example, electrostatic action-at-a-distance theory is confirmed by facts, their field reinterpretations offer alternative observations, including dielectricity and electric and magnetic waves (see my *Faraday as a Natural Philosopher*), and at times all this may indeed force us to modify our empirical observations!

A metaphysics may merge with the scientific hypotheses which conform to it. When all our hypotheses conform to one metaphysical system, and we claim in an additional – and, as it happens, testable – hypothesis that the list of these hypotheses is complete, then and only then our science entails our metaphysics.

7. PROGRESSIVE AND DEGENERATIVE PROBLEM-SHIFTS

The claim that a scientific metaphysics is possible and desired is historical. Rather, quasi-historical, as it utilizes Popper's refutability criterion of empirical character and so violates the historical fact that in the scientific tradition verification rather than mere test was the desideratum; for, as a desideratum verifiability repeatedly raised hostility to all metaphysics. In the present situation we may, if we wish, ignore altogether the fact that the framework is not testable, and take as a whole testable theories on their frameworks and call them science. Particularly conducive to such a move is Mario Bunge's claim that testability is not enough, that unless a theory fits our intellectual framework, all its testability and confirmation will not make it a part and parcel of science (*Scientific Research*). I am in great sympathy with this view since I do agree that testability is not enough (technologically important theories such as the continuum elasticity theory are testable yet not scientific). But I wish to retain the distinction of function of searchlight and powerhouse.

The question, then, is, what causes a switch from one powerhouse to another? I think this is not a pressing problem: we have very few powerhouses and we may prefer to use them all. At least this is the view I myself advocated. Yet there is a snag here. Clearly, some powerhouses simply cease to function. Why?

Here comes Lakatos' theory of the problem-shift to save the day. Lakatos proposed his theory before he allowed for any powerhouse. In his superb classic 'Proofs and Refutations' (*Brit. J. Phil. Sci.*, 1963–4) he showed that pursuing a solution to a given (mathematical) problem may raise a new problem, and the new problem may be better or worse than the original, depending on objective criteria (such as generality) or on immediate interest. He calls these progressive and degenerative problem-shift. In particular, a progressive problem-shift may generalize a given problem, or lead to a problem with far-reaching solutions. I cannot say much more on the subject since Lakatos argued chiefly with the aid of superb examples, which he largely left uninterpreted; and they may be interpreted in diverse ways.

What Lakatos brought to the discussion is the answer, I think, to the question, why do some intellectual frameworks offer no help (the Dalai Lama system) or cease to offer help (Newtonianism). His answer is, following an intellectual framework at times leads to progressive problem-shifts, at times to degenerating ones.

All this is still too preliminary, but most significant nonetheless. Problem-shifts occur on a large-scale and small. And the one who wishes to effect a progressive problem-shift may well be frustrated by all sorts of committees inspecting research projects and manned by old-fashioned, or simply old, men of science. It is a fact that these things happen and slow down progress.

The task for methodologists, of overcoming such delays, is still open.

CHAPTER 22

CONTINUITY AND DISCONTINUITY IN THE HISTORY OF SCIENCE

The problem which lies at the root of theories of historical continuity and discontinuity is the meaning of a "turning point" in history, whether it be Caesar's crossing of the Rubicon, the Fall of the Bastille, or the publication of Newton's *Principia*. The theory of continuity ideally asserts that historical change proceeds gradually by small steps, and thus denies the existence of any sudden, large-scale changes or revolutions; it therefore either denies that there exist abrupt turning points in history or asserts that they are merely convenient landmarks. There is no actual location called the equator; it is a landmark only because geographers have made it so. The approach to the equator is utterly continuous; as a landmark, the equator is a discrete, hypothetical line.

Opposed to this ideal of continuity is the more radical model which asserts that periods in history begin with sharp and abrupt turning points such as the "Scientific Revolution". In the historiography of science, two extreme models must be examined critically: (1) that there are never any discontinuities or turning points or conscious breaks with past theories of nature; (2) that there is no continuity at all in ancient, medieval, and modern developments of the sciences. We shall call these extreme views the "continuist" theory and the "radicalist" theory, but keep in mind that historians of science fall between the two extremes maintained by historiographers of science.

The monumental writings of Pierre Duhem on the history and philosophy of science illustrate for the most part the continuist view, and Francis Bacon's philosophy, in many respects, the radicalist view, despite the fact that Bacon cleverly announced that we moderns are the true ancients since we have the whole of mankind's past experience and technical information to draw upon, whereas the ancients had to start from scratch.

1. THE RADICALIST VIEW OF SCIENCE AND ITS HISTORY

Though the continuity theory has been amply adopted and almost single-handedly applied to the history of science by Pierre Duhem, it had antecedents, particularly in the work of William Whewell. Before investigating the antecedents of Duhem's continuity theory, it is interesting to notice that even among the radicalists there can be detected an unavoidable introduction of some elements of continuity. There are, by and large, three such ineluctable elements of continuity (to be discussed below), even in the most radicalist philosophy of science and its history: first, the steady accumulation of data; second, the development of stage-by-stage theories; third (and this is a latecomer to the radicalist philosophy), the idea of the spread of scientific method throughout the world.

These three elements of continuity penetrate willy-nilly any extremely radicalist framework. It is a curious fact that the nineteenth-century scholars who criticized Francis Bacon's writings, are puzzled by the fact that he says little about the inductive method and that he claims so strongly that the method he offers to the reader is so utterly novel. His chief commentator, Robert Leslie Ellis, stressed this fact. Yet, if one reads carefully the Preface to Bacon's *Novum Organum*, one can see clearly that what Bacon claims as the novel element in his scientific method is no other than the idea of radicalism, viz., that the method pursued by scientists will lead to disastrous results so long as they fail to break away from or persist in those older methods and prejudicial ideas which he calls the Idols, and allow them to persist.

Bacon was quite clear about the dangers of radicalism in other aspects of his philosophy, for in his *Essays*, he asks his reader not to apply any sort of radicalism to social studies ("Civil philosophy" or "the Law") since social life must rest on tradition and hence on some measure of fixed belief, even a superstition. Indeed one may view the development of eighteenth-century political radicalism, as well as its interest in socio-political life, as a realization of Bacon's own prophetic conservative fears. Also Bacon's radicalism is what led to the view that the renaissance of science is in fact a radical break with its past. However, when one considers the continuous growth of science as one of the more important aspects of human culture, and human culture as one of the most important aspects of human life, one cannot but derive from this what various continuist historians call the myth of the Renaissance or the myth of the Middle Ages, in short the myth of historical discontinuity. That myth has been sufficiently discredited by social, cultural, and political historians of all sorts; but it has remained, by and large, in the history of science, particularly when presented as a chapter in cultural history.

In fact there is a small element of continuity even in the Baconian approach. For however we discard theories, opinions, and dogmas about the Middle Ages, we must still stick to the hard and fast factual information gathered through the ages. Of course, Bacon adds, we have to sift carefully information from fables, but these do not matter very much. Indeed, in one place, he says one should not sift information: misinformation will be eliminated by a self-corrective process. In any case there is some continuity in the transmission of factual information from the Middle Ages to the Renaissance. It is clear to both Bacon and his followers that this carry-over is much too small to give us any clear picture of continuity.

Bacon's doctrine of the Idols helped to promote the idea that the Middle Ages were all dark, and completely subject to dogmatism. Not only did Bacon's theory explain the radical phase of Renaissance science and the success of Newtonianism, but it was also a leading doctrine that was applied by the scientific community regularly in producing the renaissance of science, that spectacular progress within one century which, at least from a distance, looks as if it covers much more ground than the preceding ten centuries. According to the radicalist, it was the result of giving up all Idols, of obliterating completely and without reserve all of the past and starting with a clean slate. Prior to the Renaissance, however, every effort to advance learning was futile, since one did not start afresh, and there were too many prejudicial elements that perverted the best of effort.

The reason for this is described by Francis Bacon in a very simple and convincing psychological theory: once a theory is accepted, for whatever reason and from whatever source, if this theory is not true, it will not be possible to avoid dogmatism sooner or later because the holder of the theory will not confess publicly that he has been mistaken. He will be unable even to see the fact which runs contrary to his theories so as to avoid the need of having to confess it in public.

Here we see the important element of radicalism in everyday scientific work. However great the scientist, once a mistake of his has been detected, it has to be treated like the plague and eliminated. Thus the importance of radicalism is not only that it demands that we set aside all previous theories and start with a clean slate, but also that we must continually try to avoid all errors, because we never know what disastrous results will ensue. The persistent quest for a way to avoid error in the future is another continuist aspect of Bacon's otherwise radicalist theory: we must proceed with scientific development slowly and by very small steps. For instance, one has to start by humbly collecting small data, and if one asks, "Has one got enough data to start developing a theory from them?" then the answer is "No." It is better to come late with one's new ideas than too early, lest one be mistaken and thereby create a new Middle Ages. The same goes for developing higher-level theories from lower-level theories. Thus Bacon presents the "ladder of axioms" or the hierarchy of theories, one built on top of others, and insists that no step of the ladder should be skipped. He criticized Aristotle very severely for having made his induction from crude facts to the most lofty part of the ladder, to metaphysics, which, according to Bacon, should be arrived at only after a very long, gradual, and painstaking process of acquiring generality by very small steps. Whewell seems to be the only philosopher of science who took "the ladder" seriously.

The step by step construction of scientific generalizations is the second element of continuity in Bacon's theory, and really the more important one. Radicalist historians of science, indeed, do present the history of science in this inconsistent manner – they show a severe break between the Middle Ages and the foundation of the Royal Society – confessing at the same time, however, that there are some predecessors to the Royal Society, particularly Copernicus, Kepler, and Galileo – and describe these minds as outstanding predecessors who were just a step ahead of their times. These historians of science tried to show, that from the days of the foundation of the Royal Society there is a perfect continuity, a perfect development step by step, of building one theoretical layer on top of the other without missing any layer in between.

In order to introduce the third element of continuity – largely due to Laplace – into the radicalist philosophy of the history of sciences one has to stress two factors. First, that the idea that Galileo and Kepler were merely forerunners of the scientific revolution is far from convincing, especially to historians of astronomy like Laplace. But second and more important is the impossibility of bending radicalism even slightly to introduce any other element of continuity. Once one takes the justification of radicalism, namely Bacon's doctrine of prejudices or Idols seriously, one can see that the slightest error in science may be seriously infectious and may lead to a new Middle Ages. The possibility that Newton's inverse square law is not absolutely true, in the sense that one may replace the number 2 in it by the number 2.00 . . . 01, was very disturbing to the radicalist school and to Laplace particularly. He tried to argue in various ways, as others did, that this remote possibility is not even a possibility – that we know for certain that even such a small modification of Newton's theory, if it leads to no possible observable results, is demonstrably impossible. Once one realizes the rigidity of the radicalist attitude, one can see indeed that there is no room for forerunners in the radicalist philosophy of science, because, however righteous a scientist is, however marvelous his results may be, if they are not all absolutely correct, they are dangerous.

This kind of problem worried Laplace when he wrote his *System of the*

World (*Exposition du système du monde*, 1796), a reconstruction of Newtonian celestial mechanics on an inductive basis. That reconstruction, Laplace admits in his preface, is not historical. In the fifth and historical part of Laplace's *System of the World*, he gives an historical sketch of astronomy and he seems to be raising the question, "Why did the history of astronomy deviate from its reconstruction as presented in the previous four parts?" He seems to be asking this question, because he is answering it. And his answer is a new continuity theory which runs as follows: the proper scientific method, as it should be practiced, was discovered before the foundation of the Royal Society and had to be spread, had to be learned by various scientists, step by step. And this is why the rise of Renaissance science is not something that took place in one day or in one year. In fact, it took two or three generations because people were fumbling and searching for the inductive method.

Once we accept this idea, we can use it as a bridge between the darkness of the Dark Ages and the enlightenment of the Renaissance and of the Age of Reason. True, once the inductive method has prevailed, the methodological continuity presented by the history of science is precisely the one postulated by Francis Bacon and is bound to be recognized in historical reconstructions of science. That is to say, if Bacon's inductive method prevailed, the history of science would be identical with its inductive reconstruction. But in the meantime there is a time of fumbling, of hesitancy, of trial and error, which leads to the discovery, not so much of scientific ideas (although there is that too), as of the true scientific method. To conclude: although Laplace is a radicalist with respect to the history of science, nevertheless he introduced a continuist theory with respect to the development and spread of the methods of science and their practice.

Whereas the seventeenth century is characterized by the spread of science and radicalism in the natural sciences, the eighteenth century is characterized by the spread of the same into the fields of the social sciences, "moral science", or "moral philosophy". The Baconian view inspired the *philosophes* behind the French Revolution. For them the most important culture is scientific culture. Otherwise, they recognized as of some significance a pre-scientific healthy common sense. The rest they denounced as error and superstition. And so, in the radicalist view, culture is either primitive with "noble savages" exercising common sense, or scientific, with recent history to exemplify it, or merely ready to degenerate, as much of Western culture illustrates. It was the failure of the French Revolution, the French Terror, and the rise of Napoleon that gave rise to an alternative philosophy of culture, viz., that of Romanticism, which was not radicalist but conservative and continuist and, in a sense, irrational. The most important spokesman of this romantic school is Hegel.

2. DIALECTICS AS REACTION TO RADICALISM

The most important aspect of Hegel's philosophy is that it contains a new idea of progress – the idea of the progress of culture within which science plays some role, but rather a minor one. Hegel created a new idea of rationality, albeit an irrationalist theory, which describes culture as deeply rooted in tradition.

Hegel saw all human history as the history of culture. Even military history, he claimed, was only an expression of cultural history. He saw history as a relay race of culture: the torch passed on from the hand of one nation to the next. It is hard, however, to decide how much continuity to ascribe to Hegel's philosophy of history as it is hard to assess the place of national (cultural, of course) heroes in his philosophy. Doubtless, heroes play a great role in history: both as expressions and as vehicles of culture. But how much a cultural hero, be he Plato or Napoleon, is an expression, and how much a vehicle, is hard to say – especially due to the ambiguity of Hegel's dialectics. However, it is doubtless true that for Hegel the hero being a vehicle of the rise of culture is identical with his being an expression of that culture: belonging to a rising culture and developing that culture are the same. And so, the hero is both active and passive at the same time. To see this better we must examine the ideas of Hegel and see what, in his view, makes a culture a rising one.

The most important element in Hegelian dialectic is the so-called first law of dialectics – namely, the law of the transformation of quantity into quality; it is the idea that, however sharp or abrupt turning points are, they come into existence through continuous development, through development by small steps. The simplest way to explain this rather obscure law is by illustration, and the simplest illustration would be the conflict between two opposite forces – one of which develops gradually, the other keeping constant, or even decaying. Although the changes of each of these forces is gradual, and by small steps, there will be an abrupt moment in which the force which was superior becomes inferior and the force which was inferior becomes superior. Therefore, there is a discontinuous element in the continuous processes of the tensions of these forces.

Indeed, there is a very obviously true element in this description, which can be even fully described in mathematical terms. When one car overtakes another on the highway, the movement is supposed to be continuous and yet, at a certain point, there is a discontinuity in which one car, which was behind the other, gets ahead of the other car, and this can be expressed mathematically with all due precision. Hegel, however, did not think that this law can be described mathematically. Indeed, he claimed that formal logic and mathematics are incapable of handling such situations, real developments, and therefore, he called his idea a new logic, so-called Dialectical Logic. Moreover, he claimed that such processes developed through contradictions.

It is Hegel's predilection for contradictions which made it possible for him to describe science and its development as a part of his theory of the rising culture. He considered that science is formally within the realm of logic and mathematics. He conceded, therefore, that natural science abides by the Baconian sort of inductive logic, and therefore, its history has a radicalist character. But, he claimed, all this does not apply to real life, especially to social life, where movement, change, development, are the salient characteristics; and so he denied that science plays a major role in the history of human culture. It was Friedrich Engels, Karl Marx's collaborator-friend, who tried to apply Hegelian dialectics to the rise of science. He had to do so since, according to Marxism, science includes technology and so it is a very major aspect of human culture. But Engels' theory has never been accepted widely, because once one endorses contradictions, one breaks away from the rules of formal logic which are considered by most scientific people essential to the development of science. This is particularly so as long as one accepts the inductive method as the proper method for science – which Marx and Engels did accept, at least partly. The inductive tradition, prior to Einstein, is one which does not permit the slightest modifications within science because of its radicalist nature, and so the philosophy of science of Marx and Engels could not, in its unmodified version, in its radicalist claim of having final verifications in science, square with the Hegelian – and their own – claim that all empirical truths are modifiable. Hegelianism, or any other claim for modification, cannot be accepted by a serious and wide audience without first having a new and a non-radicalist philosophy of science which would allow for modifications without the irrational Hegelian predilection for contradiction.

A new non-radicalist type of interpretation of the history of science arose with the philosophies of French conventionalism, chiefly of H. Poincaré, Pierre Duhem, and Émile Meyerson. The continuity theory was shared to some extent by all three, but it belongs chiefly to Duhem.

3. DUHEM'S THEORY OF SCIENCE

French conventionalism was born in the midst of the crisis in physics. In retrospect that crisis is associated with highly specific difficulties in late nineteenth-century physics – difficulties that are by now satisfactorily resolved. In other words, the current view of the crisis in physics is unphilosophical and quietist. The crisis was historically nothing short of a philosophical crisis, involving, indeed, physics, but also metaphysics and methodology. It relates to the mechanical conception and the clash between the Newtonian and electromagnetic field theory.

Newtonian metaphysics assumes the existence of atoms and forces acting at a distance (i.e., instantaneously) through the void. The void is devoid of all properties: hence forces act only where matter is; forces act between any pair of particles in a straight line, and depend on their relative distances (i.e., they are central forces). A proper explanation of phenomena must include statements about atoms and their associated central forces. Hence, when we introduce fields of force permeating space, we are bound to view these in one of two ways. Either we view fields as mere mathematical expressions (as in Laplace) or we must fill space with a thin elastic matter (i.e., the ether) with which the forces of the field are associated. It was Helmholtz who clearly declared these as the only two possibilities open to us. His disciple Heinrich Hertz first viewed the field equations of Maxwell as purely mathematical expressions, devoid of all physical meaning – in view of Maxwell's failure to endow the forces of the field with material substance, that is to say, in view of Maxwell's failure to construct models of the ether. He later changed his view and in his lecture on the significance of his celebrated experiment of detecting electromagnetic (radio) waves, he declared his experiment to be a verification of Faraday's original view of fields of force as regions of empty space in which forces freely travel.

This action at a distance, however, was questioned by both radicalist and continuist thinkers. On the one hand, radicalists like Kelvin still hoped to discover models of the ether to explain all known electromagnetic field phenomena. On the other hand, conventionalists declared all science to be mere mathematical expressions devoid of all physical meaning. The middle view that Newtonian equations have meaning but field equations do not looked more and more arbitrary and Helmholtz's view never won ground.

The new view of science, conventionalism, had much to support it. It looked especially attractive to scientists grappling with new problems. The mathematician Poincaré gave it succinct expression. All scientific theory, he said, is a mathematically succinct language and unassailable – it is true by definition. But science is not only a set of mathematical equations; it is this plus its application to physical phenomena. In the field of possible empirical applications the mathematical variables are systematically equivocal, and provide room for free play; we don't have a formal and rigid set of rules for the application of equations to experiment, but exact sciences historically have resorted to trial and error, to intuition, and to all sorts of makeshifts.

All this leaves room for much choice – indeed far too much choice. We may remember that the quarrel between Cardinal Bellarmine and Galileo concerning the status of the Copernican hypothesis was precisely concerning the physical meaning of that theory. Viewed mathematically, Copernicanism was not only allowed but even preferred by Bellarmine. But as a mathematical theory even Ptolemaic astronomy has its place. Why then is Copernicanism preferable even mathematically? Poincaré answered this question by what soon became the central principle of conventionalist philosophy: always choose the simplest mathematical system available. And Copernicanism is undoubtedly simpler than Ptolemaism, especially in view of later discoveries such as the result of Foucault's pendulum experiment (a pendulum oscillates in a plane capable of rotation which rotates once in twenty-four hours).

Simplicity is judged, since Leibniz, by two criteria, both of which are rather vague. First, the smaller the set of hypotheses (or mathematical equations, as the conventionalist will insist) a theory contains, the simpler it is. Second, the more phenomena a theory explains or accounts for (or covers, or classifies, as the conventionalist will insist) the simpler it is. Now these are intuitive criteria, since there are no rules for counting equations. It is often possible to combine two differential equations of the second order into one of the fourth order; also it is possible to view the three inverse square laws of gravity, electrostatics, and magnetism, as one with three sets of readings or three domains of applications; and to view both electromagnetism and gravity as parts of potential theory (as Poincaré sought to do). Nor is it clear how we count phenomena. Do we count the phenomena of electrolysis as separate or as parts of ionic transport (together with currents)? Do we view thick lenses as lenses though their Newtonian equations are different from those of thin lenses?

The literature on this topic is only in its initial exploratory stages. Historically, a greater difficulty led Poincaré astray. What happens when the two criteria of paucity of hypotheses and multitude of experiments clash? What if we find optical phenomena not conforming to geometrical optics? Poincaré viewed the paucity of hypotheses of geometrical optics such an asset that he felt it would be wise to retain the view that light travels in straight Euclidean lines even if some non-conforming phenomena were to be discovered, and this was just before Einstein came to the scene and took an opposite course.

Poincaré's assault on this problem no doubt led him astray, but the idea he had was not without reason: we do not effect revolutions at the first sight of a difficulty as the radicalist would have us do. The fact Poincaré overlooked was that there was an alternative to Euclidean geometry crying to be applied to physics. To build a modification of Poincaré's view then, one must study the development of alternatives to existing systems of equations, however simple these may initially be. Poincaré was somewhat blocked on this issue, believing, as it were, that somehow the peak of simplicity of Newtonianism and Euclideanism provided them with a special status.

This special status was explicitly denied by Duhem who studied and emphasized the gradual development of scientific theories.

Theories lead to predictions; these may be confirmed or refuted. Contrary to Baconianism, conventionalism claims that confirmation and refutations are not necessarily reflections on the truth or falsity of the theory – which is but a set of mathematical equations and thus formally true if consistent; but they do reflect on the extent of the applicability of that theory of empirical situations. Hence, confirmation and refutation do tell us something about the theory that they confirm or refute; not about their truth value but about their applicability. And applicability is one of the factors determining simplicity. Hence, concludes Duhem, confirmation increases the simplicity of the theories they confirm and refutations decrease their simplicity.

Here, for Duhem, is the intellectual, the ideological core of change from radical discontinuity to conservative continuity. Truth value is sharply defined in the radical view; there are no degrees of truth. Hence, if science aims at the truth of a physical theory (i.e., of a class of physical laws represented in a given set of equations), then it must separate the truth from the infinite variety of falsehoods. Radicalism is but a necessary aspect of this tendency. Radicalism, for Duhem, is an aspect of the realistic philosophy – namely, the philosophy which contends that scientific laws correspond with reality.

What are the historical roots of this realism? Why was realism, and its corollary radicalism, so popular? Because, says Duhem, of the overconfidence in mathematics as the language of Nature shown by some Renaissance scientists, because of naiveté in others; what is common to overconfidence and to naiveté, to realism and to radicalism, is – intellectually speaking – the lack of imagination.

Let us return to our scientific theory which, while fixed formally, varies in its domain of experimental application, in its partially unarticulated rules of application. When confirmed experimentally, when a new rule of interpretation succeeds in bringing under its domain older but previously unconnected facts (e.g., Poisson's identification of temporary magnetization as ordinary magnetism), in such cases the simplicity value of the theory increases. Suppose the opposite occurs. Suppose a prediction is refuted; suppose an application of a theory to a new domain turns out spurious after all (e.g., the theory of elasticity to phenomena of light), then the opposite takes place. We rescue the theory, but its simplicity is impaired, its value as a physical theory decreases.

In such cases, we can slightly alter the formal theory; we can try different ways, and increase simplicity in each of them, finally ending up with the most successful face-lifting operation. We then turn to experiment.

And so, we may conclude, the very aspect of conventionalism which made Poincaré prize traditional doctrine too much, which made him a bit too conservative, was turned by Duhem into a dual source of strength, as two sources of continuity, as a two-edged sword with which to beat radicalism.

All this is in theory. In fact, Poincaré suggested alterations in Newtonianism, and though he was wary of Einstein's revolutionism, he took him seriously and in spite of himself tended to accept his views. It was Duhem who, as an arch-conservative physicist, disliked the revolutionary aspect of Einstein's view and saw in it a serious threat to science.

One might exonerate Duhem by saying that theory and practice do not necessarily go together; that Duhem the thinker was broadminded, but that the flesh-and-blood Duhem was a chauvinist, a physicist committed to the Ampère-Weber action-at-a-distance theories, etc. This line of exoneration is too easy, too unfair to Duhem; it is much more interesting to take a closer look at Duhem and see the intellectual – scientific, historical, religious, and metaphysical – aspects of his hostility to Einstein. It is, anyway, not true that inconsistent behavior comes easier than consistent behavior.

4. THE CULTURAL BACKGROUND OF SCIENCE

Duhem's opposition to Einstein may be dismissed as not central to his theory of science, or it may be viewed in retrospect as the strongest argument against it. Since Duhem's theory is constantly gaining in popularity among physicists, philosophers of science, and historians of science, the first alternative seems obviously preferred. This choice seems to deprive Duhem's theory both of its strength and its peculiarity. For, we must repeat, no continuity theory is worth taking seriously unless it specifies what kinds and what maximum size of building blocks are needed to construct historical continuity.

Assume that we know a large step when we see one; assume, further, that we are not willing to increase the size of the elementary building blocks just to incorporate *ad hoc* this large step into our continuity theory. There is still the possibility of breaking down the large step into smaller ones. For whether Einstein was much of a revolutionary or not, Einstein undoubtedly did belong to a tradition, that of Kant, Oersted, Faraday, Kelvin, Maxwell, Hertz, Poynting, Heaviside, and Lorentz. It is equally the case that Duhem was a partisan of the conservative party, of the tradition of Coulomb, Poisson, Ampère, Gauss, Weber, Franz Neumann, Carl Neumann, Duhem (qua physicist), and Ritz. (Helmholtz, we have seen, was in a major camp all by himself.) And so one might apply Duhem's techniques to a tradition he did not notice sufficiently seriously. More than that, applying the techniques to both groups at once may yield even richer fruit – as Whittaker's *History of Theories of the Aether and Electricity* may testify in part.

Whether such an exercise as Whittaker's can be successful is a serious question. In his review of Whittaker's book, Max Born violently attacks Whittaker's application of continuity to the history of relativity, and praises his application of it to the history of quanta. Born testifies that he personally saw the revolution, that he lived on the barricades while the shooting went on. And so, he concluded, Whittaker's view of Einstein's step as a small one must be false.

And so we are back to the central question: Can the continuity theory apply to revolutions proper? And again we must answer, with the political philosophers, yes. This is the touchstone of a challenging continuity theory – its applicability to genuine revolution. Now, for a consistent continuist to explain a genuine revolution, his view must be specific; he must show which aspect of one and the same process is continuous, which is discontinuous (at least apparently so), and how the one gives rise to the other.

To show that this can be effected in the case at hand, we must first ask what aspects of science are, to Duhem, continuous, and what discontinuous. We have already noted that within science proper we have theory and empirical application and nothing more, and that both aspects evolve with continuity and close contact. Where can we have any discontinuity?

Duhem's answer is extremely odd: apart from science, all thought and cultural life he bluntly declares are not continuous with but are altogether unrelated to autonomous science. There is no contact possible between science and metaphysics, he says. Thus he blocks any possible contact between science and culture. Before explaining how this block is effected, let us explain Duhem's motivation, how it led him to endorse the worst aspect of radicalism – a total break between science and other intellectual activities – and how easy it is to graft on Duhem's view the opposite one which the historian of science, George Sarton, sought all his life in vain.

Duhem as an ardent Roman Catholic stuck to his Aristotelianism quite dogmatically. He realized that the only way for him to appreciate science – and his studies in physics and its history amply testify to his devotion to science – was to return to the philosophy of Cardinal Bellarmine, to the endorsement of Copernicanism as a mere mathematical instrument devoid of all physical interpretation, philosophy, or metaphysics (these three terms are more or less synonyms). It is perhaps unfair to speak of Duhem's return to Bellarmine, because within the hard core of Catholicism, Bellarmine never left the scene. Rather, the return was effected by the scientific tradition: Helmholtz defended Newtonianism by viewing electromagnetism as a mathematical hypothesis, and Poincaré viewed both mechanics and electromagnetism in the same light; Bellarmine won. The rift between science and faith was due to philosophic misunderstanding, concluded Duhem, thus hoping to deliver the deadly riposte to Galileo's (alleged) attack on the Church.

The radicalists declared that medieval science did not exist. They recognized the Renaissance of science to have had a radicalist history, beginning with a clean slate. This is obviously erroneous. Applying the emergence technique Duhem both discovered the medieval ancestry to Renaissance science and medieval science simultaneously.

Duhem's discovery is here to stay, no matter what else we think of him: he discovered medieval science, unlike M. P. E. Berthelot who discovered that medieval works contain merely a crop of empirical data and a few "correct" laws. Now this runs dangerously close to an admission of defeat: to discover medieval facts and those medieval theories which happen to overlap our own science textbooks is not to discover medieval science. Nor, it is also true, can we discover medieval science within medieval philosophy, metaphysics, etc. What then is there left to discover? Science stands forever between the two, says Duhem, a system of continually changing conceptual-mathematical theories yet related to observations. And incredibly, he found such science in the medieval manuscripts.

Metaphysics is discontinuous, a field of disagreements. All alleged intellectual connections between science and culture are channeled through the physical reading (and hence metaphysical import) of scientific theory. But,

Duhem insisted, scientific theory is abstract and mathematical, and hence there is no disagreement in science. Consequently, Duhem concluded, the relations of science to metaphysics, and (via metaphysics) to culture in general are spurious (to be precise, Duhem permits science to have aesthetic value; but neither he nor anyone else considers this a link between science and culture: it would be too Pythagorean and hence metaphysical, and thus everyone is glad to let sleeping dogs lie). One cannot easily escape the conclusion that Duhem separated science from all culture as a price to pay for his putting medieval science strictly on a par with contemporary science. After all, the major drastic operation he performed on medieval documents, in order to extract science from them, was that of pruning. It is amazing that with so much pruning and so much insistence on the abstract nature of what is left, Duhem's product has not entirely lost all of its original flavor.

Herbert Butterfield, a political and cultural historian, made a brief excursion into the history of science: in his monograph (*The Origins of Modern Science*, 1949) he made a major contribution to the rather arid field of the historiography of science. Butterfield dislikes Duhem's bias in favor of medieval science, but he likes Duhem's religious bias, that is, his instrumentalist philosophy. And so he has invented an ingenious variant on Duhem, much reminiscent of Laplace's variant on Baconian radicalist philosophy: whereas Laplace tempered Baconian radicalism regarding the development of science with a continuity theory of the widespread diffusions of proper (Baconian) methodology, Butterfield tempers Duhem's continuity theory of the growth of science with a radicalist theory of the invention of the proper (Duhemian) methodology. Duhem has argued that science is abstract, mathematical, amenable to modification, and has shown, in a wonderful argument against induction, that Aristotle's physics is nearer to common sense than Galileo's. Hence, claims Butterfield, science proper starts with Galileo or thereabout.

Butterfield has succeeded in modifying Duhem's theory, and in creating a new intellectual revolution – that of the adoption (in practice, not the articulation of this practice) of the Duhemian philosophy of science with its internal historical continuity and external independence of metaphysics and religion. Once this adoption took, there was no room for further revolution. True, old-fashioned concrete thinking, and its subsequent inability to adapt one's theories through modification, did not vanish everywhere. Phlogistonism was like that. Hence Lavoisier is the Galileo of chemistry. After Lavoisier, chemistry gets modified, never overthrown again. These conservative views of Butterfield are intriguing and original – but not decisive. Duhem has shown examples of medieval science – abstract and modifiable theories; Butterfield does not discuss them. Duhem has shown that throughout history metaphysical views are unrelated to the activities of scientists. Hence, a philosophical revolution can never effect a scientific revolution. Butterfield ignores all that.

The latest stage, the latest modification of Duhem's methodology, is due to Thomas S. Kuhn. His *The Structure of Scientific Revolutions* (1962) succeeds in discarding the equation of medieval and modern science. According to Kuhn science has a continuous aspect – exactly Duhem's – and also a discrete aspect: the textbook of science. The weakest element of Kuhn's philosophy, or at least the one hardest to comprehend, is how precisely quantitative changes accumulate and culminate in the qualitative emergence of a new textbook. Yet, Kuhn clearly says, a textbook does get constantly modified, its simplicity does get constantly impaired, until a moment of crisis. At what critical point do men of science spend sleepless nights until they emerge with a new textbook? That crises do emerge or lead to revolutions may be taken as an undeniable fact; the problem is, where is continuity preserved through the revolutions? If Kuhn only claims continuity between revolutions, he claims little – or less than he seems to be claiming. If he claims continuity through revolutions, he is far from clear.

Assuming, however, that science has its standard textbooks facilitates discussion of a few points. An outdated textbook is petrified science; the absence of all textbooks is pre-science, a distinct phase of quasi-scientific activity at most, scientific except for the lack of concerted coordination among scientists. Now, then, we can see that some medieval work is petrified science, some pre-science and all is well. Moreover, the textbook of science is, strictly speaking, not a part of science alone, but also a part of the sociology of science. Thus Alexandre Koyré, whose views diverge from Kuhn's (or from Duhem's, for Koyré was anticonventionalist and insisted on the metaphysical import of science as a significant component), hailed Kuhn's work as an encouragement to the long sought-for link between science and its social background based on more than merely contingent observation or on a merely technological link. How much we can count on Koyré's hopes remains to be seen. What may be said now, however, is that Kuhn views the development of a sense of crisis as a matter of public opinion, dominated by the leadership of science – thus begging the question and introducing a new traditionalist element (of continuity) which he borrows from Michael Polanyi. This much, at least, is not easily acceptable; at least not as it stands.

In conclusion, we must pay heed to the wealth of literature on the relations between science and culture which puts the whole debate in a new light. The Renaissance Platonists with their anti-Aristotelian bias, be they right or wrong, much affected the course of science. Whether determinism is a legitimate offshoot of science or not, an offshoot it is. The radicalist view leads us to the Age of Reason, to the French and Industrial Revolutions, and even to the rise of Romanticism. The cultural crisis in science has led to the revival of Bellarminian philosophy or to its popularity, and thus raised again the issues of the competition between science and religion. This creates doubts as to the intellectual (as opposed to the technological) and moral values of science and thus leads to existentialism and to a more recent new wave of (Bellarminian) fideism, on the one hand, and to the advocacy of the moral and political neutrality of science, on the other. It is thus hard to deny, as both a strict Baconian and a strict Duhemian should, Abraham Kaplan's conclusion (p. 21): "The works of the mind are all of a piece; what happens in one science is affected, not only by what goes on in the others, but also by the thoughts on the matters of religion, politics, art, and whatever" (*The Conduct of Inquiry*, 1964).

Radicalism divorces science from culture by its claims for absolute truth and thus for a total absence of any cultural dependence or cultural values of science. Duhem's replacement of truth by degrees of simplicity (which he proposed to adopt as degrees of usefulness as substitutes for truth), changes matters only internally, and only stresses the merely technological value of science. A third course, however, may be open: we may take account of degrees of falsity, or degrees of proximity to the truth (such as Karl Popper expresses in his *Conjectures and Refutations*). This accords with the radicalist concept of truth, though it is less optimistic. It allows for flexibility within science, as Duhem required, but the flexibility is that of standing corrected, and is not merely a matter of comfort. Finally, it allows for degrees of cultural independence and objectivity, and thus it allows interactions between science and other parts of human culture – whether correct or finally to be rejected. Also, unlike Duhem's and Kuhn's, this philosophy accounts for, it even encourages, the coincidence of different traditions or schools, such as the two schools in nineteenth-century theories of electricity already referred to. No other theory of science, as yet, allows for coexisting contending scientific schools.

This rivalry of schools again raises problems of continuity: there is, no doubt, more continuity within the school of Oersted, Faraday, Maxwell, and Einstein or within the school of Ampère, Weber, Duhem, and Ritz, than between these schools – at any point of history. These continuities, it turns out, are "thematic" (Holton) or metaphysical (Agassi); and so, although Kuhn's paradigms, the science textbooks, fail in their role as sets of continuity, they may still be rescued, but with a slight modification: they should be viewed not as existing textbooks but as future ones.

The problem then, ceases to be, is the history of science continuous? It turns into more specific questions; in which respects is it, in which not? There is, then, no real need to explain any discontinuity by any continuity.

Ultimately, it is reduced to the logic of the situation of a workaday scientist. What past factors should he, must he, take into account? The radicalist and the continuist alike say, only and all past science. Hence, to both science is not culturally dependent in any deep intellectual sense. Even if we assume, *ex post facto*, that the scientist's economic conditions are reflected in his scientific situation, as well as in his culture at large, things do not thereby change as far as the logic of the scientist's situation is concerned. Hegel's idea that all intellectuals are bound by their cultural conditions offers a connection between science and culture on an intellectual (theoretical or ideological) level; but it is an anti-scientific or pseudo-rationalist philosophy, and its application to science has always failed. Once we suggest that scientists may both destroy and build some cultural conditions, once we see continuity not only in the acceptance of widely acknowledged influences but also in reaction to or rebellion against them, then our picture of the growth of science will be much more varied, and the questions concerning degrees and frequencies of discontinuity will greatly change their character.

CHAPTER 23

THREE VIEWS OF THE RENAISSANCE OF SCIENCE

The Renaissance of culture has been seen as (A) an abrupt change, (B) continuous development, or (C) a rather abrupt, but not entirely discontinuous, process. Historians of culture used to accept the first, radicalist view (A), and now they generally accept the third, semi-radicalist view (C). The second, conservative view (B) was always unpopular, and defended by lovers of the Middle Ages, romantic or dogmatic; most of these few 'mediaevalists' were apologists and thus not very considerable scholars; only a handful of them are to be considered as serious, and even great, historians of culture.

We may consider three similar views concerning the rise of the Renaissance of Science, which may, then, be viewed as (a) an abrupt change (b) continuous development, or (c) rather abrupt but not entirely discontinuous. Must one who holds the radicalist view (A) concerning the Renaissance of culture also hold the radicalist view (a) concerning the Renaissance of science?

Historically, the radicalist view (a) of the abrupt rise of science was originated by Francis Bacon. It led him and his followers to the radicalist view (A) of culture, on the assumption that science is a most significant cultural factor. But this assumption can be denied. It was denied e.g. by Pierre Duhem, who invented the conservative theory (b) of continuous development in the history of science. He refused to view science as a major cultural factor: he admitted the existence and significance of philosophical, religious, and other revolutions in the Renaissance, and he detested them strongly; he admitted no significant relation between science and these other intellectual activities. Thus, he held a non-conservative view (A) or (C) about an abrupt or a rather abrupt cultural change but stuck to the conservative view (b) of science, to the total denial of the existence of any abrupt scientific changes.

It may seem, then, that the adoption of similar views concerning the rise of culture and of science depends on the assumption that science is an integral part of culture. For many generations, this assumption was taken for granted without any special study or even reference. It was indeed this assumption plus Bacon's radicalism concerning science which led to general radicalism in the eighteenth century. Today, however, this assumption – science is an integral part of culture – is not merely well known: it has acquired the status of a slogan, especially amongst historians of science. The slogan is attributed – rightly, I think – to the celebrated historian of science, George Sarton.

Now, when most historians of science view science as an integral part of culture, and when most historians of culture have shifted from Bacon's radicalist view (A) to the less radicalist view (C), one would expect the view of most if not all historians of science concerning the rise of the Renaissance of science to move in a parallel fashion from the radicalist position (a) to the less radicalist position (c). This move, however, did not yet occur. Most historians of science still adopt the Baconian radicalist view (a), and some of them adopt the Duhemian conservative view (b); the third view is still in its birth pangs. I wish to discuss now the difficulties in its way.

1. THE RADICALIST VIEW

One of the principal themes here concerns the popularity of Francis Bacon's philosophy of science, particularly amongst historians of science. The popularity of Bacon's ideas continues in spite of its having been effectively criticized by some philosophers and some historians of science, in spite of the fact that his idea of how to write the history of science has been superseded, and in spite of the fact that more and more historians of culture, social and political historians, and others, have turned away from Bacon's philosophy in general. Bacon viewed the Middle Ages with contempt, and his followers took it for granted that modern culture begins abruptly with the Renaissance. Modern cultural, and other, historians have found important mediaeval roots to the culture of the Renaissance. Even some historians of science have found important mediaeval roots to the science of the Renaissance. One might tend to expect all this to lead to the decline of popularity of the Baconian view of the abrupt rise of science – at least amongst the historians of science who see science as a part of culture. Yet this is not the case even concerning George Sarton: even though he stressed the view that science is a part of culture and even though he was a student of mediaeval science, he remained an orthodox Baconian to the last.

Bacon's philosophy of science can be briefly summarized, since it is astonishingly simple, and intuitively very convincing: once we are determined to give up all our preconceptions about nature and carefully attend to things as they really are, science will develop with great ease, by the emergence of theories from observed facts in accord with the rules of inductive logic. Baconians are, therefore, naturally prone to see the Middle Ages as the era of preconceived notions, of prejudice and superstition, and the Renaissance of science as the outcome of the successful overthrow of these prejudices and the beginning of the application of the proper scientific method, starting with the pure collection of empirical information in earnest and ending up, by induction, with the splendid results of modern science. The corollary from this is that the Renaissance of science is chiefly the outcome of the discovery of Bacon's inductive philosophy and its subsequent spread amongst, and application by, an ever increasing number of individuals. If we equate the Renaissance of science with the Renaissance of culture (which, with some qualifications seems reasonable enough), then we obtain what many historians call 'the myth of the Middle Ages'; the story, that is, of the rather abrupt transition from the dark Middle Ages to the enlightened Renaissance.

Cultural, social, and other historians have written a great number of exciting and interesting books against this Baconian myth. The defect of most of them is that they fail to do justice to the immense role which science has played in this transition or development. This defect is understandable on the assumption that these historians fail to provide a satisfactory alternative to Bacon's philosophy of science; for, consequent on this they are likely to fall prey to the Baconian myth when discussing science – so they prefer to avoid this pitfall altogether. Indeed, those few who do refer to science become pathetically Baconian all of a sudden, with some pathetically ludicrous consequences.

The myth of the Middle Ages can be traced back to Bacon's writings, especially to his brilliant though fantastically naive sketch of the history of science in his *Novum Organum* of 1620. It can then be easily traced to a direct line of histories of science written under his influence, beginning with Bishop Thomas Sprat's *History of the Royal Society* of 1667. This work is not quite a history, but an apologia for that Society. Yet it contains a somewhat more elaborate sketch of the history of science along Baconian lines. It was soon acidly reviewed by the leading opponent of the society, the notorious Henry Stubbe, in his *Legends no Histories* (1670), where both the apologetic and the radicalist character of Sprat's book come under heavy fire. Sprat's book was declared useless and frustratingly uninformative by a later historian of that society, C. R. Weld (1848); it has been recently republished and an able reviewer, L. L. Whyte (*BJPS* **11**, 1960, 264), has cursorily dismissed it saying that 'though often cited, [it] has little merit', chiefly because of its Baconian and chauvinistic character. Though it 'has little merit' it is 'often cited', I presume, because it is a milestone in the history of the history of science: it is the first history of science which presents the myth of the Middle Ages. The myth is implicitly or explicitly defended by all inductivist or Baconian historians of science, who are the majority, though sometimes with slight modifications which allow one to give more weight to the historical fact that from time immemorial there existed a semblence of scientific knowledge in the form of empirical data and simple technology. Some Baconian historians of science hardly pay any attention to the Middle Ages except in order to deride it; others try to salvage what they can chiefly by reference to factual information that existed in that period. Amongst the latter group there are authors like Berthelot and Sarton, who rebelled in almost the same way against the pitch-dark picture of the Middle Ages and studied extensively the history of positive empirical knowledge of that period in order to show that there was some ray of light even then. Yet what they found amounts to very little, and their rebellion led to no revolution of thought. Berthelot mainly tried to show that the alchemists possessed a wealth of empirical information. Sarton's great contribution to the study of the history of science in his taking up Berthelot's discovery of mediaeval Arabic science and showing its large extent, thus presenting it as having the immensely significant role as a link between the traditions of antiquity and the traditions of the European Renaissance. Incredibly, even Sarton's attitude is entirely Baconian; all he could do is salvage from the vast mediaeval literature the little which he could honestly call science; the rest he dismissed as unimportant superstition and prejudice. He fully accepted the view that science began in the seventeenth century with an 'Open Sesame!'; that there was then born a totally new being; but he compared the Middle Ages to a period of pregnancy. Thus, the picture of the Middle Ages received from all inductivist histories of science, including the ones most generous towards that period, is the picture of an age of ignorance and magic and superstition, ameliorated to some degree or other by more or less, but admittedly too little, empirical experience.

The Baconian historians of science do not bother to explain how any knowledge was obtained prior to the Renaissance of science, but they explain that Renaissance as the outcome of the discovery of scientific method – by which they mean the discovery of the inductivist view of what the method of science is. They sometimes deny that it was Bacon who discovered the principles of induction and some of them even attribute the discovery to Galileo (who explicitly opposed induction in any form); but controversial as the question may be, whether it was this or that individual who discovered the method of induction, and whether the discovery was individual or collective, the majority of historians of science are agreed in their severe condemnation of the Middle Ages as the dark period of prejudice and in their sky-high praise of the seventeenth century as the enlightened period of the rise of the proper scientific method. The contrast can hardly be sharper. Following Sarton, some of these historians have to modify the contrast when discussing some details, especially details imposed on them by Duhem and his followers. But the details do not interfere with the overall picture: in conclusion, the contrast remains sharp.

Thus, it seems to be a simple matter of fact that the radicalist Baconian theory of the Renaissance of science is based not on facts but on a preconceived notion. The preconceived notion is this: preconceived notions lead to superstition, whereas the preference for facts and induction leads to enlightenment. It therefore seems to me that the preconceived notion of the majority of historians of science is not only a superstition, but also an obvious contradiction. This criticism is, in principle, the same as the one presented by David Hume two centuries ago. The Baconian historians may answer this criticism in at least two ways.

First, they may say, the condemnation of preconceived notions applies to those preconceived notions which are descriptions of natural things, not to those which are descriptions of the methods of science. This is the famous theory of Russell on the limits of empiricism (i.e. inductivism), recently reiterated by Ayer and hosts of other empiricists. It is the view that empiricism applies to the study of nature, not to the study of the study of nature; in other words, that empiricism itself cannot and need not be empirically justified. Assuming Russell's doctrine to be true, it would follow that the Baconian method, of beginning from facts and then developing theories which are based on them, cannot and should not apply to the history of science. The history of science, then, would have to proceed on different lines. Since most Baconian historians of science accept the vast collection of historical facts as the starting point, and the basis, of all historical studies, it seems that they (implicitly) reject this Russellian reply. (I say 'implicitly' since most of them seem to ignore Hume's criticism to which Russell's theory is a reply.) For my part, I should claim that one of the severest criticisms of Russell's theory of the limits of empiricism is that it cannot at all generate a newer approach to the history of science: all of those who accept the principle of induction, as Russell does, whether or not they think that the principle of induction is based on facts by induction, cannot but apply the principle to the history of science the way most Baconian historians of science do: this is, their history will necessarily be on the preconceived idea that preconceived ideas are bad for scientists, and hence their history will include the repugnant myth of the Middle Ages – at least as far as science is concerned.

The second answer to the criticism is to deny altogether that inductivism is a preconceived notion, and to suggest instead that the principle of induction can be based either on logic or, by induction, on facts. The criticism of this alternative fills the literature of the philosophy of science from the days of Hume to date. I shall not discuss it in this place, and for the simple reason that its application to the history of science likewise leads to the repugnant myth of the Middle Ages. Hume himself was an inductivist, and his own criticism of inductivism, including his criticism of this alternative, put him in a quandary; Russell has endorsed it while offering that theory of the limits of empiricism (or inductivism) which I have already discussed. A school of thought which goes back to Kant, however, found in Hume's criticism a sufficient reason for the claim that inductivism is false, that men of science never evolve their ideas by induction.

2. THE CONSERVATIVE VIEW

Most of the philosophers who considered Hume's criticism of the doctrine of induction as unanswerable became irrationalists. Most of them rejected science or ignored it altogether. Some of those accepted science as something rather miraculous and perhaps even irrational; for instance, Whitehead and Polanyi. But all this is a very recent phenomenon. Until the Einsteinian revolution of 1905, science was too respected to be ignored or dismissed or even to be viewed irrationally. So until then most scientists and philosophers took it for granted that Hume's criticism is somehow answerable, and they viewed the irrationalist philosophers – such as the German Romantics – with great contempt. Very few thinkers followed Kant in rejecting the principle of induction while keeping allegiance to science. And even this Kantian trend became somewhat significant only with the development of the crisis in physics – the crisis which led to the Einsteinian revolution.

The majority of those who followed Kant in defending science while attacking induction were the conventionalist school, which flourished around the turn of the century. One of its leading members then was Pierre Duhem, who interests us here as one of the chief opponents of the myth of the Middle Ages. Conventionalism, and in particular Duhem's version of conventionalism, is less easy to summarize adequately than inductivism, because it is much more sophisticated and much less known, except perhaps amongst physicists. Fortunately, the following summary will do for the purposes of the present discussion. Whereas according to Bacon scientific ideas evolve from facts, according to Duhem they evolve from previous scientific ideas: every theory is a modification of a previous theory. The purpose of having theories, as well as of modifying older theories thereby attaining newer theories, is not the ultimate truth but the achievement of the simplest tool with which economically to describe and classify known facts. Thus, according to this philosophy, facts constitute an incentive to theorizing, not the basis of it, and they may at most indicate the way in which we may try to modify an older idea in order to achieve a higher degree of simplicity.

It is manifest that within this philosophy there is room for no scientific revolution; in particular for no Renaissance revolution of science. One may admire the boldness with which Duhem advocated so incredible a thesis, and the immense historical insight into mediaeval science which he showed in the process. But one can hardly accept this thesis in spite of Duhem's immense cleverness and success.

The situation regarding the contrast between the Middle Ages and the Renaissance of science has been neatly summed up by L. Pearce Williams (*BJPS* **11**, 1960, 162). 'There are still those who insist that mediaeval science was mere philosophical vapourings concerned either with the number of angels who could dance on pin heads, or as 'applied philosophy', with the magical essence of gems, fabulous beasts, and so on', he says, referring to the Baconian school. 'At the other extreme there are the followers of Pierre Duhem who view modern science as a rather extended footnote to mediaeval achievement', he continues, referring to the conventionalist school. He goes on to mention two or three exceptions (in his view) to all this, but they need not concern us here.

The situation is incredible. A glance at a history of the Middle Ages written by one who is not a historian of science reveals a strange and horrible world, hardly comprehensible to the modern reader, a world which is represented by dim-lit cathedrals with more poignancy than the modern world is represented by concrete and glass skyscrapers. It is a world of hatred of the flesh coupled with necrophilia, a world of frightened and confused genius coupled with mystical yearnings for enlightenment and divine grace and salvation. This world, the world of St. John of the Cross and his like, the spirit of which various writers have tried to capture, from the poet Heinrich Heine in the mid-nineteenth century, to the early twentieth century mediaevalist Arthur Edward Waite and his contemporary successors – this strange and disturbing world is totally unknown to almost any historian of science; or so it seems.

The Baconian historian of science is forced by his doctrine to dismiss this world as totally superstitious. Whether he writes about this topic of mediaeval superstitions and the amount of space he allots to it in his books, is a subsidiary and technical question for him, provided he registers his disapproval. The Duhemian historian of science will have to ignore this world completely, as divorced from science and as showing no continuity with Renaissance science. But then even the Duhemian historian of science must see that most mediaeval scholars, theologians and alchemists alike, were concerned largely with non-scientific intellectual activity – even by the Duhemian standard of the scientific – in great contrast to the increasing popularity of scientific thought in the seventeenth century. The Duhemian silently drops this social revolution, whose existence he must admit. The Duhemian may excuse himself by saying that he is no social or cultural historian. He may add that Baconians mistake a social and cultural revolution to be a scientific revolution; they think science is a vast collection of facts, contrary to the reasonable view that what counts as scientific is a question of quality, not of quantity.

This is the best answer I can provide in favour of the Duhemian view. But it hardly seems to me to be sufficient. Even if all that can be claimed by a Duhemian to be true is granted, the fact remains, that by and large the public turned from one kind of intellectual activity (mediaeval mystical) to another activity (Renaissance scientific). The question, what precisely has happened and why, is pertinent to the history of science and merits study. The Baconian answer that the veil of superstition fell off the public's eye is to be rejected, but the Duhemian's silence on this question is suspicious, to say the very least.

The mediaeval mystical literature which almost all historians of science ignore is evidence that in the Middle Ages a form of scholarship was alive which was very different from the scholarship they refer to; that intellectual life existed then on quite a high level with some achievements which, though different from those of science, are quite impressive all the same. Admittedly, this literature was often obscure and superstitious, yet this is no reason for dismissing it; rather it presents the problem, how is it that, during a whole period of European history, reasonable and able scholars, sometimes very impressive thinkers, were so superstitious and even obscurantist? This problem is important, and the historian of science has to cope with it if he wishes to present a theory of the transition from the Middle Ages to the Renaissance which is more reasonable than the two traditional theories. Most historians have today agreed that though the Baconian myth of the Middle Ages is false, there was some kind of transition from the Middle Ages to the Renaissance; yet historians of science are almost entirely divided between the Baconians who fully endorse the myth and the Duhemians who pretend that there was no Renaissance revolution of science at all. It may be interesting to inquire why most historians of science are so persistent on this point. I wish to present here a tentative explanation of this persistence. My explanation is this.

3. THE RISE OF NEW IDEAS

Both Bacon and Duhem have provided theories of the evolution of ideas which, however cursory and inadequate they may be, at the very least present the evolution of thought as a rather simple and unmysterious process. The general scheme is common to both Bacon and Duhem: First, there exists the smallest step of intellectual development, which is rational and unmysterious; second, all development, however large, consists of an aggregate of such elementary steps. What an elementary step is, however, allows for different answers: Bacon saw gradual abstraction as the key, Duhem saw the key in slight twists or modifications of earlier ideas. Before going into these details, a critical comment of the general scheme may be permissible.

The idea that development can be broken into elementary steps is very questionable, as well as the idea that all elementary steps will be unmysterious. Also, it is almost certain that the breakdown of a large idea into many unmysterious ones does not in itself eliminate the mystery. To consider the simplest and least mysterious kind of intellectual activity, let us take chess (too serious for a play, but too playful to be taken seriously, as Montaigne has complained). We know that every tournament game can, in principle, be analyzed and thus rendered no more mysterious than a move in the game naughts-and-crosses. Yet there is no calculator and programming available as yet to perform this analysis, so complex it is. Assuming that a step in a tournament chess game differs from the most elementary unmysterious step merely by its complexity, the mere enormity of the complexity would still suffice to make us stand in awe at the ingenious invention of a new gambit with the foresight into different kinds of strategy available to the opponent of the ingenious inventor! It is still possible to consider this unmysterious – say, by claiming that a chess-player has a super-calculator in his brains. The problems this raises need not be gone into.

Thus, even in an oversimplified artificial situation most suitable to the doctrine of unmysteriousness, the program of rendering the mysterious growth of an idea unmysterious by mere analysis, is perhaps doomed to failure. This is *a fortiori* so in fields where analysis has not even been fully proved, such as the fields of mathematics. One may claim that formalized mathematics is entirely unmysterious as each step from the axioms is done according to intuitively fully acceptable rules of inference. But this is unacceptable unless the axioms of the system can be introduced unmysteriously. The development of axiomatic geometry was never claimed to be unmysterious; on the contrary, the claim was that it took the genius of Euclid or of Peano to conceive the axioms of geometry or of arithmetic, though, once conceived, everyone can be told what they are and see, *post hoc*, how unmysterious they are. In other words, it is the *status*, not the *origins*, of axioms which was claimed – mistakenly according to Lakatos – to be unmysterious. Anyone who is even a little mathematically trained knows of methods of substitutions which obviously help solve certain classes of problems (such as some integrations) but which were invented with devilish ingenuity and fantastic insights.

Thus, there is sufficient reason to suspect that the idea of breaking down the development of thought into small and reasonably unmysterious steps is impossible: even if we could invent a method which may enable us to invent all the best methods of substitution with no effort at all, we shall have to admit that Euler had none of these and he made wild and incredibly impossible yet successful guesses which helped him solve his problems.

But let us concede that every historical development was made in small steps, that there is little mystery in piling up small steps even though the resultant big step does look mysterious – especially after the traces of the small steps were wiped out.

Let us ask now, why is the small step unmysterious. This is a question easy to ask in the twentieth century, after the fact that a few ingenious psychologists have asked it and have even tried to observe the smallest intellectual steps undertaken by children, apes, or mice. It really is no credit for myself that I can ask this question, or a discredit to Bacon and Duhem for not having discussed it in fuller detail. Still, history aside, the results are quite astounding. Kohler considered the ape's invention of the idea of using a box in order to climb and reach for a banana, or of putting two sticks together for that purpose, to be such an elementary idea. He found the invention of such an idea a very mysterious act. Gestalt psychology which he fathered is the view that there is something built in the ape's mentality enabling him to see the new complex – himself on a box under the banana reaching the banana – not as a complex but as a unit, as a single image ('Gestalt' = 'image'). This inbuilt mechanism, the limitation it thus imposes, as well as its ability to produce in the ape what Kohler called 'intelligent error', led Kohler to doubt very seriously whether Gestalt psychology has any bearing on the psychology of scientific discovery. He stressed that he declined answering that question. Recent discoveries were found to be even more puzzling: why can an octopus recognize a rectangular sheet and distinguish it from differently shaped sheets even if the sheet is rather bent? Rats in mazes raise different problems, and tend to show that what is elementary, what not, is almost impossible to say – perhaps the stimuli and the way they are administered interfere too much with the experiments to give us much indication of what is elementary.

In view of all this it may be of no more than historical interest to note that attempts were made to apply the idea of analyzing big ideas to small unmysterious ones in science and its history. Still, I should mention two of these attempts. One of the most important Baconians, Laplace,tried to sketch the breakdown of Newtonian mechanics into small steps each of which he could consider natural. I do not know what he meant by the word 'natural' and what objective criterion, if any, he introduced to distinguish the presence of naturalness from its absence; perhaps this ought to be obvious to all reasonable people. In any case, it is obvious that Laplace did not use the absence of an explicit criterion in order to smuggle unnatural ideas as natural. On the contrary, he split the introduction of Newton's Law of inertia into two smaller steps – the postulation constant speed under no coercion, and of constant direction under no coercion – and he admitted only one of these postulates as natural. The only way I can find out which postulate he did consider as natural and which he did not is to look it up – it is discussed both in his *Celestial Mechanics* and in his *System of the World* – because I do not remember what I do not understand. In any case, it seems that he was not in the least distressed by the unnaturalness of one element in his introduction of Newtonian mechanics – obviously because he still hoped that the unnatural step may be broken into smaller steps which he could claim to be natural.

The other sketch I have in mind is Duhem's sketch of development of Newtonian mechanics from mediaeval (mock-) Aristotelian mechanics by small gradual steps.

Somehow quite a few historians of science still think this could be done if the alterations are small enough and made for sufficiently compelling reasons. The process of breaking the progress of Newtonian mechanics in Duhem's way still goes on; and, doubtlessly, it is superior to Laplace's because it is genuinely historical, namely, each step is supposed to be found in the history of the subject. But for this very reason the project is so hopeless from the start. There is no doubt that the attempt is doomed to failure; we know that Newton had quite a few flashes of insight which, as Laplace himself already suspected, may remain mysterious forever.

These, and other, flashes of insight, are all too well known to be ignored. Quite a number of historians of ideas have been bothered by them and tried very hard to break them down to less mysterious steps or to explain them away. The idea that some insights should be admitted as unanalyzable, and mysterious, has always been resented, and very vehemently so: the rationalist tradition of modern science opposed this intuitionist theory of flashes of insight as a mystification and as irrationalism. The opposition to intuitionism of any kind is so deep-seated that when the rationalist philosopher and historian of science of the early nineteenth century, William Whewell, advocated a version of it, he was dismissed as irrationalist with a vengeance. And even after people whose rationality was never in question, such as Poincaré, Hadamard, and Einstein, defended some version of intuitionism – of the theory of the flash of insight – and even evinced some empirical evidence for it, the opposition is still strongly sustained. One can explain this opposition by the desire to have a rational theory of the evolution of thought, especially in order to combat the irrationalism of the mystic philosophers who claim that knowledge is achieved by the mysterious personal experience which they call 'the mystic union'. This theory is, indeed, irrationalist, as I shall now describe in detail. First, I wish to sum up my explanation of the persistence of the Baconian and Duhemian view: as against irrationalist mysticism there is a demand for a theory of the rise of scientific ideas in an unmysterious fashion, and only Bacon and Duhem answer this demand. In other words, the popularity of these views is rooted in a reaction to irrationalist mysticism.

4. MYSTICISM

The mystic union is an experience; psychologically, it is the sweet feeling of complete dissipation whereby one feels one with the universe. The novelist W. Somerset Maugham has claimed that he has experienced a mystic union; mystics would probably suggest that Maugham is mistaken, that he merely experienced the psychological side, and hence a mere semblance, of the mystic union: Maugham does not claim that his experience improved his knowledge, whereas according to the mystic's intuitionistic theory, the experience of mystic union must be accompanied by the flash of insight which in itself is knowledge. The knowledge acquired by the process is complete and hence ineffable, i.e. inexpressible in words (due to the limitations of human language); yet it is knowledge all the same. The teacher cannot transmit this knowledge since it is ineffable, but he can transmit parts of it; still better, he can teach his disciple the theory and the rituals which might lead the disciple to experience the mystic union all by himself. The theory in question – mystic intuitionism – is, thus, the view that complete knowledge is achieved in a single flash by mystic union. The mystic union is, according to this theory, only the means, though also the only means, for the acquisition of complete knowledge. The complete knowledge must be acquired first hand since it cannot be transmitted since it is not capable of being put into words since it is complete whereas language is limited and partial.

Russell's essay 'Mysticism and Logic' is a very balanced comment, from the scientific point of view, on the mystic's intuitionist theory of knowledge. Russell does not deny that the experience in question – the mystic union and/or the flash of insight – can, and sometimes does, take place. Russell also admits that the flash of insight may be very important. He does not bother to criticize the mystic's claim that the mystic experience may lead to complete knowledge – he is able to accept this claim for the sake of the argument. Nor does he repudiate the mystic's thesis that language is limited so that it is an inadequate means of stating and communicating complete knowledge. On the contrary, he expresses his contempt for those mystics who invest much time and effort to expound and defend so trivial a thesis. What he stresses is that we are not always certain that we had a proper mystic experience, that there is always the possibility of error; and, as we have noted, the mystic himself asserts this. So, in the main assumptions Russell agrees with the mystic; yet from these assumptions – that mystic insight is sometimes, but not always, attainable, and that it is always incommunicable – he concludes that the mystic insight is useless! On the one hand, accepting its outcome without question involves excessive and unjustifiable reliance on one's intuition, since uncertainty as to the mystic's success can always be entertained; on the other hand the incommunicability of the result makes it impossible to examine it; hence it must be rejected. The mystic knows of this objection. His only answer to it is this. When one has the genuine experience, one has so much genuine peace of mind that one knows that the experience was real. This reply must be rejected as too subjective to be seriously entertained. Science has smaller claims; it admits its own limitations, and even the possibility that it contains subjective elements; but as it is essentially communicable to all, it is open to examination and to development in the direction of greater objectivity and less limitation – at the same time retaining its communicability.

The mystic's alleged knowledge – be it attainable by the exercise of intuition or in any other way, be it what it may, and be its merits and scope what they may – is distinctly different from scientific knowledge. Whatever scientific knowledge may precisely be, it has always been essentially communicable: its theories are understandable in principle to anyone who takes the trouble to master the human language within which it is stated, and its experiments are repeatable in principle by anyone who takes the trouble to acquire some dexterity in experimentation. I should add to this that a great deal of effort is traditionally invested not only in the attempt to acquire new knowledge but also in the attempt to render older knowledge more easily accessible to all, in the attempt to simplify mathematics and its proofs, scientific theories and deductions from them, experiments, methods of handling tools, and even the cost of the tools. This is not always the case to the same degree. In our own days the portion of effort put into simplification, streamlining, popularization, and reducing the cost of experiments, is much too small; this only means that we may be giving up one of the chief characteristics of scientific knowledge. Yet, as long as science is recognized by its long-standing traditions, including its democratic tradition of being in principle communicable to everyone and criticizable by anyone, I for one would join Russell: I would prefer democratic imperfect knowledge to that complete one which is the prerogative of the experienced elite. Even on the supposition (which I reject, of course) that mystical knowledge is in every other sense superior to scientific knowledge, science in its openness is preferable to knowledge accessible only to the mystic.

It follows from this that it is an error to criticize the mystic's intuitionism as the root of his irrationalism: the root of his irrationalism is his view that knowledge is incommunicable. Indeed, rationalist intuitionism exists, and is much less objectionable; its having been traditionally coupled with mystic intuitionism is a confusion rooted in an exaggerated condemnation of every idea of the mystic as irrational. It is important to notice that though (by definition) all mystics are intuitionists, not all intuitionists are mystics: an intuitionist may claim that the mystic union leads to complete knowledge which can and should be stated within human language. This is rationalistic intuitionism, as opposed to the mystic irrationalist intuitionism. Plato and Aristotle advocated rationalistic intuitionism, as is well known though all-too-often overlooked. According to Plato, the mystic union is the union of the mind with the world of Ideas which leads to the axioms of science; according to Aristotle, the union is with the essence of things and leads to the definitions of science. From the axioms or the definitions all knowledge can be derived by pure logic, by deduction. Aristotle, it is true, also had another, more popular, theory of learning, namely of induction by generalization from observation. The existence of this Aristotelian scientific method and its popularity today should not obscure the fact that there also exists the intuitionistic Aristotelian scientific method; this intuitionist method was very popular in the Middle Ages amongst some leading non-mystics such as Maimonides and St. Thomas, and thus served as a link between the rationalist and the mystic traditions of mediaeval Europe.

It is clear that Russell would oppose Aristotle's rationalistic intuitionism much less than Bergson's mystic intuitionism, though he would oppose equally Aristotle's and Bergson's claims to certainly. As he claims in his 'Mysticism and Logic', all that rational philosophy requires is that the product of intuition, of the flash of insight, be put to the check of reason. This may be done if the product is formulated within human language. Assuming this to be the case, it matters little whether the product is claimed to be known with certainty or not, as long as someone takes the trouble to examine that product by the use of his reason.

It is a strange fact that even Russell, who admits the existence and significance of the flash of insight, and demands in his 'Mysticism and Logic' merely that the product of the insight be checked by reason, presents an inductivist theory of knowledge elsewhere in his writings in demanding that scientific theories be supported by, or based on, empirical findings. His reason is not that he wants a more obvious and less mystical theory of the origin of ideas, but that he thinks that checking ideas in the light of reason is enough in philosophy but not enough in science; that within science we need some positive justification for maintaining this rather than that view or theory. I discuss the question of justification elsewhere and shall not elaborate here on my point that it leads rather naturally to inductivism or conventionalism. My point here is more strict: assuming that the rise of ideas is unmysterious we are led by force of circumstances to Bacon's inductivism or to Duhem's conventionalism, and we can find no other alternative as yet. This now may be used to answer the question of the present discussion: why is there no third view of the Renaissance of science?

5. THE POSSIBILITY OF DEVELOPING A THIRD VIEW

There are, we remember, three views of the Renaissance of culture, the radicalist, the conservative, and the middle or reformist. The reformist view of the rise of the Renaissance of science hardly exists. One can show now, that the reformist view is impossible on the assumption that rationality demands the explanation of the rise of scientific ideas as unmysterious. Indeed, there are only two traditional rational views of the growth of scientific ideas which present this growth as unmysterious. Bacon's idea is that scientific ideas grow out of a wealth of factual information; Duhem's idea is that scientific ideas are the modification of older ideas, and the new ideas are designed to incorporate new factual information with greater simplicity. The Baconian idea leads to the radicalist myth of the Middle Ages, and the Duhemian idea to the conservative denial of the possibility of any scientific revolution – in the Renaissance or in any other period.

Other theories of learning have been mentioned too, which present the rise of knowledge as a mysterious event. Mystic intuitionism and rationalistic intuitionism present the process as a flash of insight accompanied by the mystic union; both declare the knowledge thus attained to be complete and final; the former declares knowledge to be inexpressible in human language, the latter declares it to be expressible.

It is obvious that mystic intuitionism cannot be applied to the history of science: as Russell has shown, the mystic's knowledge greatly differs in character from scientific knowledge. Even rationalistic intuitionism is scarcely applicable to history, or so it seems. Some Kantian or other may have tried heroically to show the rise of rational mechanics as a process of rational intuition, but one can hardly see what remains of the history of thought except a list of insights, each final and definitive.

Thus, we have only two acceptable theories on the assumption that for the rational reconstruction of the history of science we must explain the rise of scientific ideas as unmysterious events. And the assumption seems too plausible to reject. But before we fully adopt this assumption, we should examine further another theory which contradicts it – and which has so far only been alluded to. It may – and I think it will – turn out to be just the right theory to permit a reformist view of the rise of the Renaissance of science.

6. MODIFIED INTUITIONISM

The other theory, which we may call 'modified intuitionism' or 'critical intuitionism', has been presented by a few thinkers, including Galileo, Brewster, Whewell, Poincaré, Hadamard, Einstein, Russell in his 'Mysticism and Logic', and Popper. This theory describes the acquisition of a new idea as a flash of insight. But it has nothing to do with the mystic union; it contains nothing about the psychology of the person at the moment of insight except that usually the insight is a possible solution to a problem which that person has immersed himself in. Nor has modified intuitionism anything to say about the completeness or the finality of any product of the insight; rather, it declares that the function of reason is to evaluate and check critically the result of intuition or of the flash of insight. Some holders of modified or critical intuitionism, notably Galileo and Einstein, have claimed that in science finality can be achieved, but their theory of finality is not based on any criterion of the validity or authority of intuition; rather the claim that the finality of some product of our intuition can be ascertained by reason alone, by the thorough critical examination of our theories, as against facts and otherwise. Other holders of modified intuitionism, notably Russell, claim that some products of our intuition, even though not final, gain certain authority from facts by induction. Still others, notably Poincaré, claim that these products gain certain authority from their simplicity. Popper is perhaps alone in claiming that no idea outside mathematics and logic can either be final or gain any authority whatsoever. But extremist as Popper's position obviously is, it is not so extreme either in that it asserts that the source of ideas is intuition or in that it denies any validity or authority to intuition or to any of its products; on the contrary, this assertion plus disclaimer characterize modified or critical intuitionism as such, and is thus shared by Poincaré, Russell, Einstein, and others.

To conclude, all modified or critical intuitionists assume intuition to be the source, but not the authority of ideas. Most modified or critical intuitionists ascribe authority to reason when it corroborates intuition at each stage; some, notably Galileo and Einstein, ascribe this authority only to the very final stage; and Popper is sceptical even about the existence of this final stage.

It is very difficult to make the public notice that not all intuitionists hold the traditional intuitionistic views about the final authority of intuition. This has happened when Whewell's view was dismissed as a version of irrationalist intuitionism a century ago by Mill and his followers. This has happened again when Popper's view was recently dismissed by Kirk, in a controversy between the two concerning the origins of Greek thought. Kirk has rightly noticed the influence of Popper's philosophy on the way Popper writes history, and the important role of intuition in Popper's philosophy. Kirk thus feels justified in dismissing Popper's history along with his philosophy, which he considers traditional intuitionism. Kirk thus shows that he is unaware not only of Popper's ideas, but also of progress concerning intuition. If one wants to explain this strange fact one cannot do so reasonably by assuming that Kirk is ignorant, since he is undoubtedly a leading scholar. One can, however, explain it thus. There is the primary question, where do ideas originate? And the secondary question, where do ideas gain authority? The traditional empiricist answers both by reference to experience, the traditional intuitionist answers both by reference to the flash of insight. So they are lumped together in many people's minds, including Kirk's. All modified intuitionists, however, by definition, agree with the traditional intuitionists on the primary question and disagree on the secondary. The reason for lumping both questions is perhaps the mere accident of the existence of tradition of two schools which answer both questions in the same way. Possibly, however, the tradition rests on the fact that both schools share what Popper calls the pedigree theory of knowledge: the origin of a theory is its very guarantee for its validity. Anyhow, it is undeniably possible to answer the primary question with the traditional intuitionist school while answering the secondary question with the traditional empiricist school (Whewell, Russell), or in a third way (Galileo, Einstein), or while dismissing that question (Popper). By definition, modified intuitionists accept the traditional intuitionist answer to the primary question and reject the traditional intuitionist answer to the secondary answer.

This is not to say that modified or critical intuitionism is the correct view of the rise of ideas – scientific or otherwise – but merely to distinguish modified intuitionism from traditional intuitionism. Doubtless, in examining critical intuitionism one may raise hosts of problems to be met by the critical intuitionist.

7. THE DEMARCATION OF SCIENCE

Critical intuitionism, especially Popper's version of it, and the problems it gives rise to, open up a whole new universe of discourse: here I wish to conclude with the problems raised with the attempts to apply critical intuitionism to the study of the rise of the Renaissance of science.

The view of science as part of, and a special case of, the critical tradition, to wit the tradition of flashes of insights controlled by reason, raises sharply the problem of the demarcation of science from other activities, especially from activities within the critical tradition; yet, just because it so presents science as a part of the critical tradition, it also allows for a broad view of the Renaissance of science within the Renaissance of culture. The problem of demarcation is a complex affair to which a growing literature is devoted; here suffice it to present Popper's solution in a very brief outline: science is characterized by scientific criticism, and scientific criticism is discriminated from other criticism by its being empirical or experimental. Now it so happens that most of the Renaissance thought which we usually label as scientific has rather little to do with empirical criticism. It is, therefore, difficult to apply Popper's demarcation of science from other activities to the works of Brunelleschi and Leonardo, or even of Tartaglia and Galileo; and if one wishes to be tough, one can even claim that it is difficult to apply Popper's criterion systematically even to the works of Copernicus and Kepler. It is indeed very tempting to declare that Popper's criterion applies clearly to works of Newton, Maxwell, or Einstein, that they are strictly applicable to some definite theories of modern science, but reveal a great weakness when applied to the Renaissance, for example.

Undoubtedly, Popper's criterion does not apply comfortably to the Renaissance science. Bacon's and Duhem's criteria seem to apply to Renaissance science very comfortably. These two criteria tell us about fully fledged and easily recognizable Renaissance science. One of them also tells us of the non-existence of mediaeval science – the myth of the Middle Ages – and the other applies comfortably even to mediaeval science, implying thereby that there was no Renaissance revolution of science. Thus, I should say, it is a prerequisite or a *desideratum* of a criterion of demarcation of science that it apply comfortably to modern science, with great difficulty to mediaeval thought, and with some measure of discomfort to Renaissance science: the *desideratum* is that the borderline between science and mediaeval pre-science be sufficiently vague: as Duhem has said, in no time does science have a blank sheet to begin from, but as the Baconians say, the Middle Ages must be viewed as rather pre-scientific. Indeed, when reconstructing the Renaissance of culture we have a vague borderline between the uncritical and the critical, and when discussing the Renaissance of science we have a vague borderline between the uncritical and the critical, and when discussing the Renaissance of science we have a vague borderline between uncritical rational mechanics and critical rational mechanics – Galileo – as well as between empirical-but-uncritical astronomy and empirically-critical astronomy – Kepler – and between critical but unempirical theory of matter and empirically-critical one – Torricelli, Boyle, and Pascal. And in each of these wide border-strips one can expect strong interaction between the Renaissance of science and its cultural Renaissance background.

Before coming to outline this, I wish to confess that I was much too swift in the foregoing discussion, almost to the point of performing a sleight of hand: of course, the Duhemian or Baconian criterion does not fit the Renaissance – or any other period – so very well. Of course, Popper's criterion does not yield sharp results even concerning Newton or Einstein – their science also interacted with their cultural and metaphysical background. And, finally,

I do not wish to pretend that I accept Popper's criterion without qualification. Still, the fact remains that Popper's criterion allows for a better theory of the Renaissance of science.

8. THE RENAISSANCE REVOLUTION OF SCIENCE

The picture of the rise of the Renaissance of culture as the attempt to revive the ancient culture is too well known to need mention. That while reviving ancient culture experimentation and criticism developed, has been stressed by a few authors, notably the art historian Gombrich. That this is linked with the revival of ancient critical thinking has been widely noticed as well. Also, in our century authors like Burtt and Koyré noticed the influence of revived neo-Platonism including light- and number- mysticism on Copernicus, Kepler, and Galileo. Others, notably Santillana, have noticed the overflow of studies like geometry from the arts (painting, architecture) to the sciences. Others, notably Lane Cooper and Koyré, have noticed that the science of the Renaissance was a reconstruction of ancient science. But science turned out to be such an activity that, once started and let develop, it resembled its ancient origins less and less.

There remains, perhaps, the question of the revolutionary character of the Renaissance. Most authors nowadays view revolutions as chain-reactions which pass the point-of-no-return, in a quasi-semi-Marxist (or -Hegelian) fashion. This is well applicable to what was said above, and thus it allows to further the above view of the history of culture and science in the Renaissance with other histories of that period.

To conclude, the Renaissance of science connects with the Renaissance of culture first in its reconstructionist techniques which become a more extensive, inventive, and critical method. Secondly, the connection was in the reconstructions themselves – of the metaphysical and scientific ideas of antiquity as a part of the general reconstruction of antiquity.

There remain two important points concerning the Renaissance of science from the critical viewpoint advocated here. First, it has to be argued that the Middle Ages were uncritical for some reason that was removed in the Renaissance, and that the Middle Ages contained an element which, without that constraint, led to the Renaissance. Second, it has to be shown that the science of the Renaissance, and even the later period, was not as devoid of prejudice and mythology as has been claimed. These are topics for further study.

INSERTS

1. There is a need for a sympathetic-cum-critical attitude towards the Middle Ages, so as to explain that era rationally without thereby endorsing its values.
2. Such an attitude may stand behind the view that the mediaeval thinkers were misguided; but it is difficult to present such a view without falling into excessive naiveté.
3. There are three elements in mysticism: concerning the preparation towards the mystic act, the act, and its function; each of the three has local variants.
4. The most important for us is the Western variant of the theory of the function of the mystic act, namely the view that it brings salvation and revives the Golden Age.
5. The Western version of mysticism is not merely uncritical or irrational, but also it is peculiarly a reinforced dogmatism.
6. Mediaeval mysticism was the cause of the mediaeval stagnation as well as the seed of the Renaissance.

CHAPTER 24

ON EXPLAINING THE TRIAL OF GALILEO

> If a science has to be supported by fraudulent means, let it perish.
>
> (J. Kepler)

> The life of a great man ... can never be a mere record of undisputed fact ... The biographer ... must penetrate behind mere events to the purpose and character they disclose, and can only do so by an effort of constructive imagination.
>
> (A. E. Taylor)

> There are no villains in the piece ... it is what men do at their best, with good intentions, that really concerns us ... if Joan had not been burnt by normally innocent people in the energy of their righteousness, her death at their hands would have no more significance than the Tokyo earthquake ...
>
> (G. B. Shaw)

1. KOESTLER UPSETS THE HISTORIANS

Arthur Koestler's *The Sleepwalkers* is, the subtitle says, "a history of man's changing vision of the Universe". The thesis which this history illustrates, we are told, is that science and religion are essentially partners, sharing mystic intuition as their common source. Admittedly, science and religion are now divided, but this division is merely the outcome of some historical events – Galileo's quarrel with the Church of Rome and the ensuing misunderstandings – which could have been avoided; it is high time now to reunite science and faith, so as to save Man from (nuclear) self-destruction. The meat of the volume consists of two essays, one on Kepler, one on Galileo. The former is sympathetic towards its hero and was well received, the latter is critical of its hero and was frowned upon. *Prima facie*, at least, public opinion was rather apologetic.

Koestler's view of Galileo has been violently attacked by two leading students of Galileo – Santillana and Drake. Many of their strictures are just; yet even if all of them were just, I would still dissent from their overall judgment. Though Koestler is no scholar, his work is of value. It contains valid criticisms of scholary works, and the first lively image of Galileo. Koestler's great success is in managing the humanization of an idol even though he did so at the expense of committing some serious errors.

Roman Catholic writers have already claimed that the clash between Galileo and the Church was to this or that extent rooted in three faults of Galileo's. First, his difficult personality; second, his mistakes in science which were corrected by later scientific studies; and, third, his attempts to meddle with theology. Koestler, although his contempt for the Inquisition is considerable, accepts these strictures, pushes all three as far as he can, and combines them into one: Galileo fought not for the sake of truth, but because he was pathologically unable to avoid any quarrel, accept any compromise, or admit any error. Koestler has succeeded in drawing a new and very vivid picture of Galileo and of his trial, even though hardly any of the accusations he levels against Galileo is claimed to be new, with the exception of the attribution of motives. Now it is dangerous to attribute motives, especially low ones, and especially to an idol. Yet Koestler has rendered a valuable service in trying to do so, and not merely because he may be viewed as the devil's advocate. We cannot explain historical events without making hypotheses concerning the aims, interests, and motives of those who have participated in them. And after we propose such explanatory hypotheses, we can try to argue rationally about their truth or falsity, and then improve on them. Galileo held three different positions in different times: first, he concealed his Copernicanism; then – when he was about fifty – he defended it with some caution; his great battle for it took place when he was about seventy. Koestler's view of Galileo's motives and purpose is offered as one explanation of all three phases: the young Galileo's avoidance of open defence of Copernicanism was rooted in his knowledge that once he would be drawn into a controversy it would be a fierce and uncompromising battle. This self-awareness, says Koestler, first made Galileo timid; but once he fought he would not be stopped. For my part, I think the story of Galileo is not that of one and the same character: the story of the old Galileo is the story of a changed man – of a Catholic reformer who hoped to prevent further clashes between science and Catholicism, but whose plan misfired.

2. KOESTLER AND HIS PREDECESSORS

The story of Galileo has two sides to it, one scientific, one political, which are closely related but very distinct. It has other sides as well, philosophical, theological, etc. These have not yet attracted the attention they deserve. Let us, then, take the scientific side first and the political second. Most writers on Galileo have taken it for granted that scientifically he was right because he was a Copernican, and his opponents were wrong as they were defending Aristotelianism. Some Roman Catholic writers, however, have accused Galileo while defending, or at least refraining from discussing, the scientific position of the Church.

A C. Crombie's position may serve as an instance: it is apologetic, eclectic, and up-to-date. If I understand Crombie rightly, he produces three arguments against Galileo. First, Galileo was mistaken in having claimed that Copernicanism had been demonstrated, since the demonstration of Copernicanism was produced only in the early nineteenth century, with the discovery of stellar parallax [p. 323]. Secondly, as Einstein has shown, the truth is that there is no immobile centre of the world; the physicist's choice of a centre is arbitrary [pp. 327–8]. Thirdly, science concerns itself not with the search for truth but with the proposal of working hypotheses which save the phenomena [pp. 324–5, p. 328]. (Firstly, I did not borrow the bowl from you; secondly, I already returned it to you; and thirdly, it was cracked when I got it.) These arguments have little reference to the historical situation; even were they correct they should be dismissed as mere hindsights.

Koestler's attack on Galileo the scientist includes Crombie's main ammunition, and, in addition, the view that Galileo was lying when he claimed that Copernicanism had been demonstrated. This view I consider to be false, but not outrageous, and at least historical rather than anachronistic. Those who are infuriated by it are, I suppose, adherents of the principle that men of science are invariably honorable men, and *vice versa*. This principle is very widely accepted, as I have tried to show elsewhere. Even Koestler, who explicitly rejects it, abides by it throughout his work, to the point of acknowledging Galileo as a great scientist only after the Inquisition broke his pride in 1633. This terrible opinion obviously clashes with ample evidence. Since Koestler usually endorses Kepler's judgment as sane, moderate, and human, I shall mention only Kepler's view of Galileo. In his comments on Galileo's *Assayer* of 1623, Kepler makes it quite obvious that in his view Galileo was a powerful thinker [p. 355], but far from being an easy and pleasant character. Kepler himself claims there [pp. 341–2] that Galileo made statements he did not intend to make, merely in the heat of the debate. Of course, Koestler's view that Galileo's statements to the Inquisition were lies goes much further; and, unlike Kepler's, is nourished not by scientific considerations but by the Catholic hostile literature. But this is no reason to dismiss Koestler's view as false, though ultimately it has to be rejected – for other and better reasons.

Koestler's attack on Galileo the scientist is nourished not only by the hostile Catholic attitude, but also by the clumsily apologetic attitude of others. To take one example, in his introduction to his selection from Galileo's *Assayer* [p. 226] Drake speaks of him as the teacher of experimental philosophy (a phrase which Galileo never used). On the next page, he brushes aside the fact that Galileo's *Assayer* concerns the defence of a false hypothesis, according to which comets consist of (earthly) vapors, adopted in order to rescue his own version of Copernicanism. Drake claims that this is quite irrelevant to "the main point of the book, which lies not in the hypothesis itself but in its use". Santillana comes to Galileo's rescue in another way [p. 153]. It was right, as I understand Santillana, to dismiss the comets until it was possible to use them as demonstrations of Copernicanism; and this only Newton could do. Obviously, this is both apologetic and untrue. Koestler's condemnation of Galileo the scientist on the ground that he did not plot the elliptic courses of comets is faintly amusing, until one compares it with the clumsy apologies that Galileo's defenders put forward to explain his failure to endorse the ellipse. Generally, every time Galileo's defenders are apologetic, Koestler stresses the point which causes them discomfort; and every time the Catholic apologists show a weak point in Galileo's scientific views Koestler follows them. And he regularly attributes some unpleasant motives to Galileo. But let us leave the scientific side of the story for now.

The chief Catholic technique used on the political side of the story is that of pooh-poohing one document and stressing another; and Koestler follows suit. Let us see how the political evidence is handled by James Brodrick, the biographer of Cardinal Bellarmine (1928). Santillana, no ally to him, refers to him as to "no less a historian than Father J. Brodrick, S. J.", attributing to him "discriminating scholarship" and "Christian warmth". Brodrick puts the whole blame of the Church's denunciation of Copernicanism in 1616 on Galileo. This thesis, looking rather shaky, must needs be heavily supported by evidence. Brodrick quotes – from a letter written by the then (1616) Tuscan ambassador to Rome – some unpleasant words about Galileo's behavior in Rome and about the imprudence of a (young) Cardinal whom Galileo presumably had sent to talk to the Pope about Copernicanism. Copernicanism was denounced, Brodrick tells us, as the result of Galileo's pressure, which the Tuscan ambassador was describing. Brodrick adds a footnote protesting against Galileo's defenders (before 1928): "insinuating that the ambassador was a bit of a fool", and ignoring the alleged fact that he was a personal friend of Galileo, they dismiss or ignore his evidence.

On the same page where he launches this complaint, Brodrick himself suppresses a part of the same letter, which contradicts his own story. According to Brodrick's story, once Galileo extended enough pressure to get the machine going, it had to go on its own course according to established rules of procedure: the Congregation of the Holy Office had to consult experts about the status of Copernicanism, and to endorse the experts' judgment. Yet according to the same letter of the Tuscan ambassador, the question was decided not at all by the Congregation, not by any consultation, and according to no rules of procedure: the obscurantist Pope and Cardinal Bellarmine were determined to condemn Copernicanism by hook or by crook.

It is possible to defend Brodrick's omission: the ambassador's story may be questioned because of being chronologically inconsistent with the Vatican files. But this defence of Brodrick will show how much he distorts the views of Galileo's defenders when he says that they had dismissed the ambassador's evidence for no good reason. According to the Tuscan ambassador the decision against Copernicanism was taken by the Pope and Bellarmine on March 2nd 1616, whereas according to the Vatican records the Congregation had passed its verdict against Copernicanism on February 25th. The ambassador says that the Pope has told the young Cardinal who supported Galileo that "the question was to be referred to the Cardinals of the Holy Office", sometime after the question had already been referred to these Cardinals, and decided upon by them. The Pope could not have suggested on March the 2nd to the young Cardinal "to persuade him [Galileo] to give up that opinion [Copernicanism]" if on February 26th Bellarmine had already forced Galileo to be silent.

Santillana seems to explain away the ambassador's derogatory remarks on Galileo in two ways. First, he views the ambassador as "a cynical man of the world" who had deserted his acquaintance Galileo. Secondly, he thinks that the misinformation in his letter is a sign of his having been deceived by a prearranged leak designed to put the blame on Galileo. Santillana has not made up his mind as to whether the ambassador was a mere acquaintance and "a cynical man of the world", or a naive friend who was taken in by misleading inside information which he could easily have refuted by hearing from Galileo about Bellarmine's threats. (All *dramatis personae* were present in Rome then.)

As Santillana's defence is so weak, Koestler has little difficulty here; he could, and did, follow Brodrick's position rather closely, and, in addition, stress every piece of evidence concerning which the defenders of Galileo have tried to gloss over somewhat more glibly than they ought to have done. This is precisely the reason for my viewing his work as much more significant than that of other critics of Galileo, like Crombie or Brodrick. Even if we reject all of his views, and I doubt that we can do this, we must admit that he has posed quite a number of serious problems for those of us who side more with Galileo than with Bellarmine. In any case, it is to be hoped that we shall never be able to return to the old idealized picture of Galileo.

3. KOESTLER'S PEN-PORTRAIT OF GALILEO AND ITS IMPLICATIONS

Koestler describes Galileo as an extremely unpleasant and arrogant person, whose interest in science was much less a motive for his actions than his "hypersensitivity to criticism, and his irrepressible urge to get involved in a controversy" [p. 432, p. 470]. Although great men of science may be as negative personalities as Koestler's Galileo, and although one cannot entirely refute Koestler's picture of the young Galileo (it can easily be shown to contain exaggerations), one can show that the older Galileo found his personal salvation in science: he learned to rise to the occasion of defending the cause of science, and thus vastly improved his personal character. (It seems that Koestler's historical figures suffer from the same defect as his fictional figures: they never change their characters.)

According to the accepted view of the trial, the Pope and his advisers used the trial for an attack on science in general; it was a piece of self-assertion by obscurantists. Catholic historians, too, tend to accept this view. But this view conflicts with the following well-known facts. The new Pope, Urban VIII, had encouraged Galileo to write the *Dialogue* or at least let Galileo come out of six audiences with him with the impression that such was the case (which is practically the same thing). And the book won the *imprimatur*. Koestler argues that the encouragement was given to the scientist Galileo and that the trial was of the conceited and quarrelsome Galileo who took the opportunity and fought his own private battle instead of fighting for the cause of science; that by bullying ignorant censors Galileo succeeded in getting the *imprimatur* for a book of a different character than the one he was expected to write; that his bluff was called in a very short time, and he had to pay for his arrogance.

Assuming that the existence of the Inquisition and censorship are not exactly encouragements to the freedom of thought, and assuming that the

Inquisition was not created for the sole purpose of intimidating arrogant scientists, Koestler's contention that Galileo's character was the cause of the trial cannot be taken literally. It must be reinterpreted as an exaggeration of the following view. Unfavorable for science as the social and political situation was, it was not so bad to make it impossible for a man of science to defend Copernicanism reasonably and get away with it. It might have been possible – and, for the sake of science, highly advisable – to get away with a defence of Copernicanism by making a verbal concession so as to allow opponents to save face. Given the atrocious illiberalism of the times, the Church's behaviour was reasonable in view of Galileo's violent outbursts, extravagant claims, stubborn unwillingness to compromise, and immense capacity to annoy and irritate practically anybody. Thus, private affairs upset public matters and led to the divorce between Faith and Science.

It is this reinterpretation of Koestler's presentation which will be examined here. We shall take the hypothesis, examine what it can explain, what facts may conflict with it, and how it can be replaced by a better one. The hypothesis is, that Galileo was clever, vain, and hypersensitive. It follows, first, that he quarrelled obsessively, and second, that knowing his own weakness, he cleverly avoided quarrel as much as his vanity permitted. It follows, further, that he was a rather unpleasant person; not as friendless as Koestler makes him to be, but not too popular either. The historical facts the hypothesis explains are rather numerous. The hypothesis explains all the unpleasant details of Galileo's early career; in particular why, until 1613, when he was almost fifty years old, Galileo never committed himself publicly to Copernicanism: as we conclude from the hypothesis, he tried to avoid controversy whenever his vanity permitted. After he had declared his allegiance to Copernicanism, in 1613, he had, as we have concluded, to quarrel obsessively, even if the result was that he destroyed himself. The problem remains: why did he start defending Copernicanism? The answer must be, his vanity prevented him from concealing his views any longer. How? Having made some astronomical discoveries, Galileo could not but stress, not to say exaggerate, their significance (because of his vanity). In other words, he could not avoid the temptation of viewing his discoveries as demonstrations of the truth of Copernicanism.

This part of Koestler's story is logically neat. The evidence which he marshals in its favour is hardly new, but he has shown how glibly this evidence has been glossed over; yet he exaggerates its value, and makes it go a long way. Almost all his significant evidence relates to the period up to 1613. Up till then Galileo picked only quarrels which he could win and which did not endanger his career. His commitments to Copernicanism were vague enough not to cause him any serious trouble; he had a quarrel with some minor Dominicans about it, but he was confident that he could win; and he did win. But there is evidence (see below) which contradicts the hypothesis: it is a fact that in 1613 he was psychologically quite capable of ignoring a challenge. Thus Koestler's story up to 1612, though highly exaggerated in its psychology, and very incorrect in disregarding Galileo's early and passionate interest in truth, is nonetheless largely correct in describing him as an unpleasant, career-seeking, and rather touchy, quarrelsome fellow. In 1614 we find Galileo engaged in a battle with the mighty Bellarmine, in which he does endanger his career and even his life. Since Galileo was not as hypersensitive as Koestler makes him out to have been, he must have had some reason for suddenly becoming so reckless. Let us, then, consider the following modification. Galileo had been career-minded until about the age of fifty. But then he achieved great fame, and learned that truth was much more important to him than fame. Many people seek fame while neglecting to fight for the truth, deceiving themselves all the time that their motive is respectable: once they have achieved fame, they say, they will be able and willing to fight effectively for the cause of truth. It is quite possible that Galileo held the same attitude and carried out his plan sincerely. And when he started his fight he soon found out that fame is of no importance one way or another. Career-seekers normally deceive themselves in thinking that when they have achieved position and fame they will use it in their fight for the cause of truth. Yet occasionally they are sincere. Assuming that Galileo was one of them, one has to admit that he was also sincere. It must always be remembered that he was an unusual person; possibly he looked all the time for a chance of getting his ideas across without a battle; in any case, we know that he fought, and rose to the occasion wonderfully.

4. THE FALL OF GALILEO

Galileo's visit to Rome in 1611, after publishing his *Starry Message* (1610), was a great success. It was a very dramatic change from his previous status. Just before his journey to Rome he had been criticized rather sharply, and his astronomical discoveries had been declared chimerical. Only Kepler had defended them, and upon faith, not after examination of the evidence. Kepler had been worried by the fact that all the evidence he had heard was opposed to Galileo's evidence, and had asked Galileo to name witnesses confirming his evidence. Galileo could not name any witness, very much to his frustration and chagrin; and only with effort – so his letter to Kepler reads – could he turn his immense bitterness against his colleagues into scorn at the multitude, their stupidity, and their ignorance.

And then, in Rome, he was received like a king. The Jesuit astronomers of the Roman College endorsed most of his observations, and he found quite a few friends among them. He was cordially received even by Cardinal Bellarmine, who had been Professor of Philosophy, Master of Controversies, Rector of the Roman College, and remained one of the most powerful figures in Rome until his death. Bellarmine's recognition of Galileo was no small matter, especially considering that Bellarmine was a staunch authoritarian and traditionalist, and had been one of the judges of Giordano Bruno, who had been burnt on the stake but one decade before, whereas Galileo was a follower of Bruno (see below), or at least suspected of being the follower of Bruno, as even Koestler cryptically admits. (Koestler's dismissal of the thesis that Bruno was a martyr of science is unworthy of criticism.)

Of course Galileo found opposition, too, which is not surprising. But in his case one might have expected all opposition to have become weaker and weaker with the passage of time. Yet Galileo's position worsened – and very quickly too. In 1611 Bellarmine shows his esteem of Galileo; in the middle of 1612 Bellarmine writes to Galileo, expressing his affection, respect, and readiness to be of any service (the function of this most unusual letter has not been studied as yet). And then comes a change: in 1613 Bellarmine speaks against Galileo in a private conversation with a priest who is a friend of Galileo, quoting *Psalm* 19 which describes the motion of the sun; the point is discussed in Galileo's *Letter to Castelli* of December 1613, and in at least one of Bellarmine's sermons [Brodrick, p. 335], preached soon afterwards. (Like most of Galileo's early works, this one was unpublished and privately circulated; Bellarmine's sermon was published early in 1615, but, very likely, it was also privately circulated in 1614.) All is still within the bounds of civility; the rest of the story is not. In April 1615 Bellarmine creates a new nuance by launching a warning, suggesting that Galileo's opinion is opposed to a decision of the Council of Trent (this is no small threat), and contradicts King Solomon (*Ecclesiastes*, "and the earth for ever standeth"), the wisest of all men; in about November or early December 1615 he expresses displeasure at Galileo's plan to come to Rome. This is reported by the Tuscan ambassador on December 5th, 1615, but Galileo is in such a hurry to leave Florence that he never receives the ambassador's warning – he is in Rome on the 7th of December. (Koestler's report is inaccurate; almost every one of his inaccuracies can be traced back to a lack of clarity in Santillana or Drake.) The situation is so baffling that until Gebler's work of a century ago it was assumed that Galileo was summoned to Rome. As Gebler shows [p. 71f], he came to Rome voluntarily. Gebler explains this disastrous move as the result of Galileo's unawareness of the strength of his opponents' dogmatism. This explanation is a bit naive, yet Santillana accepts it as he has no better one to offer.

Two months after Galileo's arrival in Rome nothing seems to have happened; then the pace quickens considerably. On February 18th, 1616 (or a little earlier), Galileo goes to the Holy Office [Santillana, pp. 114–5]; on that day the Congregation assembles and on the next day asks for expert opinion; though the problem is most difficult, the expert opinion against Copernicanism is procured within a week; the Congregation is assembled again on the 25th; on the same day the decision is reached *and* ratified *and* Bellarmine is instructed to act; he acts on the 26th, summoning Galileo to his palace and telling him to be silent or else.

This chronology is based chiefly on the Vatican files. It is contradicted, we remember, by the Tuscan ambassador's letter of March 4th. Far be it from me to prefer the (possibly doctored) Vatican files to the ambassador's (second hand) information or *vice versa*. It is possible that the ambassador's letter had been written two weeks before it was dated, to be first delayed and then sent in a hurry; and other explanations are possible. But, obviously, there was a great rush and tumult. Why? Why?

The decision condemning Copernicanism, which was taken on February 25th, 1616, was not published until years later, and perhaps it was not intended for publication at all. It was acted upon, at least according to the Vatican files. The Vatican files report the decision of February 25th and Bellarmine's warning to Galileo on February 26th. The public, however, knew only of a decree, published on March 5th, which condemned Copernicanism in a surprisingly mild tone, and of a certificate of honor given to Galileo by Bellarmine on 26th May, which contradicts the main document in the Vatican files. Wohlwill, Gebler, and Santillana, have all argued more than convincingly that Bellarmine's inside report and his certificate of honor to Galileo are significantly different. Both these documents, however, indicate how seriously Galileo was taken by the Church authorities. Why? How could he throw the Church leaders into a panic and make them act in such a hurry and in so confused a manner? Where lay the power of that isolated sick man?

There are other points which may be explained by the assumption that some confusion resulted from the hasty proceedings of the Church authorities, and that some of the steps which were taken in a hurry were later regretted.

Santillana argues that the wording of the condemnation of Copernicanism is confused to the point of meaninglessness [p. 139]. Moreover, there is a discrepancy between two Vatican reports about Bellarmine's warning to Galileo, and between each of them and the certificate of honor which Bellarmine gave Galileo. Since the discovery of the relevant Vatican documents in 1867, the trial of Galileo of 1633 has often been alleged to be legally connected with Bellarmine's warning of 1616, and consequently much ink has been spilt on the worthless legalistic issue, which of the three documents concerning the warning is the correct one. The more interesting question, I suggest, is why were the Church authorities in such a hurry?

Koestler's explanation of the turn of events rests on a subtle error: he uses the implausibility of his own explanation of the Church's condemnation of Copernicanism to explain the Church's (alleged) subsequent attempt to forget this condemnation. This is the same kind of offence as that committed by a contractor who bills you both for a shoddy job and for the inspection and repairs of its defects. Galileo's bad temper and ability to annoy people, says Koestler, caused the move. These, obviously, could not move even a Pope, let alone a Vatican office, to a rash condemnation of an important doctrine. Therefore, says Koestler, the condemnation was soon buried.

Santillana's story is exactly the opposite. The Church was going to condemn Copernicanism – we are not exactly told why – and all that Galileo wanted was to prevent a rash action. Santillana himself admits in a way that Galileo was the person who had started the affair. Galileo wrote to Kepler in 1597, telling him that he had new proofs for Copernicanism (his law of inertia, I suppose) but would not publish; Kepler urged him, in reply, to publish – in Germany if Italy was too intolerant; Galileo remained prudent enough to say nothing until 1610. "Now [1610] that certainty [of the Copernican doctrine] had been reached", says Santillana, "the motives for silence that he had explained to Kepler no longer were valid" [p. 14]. All that Koestler has to do in order to criticize Santillana is to explain to his reader three points: first, what were the motives for silence (Koestler's answer being, Galileo's fear of criticism and ridicule); second, what was the certainty (that Aristotle was wrong, not that Copernicus was right); and third, what was Galileo's campaign (not to "build up a tidal wave of opinion" [p. 15], since the wave was building up too rapidly anyhow, but, says Koestler, to force his opponents into public admission of their errors and into public acceptance of the new and as yet unproven doctrine).

Santillana does not explain cogently why Galileo started his unfortunate campaign. He argues, with the aid of documents [p. 135], that the softening of the blow to Galileo in 1616 resulted from the strength of Galileo's (theological) *Letter to the Grand Duchess* written shortly beforehand. Yet, clearly, this letter, and its earlier version, the *Letter to Castelli*, had provided Galileo's enemies with the grounds for attack, and had made Bellarmine an enemy. When he describes how Galileo started the campaign, Santillana entirely ignores the Tuscan ambassador's letter of December 5, 1615, in which he is very opposed to the campaign; but when discussing its failure, he agrees with the ambassador [p. 117] that in December 1615 Rome was "no place to come and argue about the Moon". This is terribly apologetic. It is very hard to piece Santillana's exciting story into a simple pattern of explanation. When it emerges, it seems incredible: though Galileo started a campaign, his opponents are blamed for having started a counter-campaign. Perhaps the methods of Galileo and all his allies were more honest than those of all his opponents (this had not been shown); in any case, Galileo was endangering his opponents' faith, their social and political positions, and even their personal security. Santillana's siding fully with Galileo and against his opponents is a bit hard to endorse.

Santillana's uneasy feelings seem to show clearly, for instance [p. 53], when he explains that Galileo's *Letter to Castelli* was written in reply to Bellarmine's views as expressed in a private discussion with a priest who was a friend of Galileo. As Galileo did not accept all challenges, this is an unsatisfactory explanation. Moreover, it is incredible luck for Galileo that of all anti-Copernican passages in the Scriptures Bellarmine should quote in that private discussion *Psalm* 19, knowingly giving Galileo a chance to expound his neo-Platonist light-metaphysics (as Santillana himself notices [p. 154]), especially since Bellarmine thought he had much stronger Biblical ammunition, viewing King Solomon (*Ecclesiastes*) much greater an astronomical authority than King David (*Psalms*). This, like other facts, is explicable by assuming that it was Galileo who provoked Bellarmine to discuss Psalm 19: the person in position is seldom the more provocative party when he bears no malice and is not looking for more troubles than he already has; especially when he has plenty. (Using *Psalm* 19 Galileo was emulating Pico's *Oratio*.)

If we view Galileo as selfish and successful, we cannot explain his campaign and defeat in 1616. Koestler's psychological theory about his obsessiveness, in particular, is refuted by the evidence. If, however, we view Galileo as devoted to science, we cannot explain why he did not fight for Copernicanism in 1611, unless we assume that he shrewdly postponed the battle to a more propitious moment; which makes his battle in 1616 an incredible folly [Gebler, pp. 70–75]. There are two traditions about the case, the hostile, best represented by (Brodrick and) Koestler, and the apologetic, best represented by (Gebler and) Santillana. The easiest way to refute these two traditions is to see how the writers who belong to them struggle with the events leading to, and including, the condemnation of Copernicanism of 1616.

5. THE CHANGE IN GALILEO'S BEHAVIOUR

Koestler's story of Galileo's prudence and selfishness up to about 1612, is hard to ignore altogether. Admittedly, Koestler is mistaken in claiming that the *Starry Message* of 1610 contains no "statement in favour of the Copernican system" [p. 430, p. 431] (the mistake may have resulted from reading Drake [p. 85]). Admittedly, Koestler is mistaken in suggesting that the commitment to Copernicanism in the *Letters on the Sunspots* of 1613 is "somewhat vague in form". (In the third *Letter on the Sunspots* Galileo says, "An understanding of what Copernicus wrote in his *Revolutions* suffices for the most expert astronomers . . . to verify . . . his system" [Drake, p. 130].) These commitments, nevertheless, are not as clear and bold as his private commitments; and the absence of a sufficiently clear commitment involved another failing, namely inability to make proper acknowledgement to ones who were clearly committed. A courtier who had lunch with Kepler tells Galileo (15 April 1610) about the conversation they had: "He said concerning your book [*Starry Message*] that truly it has revealed the divinity of your talent, but that you have given cause of complaint . . . since you make no mention of those writers who gave the signal and the occasion for your discovery, naming among them Giordano Bruno . . . , Copernicus, and himself" [Singer, p. 189]. In order to explain this unacknowledged indebtedness, one has to discuss Galileo's methodology, which is cryptic, and the way he made his famous discoveries, which everyone praises but no one discusses.

Bruno's methodology is perhaps a proper starting-point. In the beginning of his first dialogue in his *On the Infinite* (and in the beginning of the last dialogue in the same book) Bruno makes it clear that he is an apriorist, who, however, does not regard observation as useless; the use of the testimony of the senses is "solely to stimulate our reason, to accuse, to indicate, testify in part; not to testify completely, still less to judge or to condemn." This passage seems to me to be of crucial importance. I do not think that Galileo kept to Bruno's methodology all his life, or that he was clear about his own view of the matter. In his *Dialogue on the Two Systems* (1632), when Simplicio asks Salviati (i.e. Galileo) whether he is an apriorist or not, Salviati refuses to answer and even bamboozles his audience [Santillana's edition, pp. 202–3, and Santillana's note there; cf. Wiener, *passim*]. But at least in his early period, and up to his *Assayer* (1623), his following of Bruno is quite conspicuous, and even in his *Dialogues*, and in his method of writing scientific dialogues, he is a Brunist.

The reason for Galileo's strong adherence to Bruno may be found in his early, mechanical works, *On Motion* (1590), and *On Mechanics* (1600), which were published only centuries later. Galileo started by accepting Aristotle's mechanics, continued by accepting Archimedes and slowly used his own Archimedeanism and clear thinking to expel his own Aristotelianism step by step. A few bits of early drafts of *On Motion* are published in the English edition, which show how slow was his progress. Nor was the process finished by 1600. Towards the end of *On Mechanics* Galileo develops his own law of inertia in order to explain Archimedes' screw, yet soon afterwards, when trying to explain what happens when a hammer hits a nail on its head, he employs an Aristotelian theory of impetus rather than his own Archimedean theory. As Lane Cooper has suggested [p. 48 and note], Galileo's early mechanical works should be viewed as part and parcel of a Renaissance attempt to reconstruct both Archimedes' philosophy and certain ancient criticisms of Aristotle. The effort involved was much greater than one may imagine: Archimedeanism was revived before Aristotelianism was thrown out, and the contrast between the two was discovered by a long and arduous process. Since his investigations were primarily conceived as logical problems [Fahie, p. 19; Koyré, b) Conclusion; Cooper, p. 48], not as empirical ones, Galileo could hardly be interested in experiments, and this is why he was so interested in methodology and laid such an emphasis on clarity. (Galileo, and to a lesser extent even Bruno, was a forerunner of Descartes in viewing clarity and distinctness – but also simplicity – as criteria of truth.)

An impressive example, which Galileo discusses in meticulous detail in the *Dialogue* as well as in earlier works, is this. Everybody (including Leonardo, incidentally) had taken it for granted that a smooth surface reflects light more strongly than a rough one. Consequently, a man on the moon should see the oceans on earth as brighter than the continents. Only clear thinking, says Galileo, can show this to be an error; no amount of experience with walls and mirrors has helped to eradicate it. And a corollary from this correction is most important since it confirms Copernicanism by showing that the moon, being bright, is a rough surface like the earth, not a crystalline body made of the pure fifth essence. Hence we may expect confirmations of Copernicanism to come more easily from clarity than from experience.

With this in mind we can easily understand Kepler's complaint. Galileo's

Starry Message contains discoveries which confirm Copernicanism: the mountains on the moon and the moons of Jupiter. To this one should add the alleged moons of Saturn and the phases of Venus, which fall in the same category. In all of these cases, there is no doubt, anticipation of the discoveries was essential to making them. As we have seen, Galileo knew, by reasoning alone, that the moon has a rough surface. In his *Starry Message* he describes how he looked for hours with a telescope at a dark spot on the edge of the light part of the moon, until it disappeared, as the shadow on an earthly valley disappears at sunrise. He could not have made such an observation without an anticipation, without, particularly, following Kepler's idea of imagining himself standing on Mars and gazing at Earth. It is hard to exaggerate the significance and novelty of this idea [Einstein, 1934, pp. 24–5; 1950, p. 225]. Kepler's indebtedness to Copernicus for it is obvious; when Galileo imagines himself standing in a lunar valley waiting for sunrise, then he is indebted to Kepler and Copernicus at once. Because he was inspired by Copernicus' theory, Galileo took his discoveries to be empirical demonstration of it (like Bruno he did not think that empirical demonstration is complete). That the same holds for the moons of Jupiter and Saturn, as well as for the phases of Venus, and that here Galileo is in debt to Bruno's speculations about the infinity of suns and satellites, is too clear to demand any further elaboration.

Now all this is largely a reconstruction. Galileo himself said little about methodology until after he wrote his *Letter on the Sunspots* (1613), probably because he was still very prudent. Koestler does not quote Galileo's style in his work. Could this be because it is unusually civil, partly even submissive, almost up to the very end? Also the *Letter on the Sunspots* is interesting because it shows that Galileo's earlier *Discourse on Floating Bodies* (1612) had not been intended to arouse opposition, and that when it did arouse opposition Galileo took it lightly, and decided not to answer his opponents, considering himself successful enough in converting judicious people [Drake, pp. 128–9]. This passage Koestler does not quote; rather, to prove his thesis about Galileo's "irrepressible urge to get involved in a controversy" he claims, in his urge[1] to condemn Galileo, that Galileo raised opposition in writing his *Discourse on Floating Bodies* quite unnecessarily.

A philosopher called Buonamici, who was probably Galileo's teacher in the University of Pisa, discovered (what Galileo did not know when drafting his *On Motion*) that Archimedes' hydrostatic theory belongs to the Platonic tradition of explaining levity (or buoyancy) as caused by the gravity of the medium, and is inconsistent with the Aristotelian tradition of assuming both gravity and levity as essential causes of motion. Consequently Buonamici rejected Archimedes' view. The chief objection to Aristotle's view is that boats made of metal float. This, Aristotle explained away (at the very end of *De Caelo*) by an auxiliary hypothesis about the resistance of water (surface tension), as exemplified by floating metallic needles and thin boards. Archimedes probably refuted this auxiliary hypothesis about floating metal boats. His own treatise *On Floating Bodies* explains why metal boats float, and without any auxiliary hypothesis; but it contains no criticism of any other doctrine. Galileo took up Buonamici's critical mode of thinking; and "speaking always without diminution of his [Buonamici's] singular learning" [p. 22] he refuted his views by reconstructing Archimedes' criticism of Aristotle. The objection to Archimedes' view, however, is the fact that metallic needles and boards do float. Galileo tried to answer this objection. This he failed to do (since the objection is unanswerable). He wriggled out of the difficulties by sheer ability to confuse a simple issue. Following Archimedes, Galileo claimed that the metal needle or board must behave like a metal boat and expel water of a weight slightly exceeding its own. Following Aristotle, Galileo's adversaries denied this. This was the crux of the argument concerning Galileo's views, and he was plainly mistaken. He adduced beautiful empirical refutations of Aristotle; and he was simply convinced that Archimedes' geometrical demonstrations were perfect.

But of course, mistaken as Galileo was concerning his own views, his criticism of Buonamici and his followers is valid. Regrettably, he was careful not to use all the ammunition which was at his disposal, judging by his earlier (unpublished) works. The first half of the *Discourse on Floating Bodies* is a discussion of the criticisms and rejoinders of both sides. The rest of the volume is a reconstruction of a Platonist rejoinder to Aristotle's criticism of the theory that the levity of a body is caused by the gravity of the medium. This goes beyond Buonamici and his followers; it is extremely interesting. Even Drake has to admit [pp. XI–XII] that the only novelty in this book is Galileo's (reconstruction of Archimedes') criticism of Aristotle. In this Galileo was continuing the job which Buonamici started. However, his superiority to Buonamici here was a source of trouble. Buonamici was not enough of a clear and critical thinker to have abandoned Aristotle's views. And in backing Aristotle, who was the accepted right horse, he was allowed by the public to use logic alone. As Galileo was backing Archimedes, the socially unaccepted horse, the public forced him (by appeals to his employer, etc.) to produce not logic but experimental facts. This pressure he may have anticipated, and this anticipation, plus his prudence, may explain why he was silent for years.

Afterwards, when he had become famous, he published only a small part of his criticism of Aristotle, and while speaking of him with great civility, saying (untruthfully, I think) "he hath exquisitely philosophiz'd" [p. 64]. But the small dose of criticism was enough to arouse dangerous opposition, of the kind which Drake and Koestler ignore, but which can be judged from the following event. Father Grienberger, a Jesuit astronomer of the Roman College, to whom Galileo referred in 1615 as to "that excellent mathematician and my very dear friend and patron", wrote in 1614 "to a close friend of Galileo to say that were it not for the deference which by the direction of his superiors he was obliged to show towards Aristotle, he would have spoken his mind clearly on the matter, in which Galileo was perfectly right" [Brodrick, p. 347; Santillana, p. 118n; surprisingly, neither gives any reference]. (Incidentally, Father Grienberger seems to have remained a friend to the last; but he could hardly be of any help to Galileo, it seems, because of his vows of obedience.)

The importance of the early (1612 and before) mechanical works of Galileo lies in his realization of the importance of logic; his considering criticism and clarity to be essential for scientific discourse. Quarrelsome as he could be, he confined his great discoveries prior to 1610 to a small inner circle, because, like most quarrelsome people, he knew with whom not to quarrel. But then, in 1610 fame had been achieved, and he found that ideas mattered to him more than wordly position. He sends a feeble feeler in the form of *On Floating Bodies* in 1612, and soon finds a wall of silence. From now on he becomes reckless and decides to take the bull by its (theological) horns not withstanding any risk to his own position. Koestler's prudent Galileo ceased to exist in 1614–5, and the reckless Galileo had an entirely different character from the one which Koestler ascribes to him. In short, the young Galileo was quarrelsome but prudent; the Galileo who got into trouble with the Church was a changed man.

6. GALILEO'S FAITH AND FATE

It is hard to imagine how much of science was a mere dream at first. That much of it was (pseudo) Pythagorean light-mysticism, has slowly transpired through works of E. A. Burtt and others. That Pythagoreanism was deeply linked then with Cabbalism and alchemy has been shown by a few scholars, such as Blau, Gombrich, and Miss Yates. How much of the methodology of the time was a mystical dream is a story which has not yet been told.[2]

In Galileo's days, hypotheses – say Ptolemy's – were claimed to encompass all known relevant phenomena, but not to describe any reality; also, usually they were over-complicated. These two points were connected: reality was always assumed to be simple, whereas the appearances were known to be complicated; hence a complicated hypothesis could be posed only as a means of calculating empirical results, not as a truth about reality. Today the term for such a hypothesis is "a working hypothesis"; traditionally, such a hypothesis was called "a mathematical hypothesis" or "a mere hypothesis" [Popper, p. 168]. One of the reasons for not suggesting hypotheses about reality was, no doubt, the idea that Aristotle had said everything about reality (though the relations between reality and appearances must then be admitted to remain obscure). The other reason for giving up attempts to describe reality, according to Popper's suggestion, was the acceptance of Plato's and Aristotle's idea that statements about reality must be demonstrable (methododological essentialism), plus the realization that demonstrations were inaccessible [p. 78ff, p. 151].

In an interesting preface to his *Three Copernican Treatises* Edward Rosen discusses Copernicus' use of the word "hypothesis". He argues that Copernicus (and Kepler) denied that Copernicus' hypothesis was a mere hypothesis or a mathematical hypothesis – it was a demonstrated hypothesis. Its being demonstrated was the same as its not being mathematical: there was no other known alternative. That it was demonstrated was shown by its simplicity. Here comes Koestler's very important point, which may have been known all along the way, but which (as Santillana's and Drake's fury illustrates) was never driven home so well: hard as he tried, Copernicus failed to show that his simple hypothesis accounts for the known facts without adding to it many epicycles which render it very far from simple. This may explain his immense reluctance to publish: his idea of simplicity was a dream; not merely a programme which he could only hope to accomplish but which he never did; it was the illusion that the programme had been (nearly) accomplished. But criticism and clarity were integral parts of the dream, and so the Copernicans had to criticize their own views even after they had claimed that these were the truths about reality, and hence demonstrated, and hence clear and simple. The contradictions and non-sequiturs here should delight Koestler. That Kepler had deceived himself in his fashion he tells us, but that Galileo could also deceive himself thus he denies. He scolds Galileo for consciously deceiving people when talking about the circular planetary orbits of the Copernican system, for not caring whether he was speaking the truth in that instance, and for being obsessed with circles.

Let us allow the accused to speak [Drake, pp. 262–3, Drake and O'Malley,

p. 279]: "it is not I" says Galileo "who want the sky to have the noblest [i.e. circular] shape because of its being the noblest body . . . . Never having read the pedigrees and patents of nobility of shapes, I do not know which of them are more and which of them are less noble, nor do I know their rank in perfection. I believe that in a way all shapes are ancient and noble" – which is an explicit commitment to Plato's doctrine of Ideas, from which Galileo shrinks at once to an almost positivist attitude:

or, to put it better [*sic*], that none of them are noble and perfect, or ignoble and imperfect, except in so far as for building walls a square shape is more perfect than the circular, and for wagon wheels the circle is more perfect than the triangle.

And yet in the same work (*The Assayer*, 1623), he claims [p. 241 or p. 197] he would accept no path for a heavenly body save a regular one, such as a circle, a spiral, or an ellipse! Clearly quite a few exciting ideas interplay here, and Galileo himself is tossed between them. That he was aware of Copernicus' epicycles, and worried about them, is also clear from *The Assayer* [p. 264], where he is very proud of having disposed of Copernicus' so-called third motion of the earth. Like Copernicus and Kepler, he has both demonstrated the hypothesis already, and is also going to complete the demonstration pretty soon. For, obviously, demonstration *is* the same as getting rid of all epicycles! This is not our idea of demonstration, but it was his: Galileo both thinks that the epicycles had been eliminated, and that he would be able to eliminate them pretty soon – more by intellectual ingenuity than by observation. It is hard to believe that the greatest logical mind of his age, and the father of scientific method, could think thus; but we should remember that such thinking occurs already in Copernicus' and Kepler's various works, as well as in Galileo's own *On Mechanics* (of 1600), when he had no possible vested interest in his blunder. It merely comes to illustrate Galileo's own point of how difficult it is to avoid inconsistencies.

Koestler contrasts the ancient Phythagoreans with modern scientists: they lived blissfully before the split between Faith and Reason (which Galileo brought about), and the modern scientists live in a world of strife, in a divided house of Faith and Reason. He views Pythagoreanism as a healthy mixture of Faith and Reason. But he scolds Galileo for his meddling in theology! Not only did Galileo declare himself openly a member of the new Pythagorean movement – his faith in science itself was a kind of religion (and still is with most of us); the Inquisition referred to him as a Pythagorean, and Bellarmine viewed him as a religious reformer. And rightly so. And with the zeal of reformers he fought, and took risks. Koestler thinks that the reform might have been implemented from within (by the Jesuit astronomers, chiefly), and that Outsider Galileo only spoiled matters by iterfering and by annoying the Jesuit astronomers until they became staunch anti-Copernicans. Yet Koestler has given not a single piece of evidence for the view that the Catholic Church has ever been reformed without a bitter struggle. And though we may easily understand, and need not resent, the official theologians' resentment of Galileo's theological writings, it is rather hard to understand Koestler's very similar resentment of the same. The way he overlooks the fact that his darling Kepler was engaged in similar theological exercises is a serious case of bias. Had Galileo's theology been accepted by the Church directly from him, he might have become a saint, rather than Bellarmine. There was a chance that this would happen, both in 1616, when Bellarmine got into a panic, and after the death of Bellarmine and of the Pope, when the new Pope, Urban VIII, encouraged Galileo to write his *Dialogue*. But even if he had no chance against Bellarmine, his sincerity and courage, as well as his important contribution to Catholic theology, ought to be appreciated (and may be appreciated in the future, even by Rome; remember Joan of Arc!).

Let us glance for a moment at Galileo the Catholic reformer. In his *Letter to the Grand Duchess* [Drake, p. 181] he offers his view for the Church to consider, he presents it neither as the known truth nor as a point of public debate. Yet the point is that we ought to separate theology from astronomy so as to enable free critical discussion amongst astronomers. And he comes dangerously close to Brunos' position, for which Bruno was burnt, and at least he attacks (Bruno's judge) Bellarmine quite clearly along Brunist lines:

> I question the truth of the statement that the Church commands us to hold as matters of faith all physical conclusions bearing the stamp of harmonious [i.e. unanimous] interpretation by all the Fathers of the Church. I think this may be an arbitrary simplification of various council decrees by certain people [Bellarmine] to favour their own opinions [p. 203].

We may remember that Bruno was willing to recant only after an argument with the Pope, not on the authority of his judges (amongst whom was Bellarmine). Koestler asserts that Bruno, a metaphysician rather than a scientist, had nothing to do with Galileo's case, and that when Galileo stood before the Inquisition in 1633 "he was afraid". Galileo's words which I have quoted, written fifteen years after Bruno was burnt on the stake, do not seem to be the words of a coward; one may fail to notice the similarity between Bruno's and Galileo's scientific views, but hardly their religious views in general, and their submission to the Pope cum defiance of the Inquisition in particular;

one may dislike their submission to the Pope, but one must admit that they were brave and sincere Catholics, even though the Catholic Church cannot as yet admit this.

To conclude, Koestler expresses more than once his desire not to be wise after the event, but he is wise after the event in Galileo's case, at least. He applies hindsight when he applies what he (erroneously) thinks is the proper criterion for judging whether Copernicanism had been demonstrated, instead of looking for Galileo's own criterion. And he applies hindsight when he takes it for granted that Galileo was bound to lose his theological campaign for Copernicanism. Galileo had a very good chance of winning it; but the point to stress is that we ought to investigate whether he had a good chance, and, what is more important, we ought to notice that he thought he had a good chance to win the battle. (Indeed, at one point he thought the battle was already won [cf. Gebler, p. 177].) Trying to explain his behavior thus may be more interesting than viewing it as irrational, as Koestler does, by saying that he was obsessed with his need to quarrel regardless of the consequences. In brief, Koestler does not attempt a rational reconstructon of the battle as it appeared before it was over.

7. THE *DIALOGUE* AND THE TRIAL OF GALILEO

The most difficult part to reconstruct is the way in which a battle can be fought from within. If one rejects an important doctrine, one becomes an outsider; and if one does not reject it first, one does not wish to fight for its official rejection. Does not one's attempt to alter the official doctrine show one's conviction that it is false? Indeed, it is universally assumed that Galileo did not believe what he was told (by Bellarmine) to believe; defenders of Galileo, like Santillana [e.g., p. 151], view his professions of faith as ironical, and Catholic apologists as hypocritical. Now (being an agnostic and a Jew) I am a person poorly qualified to explain the fact that Catholics are permitted by their Church simultaneously to believe in a doctrine and to criticize it; yet I wish to state categorically and most emphatically that such is the case, no matter what is the doctrine in question, no matter how far-reaching are its consequences. After Galileo's *Dialogue* had raised a scandal but before it led to the trial, Galileo's enemies criticized Galileo not for his defence of Copernicanism; this, for all they knew (though not according to the Vatican files), was permitted. Nor did they show that his defence was invalid; Koestler's attribution of such capabilities to them cannot be accepted without argument, and this he does not provide. They criticized him for having allowed himself to believe in Copernicanism on the strength of having demonstrated it and in spite of his having been told not to do so. In other words, they did not deny the validity of his demonstration, they did not deny his right to demonstrate Copernicanism, they merely denied his right to believe Copernicanism on the strength of valid demonstration. In a letter to Galileo from Campanella, Galileo is told that he can safely deny this allegation [Santillana, p. 191]: "Please note", advises kind Campanella to Galileo when notifying him about the oncoming storm, "that you may hold that ... [Copernicanism] was properly forbidden, without having also to believe that the reasons alleged [by Bellarmine] are good. This is a theological rule, and it can be proved", etc. Not "I believe it because it is absurd", but "I believe it although it is absurd – until I am told not to", is the theological rule; and it is a rule which Galileo did follow, while arguing that the object of belief was absurd. Both Santillana and Koestler overlook this when charging him with cynicism or insincerity.

My contention, that from 1616 onwards Galileo believed in the immobility of the earth, may be false, of course, but for a methodological reason it ought to be investigated first, especially as it explains phenomena which students of the case were puzzled about. The methodological reason is this: it is easy to attribute any motive to any person, and less easy to test such attributions; so we should normally take a person's expressed motives seriously until we can show that he did not speak the truth. Of course, when we have before us, say, the documents which Butler caricatured in his *Erewhon Revisited*, we have to reject the claim he made in his preface, that he had no intention of caricaturing Christianity [Henderson, p. 220]. And this should make us suspicious of all expressed motives, much more so in the case of Galileo who suffered religious persecution than in the case of Butler who did not. Yet as long as there is no evidence against Galileo's confessions, they ought to be taken very seriously as possibly true; and the same ought to be said of Pope Urban VIII, and of all the others involved. It is quite possible that both Galileo and Urban VIII were irresponsible rascals, as Koestler asserts; but this should not be our starting-point.

Let us glance at Galileo's preface to the *Dialogue* with the intention of believing him for a while. There he speaks "of the Copernican hypothesis, *AS IF it were to prove absolutely victorious*". This preface, Santillana comments, "was practically dictated to Galileo by his anxious friend Mons. Riccardi, the Master of the Holy Palace, who had been entrusted with the *imprimatur*" [p. 6n], the friend who was known, Santillana says, for "his immense girth and erudition" [p. 170] but whom Koestler views as an ignorant fellow whom Galileo bullied to grant the *imprimatur* without knowing what he was doing. Galileo bullied him indeed, for he was very apprehensive; but, ignorant as he was of astronomical matters, he was quite a theological authority (cf. the *Catholic Encyclopedia*); and as he considered the preface to be theologically satisfactory, it is difficult to see why Koestler should not accept this judgment. (The accusation concerning the preface was, originally, merely that it had been printed in a different type from the rest of the book [Santillana, p. 211].)

Simplicio, the Aristotelian in the *Dialogue*, says Koestler, is "the clown who is kicked in the pants". This is untrue. Even Santillana has commented on Simplicio's charming character, though he also says (*Dialogue*, xxxv): "As to Simplicio, it is reasonable that he should remain under an ancient pseudonym, for his name is legion. He is the average . . . Aristotelian professor of the universities." This is considered to be a very important question, as it was alleged that Simplicio was a caricature of the Pope, and as it was also alleged – by Galileo (in a private letter) – that this allegation is the one that led to the trial of 1633. Now Koestler has not the faintest doubt that Galileo was lying when denying that Simplicio was a caricature of the Pope. But Galileo's preface says this. "In the company of [his friends Sagredo and Salviati, after whom he has named two characters in the *Dialogue*] . . . I often discoursed of these matters before a certain Peripatetic philosopher, who seems to have no greater obstacle in understanding the truth than the fame he had acquired by Aristotelian interpretations," and who is the source of Galileo's Simplicio. So "Simplicio" is the pseudonym of a real friend of the three who had "often" argued with the three friends, and who was alive in 1632; he was neither a clown to be kicked, nor the Pope. I wonder who he was.

Assuming that Galileo had not violated the letter of Bellarmine's instructions as described in Bellarmine's own certificate of honor to Galileo, and assuming that Galileo had no intention of caricaturing the Pope, we can solve a number of problems. The story of the writing and of the publication of the *Dialogue* becomes amply clear and reasonable if we do not use hindsight, if we forget the ensuing catastrophe when trying to reconstruct the way Galileo, the Pope, and others, looked at the situation before they knew of the grave consequences of all they did. (Even the fact that they were apprehensive, all of them [Santillana, p. 216n], indicates that none of them acted unreasonably.) But it all becomes clear only when one assumes their intentions to have been honorable. An important factor in the matter may also be the possibility that some measure of secrecy surrounded the whole scheme; for we know that some Jesuit astronomers were taken by surprise when they saw the *Dialogue* on sale. It is quite possible that the Pope wanted to present them with an accomplished fact.

In the *Dialogue* Galileo says quite often that he does not believe in Copernicanism, but that soon the Pope would permit believing it, and then he gladly will do so. It is difficult to see how Galileo could have been lying, or even merely guessing, when Riccardi was the censor who read the *Dialogue*; for whatever may be said about his comprehension of astronomy, this he certainly did understand, and whether it was the truth or not he most definitely could and would examine – he could, after all, ask the Pope himself.

The whole situation points clearly at the suggestion that the people involved knew what they were doing – they were too cautious and apprehensive not to – and were acting in good faith and not from any personal motive. "If corroboration is needed, it is to be found in . . . reports" written by the Tuscan ambassador of the period, concerning the events which occurred between the publication and before the trial. "They stress that [Pope] Urban 'was so incensed that he treated this affair as a personal one', and quote Urban's 'bitter remarks' that Galileo had deceived him." This is Koestler's statement [p. 483] in support of his thesis that the case was a personal affair, whereas the ambassador says the Pope had treated it as [if it were] a personal one – because of its importance. As to Galileo having deceived the Pope, the deceptions explicitly mentioned in the letter are that Galileo did not follow instructions with rigor, which is always a matter open to a difference of opinion, and "that all is well" [Santillana, p. 192], which, (alas!) turned out to be false. And, it transpires, the Pope feared that the *Dialogue* "might bring religion very great prejudice": the success of the *Dialogue*, if acknowledged, would presumably lead to taking the authority of the Church lightly. Moreover, this Tuscan ambassador, far from trying to prevent Galileo from coming to Rome, just asks the Pope to have a discussion with Galileo; but the Pope is apprehensive, which is very understandable. A trial was much safer than a personal interview.

One of the most significant arguments against Koestler, and all the Catholics he follows, is Gebler's evidence [p. 173, p. 177] that Galileo was very surprised that his success turned into failure. Moreover, being a faithful Catholic, he cooperated with his interrogators from the start, like Bruno before him, even though to begin with he (Galileo, like Bruno before him [cf. Yates, 1939, p. 205 and note]) did not know what it was all about. Even his interrogators had to admit that from the start he behaved "like a good Catholic". But he refused to lie, and so there was an impasse. The inquisitor then suggested to Galileo in a discussion (the inquisitor received, as he says, a special permit from the Pope to take such a "bold step" as to argue instead of to interrogate), and Galileo soon agreed, that perhaps inadvertently he said things he never meant to say, and that he was probably vain in placing so much significance on his own original arguments in favour of Copernicanism. This is all: he refused to deny the strength of others' arguments in favour of Copernicanism; nor did he admit that (demonstrated as he thought it had been) he ever believed it after 1616. Koestler nowhere supports his claim that Galileo was lying and could easily have been broken; and Koestler's "understanding" of Galileo's having been afraid is therefore rather uncalled for.

But what was the catastrophe? Why did Galileo's and Pope Urban's plan misfire? Why did the Pope decide to have a trial and humiliate Galileo? Even Santillana, who has an admirably balanced and detached attitude towards the trial, confesses that he fails to understand why Galileo had to be humiliated [p. 301]. My hypothesis is that after the long silence about Copernicanism (1616–32), and with no preparation, and with the public viewing Simplicio as "the clown who is kicked in the pants", it looked as if Galileo was defeating the Church itself, not the Bellarmine sect in it. And, quite possibly (following Galileo's letter about the source of the trouble), Galileo's enemies successfully spread the rumour that Simplicio was the Pope; and then, even though the Pope knew it not to be true, he had to do something about it: this was not a period in which jokes against the Pope were harmless to the Church. So the Church had to assert its authority against Galileo. There is ample evidence for this hypothesis, which I do not wish to marshal; all I wish is that it be examined seriously.

Koestler suggests that Galileo's evidence for Copernicanism was so weak that his bluff was obvious. There is no evidence to support this suggestion. Indeed, Galileo's claim that all the planetary orbits were circular in agreement with observations is very weakly argued, as we know today. But this was not so obvious then. And as to Galileo's own original arguments in favour of Copernicanism, which in his trial he admitted to be less conclusive than he had thought, they were not as weak as Koestler suggests. The first, concerning the sunspots which show that the sun is tilted, is a criticism of the old doctrine of the crystalline spheres. Koestler claims that it does not demonstrate Copernicus' doctrine, as it does not refute Brahe's, and censures Galileo for ignoring Brahe altogether. The reason for this is simple: according to Koestler any doctrine which saves the phenomenon should be tested seriously; according to Galileo, however, this is not so: as he explains at length in his *Assayer*, there is nothing in Brahe's system to recommend itself save some errors in mechanics and some errors in theology: it has no metaphysical foundations, and thus it does not count!

Galileo's second argument for Copernicanism concerns the tides. Galileo tried to account for the tides by using his law of inertia, rather than by using a theory of force which would be more in agreement with an idea which Kepler once tossed up. Koestler criticizes this theory first by claiming that the law of inertia cannot explain the tides, and second by pointing out that even Galileo only contended that one tide per day was thus explicable whereas there are two tides per day. Now the first criticism is valid only if the initial conditions are these: the waters are at least during one moment at rest relative to the continents. Whereas Galileo's point was that the initial conditions were different from those of relative rest. This makes nonsense of Koestler's idea about the mathematical connection between Galileo's theory of the tides and the missing stellar parallax, as the initial conditions are so different in the two cases. It also shows that Koestler has missed Galileo's great idea (and its root in the metaphysical theory of simplicity): Galileo claimed that the tides should be explained not by a specially designed universal hypothesis, but by the existing mechanics plus a hypothesis concerning initial conditions, namely plus a model. And he drew great encouragement from having got some results this way. Moreover, Galileo's own (pendulum) model was the starting point of Newton's researches which led him to his theory of universal gravity, and thus it is historically very important indeed [Turnbull, p. 301]. Furthermore, Newton followed Galileo in trying to explain the tides not by a new universal hypothesis but by a model. Newton's model was criticized by Laplace on grounds similar to those of Koestler's second criticism of Galileo, and almost as vehemently [Todhunter, § 807ff]. Laplace's own model, incidentally, has meanwhile been rejected as well, and again as one which is not even a tolerable approximation to known facts. And this shows how dangerous it is to be wise after the event and be indignant about the shortcomings of our predecessors. Quite possibly, Galileo was too humble in confessing that he had been too vain when thinking highly of his theory of the tides.

But this is not to commit myself to Galileo's methodology or physics. Though nearer to the truth than its predecessor, Copernicanism is false: the sun is not the immobile centre of the universe. And scientific hypotheses are either tentative or refuted; they are neither "mathematical" nor demonstrated. But Galileo's views on criticism and clarity and on freedom of thought are admirable, and so was his battle for these, which he lost neither entirely due to his own character nor entirely due to his opponents' malice or stupidity. It was touch and go. It is a great pity that he lost the battle, even though in the long run his ideas were taken seriously both by the world of science and by Catholic theologians.

8. CONCLUSION

"We should be grateful" says Copernicus [Rosen, p. 93] ". . . to those who have spoken incorrectly, because to men who desire to follow the right road, it is frequently no small advantage to know the blind alleys." This holds for Koestler too. Although most of his points are not essentially new, it is the first time that a non-Catholic has asserted them, and with such force and vividness. Even Santillana's Geblerian *The Crime of Galileo*, which wonderfully conveys the electric atmosphere of the times, and has in it crowds of real people made of flesh and blood, has for its Galileo and Bellarmine cardboard figures with little cardboard wings instead of real people. Koestler is not half as scholarly, he has no atmosphere, and his Bellarmine is the cardboard figure borrowed from Santillana (borrowed from Brodrick?), and his Galileo has no family and barely any everyday life; yet his Galileo is nonetheless alive and kicking – especially kicking.

Some commentators have suggested that only his Kepler is alive, and they have explained it by his sympathy for Kepler. This is unfair, not to say highly unimaginative. But alive as Koestler's heroes, Copernicus, Kepler, and Galileo, may be, regrettably they cannot be the true historical figures he wants them to be, because he is too nonchalant in his statements about the working of science to explain what a scientist is supposed to do, and too unappreciative of the difficulties which scientists encounter. The following is an example I chose because it wants little comment. Indeed, it is a collector's piece. "Instead of proceeding by observation and measurement, as the Pythagoreans did," Koestler tells us [p. 108], "Aristotle constructed, by that method of *a priori* reasoning which he so eloquently condemned, a weird system of physics 'argued from notions and not from facts'." Yet the whole of the book, including the passages about "the Pythagoreans", comes to illustrate, Koestler also tells us, "one of the points that I have laboured in this book," namely, "the unitary source of the mystical and scientific experience" [p. 426] – that is, that the process of developing scientific theories is that of mystic intuitions and not of observations and measurements. A person who speaks thus about science cannot be expected to do justice to the difficult problems involved in Galileo's method and methodology. It has not even occurred to him that to Galileo scientific demonstration is "geometrical" in the sense that Archimedes' *On Floating Bodies* is, and definitely not in the modern sense. This raises problems concerning observation which Galileo could not solve; nobody has solved all of them yet. I know of very few philosophers whose views on observations ought to be taken seriously. Seeing the situation thus, I cannot even start understanding how we can judge Galileo's behavior by the yardstick of whether and when Copernicanism has been demonstrated. All we have to remember, I think, is that no one in modern times had even thought about tentativity before Pascal and Boyle. And even after tentativity had been invented, or reinvented, Newton could not stomach it, so that it did not receive any popularity before the Einsteinian revolution, and before Popper presented a methodology – a false one, I think – based on the true idea that tentativity is an essential feature of all empirical science. But Koestler takes no account of all this, as we can see from his claim that Einstein has not yet influenced man's changing vision of the Universe (not to say his – rather funny – attribution of the idea of tentativity to Kepler). His chief accusation against Galileo, his claim that Galileo ought to have held Copernicanism as a tentative hypothesis, is thus answered. Koestler wants Galileo to have done the right thing, though this even Newton and Maxwell were unable to accomplish. Koestler's chief weakness seems to be taking great ideas for granted (quite against his intentions). In particular, he does not appreciate the greatness and novelty of the various ideas invented by Galileo, and of ideas invented much later, including the idea of tentativity in science.

NOTES

[1] It is worth quoting Koestler in full on this point, to allow the reader to notice his following three striking errors. (*i*) Koestler himself, though he has published some interesting thoughts about Archimedes, completely fails to restate the contrast between Archimedes and Aristotle. (*ii*) Yet he views Galileo's opponents as unworthy of being criticized. (*iii*) He can be unbelievably unfair to Galileo, to the point of viewing even his (alleged) anticipation of criticism – the thing which every good author does – as a vice. "After his return, in the summer of 1611, from his Roman triumph to Florence, Galileo became immediately involved in several disputes. He had published a treatise on "Things that Float on Water" – a title that sounds harmless enough. But in this pioneer work on modern hydrostatics Galileo had embraced Archimedes' view that bodies float or sink according to their specific gravity, against the Aristotelian view that this depends on their shape. The backwoodsmen were out at once in full cry, swinging their stone axes. They were the more irate as Galileo, instead of letting the facts speak for themselves, had employed his favorite trick of anticipating the peripatetics' arguments, building them up in a mock-serious manner, and then demolishing them with glee" [p. 428]. The absurdity of Koestler's condemnation, however, is no excuse for the absurdity of Drake's praise;

on the contrary, Koestler's indebtedness shows how easy it is to change the nuance in a passage from absurd praise into its opposite, while leaving the key points untouched. "Shortly after Galileo's return to Florence in the summer of 1611 he found himself once more in collision with the followers of Aristotle . . . Now they were confronted with a record of experimental data that anyone could verify at will, and the only thing open to question was the matter of interpretation. But in this they were no match for Galileo, whose specialty was the study of experimental results; and even in their own field of constructing ingenious arguments they were hopelessly outclassed for once. Galileo had in fact all their arguments, strengthening these, adding others that had not occurred to them, and then demolishing the whole structure with his own demonstrations and proofs. It was a device which he was to employ extensively in his later works, and one which accounts for his vast influence with nonprofessional readers as well as his extreme unpopularity with the targets of his polemic compositions" [pp. 79–80]. This praise is reminiscent of the tragedy of Nijinsky who was hurt by public enthusiasm for his jumps; he wanted to be appreciated as a dancer and not as an acrobat. Drake admires Galileo as an acrobat-polemicist rather than as a teacher of critical thinking. Incidentally, what Drake and Koestler say about Galileo's new experimental facts, about the ease with which Galileo demolished his opponents, etc., are sheer fantasy [cf. Fahie, pp. 143–5], as explained on p. 336.

[2] Wolfson sees a continuity of method from Antiquity to date [a) p. 25; b) pp. 106ff.], whereas Popper considers the Middle Ages as the outcome of the death of the Greek methods and the Renaissance of science as their revival [p. 151]. The scholastic methods, so characteristic of the Middle Ages, can be traced to the Talmud, Philo, early commentators on Aristotle, and Aristotle himself. The scholastic method is the critical method as employed elsewhere, in Antiquity or in modern science and scholarship, but with the proviso that the fundamental tenets remain unchallenged plus the technique of inventing *ad-hoc* hypotheses in order to protect them. The criticism of the fundamental tenets was offered by the mystic irrationalists who thus entrenched the identification of rationalism with Aristotelianism. Moreover, since the content of their criticism was identical with (and borrowed from) parts of the Aristotelian commentaries, what distinguished them was their method: by forbidding *ad-hoc* ripostes they turned innocuous flashes of debates into deadly hits. This may explain why the principle of simplicity was of such great methodological import and mystical excitement at the same time (and violating it was so sinful). The peak of the mystic criticism is achieved by Al Ghazali and Crescas [Wolfson, a) pp. 11ff.], both of whom view their attack on Aristotle as an attack on rationalism. Crescas, however, being a Jew (and thus an adherent to the commandment to study the Law), finds a limited role for reason. The fact that criticism rises in the Renaissance together with mystic irrational cabbalism or Pythagoreanism or Neo-Platonism (e.g., Boccaccio, Pico, Cusanus) is thus no accident. [See also Yates, b) especially conclusion.]

BIBLIOGRAPHY

Agassi, J., 'Duhem versus Galileo', *Brit. J. Philos. Sci.* 8 (1957).

Agassi, J., *Towards an Historiography of Science*, 's-Gravenhage, 1963. Fascimile reprint, Wesleyan U. P., Middletown, 1967.

Agassi, J., 'Can Religion Go Beyond Reason?' *Zygon, Journal of Religion and Science* 4 (1969); reprinted in *Science in Flux. Boston Studies*, 1975.

Blau, J. L., *The Christian Interpretations of the Cabala in the Renaissance*, New York, 1944.

Brodrick, J., *The Life and Work of Blessed Robert, Cardinal Bellarmine, S. J., 1542–1621*, vol. II (first edition), London, 1928.

Burtt, E. A., *The Metaphysical Foundations of Modern Physical Science*, London, 1924, 1932.

Clagett, M., *The Science of Mechanics in the Middle Ages*, Madison, 1959.

Cooper, Lane, *Aristotle, Galileo, and the Tower of Pisa*, Ithaca and London, 1935.

Crombie, A. C., *Augustine to Galileo, The History of Science A.D. 400–1650*, London, 1952.

Drake, S. (editor and translator), *Discoveries and Opinions of Galileo*, New York, 1957.

Drake, S. and O'Malley, C. D. (editors and translators), *The Controversy on the Comets of 1618, by Galileo, Kepler, and others*, Philadelphia, 1960.

Einstein, A., *The World As I See It*, London, 1934.

Einstein, A., *Out of My Later Years*, London, 1950.

Fahie, J. J., *Galileo, His Life and Work*, London, 1903.

Galileo, *On Motion*, translated with introduction and notes by I. E. Drabkin, and *On Mechanics*, translated with introduction and notes by Stillman Drake; Wisconsin, 1960.

Galileo, *Discourse on Bodies in Water*, translated by Thomas Salusbury, with introduction and notes by Stillman Drake, Urbana, 1960.

Galileo, *Dialogue on the Great World Systems*, Salusbury's translation, revised and annotated and with an introduction by Giorgio de Santillana, Chicago, 1953.

Galileo, *Dialogue Concerning the Two Chief World Systems, Ptolemaic and Copernican*; translated by Stillman Drake, foreword by Albert Einstein; Berkeley, 1952.

Galileo, *Dialogues Concerning Two New Sciences* – translated from the Italian and Latin into English by Henry Crew and Alfonso de Salvio with introduction by Antonio Favaro; New York, 1914.

Gebler, K. von, *Galileo Galilei and the Roman Curia*, London 1879.

Gombrich, E. H., 'Icones Symbolicae: The Visual Image in Neo-Platonic Thought', *J. Warburg Inst.* 11 (1948).

Henderson, P., *Samuel Butler, the Incarnate Bachelor*, London and Bloomington, 1954.

Koestler, A., *The Sleepwalkers, A History of Man's Changing Vision of the Universe*, with an introduction by Herbert Butterfield; London, 1959.

Koyré, A. A., *Études Galiléennes*, Paris, 1939 (*Actualités Scientifiques et Industrielles*, 852–4).

Koyré, A. A., 'Galileo and Plato', *J. Hist. Id.* 4 (1943); reprinted in Wiener and Noland, *Roots of Scientific Thought*, New York, 1957.

Lovejoy, A. O., 'The Dialectic of Bruno and Spinoza', *University of California Publications in Philosophy* 1 (1904).

Popper, K. R., *Conjectures and Refutations*, London and New York, 1962.

Price, Derek J. de Solla, 'Contra Copernicus: A Critical Re-estimation of the Mathematical Planetary Theory of Ptolemy, Copernicus and Kepler' in Marshall Clagett, *Critical Problems in the History of Science*, Madison, 1959.

Rosen, E. (editor and translator), *Three Copernican Treatises, the Commentariolus of*

Copernicus, the Letter Against Werner, The Narratio Prima of Rheticus, translated with introduction and an annotated Copernicus bibliography 1939–58; New York, 1959.

Santillana, G. de, *The Crime of Galileo*, Chicago, 1955.

Santillana and Drake, 'Review of Koestler's *The Sleepwalkers*', *Isis*, (1959).

Singer, Dorothea Waley, *Bruno: His Life and Thought*, New York, 1950.

Todhunter, I., *A History of the Mathematical Theories of Attraction and the Figure of the Earth*, Cambridge, 1873 and New York, 1962.

Turnbull, H. W., (editor), *The Correspondence of Isaac Newton*, vol. 1, Cambridge, 1959.

Wiener, P. P., 'The Tradition Behind Galileo's Methodology', *Osiris* 1 (1936).

Wohlwill, E., *Galilei und sein Kampf für die Kopernickanische Lehre*, Hamburg and Leipzig, 1909.

Wolfson, H. A., *Crescas' Critique of Aristotle. Problems of Aristotle's Physics in Jewish and Arabic Philosophy*, Cambridge, Mass., 1929.

Wolfson, H. A., *Philo. Foundations of Religious Philosophy in Judaism, Christianity and Islam*, Vol. I, Cambridge, Mass., 1947.

Yates, Frances A., *Giordano Bruno and the Hermetic Tradition*, London and Chicago, 1964.

Yates, Frances, A., 'The Religious Policy of Giordano Bruno', *J. Warburg Inst.* 3 (1939–40).

CHAPTER 25

THE ORIGINS OF THE ROYAL SOCIETY

1. THE HISTORICAL SIGNIFICANCE OF BACON'S TEACHING

Trevor-Roper's introduction to Miss Purver's work[1] sums up the book adequately. There are two stories about the antecedents of the foundation of the Royal Society of London, overlapping yet different. Sprat's *History of the Royal Society* of 1667 declares the Oxford group to be its antecedent, whereas most other writers assume the London group to be the one. The prejudice against Sprat, namely that he was himself prejudiced in favour of Oxford, is dispelled by showing that his *History* was the semi-official one. The question, however, is ideological. The Royal Society was Baconian. So were both the Oxford and the London groups. But whereas the London group held vulgar Baconianism, the Oxford group and its successor the Royal Society were purist Baconian. Thus, the true predecessor is the Oxford group as Sprat has claimed, not the London group as his successors have claimed.

The London group's vulgar Baconianism, to continue Trevor-Roper's summary, was the weaving into Bacon the ideas of pantheism, social radicalism, millenarianism; the nineteenth century followed Macaulay and read Utilitarianism into Bacon. Pure Baconianism ("the new philosophy") replaces the idols of the theatre and the market-place with the truth.

So much for Trevor-Roper's summary. Regardless of how well he represents Miss Purver's views, the question may be asked, is his summary acceptable prior to considering the new evidence? I start here because Trevor-Roper concludes his introduction frowning at those who, like myself, tend to resist Miss Purver's conclusion before examining her new evidence. (She had published her conclusions without the evidence some years before; see Appendix below.)

It is trivially true that antecedents in intellectual history have to be judged ideologically in the first instance. But there are, or may be, other kinds of antecedents which may be of interest to a historian, even to a historian of ideas and ideologies. In the case at hand, in particular, the concern may be with a scientific organization. It was the seventeenth century which developed the very idea of scientific societies, and almost all important firsts, particularly the Paris, London, and Oxford groups, were inspired by Bacon, the inventor of the idea of the lay university, with its research laboratories – the non-monastic monastery. The Society was, in a way, a lame substitute for the lay university. To be more precise, the idea of the Society itself, as opposed to the groups, surely belongs to Evelyn, Boyle, and Wilkins, not to any group, the Oxford or the London. Evelyn was for a lay university, but not Boyle. The antecedent events leading to the formation of the Society, the various abortive efforts to organize something of a scientific institution, surely belong to the vulgar Baconians of the Paris and London groups. It will be interesting to see what Miss Purver has to say about the origins of the theory and practice of building a scientific community.

So much for organization as an additional dimension of the problem. Confining ourselves to ideology, then, we have one more problem. How Baconian were the London or the Oxford groups? What exactly is true Baconianism and what is vulgar Baconianism? Trevor-Roper says, true Baconianism is the idea of replacing the idols of the theatre and the market-place by a "true model of the world". This is what philosophers call the (empirical) verification of scientific theory. Was the London group against it and the Oxford group for it? This is hardly conceivable. Amos Comenius, an idol of one London group (for there were two or three of those, naturally) regretted that Bacon had discovered the key to the secrets of Nature, yet failed to use it. It is also true, of course, that various people had various millenarian ideas of social radicalism, and that the Society confined itself to intellectual radicalism (as described by Trevor-Roper) and tabooed all other radicalist ideas, social or religious. How much this exclusiveness belongs to the Oxford group, how much to the Restauration (as noticed by Macaulay), is an open question.

As to Bacon himself, it is hard to say *a priori* how much he was a pure Baconian, how much vulgar. The vulgarization of the seventeenth century may be in part an expression of immediate needs of the second quarter of the century which had nothing to do wih Bacon, who wrote in the first quarter of the century. Alternatively the immediate successors of Bacon may have shared with him much background knowledge and so naturally read him nearer to his intention than we do. This second idea is not mine, but that of James Spedding whose own small odyssey is not without interest.

Spedding began his career as a Bacon scholar by responding to Macaulay's *Essay on Bacon*. In that essay Macaulay debunked Bacon the philosopher to the extent that he gave rise to a problem: why had Bacon been so revered throughout the eighteenth century as a philosopher? This problem has engrossed most Bacon students since, and is the chief problem in the centenary lecture *Bacon's Philosophy*, 1926, by C. D. Broad, and even more recent works. Sir David Brewster, the biographer of Newton and the originator of the problem, explained Bacon's fame by reference to his political and literary career. Justus von Leibig, the greatest and most influential of Bacon's debunkers, later endorsed this solution; Macaulay, the utilitarian, was more charitable, and praised Bacon as the father of modern utilitarianism.

At this junction of Macaulay's essay, the historian recedes to the background and the propagandist takes over. Had Macaulay been challenged in his attribution of modern utilitarianism to Bacon, it seems he would have preferred to discuss the truth or falsity of the doctrine than the truth or falsity of his attribution of it to Bacon.

Spedding read Macaulay's book as debunking, mainly Bacon the politician, since it hardly treats Bacon the philosopher. Spedding wrote a very long defense of Bacon against Macaulay's attack. At about the same time Robert Leslie Ellis was working on a complete edition of Bacon's works. Ellis was a mathematician and a biologist, in addition to being somewhat of a classicist and a surprisingly well-read scholar. He was a consumptive who died before he was forty. He asked Spedding to join him in his work, and died soon after, leaving it to Spedding to do with the uncompleted task whatever he found fit.

Spedding found the editorial work painful. He followed Ellis' change from admiration to puzzlement, and found even more puzzlement. He found, with Ellis, that Bacon had no scheme for a new philosophy, no idea about induction, no willingness to accept, even tentatively, induction by generalization, no suggestion as to how science can be built on solid foundations. In addition he found that Bacon's immediate successors were right in reading Bacon's mythological and utopian writings not as mere fables, but as serious works intended to be taken as serious.

Trevor-Roper says (p. xv) that

> Miss Purver has re-created the 'new philosophy' of Bacon, redeeming it not only from the Puritan vulgarization of Hartlib and his friends but from the Victorian vulgarization of Macaulay.

He does not mention Ellis, nor Spedding, but implies that no Victorian commentator on Bacon is better than Macaulay. One cannot but consider this a bit below the dignity of a scholar – particularly so since a page later Trevor-Roper admits that Sprat's reading of Bacon, which is the same as the Oxford group's reading and as the semi-official reading of the Royal Society, is an "idealized" version; that is, he admits that Bacon himself is a bit vulgar. It is one thing to say that a reading of a text is unhistorical and another to say that it is vulgar; as it happens, Spedding's reading is both vulgar and historical, and he also offered an idealized and frankly unhistorical reading akin to Sprat's, which he recommended should be tried out.It never was tried out. In his famous review of Spedding, William Whewell criticized this proposal quite sharply, and it was forgotten.

This much for extended comments on Trevor-Roper's seven page introduction. Let us examine the detail of the book, but with an increased pace.

2. BACONIAN RADICALISM IN ACTION

The problem Miss Purver comes to solve is, how reliable Sprat's *History* of 1667 is. The current view is that his claim that the Oxford group was the one which led to the foundation of the Royal Society in 1660 is suspect as he was the mouthpiece of Wilkins of Oxford; Boyle had spoken of "the invisible college" in London of the 1640's, and John Wallis, the mathematician from London, made a similar claim. Then there are accounts of activities of other personalities in London in the period in question, especially Samuel Hartlib whose invitation had brought Jan Amos Comenius to London. After 8 pages of thus presenting the problem, Miss Purver devotes about 150 pages to Part One, where she expounds Sprat's and her own view about the Oxonian origins, and about 80 pages to Part Two on the alternative views.

Part One, Chapter one, allegedly on the validity and significance of Sprat's *History*, but in fact largely (ample) evidence that Sprat's work was officially declared the semi-official view of the society. This explains the delay in publication from 1664, when we know it nearly went to press, to 1667. This also explains why, as Charles R. Weld, a mid-nineteenth century historian of the Royal Society has observed, and bitterly complained, Sprat's *History* is so very uninformative. (Incidentally "history" in Sprat's title may be read as in "natural history" or as in "the history of England." Anthony à Wood's *History and Antiquities of the University of Oxford*, for example, written at the same time, will now read, description and history, etc.)

Chapter two on Francis Bacon's philosophy, without which the Royal Society could not be as important as it was. Though a number of societies of similar character had come and gone, the Royal Society was a real first. Evidence: Sprat says it succeeded to bring about in six years more than others have in six thousand (i.e. since Creation).

This is a very strange thing. It is not easy to declare that Miss Purver agrees with Sprat and accepts his testimony as final. But I am afraid I could not find another reasonable reading of her text. This has something very nice and commendable – her taking Sprat's claims seriously and literally – as well as something very partisan and intolerable – her uncritical acceptance of so much on his mere say-so, and contrary to so much extant evidence as to make one wonder. Indeed, as we shall soon see, she may misread his claim for an achievement to mean not scientific but social achievement. This, however, is plainly missing his point! He says the proof of the importance of the new organization is the enormous scientific achievement it had to its credit already in 1667. This claim is just clearly not acceptable at all.

Radicalism is quite a thing to contend with. When one reads Bacon's claim that he has followed no one's footsteps and is the very first of his kind, one cannot but be moved. Robert Leslie Ellis was so moved that he was determined to attribute some valuable idea to Bacon. He could not attribute to him any idea about induction, because he said little about the technique of induction, because what little he did say was contingent on the questionable assumption of a very high degree of simplicity and comprehensibility of nature, and because he explicitly declared induction by generalization childish. Ellis finally attributed to him a version of atomism and an important idea in the psychology of learning, both of which he found in Bacon's myth of Cupid. Ellis was an immensely learned man and so he could find the source of almost any idea in Bacon which he considered valuable. Yet somehow even his scholarship was not broad enough. He had overlooked Natalis Comes or Conti, though this student of myths was fairly well known in the early seventeenth century (see Paolo Rossi, *Francis Bacon*, Chicago, 1968, p. 80). As C. W. Lemmi has shown (in his *Classical Deities in Bacon*, Baltimore, 1933), what little original material Ellis had found in Bacon was material lifted by Bacon from Comes.

In his by now classical *Ancients and Moderns*, R. F. Jones quotes one Hakewill, probably a disciple of Bacon, to say of himself that he is utterly original, in almost the same words as Bacon. Later on Lynn Thorndike, in an essay in *Isis*, quoted a long list of important Renaissance figures, all claiming utter originality in accord with what obviously was the fashionable formula of the day. In the light of this it is hard to take radicalist claims as seriously as Ellis did. Meanwhile the archconservative Michael Oakeshott in his *Rationalism in Politics* (Cambridge, 1959) has shown this to be a standard feature of radicalism (which he identifies with rationalism so as to arrive from conservativism to irrationalism), quoting even Bernal to say in our century that since by comparison all science prior to ours is microscopic we may well view science as more or less an ultra modern creation. And Imre Lakatos has quoted Bertrand Russell (*Mysticism and Logic*) to say that perhaps George Boole was the first mathematician, but more likely Russell himself was. And

Lakatos himself called himself, in a debate ensuing a public lecture in Boston, the first historian of mathematics.

All this takes us far afield from Miss Purver's study. She indicates her radicalism by approving of Bacon and of Sprat. Ellis is for her but a follower of Macaulay who distorted Bacon and presented him as a utilitarian. She makes no mention of R. F. Jones or of Lynn Thorndike. Since she begins her chapter by a survey of the history of scientific societies I had hoped to find a reference to Martha Ornstein's *The Role of Scientific Societies in the Seventeenth Century* (Chicago, 1920), one of the earliest recognitions of the place of radicalism in the seventeenth century. But no. Miss Purver, says H. R. Trevor-Roper (p. xiv), "has presumed nothing. She starts from the beginning, and tests every piece of evidence before using it." He does not say what she does or ought to do with the evidence she does not use.

Bacon's philosophy begins with the maxim, presume nothing, collect all the available evidence. Let the evidence lead you to the formation of a theory. With enough labour and patience you will arrive at the true theory. The basis of this process is radicalism: assume nothing; first, destroy all error. Somehow, all of Bacon's debunkers, from Brewster to date, being radicalists like all debunkers, took radicalism for granted, took for granted the idea that it is best to take for granted no idea; they therefore could not see Bacon as the great innovator, as the inventor of radicalism. Miss Purver is a radicalist yet will not debunk him.

Miss Purver, at least, finds Bacon's greatness in his radicalism; and though not original, she is quite right. (Her view is expressed in Paolo Rossi's *Bacon*, opening of Chapter 6.) Also, this is for her the rationale of the founding of a society: the process of collecting vast data requires collaboration (Rossi) and hence organization. Also, she endorses Bacon's radicalism. It is unusual to endorse Bacon's radicalism yet praise him; for a radical it is hard to forgive Bacon's failure to abide by his own radicalism, his erroneous acceptance of magic, alchemy, and geocentricity, his calling Copernicus a charlatan, his poking crude fun at Gilbert while plagiarizing his manuscripts, his inability to understand or to take notice of Galileo's headline-making discoveries, etc. Miss Purver does not meet the difficulty: she says,

> The celestial bodies were, as Bacon scornfully remarked, 'supposed to be fixed in their orbs like nails in a roof' (p. 28). Bacon was far from being the only one to see that such a concept of the natural order [the 'Aristotelian'], even if considerably modified, presented grave obstacles to scientific progress. (pp. 29–30)

and even "In this context Bacon's own resistance to the Copernican hypothesis is not only reasonable, but scientifically impeccable." (p. 40) – but not a hint at Bacon's magic, alchemy, staunch geocentrism, and tirades against Copernicus. Miss Purver, says Trevor-Roper, "tests every piece of evidence before using it".

Miss Purver's chapter on Bacon comprises over forty pages, most of which are devoted to a general exposition of Bacon's works and thoughts. Ellis' classical summary is not much longer and C. D. Broad's (Cambridge, 1926) is shorter. Both are more accurate, more informative, more interesting. Neither is up-to-date, to be sure, but their errors are at least understandable within the terms of the accepted standards of scholarship.

In other words, the scholarly world, even when spouting the pure milk of Baconianism, practises a different, and non-radicalist, standard: the scholarly world recognizes that certain errors are permissible within scholarship yet other errors disqualify their perpetrators as scholars. Miss Purver disqualifies Macaulay, whose *Essay on Bacon* is a paragon of beauty and of scholarship, because his presentation of Bacon's philosophy is (undoubtedly) scanty and erroneous. Standards much more lax than hers are violated by her, such as those which include the counsel to people who live in glass-houses to be sparing with the throwing of stones.

3. RESERVATIONS REGARDING BACON

We now come to the chapter on the Royal Society's Baconianism. "That the movement originated in the University of Oxford is not very surprising" we are told (p. 63).

> Its leaders, knowing that academically they were in hostile country, were conducting themselves with caution and tact, for nothing was to be gained by antagonizing the main body of their own university or of academic opinion elsewhere. So when, in that year, Thomas Hobbes, in his *Leviathan*, attacked the whole range of Aristotelian learning in the universities, the club did not welcome his efforts. (p. 64)

I like the juxtaposition of these two quotes – less than a page between them!

It may intrigue one that Miss Purver has chosen this line as an opener for the support of the thesis that the Oxford group comprised true Baconian radicalists, the followers of the one who – to date – is the severest critic of academic and Aristotelian practices. Miss Purver has an explanation: "In a witty rejoinder" she continues (this is a slip of her pen, "rejoinder" signifying a counter-offensive), the group criticized Hobbes as one who wished to replace Aristotle. This is a variant of Bacon's attack on Copernicus which soon became traditional. Huygens said the same of Descartes, and Dr. Thomas Thomson said something similar of Lavoisier: all radicalists must explain failure to implement the radicalist formula; the absence of pure intentions is the easiest available, and the one which Bacon had found in the Cabbalistic and alchemical literature (including the works of Comes) and expounded in his various works.

Another attack on Aristotle took place in the mid-fifties, this time by a frank Baconian (Hobbes, too, was influenced by Bacon, and even a personal friend; but not quite a disciple). It was, however, aberrant: its author believed in astrology and alchemy. Miss Purver forgets that even Boyle and Newton were believers in alchemy, and so equally aberrant. Another member of the group joins the comments this time, and with a "tactful statement" (p. 65) defends Aristotle's scholarship, "implicitly" endorsing some of the criticism (p. 65).[2] The interesting part of the group's counter-attack on the poor frank Baconian was an expression of mixed feeling towards Oxford. Miss Purver quotes but does not comment. She does not say why the Oxford group moved (1658–9) to London, even though Oxford was "Oxford of this enlightening and ameliorating influence". By "this" she means merely the Oxford group – now departing and taking the amelioration with them. Why did they depart? Was anything amiss? No answer.

In 1661 Glanvill attacks astrology. In 1665 he revises his work and dedicates it to the Royal Society. He is soon elected fellow of the Society. (The story of his being forced to revise his book is told by R. F. Jones. We find no explanation here as to why he revised it.) In 1667 Sprat's *History* appears, and in 1668 Glanvill's *Plus Ultra*, both apologies for the Society. Miss Purver gives the impression that Glanvill is a Baconian. Those interested in him may read Professor Richard Popkin's exposition of his skeptical philosophy.

In the fifties Oxford was the birthplace – a "not very surprising" (p. 63) fact since Oxford had "enlightening and ameliorating influence" (p. 67) – even if all this required some compromising. Things got better with the rise of the Society and its defense by Sprat and Glanvill. So, in 1669, it all led to an open clash between Oxford University and the Royal Society. This may all be very clear to Miss Purver; for my part I wish she had explained the trend more clearly.

Anyway, clearly, the universities (for Cambridge joined Oxford) feared competition (p. 72–3), and competition not from a new university or its like, but from the new experimental Baconian ideology (p. 75–6). Is this "enlightening and ameliorating"?

The evidence is from Sprat's *History*. The history is of 1667, the quarrel from 1669. It is clear, however, that not Sprat was on the attack but the universities: clearly, when he said the Society did in six years more than the whole world since Creation, he was just stating the facts. What, however, has happened to tact? Was the Society so sure of the oncoming attack that it decided that 1667 was no time for niceties? Miss Purver does not say.

Nevertheless, Miss Purver is right on the major issue: the quarrel with the Universities was ideological: it was the Baconian radicalist ideology which made the Royal Society declare through Sprat that the universities were worse than nothing.

We are now in the midst of Miss Purver's exposition of Sprat's Baconian radicalism. The main point is Baconian indeed: the Society insisted on experimenting first, leaving theorizing to a later stage (so as to avoid error and dogma). Miss Purver admits, however (p. 84), that in some cases the society expounded theories, and, moreover, theories that were quite mistaken and that some of the reports of observations of facts the Society expounded were silly or at least irrelevant. And she takes as a silly example Boyle's report that he had been informed that excessive coffee drinking causes palsy.

With so many silly examples around, Miss Purver had to choose a clever one. First, it is not to be doubted that Boyle's report is true . . . Second, that suppressing it would have been irresponsible, since he could not know *a priori* whether there was anything to it. Third, at least in view of the social unacceptability of coffee, possibly even the theory Boyle reports to have heard was also true (assuming the paralysis reported to be hysterical). But this only refers to incidental and to unknown facts (was the paralysis hysterical?), where principle matters much more; to which I now revert.

A scientific fact must be repeatable. This was instituted by Boyle in 1661, in his essay on 'The Unsuccessful Experiment' (the expression is Bacon's, but he meant it – in his *Advancement* – to denote an unsuccessful attempt to build a useful machine), under the influence of Galileo (in *The Assayer*). Boyle also suggested to declare any unrepeatable experiment unsuccessful rather than a distortion. Now a delicate matter, both politically and philosophically, is hidden in this suggestion and I wish to discuss both.

Henry Stubbe, the leading enemy of the Royal Society, attacked its radicalism most. In his attack on Sprat (*Legends No Histories*) he says, the Society should make experiments instead of trying to remove all the rubbish of the past. Indeed, he adds, everyone knows that Bacon himself made a lot of mistakes, especially in gardening, as even fellows of the Society admit. The reference is to the fact that Bacon had transcribed from Pliny about gardening, especially roses, forgetting that the climatic conditions in Italy and England are different, and to the fact that Boyle himself uses this as an example of an obstacle to repeatability though without explicity asserting that Bacon had transcribed what he had professed to report.

Bacon must have embarrassed his followers quite a bit. John Evelyn, for example, shows this in his letter to William Wotton on Boyle (Wotton was going to write Boyle's life but never did). Evelyn says there, Boyle always performed his experiments, unlike Bacon, though the fact about Bacon need not be broadcast. (Miss Purver quotes from this letter only the passage about the early days of the Society.)

Bacon's most Baconian work was his *Sylva Sylvarum*, ten books of one hundred facts each, put at random and full of superstition. Boyle wanted to write a book to replace it, and called it *The Promiscuous Experiment*. The fact that he advertised it, yet never published it, is quite remarkable since the man published voluminously and regularly. John Beale, his old school-mate from Eton, regularly urged him to publish the book. He even reminded him how grateful they were to Bacon, how impressed they were when, for the first time Henry Wotton (the founder of Eton and the father of the above mentioned William) placed Bacon's work in their hands (when they were in their teens; Wotton was the first Baconian who even performed experiments such as conceived by Bacon – see his posthumous *Reliquia Wottoniana*).

The reason for Boyle's reluctance to publish his promised *Promiscuous Experiment* can be found in what seems to be the substitute for the *Promiscuous Experiment*, which is Boyle's posthumous *Experimenta et Observationes Physicae*. Boyle wrote in the preface to this work that he tried to describe the fact in it as circumstantially as possible, but he feared that nevertheless it is possible that in some descriptions some circumstances necessary for repetition were inadvertently omitted; and he calls on his old friend Oldenburg (the secretary of the Society) to testify that he had performed even the experiments which the reader may find unrepeatable. When a dying man calls a dead man to testify in his favour he must be talking in earnest, and on a disturbing point.

This indicates how aware Boyle was of the philosophical difficulty involved in the philosophy of induction: we cannot decide what is an observed fact without deciding what of the observation is part of the observed fact and what is incidental to it. To decide this is to rely on theory, and to rely on theory prior to experiment may be a prejudice . . .

Back to Miss Purver, who accepts the maxim to begin with facts yet insists that they all be relevant. She is now recounting the list of the experimental projects reported by Sprat – with great approval. Those interested may be well advised to supplement her review of Sprat's history with L. L. Whyte's review of the facsimile edition of the work in the *British Journal for the Philosophy of Science*. Miss Purver quotes Glanvill's *Plus Ultra* and other sources to prove that the Society was orthodox Baconian. She then devotes a page to represent Sprat on language, concealing his chauvinism and his mention of Milton as the only English poet of any significance. She ends by declaring that the Royal Society alone put Bacon's vision into practice. The vision, we remember, is of removing past prejudices and of organizing a vast search of data. The vision was quite reasonable in the seventeenth century, when only geniuses like Boyle could criticize it. Centuries later, after criticisms by philosophers, psychologists, and Bacon scholars, Miss Purver endorses it with the same naïve freshness.

4. THE RISE OF THE ROYAL SOCIETY

Chapter 4 takes us back to Oxford – Sprat on Oxford, others on Oxford, some biographical data. The Oxford Club was founded in 1648; this is a bit of an exaggeration: there was no foundation and no club, only an informal colloquium. Anyway, the first public reference to it is by Ward in 1654, "declaring that Aristotelianism was being combatted" (p. 113). A footnote refers us back to pages 64–7, where all that we are told is on page 65 that Ward wrote a rejoinder to Hobbes's attack on Aristotle and on page 66 that Ward said in Oxford they were teaching not only Aristotelianism but also modern versions of Copernicanism "either as an opinion, or at leastwise, as the most intelligible and most convenient hypothesis." This is not exactly evidence that "Aristotelianism was being combatted". Miss Purver takes some liberty with her own crucial point of evidence.

The chapter can hardly be summarized; at least I cannot summarize it. Again, no explanation of the movement to London, or to Gresham College. They referred to themselves as "the Society" or "the Company" or "the illustrious Company that meets at Gresham College". And soon they received the Royal Charter.

Chapter 5, The Royal Charter. The Society was founded in 1660 and ran into financial difficulties. C. R. Weld, a later historian of the Society (1848), discusses these difficulties; Sprat does not, nor does Miss Purver. The farthest she goes is to quote in a footnote Sprat and Birch to say that some fellows were researchers, others were financial contributors. On a previous page (109) she mentions that in 1654 Wilkins "had given 200 pounds towards a College of Experiments and Mechanics to be set up" in Oxford. Also, she quotes

(p. 113) Seth Ward to speak then of "a conjunction of both purses and endeavours of several persons". It now seems that something had changed: Wilkins was broke (p. 130). There were people in better financial shape, especially Boyle. Weld complains. Even a biographer of Boyle, L. T. More, is not very approving. Miss Purver is reticent.

It is clear that Boyle's friends, particularly Oldenburg, tried hard to get Boyle to finance some scientific activity or another, preferably to found a secular college on the Baconian line. But Boyle never did. Even when his friends procured for him some confiscated Irish land (1662) so as to enable him to support science without loss he was adamant: he said since his friends had not consulted him he was not bound by their intents and spent the money on charity and on missions. (This is the source of the complaints.) In his important early work, *The Spring of the Air* (1660), in the introductory part he says, a philosopher needs a purse as well as a brain (in obvious contrast to Ward's above quoted remark); in his will he bequeathed all his scientific materials to the Royal Society, including his stones but excluding the gems. All this was deliberate, it seems, and systematic. It even agrees with Boyle's philosophy of mind: whereas Descartes assumed the mind to possess reason alone, Boyle assumed it to possess reason and emotion. To reason he ascribed natural religion, which includes natural theology and experimental philosophy (as doctrine and ritual respectively); to emotion he ascribed Christianity (including revelations and miracles) as ancillary doctrine and as second chance for those who jettison reason. And so charity becomes religion but not science. Also, of course, science is rationally superior to religion as it is rational and so when science and faith clash science must win, and the Bible must be understood as a mere system of ethics, etc.

Without discussing Boyle's role in the Royal Society, we can take it for granted that he was a prominent member of the Society, and that he wanted it to function as a means of bringing amateur scientists together. If so, it cannot be suggested that he would oppose the idea of admitting to the Society people whose contribution was only financial, though he must have coveted their brains more than their purses. Nor could he have objected to the Royal Charter and such, and for similar reasons. We must remember that though he rejected peerage, bishopric, presidency of the Society and provostship of Eton, though he was proud of not being a college fellow, he could not resist an Oxford degree as this eased the tension between the University and the Society. This last point was noticed in the thirties by J. F. Fulton, the renowned Boyle bibliophile.

And so, Miss Purver's explanation of the foundation of the Society from

Ward's viewpoint does not quite clash with any explanation from Boyle's viewpoint – on the condition that we notice that they differ, Ward liking better the idea of the man with a purse supporting the study of the man with the brain, and Boyle liking better the man with a purse pursuing his own researches.

Anyway, the story according to Miss Purver is sufficiently straightforward. The Royal Charter was given in 1662, allowing the Society a few privileges. It was revised and implemented in 1663; the revision did not offer new privileges (contrary to what historians say), but a coat of arms, the full name – the Royal Society for the Advancement of Natural Knowledge – and the statement that the King was its Founder and Patron. Some details about membership. I remember having read that one founding member was expelled. Miss Purver's talent could be put into use in search of the story; her disposition lies elsewhere.

Chapter 6 on the religious policy of the Society, and the end of Part One. "Bacon's vision of new sciences was down to earth", it begins. Before one stops to gasp she adds, "the facts of nature were the subject of his study. Yet the impulse behind it was essentially a religious one". Before one stops to congratulate Miss Purver on her perceptive notice of the religion of science, she adds, "and the Royal Society, as a body, followed his precepts on religion in its relation to science." I have now quoted the whole first paragraph of Miss Purver's chapter on religion. I can only say I am at an utter loss.

That Bacon wanted people to study facts and find natural laws is uncontestable. Does this make him "down-to-earth"? Miss Purver analyzes Bacon's utopia, *The New Atlantis*, in detail. In particular she notices that *New Atlantis* is Christian but religiously tolerant. But this sounds more pedestrian than visionary. She does not state clearly enough to my taste that there was a revelation particularly for the benefit of the inhabitants of *New Atlantis* (which is isolated from the outside world, though somehow it has all sorts of immigrants, including Jews), and she hardly presents the place in all its true colors. In *New Atlantis* the lay college which engages in research has the power to decide which of its inventions to make public, which to make state secrets, and which to withhold (as too dangerous) even from the state. The Oppenheimer case shows that technocracy has not yet developed to the height of Bacon's vision. Neither Einstein nor Bohr, nor the President of Harvard, ever entered town in a procession the way the College President in *New Atlantis* did. Bacon even tells us he found it impossible to get a ticket for the stand, and he only got one through his Jewish host. The host even arranges for an audience and the president tells Bacon all about the college, including their statues of discoverers and inventors and including the prayers to God to assist them in their researches.

Miss Purver is right: Bacon's view of science as a mode of worship separated from established religion, as well as Bacon's notion of religious tolerance, were central to the Society, which had Catholic members and somewhat low-Church (not really) Protestants. This, says Miss Purver, disproves the thesis of Merton that the Royal Society was an expression of Protestant ethics (in Weber's sense of the word). And since Protestant ethics is utilitarian, she adds, surely the Society did not accept this ethics.

The interested reader may find a summary of the literature on the topic in Richard L. Greaves' 'Puritanism and Science' in the *Journal of the History of Ideas* (1969). Here let me only add this. In Weber's sense Protestant ethics represents the idea of the virtue of work, and this certainly is something which Bacon had preached. Also Weber assumes that Protestant ethics is Calvin's invention, which may be true for the business world (though this has been questioned too), but is certainly not true for the learned world where good works and ritual were parts of purification processes of the mystic scholar, as expressed in the cabbalist and alchemical literature and echoed in Comes and in Bacon. Boyle, expounding similar views in his *Seraphick Love* of 1659, ascribes its origins to Philo Judaeus! Unless we make clear what is new in "Protestant ethics", we can scarcely decide its influence on the "new philosophy".

Miss Purver quotes some details about the toleration of the Royal Society which had even led some of its opponents to view it as an instrument in the hands of the Catholics. The presence of this typical intolerant argument might be expected *a priori*, though how weighty it was considered, or how large was the intolerant group amongst the intellectuals is very hard to assess. Miss Purver does not raise the question, how significant her evidence is. Miss Purver mentions that in the House of Lords Wilkins openly criticized the King's attempt to pass an intolerant law. But this has almost nothing to do with our topic. Macaulay has noted that the political significance of the foundation of the Society is that it kept some important minds off politics; now religion was at the time a major political item. There is little doubt that the King could have his cake and eat it, allowing the Society to fight for tolerance and impose tolerance on its members, not on its founder and patron. I think Miss Purver should have told her readers clearly that we have ample evidence against the notion that Charles II had any weakness for either enlightenment or toleration.

5. DISPUTING ANCESTRY CLAIMS

We now come to Part II, on the London group, pp. 161–234 in four chapters. First John Wallis's account of the origins of the Royal Society, second on Gresham College, third on Boyle's account on the Invisible College, and fourth on Hartlib and his pansophia.

Wallis. He reports that before Wilkins went to Oxford he belonged, with Wallis, to a London group which was interested in the new philosophy. Now, first of all, we are told, Wallis was a plagiarist and a hot-head. True, but unimpressive. Secondly, Wallis uses the term "New Philosophy" for ideas preceding those of the Baconian radicalism of the Royal Society. He considers Harvey's views on the circulation of the blood, which he had studied in Cambridge, as an example. Now, clearly, absorbing a new idea into the old system is a sin by any radicalist standard, of Bacon or of Miss Purver (p. 169); and Wallis' acceptance of his Cambridge teacher's non-radicalist practice is really bad: "it seems to indicate", she quietly chafes, "that he never did fully appreciate the state of affairs which John Wilkins and . . . his group at Oxford sought to remedy" (168–9).

This is a tough spot for a critic, and I wish I had the tact which Miss Purver ascribes to Wilkins and the Oxford group when she explains why they did not act in Oxford as good radicals should. But let this ride. Wallis' list of sins is not here exhausted: he calls "the New Philosophy" not Bacon's ideas, but those "which, from the time of Galileo . . . and . . . Bacon hath been much cultivated . . . abroad, as well as in England." He gives as examples a list of topics discussed by the London group. This is no evidence that the group made new discoveries. Indeed, the same list had been presented in Glanvill's *Plus Ultra* as examples of individual contributions, in contrast to the Baconian collective projects of the Society.

Thus, Wallis' evidence, though acceptable as factual, is rejected as an interpretation. Indeed, Wallis reports that Theodore Haak, a foreign resident, initiated these meetings. As Harcourt Brown has suggested (*Scientific Organizations in Seventeenth Century France*, Baltimore, 1934), Haak was influenced by Father Mersenne, the founder of the Paris group, with whom he corresponded. Mersenne was to a large extent a Baconian, who made even Descartes express approval of Bacon's attitude towards experiments. Miss Purver, however, has no difficulty showing that he was first an Aristotelian of sorts, then a Cartesian of sorts. (There is no difficulty showing this of Galileo, Bacon, Boyle, etc., etc.) She even sees (p. 174) in Mersenne's suggestion to found an academy of science in France an attempt "to ensure what he considered to be the proper intellectual control of knowledge, and no doubt to offset Bacon's proposal of colleges on an international scale". Even if her quotation from Bacon were in agreement with her statement, even if Mersenne's (Cartesian) mechanical philosophy were not the one also endorsed by the Oxford group and the Royal Society, even then, Miss Purver's reading of nasty motives in Mersenne may make me withdraw the wish that I could be tactful.

Miss Purver is right in dividing Wallis' account into the factual and the interpretative; she is right in saying he later withdrew his own interpretation and said, the Society had originated in Oxford; she is right in saying, the official version says Oxford, not London. She is even right in saying the official version is not a small matter since it is the radicalist version. There is only one snag; radicalism is false.

Up till now Sprat was the authority, not Glanvill. The Society even recognized, we are told (p. 14), differences between the two, and endorsed only the former. Some of Miss Purver's evidence against Wallis is from Glanvill. Of course, the reason is that both Glanvill and Wallis mention the same list of discoveries – Wallis to prove that the London group was the original one, Glanvill as a mere admission that some pre-Baconian individual discoveries – of Copernicus, Galileo, Harvey, etc. – are quite important. Perhaps we have here some clash between Glanvill and Sprat who said, we remember, that the Society, as a group of Baconians, did more in six years than the rest of the world in six thousand! If so, Glanvill must yield to Sprat; by Miss Purver's own standards.

Miss Purver's interpretation is not very convincing. She should not follow Glanvill and say, as she does, the discoveries Wallis mentions are not new; she should follow Sprat and say, as the good radicalist she is, the discoveries are not important! If they are important, as Glanvill but not Sprat admits, then those who met to discuss them may be seen as some beginning of the Society. Radicalism permits no ancestry to radicalism, as Bacon declared, as many others did (see above p. 356). Miss Purver's study of the evidence is somewhat coloured by her radicalism. Take away her radicalism and Wallis' reading may sound much more congenial.

The next hypothesis Miss Purver refutes in the second chapter is that Gresham College had anything to do with the foundation of the Society. I really find it too tedious to go into details, where the all-or-nothing attitude of Miss Purver leads her to an ever easier victory. It is a real pity. For, though not a real college, Gresham could have developed into a secular college proper, indeed in accord with Bacon's wishes, and the wishes of almost all its founders.

I do not wish to quarrel with Boyle's insistence that the Company found a Society, not a college. His idea of amateur scientists was also inspired by Bacon, and made better sense to him as the basis of a disinterested activity. There is evidence that when Evelyn, Boyle, and Wilkins, called the founding meeting, a least Evelyn, and probably also Wilkins, wished for a college, but Boyle was adamant and only he could afford the founding of a college. It was no doubt his privilege to refuse and his alternative idea did prove useful. Yet the connexion with Gresham for about half a century is some indication of the retention of some vestige of hope to establish a college. The University of London was formed only in 1830, partly because the scientific societies played a significant intellectual role amongst those debarred from Oxbridge and partly because even in the 19th century Oxbridge was not very tolerant and debarred nonconformists, Jews, and agnostics, not to mention the poor. As to Miss Purver's details of the weakness of Gresham College, they are misleading: the other universities were terrible at that time, and showed less hope. The hope, finally, fizzled out; but it could have materialized even after the college's demise – just as the Society could have disintegrated but did not, after Boyle died and before Newton revived it.

The third chapter deals with Boyle's report, with his famous "invisible college" – this is his label; theirs was "philosophical college". Though it had been identified with Theodore Haak's group, it is clearly the group of Samuel Hartlib – another foreign resident of London. (This was first noted by Miss R. H. Syfret; see note on p. 200.) Miss Purver also argues from the fact that Hartlib and his group wanted a college proper; but so did Evelyn, Oldenburg, Petty, and others.

There is also the question, how distinct were the three groups. From all we know the overlaps were small; but then this may be due to a division perceived more than practised. And Boyle may have felt the need to view all groups as essentially one.

Also Boyle, in 1646, at the age of 19, says the invisible college is in principle utilitarian. This, says Miss Purver (p. 194), shows it belonged to Hartlib. But again I am uneasy. I do not think anyone was a utilitarian then, not even Boyle. A private letter of a 19 year-old, even a genius, is not exactly clinching evidence.

Also, Hartlib's college was supposed to preach "reformation of church and state" (p. 201). Now this is not a utilitarian reformationism; even Miss Purver notices that much. The whole adventure was one muddled turmoil, where Copernican and Cartesian and atomic theories mixed with Baconian condemnation of all speculations; where looking backward to antiquity mixed with looking forward to new horizons; where extreme radicalism in philosophy mixed finally with Restauration moderation and toleration in politics and religion.

There is no need to go into much detail of the discussion in the next chapter on Hartlib, Comenius, and "pansophia". Of course Comenius admired Bacon most and was influenced by him both as a philosopher and as an educationist. Of course "pansophia" averts to the fact that Bacon had taken the whole of Nature as his province. Yet Miss Purver sees only "a superficial resemblance" (p. 210) between his aphorisms andBacon's. This is a superficial impressionism, and of an apologetic brand. Of course, Comenius was also influenced by others, including one Johann Valentin Andreae,[3] the inventor of the word "pansophia" who, too, was a Baconian, though even less than Comenius (p. 211). True, Andreae was much influenced by other utopians; so was Bacon, to be sure. To me, clearly, Bacon's dream of a technocratic society is part and parcel of contemporary utopianism and an offshoot of a remark of More, perhaps indirectly related to the dreams of Roger Bacon. Hartlib even saw both More's and Bacon's utopianism as a symptom of the period's (neo-neo-) Platonism (p. 218). Miss Purver, however, puts a wedge here: Plato in his *Atlantis*, More in his *Utopia*, etc.,

> saw his ideal society as an end in itself. Bacon, on the other hand, had a specific aim in his New Atlantis. Although his society was to have the spiritual and social virtues which he considered desirable, the actual purpose of his proposed institution was to build up a new system of natural sciences (pp. 225–6).

This hot air should read, More wanted mainly justice, Bacon wanted justice too but stressed efficient technocracy. For Miss Purver that sets him apart, for me that sets him well within, the group of utopians – though admittedly with the merit of an added variant which proved very ingenious indeed.

In the last pages of her book, though, Miss Purver offers a pleasant surprise, a hitherto unpublished letter from John Beale (Boyle's schoolmate mentioned above) to Samuel Hartlib, concerning a hitherto unpublished plan. The same Andreae who had influenced Comenius had also influenced a Swedish nobleman who developed a plan about a Royal Society which made Beale suggest that King Charles II should be the patron and founder of the Royal Society. (See p. 228 and p. 229 and notes there). And so, Andreae, Hartlib, and others, somehow managed to enter into the act. This should warm the heart of an anti-radical like me, but I am not so much at home in the Establishment either, and find the whole business of "Royal" in the Royal Society not over-exciting.

6. BACON VERSUS BOYLE

In conclusion, Miss Purver has rendered us a service she has not intended to, and at least I am glad I have studied her book – though this is only a retrospective feeling. She did force me to reexamine the known documents, and she did impose on me an image of a radicalist group which knowingly suppressed their origins in earlier groups which had understood the term "the new philosophy" in a less radical sense; a group which was embarrassed by the fact that it had to pay homage to some thinkers of previous generations other than Bacon; a group which functioned as a group and with a radicalist ideology which justified just this new facet.

Yet science is not a group activity or a collective activity. Contrary to the Baconian ideology of the founders of the Royal Society, we still see the foundation of their Society as no more than a landmark: Miss Purver is quite right in the factual part of her complaint (p. 3 *et passim*), though I do not see that we need change our appraisal.

This being so, one may wonder how the Society could function and contribute so much to the advancement of learning. That it contributed to human welfare in general is neither problematic nor questionable: in addition to their Baconian toleration and drive for enlightenment in general, their stress on the mechanical and agricultural practices, from shipbuilding and gunpowder to milking and gardening, this had a lasting democratizing effect; and their anti-Baconian stress on natural knowledge in a period renowned for its witch-hunts is of supreme significance too. Even within the commonwealth of learning their influence in the arousal of interest and hopes, as well as their offering a platform for scientific encounter, publications, and the like, could not but be beneficial. Yet the Society did more as an instrument for the advancement of learning: contrary to its own ideology, it encouraged the development of hypotheses and controversies, and contributions of individual thinkers as individuals head and shoulder above their colleagues.

There is little doubt that the Society's ideology was somewhat tempered with common sense from the start. The only staunch anti-radicalist in the group was Robert Boyle. His *Seraphick Love* of 1659, which moved Evelyn to tears and sent him first to Boyle and then to Wilkins and thus to the foundation of the Society which soon became Royal, spoke of natural religion as encompassing experimental philosophy as a ritual and as sublimation of unrequited love. His 'Proëmial Essay' to *Certain Physiological Essays* does endorse a quasi-Baconian philosophy, but staunchly rejects all radicalism and all hostility to hypotheses (such as preached by writers from Bacon to Miss Purver). This 'Promëmial Essay' is well reflected in the constitution of the

Society which was proposed by Lord Brouncker, the light-weight first president of the society, and seconded by Boyle.

The tradition of science still reflects a double-standard, a shopwindow image which is radicalist and neat and devoid of all problems, and a workshop image where all is in constant mess. In this century, for some time, men of science tried to break away from this tradition and expose the workshop to lay inspection; but old traditions die hard, and Miss Purver's volume is but an instance of this.

NOTES

[1] Margery Purver, *The Royal Society: Concept and Creation*, with an introduction by H. R. Trevor-Roper; Routledge and Kegan Paul, London, 1967, pp. xvii + 239.

[2] Since this was first published, a detailed study has appeared on the Baconian radicalist critique and the defense of Oxford by Wilkins and Ward: Charles Webster, 'William Dell and the Idea of University', in M. Teich and R. Young (eds.), *Changing Perspectives in the History of Science: Essays in Honor of Joseph Needham*, Heinemann, London, 1973, pp. 110–126, esp. 125. The defense appears anti-Baconian.

[3] Further detail about Johann Valentin Andreae can be found in a most interesting review by Dashiell Hammett of Arthur Edward Waite's book on the Rosicrucians reprinted in his "Tulip", an autobiographic fragment, published posthumously in his *The Continental Operator*.

APPENDIX

The Beginning of the Royal Society, Oxford, 1960, by Margery Purver and E. J. Bowen, F.R.S., 16 pages, recounts Sprat's story in brief, supplements it, adds some biographical data, and such. It is a publication devoid of any merit except that it announces Miss Purver's conclusions to the world. In view of this it is hardly surprising that, as Trevor-Roper complains, her conclusions were resisted before her evidence was heard. Why there should be no resistance to the conclusions when they are not argued for? But Trevor-Roper finds comfort in the fact that already in 1683 Wilkins said it is the fate of new truths to be derided by the ignorant and rejected by others who are perverse. This is nice: if I endorse your view it is because you are right, if I reject it, it is because new truths are resisted. Is it possible that some allegedly new ideas are resisted because they are old hat? The combination of scientific radicalism with Establishment social attitudes is one we may call *passé*. For my own part, I neither accept nor reject Miss Purver's solution, as I reject the presuppositions of her problem. Antecedents are never as clear-cut as to allow us to pose the question. Those who did pose it in the 17th century wanted a neat and true shop-window picture of the antecedents of the Royal Society. This is neither possible nor interesting.

CHAPTER 26

THE IDEOLOGICAL IMPORT OF NEWTON

1. BACKGROUND: THE ENLIGHTENMENT AND NEWTON

Considering Newton and the Enlightenment, the standard topic as I understand it is the contribution of Newton's specific achievement to the culture of his times and the following generations, known as the Age of Reason, or the Enlightenment. It is not clear what characterizes that age, nor even how the *avant garde* thinkers of that age saw themselves. But, to begin with at least, we can take the central thesis of Immanuel Kant's *Was ist Aufklärung?* Kant says Enlightenment is two things: the autonomy of the individual and rational thinking. The recognition that each individual is responsible for himself and that thinking is the major factor in human affairs leads to the immediate corollary that every individual person must be his own light. Hence, science as a specialism is impossible since the specialist, however, clever and intelligent and learned, cannot have authority over anyone but himself. Thus, rather than having the universal man as the image of the age, as Leonardo and Michelangelo still are the images of the Age of Humanism, the Age of Reason has as its models or symbols of the age, people like Franklin, Priestley, Lavoisier, Boscovich and even Voltaire. In a later age a joke was invented about Euler's poking fun at Diderot's ignorance of mathematics. This joke has provoked historians of ideas to show that Diderot was a mathematician; and everyone who knows D'Alembert's preface to the *Encyclopedia* knows that he was a rounded man too.

Of course, this picture is, like all symbols, but a myth. William Whewell says of Newton's *Principia* that when it was published only a few people could understand it. He compares the mathematical tools in it with heavy weapons that only giants can lift and few of these could move about freely so as to be able to use them properly.

This is not the only exception, though it is a very conspicuous one. Kant's *Critique of Pure Reason* is another. Kant thought Newtonian mechanics is self-understood; *a priori*! But as Fichte said, the *Critique* itself was too difficult for simple people. Kant was extremely upset by this charge, denied it, and cited the famous Italian proverb, God protect us from our friends. One more example: Laplace wrote his *Essay on Probability* twice, once as

Analytic Essay etc., once as *Philosophic Essay* etc. Isaac Todhunter, the leading historian of probability in the nineteenth century, found a formula put in mathematical symbols in the one and in ordinary French in the other, and said he understood it when put mathematically but did not understand it when put in ordinary French.

Here, then, is the uncritical point of the Age. True, the attempts to popularize Newton, including the celebrated Italian *Newton per le donne*, including Laplace's excellent *System of the World*, and other masterpieces, only lessen the problem but do not overcome it.

I cannot stress enough the importance and generality of the problem, but let me hint. Paul Hazard mentions that scientific research was the most popular hobby in the Age of Reason. This hobby was first advocated in Sir Francis Bacon's *Advancement of Learning* in 1605. In his *Novum Organum* of 1620 and elsewhere Bacon says, the method of induction is like a ruler and a compass that takes talent out of the draftsman's job: with these tools everyone can learn to be a draftsman, and likewise the inductive method makes everyone a scientific researcher. When the historic meeting of modern philosophers took place in Restoration London in order to organize scientific research, the leading members wanted to found a non-monastic monastery, i.e. a lay or a secular university. But one of these philosophers was Robert Boyle, a student and disciple of the amateur scientific researcher William Wotton who was a friend of Bacon; and he decided for a society of amateurs, and so the Royal Society of London was founded. As Robert Leslie Ellis, the leading Bacon scholar, suggested, Bacon was so influential because of his new idea: everyone can be a philosopher; all he need do is perform some simple observations and experiments.

So much for the problem. The solution was that since science is cooperative and cumulative, we must all accept others' contributions, but understand them and accept them after each of us examines them personally. This is an excellent solution, and many members of the Enlightenment movement read scientific papers while repeating the experiments they describe. Not knowing this fact many historians misread these papers. But I will leave this point now. Rather, let me say, this solution will not do. Already Laplace noticed, in his eulogy of Newton at the end of his *System*, that something is amiss here. Newton, he said, was both the luckiest and the cleverest man ever. Luckiest, because he was born just in time for the facts to have accumulated and stand ready for the grand generalization; and cleverest, since of all his contemporaries he alone made the generalization.

But if science needs a genius, then not everyone can be a scientist, not all minds are equal, and the very recognition of a great mind is the acceptance of it as an authority of sorts.

This is why verification was so important for the Age of Reason. Many seventeenth-century thinkers suggested that the highest probability of a given theory should suffice for according it scientific status, for thinking it is true. But this is true only if there is no problem of authority at all, if every individual comprehends a theory, correlates it with facts, knows all the facts, can compute his own probability. This idea was accepted by the eighteenth-century Italian thinker Bruno de Finetti who was born in the twentieth century. His disciples were very uneasy about the fact that common people compute probabilities wrongly, and even simply guess them instead of computing them. But then these disciples do not believe in the Enlightenment of each individual.

In the age of Enlightenment, then, the idea was and is of supreme importance that everyone can add to the human stock of knowledge by making experiments and can at least repeat experiments that verify important theories. And the prime example – the paradigm – of a verified theory was Newtonian gravitational theory. And so the very attempt to bridge the gulf between the common man and Newton only broadened the gulf and made Newton the paradigm, the model, the symbol. The ambivalence only deepens.

2. THE PROBLEM SITUATION AT PRESENT

An incongruity regarding Newton's own works still exists and is presented in the opening pages of I. Bernard Cohen's masterly preface to his and Koyré's edition of Newton's *Principia*. That work was prepared most carefully by its author and published in several editions; yet posterity neglected it. Newton's attention to details of his *magnum opus*, his careful revisions through its three editions and more, equal those given to no other scientific text unless it be Charles Darwin's *Origins*. Yet there was a remarkable neglect by scholars of the changes in the texts. Even the most conspicuous change in the opening Book III of the *Principia*, from *Hypotheses* to *Regulae Philosophandi*, was not noticed; Cohen says that quite accidentally both Koyré and he himself discovered this and it was the pivot of their joint venture. Now, how come, asks Cohen, that such a careful attention by Newton to details of his *magnum opus* remained unnoticed? He says there is almost no scholarly work on scientific texts – history of science is a relatively young discipline – and the task was, anyway, quite formidable. Fine; but why no scholarship of the *Principia* comparable to that which Kant's *Critiques* have attained?

The question rests against a complex background. The fact that the thinkers of the Enlightenment were little concerned with history and with scholarship is fairly well known. So is the fact that when historical scholarship developed in the period of reaction to the French Revolution most natural scientists followed the ideology of the Enlightenment – either along the whole line or at least as far as science was concerned. And their interest in the history of science was marginal. A lecturer and writer like James Clark Maxwell – of the famed Maxwell's equations – used to spice his scientific presentations with historical anecdotes and thereby drew much admiration. So it would not be surprising to conclude that throughout the eighteenth and nineteenth centuries there were scarcely any histories of science and thus to explain I. B. Cohen's claim that the field of the history of science is relatively young.

Yet this explanation is false or at least misleading, since histories of science and of the diverse sciences were published ever since the foundation of the Royal Society of London, of course. But then, consider what Robert Schofield, a modern historian of science, had to say (in his appendix to Cohen's collection of papers and letters by Newton) concerning Thomas Birch's famous history of the Royal Society: "strictly speaking, it is not a history at all, though it may serve as historical raw material". I wish to emphasize that this very characterization, made to condemn Birch, would certainly seem to him nothing short of a compliment. History, for all followers of Sir Francis Bacon, meant, as Bacon's expression had it, "a just history of Nature", or "history as she really is", to use a latter expression, or "facts uninterrupted by thought" as Goethe and Lavoisier admiringly described Joseph Priestley's reports on his experiments. I do not mean to adjudicate between Robert Schofield and Thomas Birch, though clearly my prejudice is with Schofield, of course. I cannot, nevertheless, accept Schofield's dismissal of Birch as a historian in so ahistorical a fashion: a historian might benefit from the views on history that an older historian may hold. Moreover, by whatever standard we judge history, we must observe that there are all too few histories of science even in the age when writing histories became popular. Moreover, contrary to Cohen's claim, Newton's writings fare worse than one might expect. "Why should England have published the monumental editions of Cayley's and Sylvester's works", asks David Eugene Smith in the 1927 bicentenary Newton Volume, "and yet have so neglected Newton and have allowed the Portsmouth papers to lie practically in oblivion?" Now the expression "the Portsmouth papers" is ambiguous. It may refer either to the bunch of papers and books of and by Newton, as well as unpublished manuscripts of Newton, that were offered to Cambridge University by the Portsmouth family, most of which manuscripts the University rejected, believe it or not. And it may refer to the part of that bunch which the University kindly accepted and which is now known by the descriptive title "the Portsmouth Collection" given to it by the University Library. And, since the chemical papers of the Portsmouth Collection were mislaid, very conveniently, and returned to the collection only after David Eugene Smith's death, he may have referred to only a part of that collection, especially if he meant to refer only to Newton's scientific manuscripts, thereby meaning to exclude his alchemical papers. In 1958 Marie Boas Hall and A. Rupert Hall still insisted that "Newton was not in any admissible sense of the word an alchemist" – even though in depriving his experiments of their alchemical rationale they had no alternative rationale for them. This has been noted by Betty J. T. Dobbs in her 1975 study of Newton's alchemy, in which she notices that Boas and Hall did not bother to examine Newton's alchemical manuscripts so as to check their contention.

More generally, there is the problem of Newton's scientific works, and these are now studied and published often by some old-fashioned historians, who are still embarrassed by Newton's deviation from the Enlightenment's ideal of the rational man and his devotion to alchemy and theology, and either choose to ignore his failings or condemn him for them or vacillate between the two. And there is the new historical school of students of the intellectual framework of Newton's time who try to correlate Newton's science, alchemy, theology, and eschatology, in a manner agreeable to that framework, among them Herbert McLachlan, Frank E. Manuel, and the already-mentioned Betty Dobbs. Their work, I feel, could not be advanced without the deeper study of the question, does science need an intellectual framework? – was the Enlightenment right in its claim that no framework can be accepted by a rational investigator unless it be proven? Amongst the students of this question were many fine thinkers; I must mention Émile Meyerson, E. A. Burtt, and Alexandre Koyré.

All these developments, to return to my quotation from David Eugene Smith, are contemporary ones. John Maynard Keynes, who purchased many of Newton's papers originally among the Portsmouth papers but rejected by Cambridge University and so not in the Portsmouth Collection, has given full expression, perhaps an exaggerated one, to the sense of embarrassment they caused to those who had seen them and acted as though they did not exist. Keynes himself went after the papers, he said, only out of a speculative bent; but once he possessed some of them, he looked at them, was surprised, studied their history, and published his memorable 'Newton, the Man' of 1946, where he argued that Newton's papers on theology and alchemy were suppressed because they present Newton to us not as the first of the scientists, of the men of the Enlightenment, but as the "last of the magicians".

And this "last of the magicians" embarrassed his followers not only in his unpublished alchemy but also in his published science. This is not to say that Newton was not the chief scientific influence, of course; merely that orthodoxy changed and so his writings became embarrassing. Even Newton's theology, which was largely successfully suppressed, was of much consequnce: as G. S. Brett, the famous Toronto philosopher-psychologist, has pointed out in the 1927 Newton bicentenary volume, Newton's reconciliation of science and religion was an important step in the process of science ousting religion. Also Frank E. Manuel recommended in his excellent *Isaac Newton: Historian*, and *The Religion of Isaac Newton*, that we approach Newton's work in a proper historical perspective, and thereby avoid embarrassing ourselves with his naive views and outlook and see his influence as a part of an *avant-garde* stream of histories. This stream was, by and large, characterized by attempts to modernize, to make more scientific, the writing of the history that religion shrouds in myth and in the reduction of the mythical component of religion by the techniques known as Euhemerism, after the ancient Euhemerus who viewed many pagan myths as parables, metaphors and such. Nor was Newton's alchemy, though it was entirely suppressed, it seems, of no consequence. J. E. McGuire argued that it helped Newton shape his opening to Book III of the *Principia*, which was one of Newton's most influential passages, in the *Hypotheses* of the first edition and the *Regulae Philosophandi* that replaced them in the third edition. Moreover, Newton's speculations on the nature of matter and subsequent doctrines of the eighteenth century in the same vein, all had a central idea, the unity of all matter, i.e. the claim that the elements are not truly elementary but must be transmutable – an idea now known as Prout's hypothesis, though Prout himself claimed in his famous 1915 study that it was very much in the air.

All this comes to illustrate what a great shift in attitude is to be found in the recent story of the history of the physical sciences in general and of Newtonian studies in particular. Yet the shift is far from complete and is not in the least explicit. The more old-fashioned historians of science simply undertake the task of studying and publishing Newton's scientific work and try hard to avoid the question "Why was Newton neglected till now?" and they also try hard to delay attention to his unscientific material. And even the more modern historians of science do not like to record the shift. Thus,

I. B. Cohen prefers to explain the shift technically rather than as a result of a changed attitude towards intellectual frameworks – even though he is one of the scholars who have contributed towards the shift quite considerably. Cohen can explain the delay of the venture of issuing a proper *variorum* edition of Newton's *Principia* technically – by reference to the fact that the venture is tremendous; and indeed his and Koyré's monumental work is testimony to that. But even a collection of Newton's published scientific papers, akin to Maxwell's, say, was not published, nor even his published optical papers – until very recently, of course. In brief, there is no doubt that even by the standards accepted by given communities at given times Newton's works suffered from neglect. Even if the Cohen and Koyré *variorum* could not be attempted before, surely Cajori could easily have improved his edition of Motte's translation by adding here and there the passages from earlier editions which most conspicuously differed from the third and authoritatively last – at least he could have published the different openings to Book III of the *Principia* with the *Hypotheses* of the first edition and the *Regulae Philosophandi* that replaced them in the third. He chose not to. There was, in other words, much embarrassment not only concerning Newton's theological writings, but also concerning his scientific ones. I wish to dwell a bit on this.

No doubt physicists today are not adherents to any doctrine that may justly be called Newton's own or even Newtonian in any sense, however loose. They may feel uneasy about this, but then only when pressed; and then they have some ready-made excuse to soothe their conscience. Not so the physicists – and other thinkers, of course – of the pre-Einstein generations. Thus, E. A. Burtt in the famed *Metaphysical Foundations of Modern Physical Science* of 1924 described Newton's authority as surpassing the heights of Aristotle's authority. And Brett, in the already-mentioned 1927 contribution, said that so great was Newton's "undisputed dictatorship which he held in the kingdom of science that some importance was inevitably attached to any works or writings which carried his authority". This is an understatement: since Newton's authority lay in his supreme mastery over the inductive method, which was supposedly infallible, anything he said that was anchored in fact had to be undisputed. And so, when we read in the terrific *Newton the Man* of 1931 or so by Lt. Col. Richard de Villamil, the concluding note concerning the hero-worship accorded to Newton, as expressed by the now famous remark attributed to Marquis de l'Hôpital, in which he allegedly found it hard to believe that Newton ate, drank, and slept like ordinary mortals, then we must approach all this with mixed feelings. It is a breath of fresh air, and a relief from an oppressive authority, but it is also an authority voluntarily undertaken in the name of Reason.

And this brings me to one concluding observation as well as to the heart of my problem. We may find many works, such as physics and philosophy textbooks, and even texts on the social sciences, with scores of references to Newton, yet with no Newton and no Newtonian ideas in them: Newton serves there as a sort of symbol, a generic name, like Quixote or Sancho or Rosinante or Donna Dulcinea. In older books the opposite is true. The name is not mentioned, but the phraseology chimes all the way through all the writings of the Age of Reason. If someone reads Ben Franklin saying he knows not the cause of electricity and does not hear in that the chime of Sir Isaac's confession of not knowing the cause of gravity, then that person's historical knowledge is wanting. But the allusions are so many and profuse that I suppose even Newton's contemporaries may have missed some of them. This is why studies such as I. B. Cohen's *Newton and Franklin* of 1956 and Perry Miller's *Jonathan Edwards* of 1948 are of so much value. But enough of that for now. What all this illustrates is the authority Newton had on the commonwealth of learning – the same commonwealth that held the autonomy of the individual's judgement as a supreme value. The problem, then, shifts from Newton scholarship or its absence, to its roots in a deeper problem. That deeper problem is, how can one and the same milieu profess both the autonomy of the individual's reason and the authority of one particular man's reason?

The question looks facile: we can all see the Truth when she is revealed, but Truth may hide until we perform the proper ritual that forces her to appear. After all, Bacon presented this view and said the ritual is the observation of the smallest and minutest facts of Nature that are made in an act of homage to Nature, in an act of submission designed for the resultant conquest: Nature only reveals herself to her humble worshippers.

Therefore, one must observe the fatuity of this answer: not the question, but the traditional answer is facile. Why did Truth, or Nature, reveal herself only or primarily to Newton? Newton, said Laplace, we recall, was both the cleverest and the luckiest fellow: he was born just when sufficient amounts of minute details had accumulated to open the way for the grand generalization. Yet no one felt the absurdity better than Laplace: What is now left for us to do? he asked. And he answered, just develop the mathematics, improve the observations – in brief, exegite, though in a new style. Are we all doomed to mere exegesis? Is this what intellectual autonomy amounts to? Can Truth reveal herself only once and to our Great Leader only? Where, then, is intellectual equality? Was Newton *only* unusually lucky or also unusally clever? We do not need both qualities and so can ask, which is the crucial one?

No doubt, Newton's authority was problematic. Otherwise people would have mentioned him rather than emulate his idiom in allusion. The problem is still alive, nowadays with newer geniuses, even though Einstein, for one, claims much less perfection and finality and infallibility than did Newton.

So much for the problem at hand. This concludes my presentation. Let me only round it off by returning to the problem of Newton scholarship. Traditionally, scholarship was authoritarian: the objects of the scholars were the worthy and hence the authoritative texts: the authority of the text justified practically all investment of energy in scrutinizing the sacred text and its worth justified it intellectually. Modern scholarship is critical and egalitarian at once, and one of the chief burdens of the modern but not the traditional scholar is to explain the import of his exegesis. When he fails to do this he becomes a scholastic yet without the vitality of the old scholasticism. Newton was both an authority and an equal. And so his admirers – who of course would not admit him as an authority – could not be openly critical of his scientific output – they did have criticism, and Newtonianism kept moving, but like many a school dogma, the moves were all surreptitious – to echo Popper's observation on schools. And so, all historians of science could do at first was to record some indifferent facts. But they soon ran out of these, especially since in the course of time even a seemingly innocuous fact may begin to seem odd. So next they shifted to personal problems: Was he right to go to the Mint, and what were his sexual habits? Critical history caught up with Newton only after his dictatorship ceased to be an open secret and began to be a mere historical fact. Which, we remember, is after Einstein's theory of gravity won over Newton's, with E. A. Burtt and G. S. Brett and D. E. Smith in the 1920s, and Koyré in the 1930s, who, in his *Etudes Galiléennes*, inaugurated the new style of history of science. Every historian is a bit of a hagiographer, Koyré confessed; he thereby concluded the period of the open secret of the authority of the man of science and began the tradition of the explicitly critical, however admiring, history of science.

Back to my final and central problem, then: how could the autonomism of the Enlightenment and Newton's absolute authority live side by side?

3. THE PERSISTENCE OF THE TENSION

The tension between individualism and the authority of science is one that sensitive people were always aware of: they resolved it with sweet reasonableness: logic is universal; scientific method is universal; great minds think alike; it is not that you must agree with What's-his-name because he is great; you must agree with him because of the dictates of Reason – mine, yours, his, anyone's. Anyone reasoning under similar conditions accepts the same conclusions. For example, two able arithmeticians doing the same arithmetical exercise must come up with the same result. Indeed, we like a second opinion exactly for this reason: we want the voice of reason, not an eccentric idiosyncracy. This is why, since the rise of the Royal Society of London, each discovery had to be reported by at least two independent observers before it was endorsed.

This is also the case of multiple discovery: a few people in the same situation come up – quite rationally unless they share an idiosyncracy – with the same idea. At times the multiple discovery is done independently and is a multiple discovery proper. Less impressive but much more relevant is a shared discovery – shared by a team or by a large group of researchers who publish frequent progress reports.

Newton's case was nothing like that. Even the possibility that Robert Hooke made a slight contribution to Newton's thought by such a progress report, by the mere posing of a question or a hint of an answer – this possibility was violently repudiated. Newton jealously guarded his claim for isolation, and for generations his claim was endorsed. The normal solution to the conflict between the autonomy of the reasoner and the authority of reason does not obtain here. Let me rub this in a little.

What both Bacon and Descartes said, and many other authors, including Kant, avidly repeated, was that they did not want to convince their readers of their conclusions but to provoke them to think for themselves: if you and I both use the same method, and it had better be the right one, then we both attain the truth, and the truth is one – so in essence you do not need to read my writings, but to mediate sagaciously over the situation and you will see what I now see. But expediency suggests you allow me to save some of your time so that you can go further than I and have the opportunity to progress.

Assume this to be so, and ask what are the chances that due to faulty communications and such there might be another Newton or another Kant, somewhat lesser perhaps. This incredibly silly problem could only arise because of an error in the background that has led us to pose it. Hence the problem – Could there be two Newtons? – emerges from the error of juxtaposing a universal method with the undisputed uniqueness of "the incomparable

Mr Newton", to use the phrase which John Locke has immortalized, whose uniqueness was the over-employment of that method.

It is not that I have seen the problem stated anywhere. The nearest to its statement of which I am aware is Bertrand Russell's, who says in his *History of Western Philosophy*, in the chapter on the Rise of Science:

> It appeared later that even the minute departures from elliptical orbits on the part of planets were deducible from Newton's law. The triumph was so complete that Newton was in danger of becoming another Aristotle, and imposing an insuperable barrier to progress. In England, it was not till a century after his death that men freed themselves from his authority sufficiently to do important original work in the subjects of which he had treated.

Yet this is hardly adequate. How come that a century later men – which, and how many? – got the temerity to break Newton's spell?

The allusion is to optics, and the men are Young and Fresnel. They encountered hostility, we know, and won their victory: even Laplace conceded despite his earlier dogmatism and hostility. How they came to rebel against Newton's authority is quite a story, and so is the story of how the Newtonians absorbed the shock. They blamed Newton's discipline for holding dogmatically views he had offered quite tentatively – especially Sir John Herschel, who wrote a book of much influence in which this point is made. And William Whewell evolved the view on methodology that became official: whereas Newton's optics had never been properly tested, his dynamics, especially his theory of gravity, was secure and still unchallengeable.

Anyway, Russell only speaks of a danger, and one which was averted. I am concerned with the root of the danger, with the intellectual consistency of the framework which gave rise to it.

Nevertheless, Russell, and likewise also Popper, does solve part of the problem: not "How do autonomy and authority dwell together in the Age of Reason?" but "Why are there no references to all this in that age?" The answer is, the scientific community was both militant and triumphant; its militancy permitted oversight of difficulties and its triumphs justified the venture and goaded it on. It is no accident that social scientists kept complaining that though scientific method was known, since Bacon, social science still awaited its Newton. As Paul Hazard has noticed, Bacon's name was the only one permitted to be mentioned in conjunction with Newton's during the whole of the Age of Reason without it being a blasphemy: they had to, since they lamented the lack of similar success in the social field as in the natural one. Nor was there a dearth of takers. Hume, Bentham, St Simon,

Marx and others, were claimants, but their claims failed to bring recognition. And as long as the struggle continued, no matter how many problems existed, the façade had to be kept up of a success story uninterrupted and unmarred by failure.

No doubt all this is true and supportable by many details; yet it is hardly satisfactory. Indeed it is nothing but the problem itself restated: why was the Age of Reason the age when so much success was rightly claimed that the very existence of some problems had to be concealed? Success does not usually need concealment and concealment usually is not conducive to further success but rather covers up stagnation! Yet even these days, when we are much freer of the authority of Newton and the problem is felt by so many scholars, it still lies in relative obscurity. Why?

What I wish to indicate is a difficulty we still have with us. Every historian of science has some idea of what is rational, what not; and of how binding rationality is, and to what extent; and what is the nature of the moral and intellectual autonomy of his heroes. And so he must face squarely the authority of Newton, in its historical eighteenth-century setting and in its modified twentieth-century setting. For, in our twentieth-century setting we see in Einstein an authority – less than Newton's but still not inconsiderable – and we recognize in Newton in retrospect the same authority that we recognize in Einstein now.

This is a point in contemporary philosophy of science; and since it is my field of expertise, I will not dwell on it. Some philosophers of science say Einstein's theory of gravity is today's most probable, and yesterday it was Newton's that was most probable. Others, with Popper and his disciples, come closer to Einstein's philosophy and say, Einstein's theory of gravity is nearer the truth than Newton's, as Newton's is nearer than Galileo's. (I. B. Cohen, incidentally, has contributed a lengthy and detailed historical critique of the writings of Popper and his followers in Yehuda Elkana's *The Interaction Between Philosophy and Science*. I am not even clear in my mind whether Cohen intended his critique to repudiate or to rectify the Popperians on this point. I suppose he is still undecided.) Anyway, the question, as I have argued elsewhere, is not what kind of rationality is it that imposes the ideas of Newton or of Einstein on their relative contemporaries, but rather what is the degree of that rationality. And all canons of rationality before our own age were (a) of complete rationality admitting of no degrees and (b) compulsory rationality for all thinking persons as such. Hence the tyranny of reason stands behind the tyranny of the ideas that manage to best comply with the canons of rationality.

But perfect rationality is but a myth. What I wish to argue is that rationality is not given to us full-blown, and is not fully obligatory anyway.

4. THE ENLIGHTENMENT MYTHOLOGY

To plunge straight to my thesis, I wish to apply the doctrine of Claude Lévi-Strauss – of myths coming in pairs and being mixed to varying degrees, depending on circumstances – to the movement of the Enlightenment – which endorsed the supremacy of Reason in all circumstances of the life of the individual. Let me first say that I do so without the slightest authorization. In the conclusion of his *Savage Mind*, for example, Lévi-Strauss makes it quite clear that scientific and mythical thinking are poles apart. For my part, though, I do not in the least deny the science-myth polarity; I see in the scientific world in general and in the ideology of the Enlightenment in particular, the prevalence both of science and of myth, the myth being that proper science is divorced from all myth. In places Lévi-Strauss allows for the view of parts of the modern world as myths, including his own theory which, in the beginning of his *The Raw and The Cooked* he calls the myth about myths, and he recognizes the French Revolution as a myth in quite a few places. But never mind that: the responsibility for my application of Lévi-Strauss to the Age of Reason is mine.

My application, then, is simple. *The Enlightenment had both authority and reason as two poles, and the authority of reason as the mediation; Newton was the symbol of all this*. Let me explain.

Lévi-Strauss finds everywhere polarities – i.e. pairs of conflicting qualities – that are given *a priori* within an intellectual framework. He considers given polarities within a given framework much more revealing than who is supposed to occupy which pole, or any other data within that framework! The questions, who is a magician or who is a scientist, differ from each other more, and so are *a priori* more informative, than the answers. Merlin is the greatest magician; Newton is the greatest scientist; or, as Keynes would have it, Newton is both; or even, Newton is the greatest philosopher as contrasted with Newton is the greatest scientist. One reason our age is so unable to comprehend Newton, all the magnificent Newton scholarship notwithstanding, is just the fact that so many of us insist on our framework and call him a scientist even while discussing his theology and his methodology, rather than calling him a philosopher, in accord with his framework.

Now, in any myth framework, only extreme positions are marked; only polar characteristics count as really important. But in real life these are rare and so the users of myths mix the polar opposites, and in varying proportions. Moreover, the polarities, being extreme, create tensions – emotional as well as intellectual – often leading to mediations or compromises. So much for Lévi-Strauss.

To apply this to contemporary historians of science, some of them claim to have and need no framework; of course they are mistaken. The historians who insist on framework are all too often irrationalists – agreeing with the Enlightenment that taking a framework as given is uncritical and thus irrational – and so they do not usually pay much attention to science. The few historians of science who do write about frameworks, whether Newton's, the 17th or the 18th century, or any other, are struggling with the problem of rationality.

It is futile to have a historian who claims he has no framework and one who claims he has, debate between them a historical detail, unless it concerns the question of whether historians of science have to use a framework. Unfortunately, just this major topic is seldom discussed by historians of science, and then more usually their discussion is glib or even downright contemptuous.

There is, for example, the celebrated volume on Newton's *annus mirabilis*, 1666, (M.I.T. Press, 1970), in which the old-fashioned attitudes are expressed with a vengeance. Asger Aaboe, for example, dismisses there the social and psychological components of the history of science as rooted in their students' scientific incompetence; the history of science, he says, is scientific and so indubitable and undebatable! Aaboe thus makes it clear he could barely be bothered with the essay of Costabel in the same volume, since it is on Newton's concern for the unity of science.

Going through such a volume as *Newton's Annus Mirabilis* that is so representative of Newton scholarship today, is hard work and may, all too easily, lead one to lose the forest for the trees. It is so much better to take a representative judicious work that offers only an impressionistic view. I chose Russell's *History of Western Philosophy*, where, in the chapter on the rise of science, Russell notices that Newton's influence was regularly coupled with that of Locke; "quite illogically" he adds. This puzzles me. It is one thing for a historian to be blind to connections between minute scientific and methodological views – all he need do is to dismiss them as incidental, uninteresting, irrelevant – and quite another for an author of an overview who misses a needed connection, as is clear from the facts – which Russell reports – that Newton's success was a major factor in the Age, as was the demand for a Newton in the social sciences. On what ground, then, do we have the right to expect a repeat of Newton's success? Russell ought to know: it is on the ground of the theory of induction, of course: success must follow proper induction. Hence, the methodologist Locke must be coupled with the philosopher Newton – not illogical at all. It was Newton's inductivism, not his genius, that made so rational both attempts to follow him and efforts to emulate him. This pole is all too clear. It is the other pole, Newton's genius, that we usually associate with the Romantics and we refuse to associate with the Age of Reason. We do so because it runs against their ideology, not because of want of evidence, if I may repeat myself.

The problem is still here. In the *Annus Mirabilis* volume at least one writer tries to answer it, though none of them is so bold as to ask it explicitly. Robert Palter, scathingly contemptuous of hypothetico-deductivism, adduces inductivism from the writings of Newton. But he offers room for Newton's genius: he was great at discovering generalizable facts. It is not my custom to criticize ideas which I cannot praise a little as well, and so I shall not comment on this one. I shall only mention that at least one other author in this volume is a real conservative – Establishment as well as backward-looking – inductivist, namely Truesdell. But there also lurks another kind of conservatism here: the chief corollary to inductivism that was retained by the opposition – by the conventionalists-instrumentalists – is positivism or anti-metaphysics. Newton's metaphysics is examined by both Howard Stein and Dudley Shapere. Stein declares Newton's views of space and time as "a classic case of the analysis of the empirical content of a set of theoretical notions" which he deems almost correct; and Shapere concedes that Newton had a conceptual framework that may be called metaphysical, but not in the classical sense in which all scientific propositions (allegedly) follow from some metaphysical principles, he says. I know of no classical metaphysician, Plato or Aristotle. Descartes or Gassendi or Boyle, who made the claims that Shapere ascribes to "classical philosophy" with no comment or evidence, except to say that the framework he ascribes to Newton may be reconciled with positivism.

What I want to say is that positivism, inductivism, pure rationality, scientific proof, and all that, are parts of a myth, a myth that rested on Newton's success. And the opposite pole was Newtonian World-View as described in both Perry Miller's excellent *Jonathan Edwards* of 1948 and I. B. Cohen's excellent *Newton and Franklin* of 1956, both of which totally ignore Newton's positivism just as the positivists ignore or combat Newton's metaphysics. It is time to recognize both Newton's exciting metaphysics and his hostility to metaphysics.

To repeat Lévi-Strauss' thesis, we use both myths in proportion when doing history though, in contrast to Lévi-Strauss' savages, we demand consistency of ourselves and may achieve it by swearing allegiance to only one pole. Yet the tension between poles, to return to Lévi-Strauss, does exist, all the more in a rationally minded society, I should add; and it requires mediation. Newton was both poles – both inductive rationality and genius; and so he was the best *prima facie* candidate for mediation: he was the compromise solution, as hinted by Bacon already: everyone who reasons properly may find himself a genius.

I confess my bias is rationalistic. Also, because of this I prefer my position tempered rather than to its incorporation of extremes that call for their opposites and cause tension. I therefore endorse a tempered or gradualist view of rationality. For example, the Age of Reason was by far more rational than any other period of Western history, Greece not excluded; yet it was still mythridden; and we can do better then the Age of Reason by noticing that utter rationality is a myth creating its polar opposite, that the Age of Reason makes way for Romanticism – this quite against all it stood for.

If we agree that one major component of Romanticism is hero-worship, then we can see how Romantics made use of scientific hero-worship as a kind of *reductio* of Rationalism. And this ought to be admitted. Can we admire Newton without hero-worship? Can we read his texts critically? Even inductivists ought to be able to do so, and indeed Truesdell does so in his already mentioned contribution. But in an inductivist mood he ridicules Koyré's desire to see Newton's theory as a whole and wishes to sift the correct from the erroneous. This simply cannot be done. We can admire Newton without uncritical worship and see his ideas as a unity, with its strong and weak sides inseparable, and as such a part of our own heritage: physics, rationalism, cosmology, theology. This task is still open; indeed, it was just begun by Koyré, whose assessment of Newton in his classical 'The Significance of the Newtonian Synthesis' of 1948, reprinted as the first paper in his *Newtonian Studies*, I have ventured to echo.

CHAPTER 27

SIR JOHN HERSCHEL'S PHILOSOPHY OF SUCCESS

This is an extended critical book-report on Sir John Herschel's *Preliminary Discourse on the Study of Natural Philosophy* (1831).[1] The book is not such a masterpiece as to deserve detailed study, but it contains one important discovery and a few interesting passages which I shall comment on. It expounds a popular philosophy of success which will become amply clear from the present summary of its presentation and ideas and which will be discussed in the concluding section. It expresses the atmosphere of the time in which it was written, as I shall endeavor to illustrate chiefly in my introductory and concluding sections. It influenced the literature considerably, if for no other reason than that the writings of both William Whewell and John Stuart Mill follow in its wake. Its very conception as an updated version of Bacon's *Novum Organum* is a forerunner to Whewell's *Novum Organum Renovatum*. But I shall not discuss Herschel's influence on posterity here, since this is a topic for a separate essay.

Herschel's *Preliminary Discourse* echoes, of course, d'Alembert's *Preliminary Discourse to the Encyclopedia*; indeed, Herschel's work, too, is a preliminary – to the Cabinet Encyclopedia. However, that encyclopedia falls far short, even in ambition, of its French predecessor, this quite in contrast with the fact that its preliminary (as we shall see) was very much more ambitious and pompous. Whereas d'Alembert had his encyclopedia as his main reason for writing his *Preliminary Discourse*, for Herschel the encyclopedia was only an excuse for writing his. Herschel's main reason for writing it lay elsewhere.

Herschel's work was much more successful than d'Alembert's – partly from the fact that d'Alembert's preliminary stood in the shadow of his monumental *Encyclopedia*, and partly because d'Alembert merely reaffirmed an accepted philosophy, whereas Herschel reaffirmed a philosophy whose foundations were shaking and whose tenets were thrown into serious doubt and confusion. Not only did philosophers like Whewell and Mill take it very seriously, but an original scientist like Faraday could write Herschel a fan letter telling him that it had encouraged him to pursue his scientific work. No doubt Faraday was a good friend, but no one even superficially familiar with him can doubt the sincerity of his gratitude.

It is therefore particularly intriguing to find out the core problem which Herschel and his contemporaries were facing. The choice of *prima facie* hypotheses is rather obvious. In Herschel's day the great events were: (a) the rise of electrochemistry, Davy's overthrow of Lavoisier's doctrines, and Dalton's atomism; (b) the rise of electromagnetism; (c) the overthrow of Newton's optics by Young and Fresnel. Now chemistry did not particularly engage Herschel. Since his book is rather general, he refers in it only to atomism, and rather as to a success than as to a problem. Electromagnetism was a serious problem, but conviction was growing that Ampère's electrodynamic theory was an adequate and satisfactory solution. Herschel's book was written under the conviction that Ampère's success was unshakeable; this conviction, incidentally, was shaken very soon after the publication – by Faraday's discovery of magnetoelectricity of 1831 (and it took the scientific world many years to recognize this fact). This leaves us with the revolution in optics. I shall argue throughout this report that the optical revolution proved Newton to have been not infallible and thus threw serious doubt on the widespread metaphysical belief in the existence of a foolproof scientific method. Herschel, as Whewell after him, tried to show that there exists such a method and that therefore there is no ground for the fear that Newton's mechanics will ever be superseded as Newton's optics had been. Herschel's book, in brief, comes to reinforce faith in science, in induction, and in Newtonian mechanics.

N.B. In all ensuing quotations, emphases are mine, except where otherwise noted.

1. THE PUBLIC SITUATION

Sir John Frederick William Herschel was the son of a poor Hanoverian musician, William Herschel, who settled in England as a court organist to King George III, and who became famous for his astronomical discoveries and cosmological speculations (not to mention his discovery of radiant heat). Sir John himself was a famous scientist and a fellow of a college in Cambridge where he had studied. He later became President of the Royal Society. During his first years in Cambridge he was one of the three heads of the undergraduate revolutionary club which ventured to overthrow Newton's symbolism and replace it with Leibniz'.

This dispute concerned not merely symbolism but an idea too. Newton introduced calculus as a branch of geometry. Now today we know that even Lagrange's analytical mechanics can have a geometrical representation, but in the early nineteenth century this was not known, and Lagrange's novelties could not penetrate the wall of British conservatism. It may sound strange that a club of undergraduates can make such an easy and successful revolution. The need for the *new* mathematics was very strongly felt; in *Thomson's Annals* (1815), for instance, there was a long discussion over the British backwardness in mathematics, but no one before Babbage, Herschel, and Peacock dared to create the precedent.

The most revolutionary of these three was Herschel. Babbage is now remembered chiefly for his mathematical calculation-machine, and Peacock for his life of Thomas Young. Herschel is known as the one who introduced the wave theory of light (of Young and Fresnel) into England. The new wave theory of light had already been advocated in Britain by Brewster since 1814. In 1819 Herschel followed him, being the first who brought respectability to it in England.

One might have expected Herschel to be above the vulgar identification of all error with prejudice, as Brewster was; but this was not the case. On the contrary, Herschel, the revolutionary son of the great speculative thinker, tried to return to the old view that all errors are culpable and hence that all speculation is dangerous and to be avoided.

Nowadays, one can hardly imagine what a shock the refutation of Newton's theory of light produced. Newton, the infallible, was found to be mistaken, and hence, allegedly, prejudiced; his theory of light, which was much more widely known than his theory of gravity, and which had been viewed in 1810 as doubly demonstrated beyond any shadow of doubt, was entirely deserted in 1820!

I cannot discuss all the repercussions of this shock wave. But it surely was a unique chance to dislodge the old Baconian myth and reject once and for all the identification of all error with prejudice. I quote only one passage from Captain Forman RN, who was neither a Newtonian nor a Huyghenian. Let me contrast the calm and far-seeing character of this passage from this elderly gentleman with the reaction of a young contemporary of his, Michael Faraday, a reaction dominated by a sense of disaster and by the urge to avoid wild speculations. Let me first quote the remarkable introduction to Forman's paper:

> Truth is the only legitimate object of philosophical research; and whoever believes his own opinion to be true, and fancies that he can add to the general stock of knowledge by imparting some new discovery, *has not only the right but is bound in duty to make his opinion known.* If among a number of erroneous opinions, he has afforded but one hint which in the hands of a wiser man may lead to important results, he has conferred

a real benefit on society; while *his errors*, though they may outlive his own time, *will finally be dispelled* by the light of true philosophy, to which his own hint has so materially contributed.

My design in writing this essay is to show the true cause of reflection and refraction of light; and as *Sir Isaac Newton* has already accounted for these phenomena by his hypothesis of alternate fits of easy refraction and reflection in the medium, it follows of course that *I intend to oppose his opinion* .... For this opposition however I shall offer no apology, because I am only excercising the right which he excercised before me ....

For a great many ages the history of natural philosophy was little more than a record of errors .... Here and there we meet with a transient gleam of philosophy ... which serves a traveller a few steps further on his way, and then left him to grope in the dark, perhaps for another generation .... *It is chiefly to Lord Bacon that we are indebted for the principle of making experiments the basis of philosophy* .... Before his time the *ipse dixit* of an acknowledged philosopher was sufficient to establish an opinion however absurd: but with this test, like another Hercules, he has cleansed the Augean stable of all its impurities, *and no hypothesis now can long hold its ground whose foundation is not laid on experiment* .... There can now be no danger of my propagating error, even if my opinions should be wrong, because, in that case, they will not stand the test of experiment.[2]

As this passage indicates, the refutation of Newton's theory of light provided an opportunity for passing to a more explicit mode of arguing. Forman's arguments in favor of his proposal to endorse the explicit argumentative method of presentation are very strong. In his view scientific progress is slower than was generally believed; progress is made by presenting hints and rudimentary theories, and errors need not be feared since, being refutable, they will sooner or later be eliminated.

But Forman missed his target, and, I think, for an obvious reason. He refused to declare boldly that Newton's theory of light had been refuted. And he expresses his indebtedness to Bacon in the very passage in which he only implicitly rejects and severely criticizes Bacon's philosophy. Although he declared that refutations are not to be feared, but should, on the contrary, be considered as the means of improvement, he was not ready to declare both Newton's theory of light and Bacon's doctrine of prejudice as simply refuted errors.

That the idea is hard to concede is not hard to imagine, though documentation may not be easily available. There is a remarkable record of a shocked reaction in Faraday's lecture relating to the nature of light, in a series of lectures to the City Philosophical Society, delivered in 1819 (when he was twenty-seven years old), as quoted in his *Life and Letters* by Bence-Jones:

In the constant investigation of nature pursued by curious and inquisitive man, some causes which retard his progress in no mean degree arise from the habits incurred by his exertions; and it not unfrequently happens, that the man who is the most successful in his pursuit of one branch of philosophy thereby raises up difficulties to his advancement in another . . . .

The evil of method in philosophical pursuits is indeed only apparent, and has no real existence but in the abuse. But the system-maker is unwilling to believe that his explanations are not perfect, the theorist to allow that incertitude hovers about him. Each condemns what does not agree with *his method*, and consequently each departs from nature. And unfortunately, though no one can conceive why another should presume to bound the universe and its laws by his wild and fantastic imaginations, yet each has a reason for retaining and cherishing his own . . . .

As it regards natural philosophy, these bad, but more or less inevitable, effects are perhaps best opposed by cautious but frequent generalisations . . . .

Ultimately, however, facts are the only things which we are *sure* are worthy of trust. All our theories and explanations of the laws which govern them, whether particular or general, are necessarily deduced from insufficient data. They are probably most correct when they agree with the greatest number of phenomena, and when they do not appear incompatible with each other. The test of an opinion is its agreement in association with others, and we associate most when we generalise.

Hence I should recommend the practice of generalising as a sort of parsing in philosophy. It occasions a review of single opinions, requires a distinct impression of each, and ascertains their connection and government . . . .

Matter classed into four states – solid, liquid, gaseous, and radiant – which depend upon differences in the essential properties.

Radiant state. – Purely hypothetical Distinctions.

Reasons for belief in its existence. Experimental evidence. Kinds of radiant matter admitted . . . .

Nothing is more difficult and requires more care than philosophical deduction, nor is there anything more adverse to its accuracy than fixidity of opinion. The man who is certain he is right is almost sure to be wrong, and he has the additional misfortune of inevitably remaining so. All our theories are fixed upon uncertain data, and all of them want alteration and support. Ever since the world began, opinion has changed with the progress of things; and it is something more than absurd to suppose that we have a sure claim to perfection, or that we are in possession of the highest stretch of intellect which has or can result from human thought. Why our successors should not displace us in our opinions, as well as in our persons, it is difficult to say; it ever has been so, and from analogy would be supposed to continue so; and yet, with all this practical evidence of the fallibility of our opinions, all, and none more than philosophers, are ready to assert the real truth of their opinions . . . .

The history of the opinions on the general nature of matter would afford remarkable illustrations in support of what I have said, but it does not belong to my subject to *extend upon it*. All I wish to point out is, by a reference to light, heat, electricity, &c., and the opinions formed on them, the necessity of cautious and slow decision on philosophical points, the care with which evidence ought to be admitted, and the continual guard against philosophical prejudices which should be preserved in the mind. The man who wishes to advance in knowledge should never of himself fix obstacles in the way.[3]

There is a general pattern which the responses of Forman and Faraday illustrate. When a radical measure proves to be ineffective, one may become less of a radicalist and more of a reformist; alternatively one may become more radicalist. I have argued elsewhere[4] that this general pattern is illustrated both in social and political history and in the history of ideas. Here, however, I must add that each of these two responses is intellectual in its own way; and one need not be an intellectual. Instead one can respond with a mixture of a cocksure manner, pretending that nothing has ever gone wrong as far as the principles are concerned, and a shocked manner as far as the deterioration of people's application of them is exposed.

One of those cocksure responses to the crisis in optics can be found in the works of Sir John Herschel, who introduced his low standard of discussion even to the *Philosophical Transactions of the Royal Society* (though, it is true, only in a footnote):

> The author of the article on Polarization, in the 63rd Number of the Edinburgh Review, just published [1919], *is guilty of a most unpardonable mistake*, in asserting, (p. 188), as deducible from Dr. Brewster's experiments, that the Huygenian law is *incorrect*, for carbonate of lime. Dr. Brewster's general formulae for crystals with two axes resolve themselves into the Huygenian law when the axes coincide, of which case it is only an extention. That excellent philosopher, if I understand English, in the paragraph which gave rise to this strange assertion, only means to declare his opinion that it remains undemonstrated.[5]

For the sake of completeness, or for the curious, the passage which enraged him is: "We shall show, from the experiments of Dr. Brewster, that the Huyghenian law *is not general*; that it *is not even correct* for the phenomena of calcareous spar [= carbonate of lime]; and that the explanaton which Laplace attempted . . . falls to the ground". Both Herschel and the writer whom Herschel declares unpardonably mistaken agree that Huygens' law explains certain phenomena of double refraction but *not* all of them (as it was originally intended), and that Brewster generalized it into a law which explains more phenomena. The author, then, is blamed for an "unpardonable mistake" even though he supports the right views by the right arguments. The unpardonable mistake is purely stylistic – the right argument for the right theory was put in the wrong language, not in the usual clichés. What exactly these clichés were, how exactly one had to apply them, and what one had to do when they were inapplicable or applicable but not to one's taste – these were difficult and vexing questions. Consequently, sensitive people found it ever harder to find the right words before submitting their works for publication. (Even in our own relatively very permissive age the difficulties still exist and often cause much tension between author and editor.) Postponing publication, however, could incur a different kind of censure – the impatient demand for publication (publication pressure). Let me quote Herschel again, this time a footnote published in 1845: "This memoir [of Fresnel] was read in the Institute, Oct. 7, 1816. A supplement was received Jan. 19, 1818. M. Arago's report on it was read June 4, 1821. And while every optical philosopher in Europe has been *impatiently* expecting its appearance for seven years, it lies as yet unpublished, and is known to us only by meagre notices in a periodical journal".[6]

I do not wonder that people preferred to communicate their views verbally rather than in writing. The method of writing was standardized in the inductive style which was out of date. In England two people were in the authoritative position to suggest changes of standards, the two friends Herschel and Faraday. I have discussed Faraday's activities elsewhere.[7] Herschel was a conservative *par excellence*; he demanded strict adherence to the inductive style of presentation.

In another essay[8] I have discussed the standards of publication of scientific essays as proposed by Boyle, instituted by the Royal Society, and rubber-stamped by Newton. With the refutation of Newton's optics all became fluid. Doubts about Newton, about science, and about modes of research and modes of publication all found their way to public attention. Herschel tried to handle them and restore an order of sorts in the scientific community. He did so, I propose, by publishing his views in a definitive essay about science in general.

2. A GENERAL VIEW OF SCIENCE

Herschel's *Preliminary Discourse* is composed of three parts. The first is on the nature and advantages of physics; the second is on prejudices and induction, illustrated by historical examples; the third is about the classification of physics, illustrated by further historical examples. Here is a brief summary of the work, starting with part 1, 'The Nature and Advantages of Physical Science'.

In the struggle for survival man has proved the fittest. This is due to his rationality. Man is driven to learning both by utilitarian motives and because of intellectual needs. Herschel's story of man's development of his intellectual powers sounds to modern readers a bit naive: it seems that God created man as a nineteenth-century thinker, but without nineteenth-century knowledge. I do not know how historical Herschel considered this picture; its basis is surely in the classical theory of the intellect, whose *locus classicus* is in Buffon's work and whose origins, as Paul Hazard has observed, are traced to Bacon's philosophy. Yet, it may strike one as odd that Herschel is more of a Darwinian-before-Darwin than the eminent naturalist Buffon; this may be related to the fact that the struggle for survival first occurred in the economic literature, especially in Malthus, to whom, of course, Darwin acknowledges his great debt.

Having established the value of science, Herschel now launches an attack on those who oppose science as anti-religious and on those who support science from purely utilitarian considerations. Science is opposed to the religion of Galileo's persecutors, not to religion as such. Science can be of great utility, but to say that it is merely useful is humiliating to the rationalist and the speculative philosopher. Besides, the utility value of a theory cannot be predicted. The theory is remote from everyday life at first (though it will sooner or later be connected with simple facts no less than with uncommon phenomena).

In Chapter II Herschel tells us that science is composed of abstract and empirical components. The abstract component seems, at least casually, to be nothing less than a Kantian *a priori* scheme of language, logic, and mathematics – the scheme of *possible* science. Man is rational, and he would not accept the authority of science unless he understood what is behind scientists' seemingly mad assertions. Only when one understands science, one sees the certainty of these assertions. The certainty is achieved by verification, by the occurrence of the results predicted by the theory, whether it be Newton's theory of gravity or Fresnel's wave theory of light.

This, however, was *apropos* of *abstract* science, the *a priori* framework. Next, in Chapter III, comes the law of causality, which, Herschel says, is "the first thing [which] impresses us from our earliest infancy". This statement seems puzzling not so much because one may doubt it as because Herschel does not need it at all: he deliberately admits the law of causality to be *a priori* valid, and so he need not claim it to be provable by observations, from infancy or otherwise. The law, or any specific causal law, is not written by God in a way in which human legislators write laws; it is not isolated but a part of a system, a part of the whole truth about matter. Thus God comes *ex machina* to help Herschel in his traditional Cartesian role of the defender of scientific truth.

Atomism, for example, incorporates, in its modern form, various laws. But "the ascent to the origin of things", the finding of such laws, must not be through the act of speculating about the nature of things, but the empirical discovery of these particular laws (p. 29). Herschel has by now forgotten that he had previously (p. 12) spoken favorably of the philosopher as the person who speculates on everything. He now claims that the business of the natural philosopher is decidedly not to speculate but humbly to search for particular causes. The ancients, who lived in the infancy of mankind, had good reasons to study the problem of whether or not there exist fixed laws of the universe;

> but to us, who have the experience of some additional thousands of years, the question of permanence is already, in great measure, decided in the affirmative. The refined speculations of modern astronomy . . . have proved to demonstration, that . . . the force of gravitation . . . has undergone no change in intensity from a high antiquity (p. 30).

This dual allusion to Laplace's inductive investigations and to his own father's speculations makes it very hard to decide whether to Herschel causality is *a priori* or *inductive*, and whether speculations are permissible or forbidden. Be it as it may, Herschel expresses explicitly the view that *prior* to science there exists a twofold framework. The one side of it is "abstract science" and it is composed of language, logic, and mathematics. Herschel's abstract science corresponds roughly of Kant's *analytic a priori* knowledge. The other part, quite different from abstract science, includes causality, the immutability of natural laws, and the indestructibility of matter. This corresponds to a very substantial part of Kant's *synthetic a priori*, provided that the seeming inconsistencies I have alluded to be removed in one way or another. (Even if causality and such were considered *a posteriori* in the *epistemological* sense, no interpreter would deny that for Herschel they have *a priori* status, at least from the *methodological* point of view: they are prerequisites for induction. Indeed, this very distinction may offer the clue to the resolution of the seeming inconsistencies.) I do not wish, however, to go further into this topic, or even to suggest that Herschel, who had but little epistemological interest, had ever read Kant, whose works after all, were rather unknown in England before the efforts of De Quincy to popularize them. On the whole, Herschel's view is much too empiricist to be considered even mildly Kantian; yet *apriorist* arguments are contained in it all the same. This seems to me to represent a general trend which historians of philosophy unjustly overlook, and which I have discussed at length elsewhere.[9] Let me only add here that Herschel studied Dr. Isaac Watts's once very popular *Logic* (1724), where the distinction is drawn between essences which are grasped *a priori* (Platonic ideas) and ones which are arrived at by induction (Aristotelian essences proper).

From the *a priori* we come to the *a posteriori*; Part II deals with the empirical aspects of science.

3. THE DOCTRINE OF PREJUDICE

Herschel is one of the last exponents of the orthodox Baconian doctrine of prejudice. In line with Bacon, Watts, and many other thinkers, he sees the doctrine of induction and the doctrine of prejudice as merely two sides of the same coin. After Whewell's defense of the role of the imagination in science, a change took place in the inductive tradition. But for Herschel, still, the theory of induction is the same as, or the contrapositive of, the theory of prejudice. The ultimate source of knowledge, he says, is experience: observation and experiments. From this, he concludes, the doctrine of prejudice follows:

> Experience once recognized as the fountain of all our knowledge of nature, *it follows* that, in the study of nature and its laws, we ought *at once* to make up our minds to *dismiss* as *idle prejudices*, or at least *suspend as premature, any preconceived notion* of what might or what ought to be . . . the case, and content ourselves with observing, as plain matter of fact what *is*. (p. 60)

But what should we observe first? Of course preconceived notions are premature and might be refuted by fact; but why dismiss them? Merely because they may be false! The science which Herschel advocates is infallible! It is infallible, however, only if we make "one preliminary step", which is "the absolute dismissal and clearing the mind of all prejudices, and the determination to stand and fall by the results of a direct appeal to facts in the first instance, and of strict logical deduction from them afterwards" (pp. 60–61).

This is absurd. If we dismiss all prejudices what remains "to stand and fall by the results of a direct appeal to facts in the first instance"? And how, for example, can atomism, to which Herschel had already committed himself, be obtained by "strict logical deduction" from facts? But Herschel proceeds undisturbed. In accord with Bacon and Watts, he now classifies prejudices into two groups: prejudices of opinion and prejudices of the senses. He also defines them echoing Watts. At least five of his six examples were endorsed by Bacon as truths. I do not know whether he did this on purpose; he may have merely tried to bring examples of unambiguously refuted theories, for which it was safer to go to the distant past. And of course, Herschel stresses, a refuted theory never was a part of science; it was always a prejudice.

Herschel discusses the method of avoiding prejudices of opinion, which shows again the Wattsian origin of his view:

> To combat and destroy such prejudices we may proceed in two ways, either by demonstrating the falsehood of the facts alleged in their support, or by showing how the appearances, which seem to countenance them, are more satisfactorily accounted for without their admission (p. 61).

Again we see that we may have to refer to a theory in order to show that it is false and thus a prejudice. But then those who had held it prior to its refutation may have been in simple and honest error; why call it, then, a prejudice? Herschel continues the above quoted passage, as if to answer my question:

> But it is unfortunately the nature of prejudice of opinion to adhere, in certain degree, in every mind, and to some . . . [even] after all grounds for their reasonable entertainment is destroyed. Against such a disposition the student of natural science must contend with all his power. Not that we are so unreasonable as to demand from him an instant and peremptory dismission of all his former opinion and judgment; all we require is, that he will hold it without bigotry, retain till he sees reason to question them, and be ready to resign them when fairly proved untenable, and to doubt them when the weight of probability is shown to lie against them. If he refuses this, he is incapable of science (pp. 61–62).

It is strange that Herschel suggests that one should try to refute a prejudice, and then he admits that of necessity the refutation "to a certain degree" fails to dissuade: either the "certain degree" is marginal and one may ignore it, or it is substantial and then perhaps investing efforts in refutations is not productive enough to be commendable. It seems obvious to me, but I shall not insist on the point, that the ambiguity here reflects a real ambivalence: on the one hand Herschel adumbrates the Baconian doctrine according to which it is futile to try to make people change their minds by criticizing them, and, on the other hand, Herschel, only one decade before, when he was young and without authority, had succeeded in convincing the whole of his English public that the infallible Newton, the greatest scientist on earth, had been wrong. He never expressed a word of appreciation for such a "candid and sincere" public (to use Boyle's idiom). Nowhere did he mention that the success of science of which he is so rightly proud is at least partly due to the good will and curiosity of the general public. Instead, one decade after his victory, he tells his public *ex cathedra* that no matter how strongly and sincerely they discard their past views, they are still prejudiced "to a certain extent" somewhere in the corners of their minds.

What, then, can we do if our natural tendency is to be prejudiced? The student of natural science must fight this tendency with all his power, Herschel says, echoing the puritanic Watts again. "Not that we are so unreasonable to demand of him an instant and peremptory dismission of all his former opinions and judgments", Herschel adds by way of qualifying Watts's doctrine of empty-mindedness. "All we require" from our student, Herschel says in all modesty, is non-dogmatism; let him by all means "retain" his past judgments "till he sees reason to question them and be ready to resign them when fairly proved untenable and to doubt them" in proportion, as Watts had demanded.

By now we have before us an admixture of Bacon's, Boyle's, and Watts's views together within some new adjustments – which is not easy to put together cogently. Now we come to the method of eliminating prejudices of the senses. This elimination in no way advocates any degree of mistrust of the senses; rather, it aims at destroying unconscious judgments by appealing to the strict use of the senses themselves. A prejudice of the senses can only be destroyed by the highest court of appeal, namely, the senses themselves, in all their purity. The senses may, it is true, be prejudiced, or influenced by faulty judgment, by an admixture of opinion and sensation; we may, therefore, receive from the senses some contradictory evidence; and we must in such cases discard a part of the evidence. But once we realize that the error is originally that of judgment, we shall not fail to pin the fault on the sensation which is less direct and which is less purely sensational.

I need not discuss this view at length: I have discussed it and explained its historical significance elsewhere.[10] Rather, let me mention that Herschel had in mind specific instances of prejudices of the senses which intervened with scientific research; I will illustrate with an example. In his *The Decline of Science In England* (1830) Charles Babbage concludes with a comparative sketch of Davy and Wollaston. He refers to Wollaston's refined senses and the allegation that these made him a good scientist, and dismisses this allegation with disdain. To reinforce his view he tells the story of how Herschel had showed him solar absorption spectra: I shall put the instrument before you so that you will be able to see them, but, Herschel predicted, you will not see them from not knowing how to see them; then I shall tell you how to look at them and you will be unable to comprehend how they had eluded you before. And, Babbage adds with satisfaction, Herschel's prediction came true.

Having described some sense illusions as examples of prejudices of the senses, Herschel concludes (p. 63, italics in the original) that the "*sensible impressions* made by the external objects on us" depend on the circumstances and therefore we must be careful; we must lay confidence in our observations only in a limited degree at first, correcting them appropriately before being certain of them. He does not tell us how to do all this, but in this respect he is no worse than Bacon and Watts. Next comes his theory of observation. As we do not sense with our mind, he says, all sensations are in fact signals, the mechanism of which is inexplicable to us. This, it seems to me, is, in a humble way, a statement of the problem of observation: all signals, observations included, need *interpretation* before they convey a message; how do we interpret them? Herschel's solution is stated as briefly and psychologistically as the problem: the interpretation is made by *association*, which means, I understand, induction by simple enumeration: "we can only regard sensible impressions as signals conveyed . . . to our minds . . . which receives and reviews them, and by *habit and association*, connects them with corresponding qualities . . . of objects; just as a person writing down and comparing the signals of a telegraph might interpret their meaning" (p. 64).

I shall not discuss the psychology involved. It is hypothetical and thus in need of testing; it was actually tested, and it was promptly refuted. But as Herschel had no access to this refutation we should overlook it here; Herschel's linguistic simile, however, is rather unfortunate, since in his own view language is *a priori* valid, and so should be, then, the language of the signals, i.e., the rule of association. But, the signals we receive from our senses, he also says, should convey messages which do not in any way depend on our theories!

Immediately following the simile, Herschel brings in an example:

> As, for instance, if he had *constantly observed* that the exhibition of a certain signal was *sure to be* followed next day by the announcement of the arrival of a ship at Portsmouth, he would connect the two facts by a link of the very same nature with that which connects the notion of a large wooden building, filled with sailors, with the impression of her outline on the retina of a spectator on the beach (p. 64).

This is more puzzling than helpful: can one *observe* that a signal is "*sure to be* followed" by another signal? Why is it so certain that if two signals are always consecutive one would connect them in the same way as an impression of an outline of a ship is connected by an observer with the notion of a ship? Is the notion of a ship identical with "the notion of a large wooden building, filled with sailors"?

I have an answer only to the last question. It is perhaps not very satisfactory, but it is the only one I found: in Watts's *Logic* there is an arbitrary example of a definition: "A ship may be defined as a large hollow building made to pass over the sea with sails" (Pt. I, Chap. VI, Sec. 6). I admit that Watts's "hollow building" is different from Herschel's "wooden building" just as his "sails" differ from Herschel's "sailors"; but the latter is probably a misprint and the former is an insignificant variation. Moreover, as Watts is not dogmatic about his definition, and presents it only as a possible example, Herschel may have taken the liberty to deviate a little.

What matters, however, is not Herschel's problem but rather his claim that we can actually solve it with *certainty*. Hersehel ends the chapter with the beautiful image of a guide who knew, by the sight of the flight of condors soaring in circles miles away, that below them lay a carcass over which stood a lion.

4. INDUCTION

From the doctrine of prejudices of thought and of sensation Herschel goes over to discuss "the analysis of phenomena," the creation of experimental or observational set-ups in which complicated phenomena appear to have a simplicity, with fewer ingredients than in everyday life, so that *causes* of fact can be found with their aid:

> But, it will be asked, how we are to proceed to analyse a composite phenomenon into simpler ones, and whether any general rule can be given for this important process? We answer, None. . . . Such rules, could they be discovered, would include the whole of natural science. . . . However, . . . the analysis of phenomena . . . is useful . . . as it enables us to recognize, and mark for special investigation, those which appear to us simple; to set *methodically* about determining their laws, and thus to facilitate the work of *raising up general axioms* . . . which shall, as it were, transplant them out of the external into the intellectual world, and enable us to reason them out *a priori*. (p. 73).

This, he continues, enables us, by "reasoning back from generals to particulars," to devise more simple set-ups, and come to more discoveries. While the preparation of the original simple experiment is unaided methodologically, the *a priori* reasoning is; the method is Bacon's "true ladder of axioms." Any abstract theory can have a *class interpretation*, and when we find a similarity between two laws *we search for the class to which this similarity applies*, thus arriving at a higher level law.

This is the whole of Herschel's methodology, a version of refined empiricism; it can be briefly summed up thus. We give up all prejudices and start afresh. We see a complicated world. Somehow, unaided by any method, we isolate a set-up *A* which leads to a result *B* invariably, i.e., repeatedly. This we generalize into a law: all *A* is *B*. We meditate over many such laws and by a method of similarity we deduce new laws from these, of a higher level of abstraction. To put it more sharply, Herschel has three stages of induction: (1) fact finding (and temporary classification), (2) finding immediate causes, and (3) finding the higher causes until we find the very first cause. Which of these stages is induction or the "analysis of the phenomena" in Herschel's terms? Induction, it seems, is alleged to be the *finding of causes*; hence, it belongs to, if it is not identical with, part (2), which raises the problem: why does Herschel present induction before (1)? The answer is imposed on us by the logic of the situation: only those facts the causes of which we already know (or at least suspect) come under (1) [if (1) will include all observed phenomena it will be a mad collection]; so (1) is in fact only a Baconian retrospective myth. Herschel's claim is in fact this: we start with (2), with causes, and *then* pretend to have started with (1), with facts. And, when discussing (2), the causes, Herschel does offer, to use Bacon's term, some "aids to the intellect": he recommends the employment of what he calls "proximate causes" and others customarily call "hypotheses." But before coming to this, we should stop to notice Herschel's explanation to the reader of why he must accept induction. Induction, Herschel says, is the basis of science, as we can see from the "fact" that before Bacon there was no science.

5. A EULOGY ON BACON

It is to our immortal countryman Bacon that we owe the broad principle and the fertile principle; and the development of the idea, that the whole of natural philosophy consists entirely of a series of inductive generalizations, commencing with the most circumstantially stated particulars, and carried up to universal laws, or axioms, etc., etc. (p. 78)

So opens Chapter III of Herschel's *Preliminary Discourse*. It is the only passage in the book that was quoted repeatedly and highly approvingly by a variety of authors in the nineteenth, and even twentieth, century. It was even quoted in Whewell's *Novum Organum Renovatum*, Whewell's sharp criticism of Bacon's notwithstanding. This shows us that one only has to make a broad and unqualified statement in favor of a thinker in order to be quoted by all the adherents of this thinker. Of course, the merits of Bacon are not those which Herschel enumerates: the ladder of axioms is credited by Bacon to Plato, and the theory of circumstantial descriptions is Boyle's not Bacon's, and inductive generalizations are Aristotle's and Newton's (Bacon viewed them as puerile). Not one of those who quote this passage says where in his writings did Bacon ever mention, or allude to, circumstantial description. It is more a matter of rhetoric than of serious history of ideas.

Herschel's proof of the fertility of Bacon's teaching, which is the thesis of Chapter III with which the above quoted passage opens, is very surprising indeed: "Previous to the publication of the *Novum Organum* of Bacon [1620], natural philosophy, in any legitimate and extensive sense of the word, could hardly be said to exist" (p. 79). The Greek philosophers were great abstract thinkers and excellent arguers, (see *Novum Organum*, I, Aph. 71 and 79), but as empirical scientists they were defective in the manner described by Bacon: they were prejudiced. Yet, even Herschel had to admit that at least Aristotle had displayed some knowledge of nature: he had a system of physics, and he happened to make some observations. But his physics was prejudiced by his biological approach and led to dogmatism (see *Novum Organum*, I, Aph. 77). Archimedes is mentioned only in other parts of Herschel's book; in the discussion about Greek science he is conspicuously absent. In another passage (p. 72) he states that Archimedes was "too late," which is an echo of another ploy of Bacon's. Next comes the routine mockery of scholasticism and alchemy, with Arabic science not even mentioned (see *Novum Organum*, I, Aph., 71). True, prior to Bacon there were Roger Bacon, Paracelsus, Agricola, and Gilbert; but these were only rudiments of the beginning. The real beginning was Copernicus, Kepler, and Galileo. But even they can be belittled, and Herschel does belittle them – not to spite them, I am sure, but in order to show that science begins with induction by showing that it begins with Bacon:

> By the discoveries of Copernicus, Kepler, and Galileo, the errors of the Aristotelian philosophy were effectually overturned on a *plain appeal to the facts of nature*; but it *remained to show on broad and general principles*, how and why Aristotle was wrong; to set in evidence the particular weakness of his method of philosophizing, and to substitute in its place a stronger and better [one] . . . . (p. 85)

If we turn back to Herschel's discussion of Aristotle (p. 110) we find in a footnote that it was Galileo who "exposes unsparingly the Aristotelian style of reasoning"; he gives there an example of an Aristotelian "string of nonsense" and, ironically enough, this same example of Aristotelianism can be found in Bacon's *Novum Organum* (Book II, Aph. 48).

There is little doubt that Galileo did more than expose details of Aristotle's errors. But we should not go into needling particulars. Herschel sticks to generality and ignores Galileo's methodology as most of his contemporaries do. He wants to praise Bacon for his invention of modern inductivism and here he is on safe ground, except that his thesis (science = induction) is not proven by historical evidence, but rather is imposed upon history; accepting induction, he is determined to praise its inventor:

This important task was executed by Francis Bacon, Lord Verulam, who will, therefore, justly be looked upon in all future ages as the great reformer of philosophy, though his own actual contributions to the stock of physical truths were small, which were the fault rather of the general want of physical information of the age than of any narrowness of view on his own part; and of this he was fully aware. (p. 86)

Here Herschel speculates about Bacon's place "in all future ages" and about Bacon's own state of mind. This is sheer hero-worship. Was Bacon so unsuccessful because of the lack of information? Robert Leslie Ellis states in his general introduction to Bacon's *Works* that it was the insincerity of Bacon which stood in the way of his scientific achievement. Charles Singer argues in the *Encyclopedia Britannica* (Art. "Bacon") that empirical refutations of many of Bacon's views as expressed in the *Novum Organum* were known prior to the publication of that work.

An immense impulse was *now* given to science, Herschel's eulogy continues, and everybody "rushed eagerly" to observe "matters of fact"; art and even nature herself helped by kindly supplying the telescope and microscope, and by displaying two rare astronomical phenomena (the two novas). The word "now" makes sense only if it means "after 1620"; but then the claim is to be interpreted as implying that the novas appeared after 1620. To repudiate this silly implication, perhaps, Herschel gives the dates of the novas in a footnote. It is not that he is ignorant of the history of astronomy, of course; he was merely misusing the word "now." As I have discussed this misuse twice already, I shall not go into that further here. Let me simply state that Herschel did not explicitly state that the "immense impulse" to observe and experiment was *caused by* Bacon; he only mentions it, opening with "now," and ending with the phrase "lifetime of Galileo." But all this eulogy to the Renaissance of science follows the eulogy to Bacon which follows the abuse of the peoole who made this Renaissance. To mislead the reader still further, Herschel goes on to judge "the immediate followers of Bacon and Galileo," namely, Boyle and his associates, who observed facts, but were still under the spell of "natural magic"; they were not sufficiently theoretically minded.

The "immense impulse" peters out into "natural magic"; both, however, relate to observation: the impulse was to observe, and the magic kept within observation and away from theory. The general impression and thrust is rather clear, though nothing else is: Bacon's star shines in a dark period; his faults are few and due to others, while others' stars are dim in comparison with his and only reflect his light. Copernicus, Kepler, Galileo, and Boyle all found some facts, partly due to his own inspiration, but he, Bacon, was the reformer of philosophy, the herald of Newton's success!

Herschel wrote his book at the time when the claim that Bacon had greatly contributed to philosophy was highly contested in some quarters. This, again, is the worst thing about Herschel's mode of arguing: he does not mention the contrary opinion which he was exorcizing. He did not mention, in particular, that both Bacon's alleged greatness and his methodology were equally challenged. Herschel seems merely to reiterate naively Bacon's view that no real science existed before Bacon, and to present the rest of the history of science as the history of inductive reasoning.

Herschel's *Preliminary Discourse* seems to be a modernized version of Bacon's *Novum Organum* – the same thesis with more modern illustration, and perhaps with some additional new ideas. In order to know what Herschel accepted from Bacon and what he did not one has to compare these two works; but I cannot say how to find out Herschel's reasons for having selected what he did. In the end of the *Novum Organum* there are the twenty-seven so-called "Prerogative instances." Herschel selected only six of them, though without telling his selection rule. Besides the *crucial instance*, his instances share with Bacon's only the labels. I shall choose one instance in order to show how apologetic Herschel could be.

Take the "travelling instances" (*Novum Organum*, II, Aph., 23), which are instances of generation and corruption, or of increase and decrease of certain qualities. I do not think that it is possible to understand Bacon's muddle unless we see that he speaks of Telesio's theory in an Aristotelian idiom. Telesio tried to explain the variety in the phenomena by reducing them to series of dualities. His hot-cold duality was perhaps the better known; but as Bacon's discussion of induction is mainly illustrated by Telesio's theory of heat, he had to use for an example of his "travelling instance" something else. He chose Telesio's theory of light, and his examples, like most of his factual optical examples, from Telesio's *Consentini De Colorum Generatione Opusculum* (Naples, 1570). In Telesio's theory of light, whiteness and light are opposites. Transparent things like glass and water can be turned white, by chopping and foaming them respectively. Telesio, and likewise Bacon, explained these two facts by making air responsible for whiteness. This explanation Herschel knew to be false. The explanation is confirmed by the fact that white paper is rendered less opaque when wetted; the assumption is that water replaces air in its pores. This confirmation is true, of course. We are now ready for Herschel's quotation from Bacon; it is one of six which he chose because he wished to praise Bacon and which I choose in order to poke fun at his uncritical apologia:

Bacon's "travelling instances" are those in which the *nature* or quality under investigation travels in degree; and thus affords an indication of cause by a gradation of intensity in effect. One of his instances is very happy, being that of "paper which is white when dry, but proves less when wet, and comes nearer to the state of transparency upon the exclusion of air and the admission of water." In reading this and many other instances in the *Novum Organum*, one would almost suppose (had it been written) that the author had taken them from Newton's *Opticks*. (p. 141)

6. AIDS FOR THE INTELLECT

After the eulogy on Bacon and the unfair attack on all the rest of Newton's predecessors there remains little for Herschel to do but to expand his remarks on induction and to give some examples. Chapter IV is on collection of facts. Scientists make many observations, measure what they can, and make, by the way, some accidental discoveries. Chapter V is on classification of this multitude of facts. The classification should not be held as a theory but merely as a temporary nomenclature. Chapter VI is on the first stage of induction, on lowest level generalizations and their verifications. Here the concept "force" is introduced (p. 108): Herschel identifies "force" with "cause" and alleges that we can have a "direct perception" of forces! But forces are only sometimes visible, not always; and so, when the cause of a phenomenon is not obvious, a crucial experiment should be made to decide between the possible causes. Moreover, observability is not an essential matter, but a matter of external conditions: usually forces are observable only in simple experimental set-ups.

How do we arrive at the idea of preparing an experiment which makes forces manifest? First of all we postulate "proximate causes," in fact hypotheses, to the effect that under certain circumstances certain results occur. The name "proximate causes" shows that Herschel considered the possibility that the result would not be quite the same as what was conceived. Yet he ignores the fact that most experiments lead to no new results.

So we suppose that by the method of "proximate causes" we have an experiment with a cause captured in it. How do we observe or find it? There are four or five necessary and sufficient clues. A cause is found (1) by the presence of the effect whenever it is present, (2) by the absence of the effect in its absence, (3) by the increase and the decrease of the effect with the increase and decrease of the cause (even in strict proportionality!), and (4) by the reversal of cause and effect. This is an improved Baconian theory of causation, which is in origin medieval and draws on Aristotle. It is most questionable.

For one thing, it prohibits all cases of delayed effect, such as phosphorescence. One may argue that cause and effect are contiguous, and delayed effects are linked by causal chains; this, however, will be a more Renaissance and less medieval theory than Herschel presents, where cause and effect are not contiguous but coincidental. Similar arguments may be easily found against strict proportionality, or even monotony, of cause and effect: already Hume used the case that increase in heat increases comfort up to an optimum and then increases discomfort. There are many equally incontestable instances in which an optimum effect is reached while cause continues to increase. Herschel even takes it for granted that all causes are strict, quite contrary to commonsense; who ever doubted that a gun is caused to fire by the finger's pressure on the trigger? – least of all the soldier who has experienced hundreds of cases in which the cause was present and the effect absent. And, need one add, all too often the effect is unfortunately present in spite of the absence of the cause; the gun fires by itself, as they say.

The above rules are followed by ten "observations" or "rules of philosophizing" (the phrase is due to Newton, who opens Book III of the *Principia* with his four famous rules of philosophizing), which, unlike the previous rules, are not general – indeed they are mere rules of thumb.

Next comes Chapter VII, on the method of verification (p. 142). It involves checking that all cases are observed, and that all empirical exceptions to the conclusion to be verified are carefully re-examined: if the exception is explicable by finding a different cause, the induction will be saved, but if there are too many exceptions "our faith in the conclusion will be proportionally shaken" (a splendid Pickwickian sense of "verification"!).

No wonder that inductivists need the doctrine of prejudice. Since they teach that all scientific theory is verified, they do indeed need extra admonition to accept refutations of allegedly verified theories. To add confusion Herschel uses – in places but not systematically – "verification" and "confirmation" as synonymous. There also exist as yet "unverified inductions," which are none other than testable but as yet untested hypotheses. We should not place confidence in them, we are admonished, until they are verified. Kepler's laws are Herschel's example. He does not mention, even in this context, a testable but refuted hypothesis: all refutations are dealt with along with the doctrine of prejudice: all refuted theories are, and always were, sheer prejudices.

7. HIGHER LEVEL GENERALIZATIONS

We have thus far observed facts, analysed them, based causal laws on them, and now we must proceed since we should not rest content with mere facts. We are now willing and ready to develop theories. How? Having been warned up to this point to be on guard against the method of speculating, we now learn the secret of theorizing – speculate. There is no other method of theorizing than that of inventing a hypothesis, and Herschel has a reason for not telling us until page 143: "*The liberty of speculation* which we possess in the domain of theory is not like the wild licence of the slave broke loose from his fetters, but rather like that of the freeman who has learned the lesson of self-restraint in the school of just subordination." Though I much disagree with the sentiment of this passage, I should only draw attention to the following. At least *prima facie* this passage contradicts most of what precedes it. It is, of course, possible to reconcile the seeming contradiction, but how to do so satisfactorily I, for one, do not know. And even if this were no serious problem, the fact remains that Herschel conceals the idea of this passage from his reader for a long time, thus treating him as a bad schoolmaster treats an unruly schoolboy.

Let us, however, observe the methods of self-restraint of the "freeman," the scientist in action. We have to know the mechanisms of the universe; regrettably, however, these are often either too big or too small. "Yet we are not to despair since we see regular and beautiful results brought about in human works," which are incredible, like printing and steam engines – even though *why* a steam engine works we shall not know for a long time. In the meantime there is a ray of light: we may be able to make *hypotheses* (p. 145).

This may invite trouble: we may have too many hypotheses on our hands; there may be two, or even more, hypotheses with which to explain the same phenomena. Now, Herschel says, "are we to be deterred from framing hypotheses and constructing theories, because we meet with such dilemmas, and find ourselves frequently beyond our depth? Undoubtedly not" (p. 147). Hypotheses, he strongly claims, are not theories. This is clearly so, since theories are certain, while hypotheses are uncertain forever. The hypothesis can be refuted in the course of research, and theories are arrived at by generalization. What Herschel means, but does not say, seems to be this: after having *tested* the hypothesis, if it is not refuted, we give it a *class interpretation* which then becomes the (causal) generalization and thus the irrefutable theory. "A well-imagined hypothesis," Herschel continues, is very helpful, though he does not tell us how a hypothesis is "well imagined." Anyway, it must lead to a theory, and if it is verified, it is of the highest importance (p. 148). Therefore, Herschel says, hypotheses must concern the *agents* of the phenomena: "These agents are *not to be arbitrarily assumed*; they must be such as we ALL have good inductive grounds to believe they do exist in nature" (p. 148).

This stress on unanimity may be legitimate, even appropriate. But one cannot help asking what is the place of a rebel like Thomas Young; and the answer seems inescapable. Science, it is true, is like the Polish Sejm; accepting a law is like voting there: every member has the right to veto. But, to be a member you must be unprejudiced. The unprejudiced bases his judgment on "good inductive grounds." The unanimity, then, is quite spurious; what matters are the good inductive grounds.

Here the whole of Chapter VII stands or falls. For, it is here that Herschel permits us to speculate for the first time, provided that we speculate with restraint, and restraint is a matter of having good inductive grounds recognizable by all who are unprejudiced. It is somewhat distressing, I confess, that such strong claims are made within a rather obscure passage concerning a very obscure entity – the agent – whose entry into the scene was unheralded.

Herschel continues: we made a hypothesis concerning the agents, and "we have next to consider the laws which regulate the actions of these primary agents" (p. 149). So now the agents have become primary. This rule, it seems to me, is quite unscientific. If we first make a hypothesis about the agent and then about its mechanism, laws, or properties, then the first hypothesis is untestable. I have no objection to such a procedure, having repeatedly claimed that the roots of scientific hypotheses are in metaphysics; but Herschel should object. It seems rather obvious to me that Herschel implicitly suggested accepting Newtonian dynamistic metaphysics as prior to any hypothesis concerning the particular mechanism or force acting under certain circumstances. Herschel demanded first that we assume the existence of an agent, a force or a fluid, and *then* assume its mode of action, its mechanism, its specific laws of force and motion; this is because the science of his day partly proceeded on such Newtonian lines, and he did not notice that far from having been verified, the first assumption concerning the agent cannot be tested as long as it is not supplemented by a hypothesis concerning the specific mode of action of that agent. If one first assumes the existence of the luminous ether or, to take Herschel's example (in slight modification), the electric fluid, and only generations later find an adequate theory of the mode of action of the ether (Young) or of the electric force (Coulomb), then only the second step is testable. But in Herschel's view not only can the mode of action be verified, but the existence of the medium or substance is usually verified first. He knew very well, but simply did not consider relevant here, that although the assumption concerning the existence of the ether as the agent of light was pre-Newtonian, only Young's assumption concerning its mode of action as being *transversal* waves convinced people to try out the ether theory again.

Be that as it may, by now we have verified the theory concerning the existence of the agent and we wish next to find its mode of action. Three ways are open to us: (1) "inductive reasoning," (2) "a *bold* hypothesis," and (3) a combination of these two: a bold hypothesis supported by much previous knowledge (p. 149). This is a novelty in the inductive literature. Until then the prevailing idea was that a mild hypothesis, or, to use Laplace's terminology, a "natural" hypothesis, is preferable to a bold one. This novelty, I think, is thanks to Young and Fresnel. Young was bold enough to present his hypothesis as a hypothesis, and Fresnel was too. It seems that these facts were too remarkable for Herschel to ignore; he had to give bold hypotheses a prominent place in science, even though only as tentative tools until verified and converted into theories proper. Here is how Herschel views the relation between theory and hypothesis:

> In estimating, however, the value of a theory, we are not to look, *in the first instance*, to the question, whether it establishes satisfactorily, or not, a particular process or mechanism; for of this after all, we can never obtain more than that indirect evidence which consists in its leading to the same result. What . . . is far more important for us to know, is whether our theory truly represents *all* the facts, and includes *all* the laws, to which observation and induction lead. A theory which did this would, no doubt, go a great way to establish any hypothesis of mechanism or structure, which might form an essential part of it: but this is very far from being the case, except in a few limited instances; and till it is so, to lay any great stress on hypotheses of the kind, except in as much as they serve a scaffold for the erection of general laws, is to "quite mistake the scaffold for the pile." Regarded in this light, hypotheses have often eminent USE: and a facility in framing them, if attended with an equal facility in laying them aside when they have *served their turn*, is one of the most valuable qualities a philosopher can possess; while, on the other hand, a bigoted adherence to them, or indeed to peculiar views of any kind, in opposition to the tenor of facts as they arise, is the bane of all philosophy. (p. 153)

The quotation comparing hypotheses to scaffoldings is very common and, I think, belongs to Goethe.[11] Notice that the chief preoccupation of the whole passage is typically classical inductive: it is that of order of procedure: which comes first, the hen or the egg?

Herschel had earlier suggested that we should make a hypothesis first about the agent and then about its mechanism: namely, from the more general to the more particular aspects of the specific explanation. Now, without any indication of a change of mind, he suggests we start with the more particular aspect, the better testable part of the theory (the more directly observable, to use the inductive idiom). The mechanical model is only the scaffold which may serve to find a theory. A theory is verifiable; a hypothesis is a refutable instrument.

By viewing a mechanical hypothesis as an instrument Herschel shows that we may arrive at a theory by a "hypothesis." The "theory" is nearer to facts (more testable) than the hypothesis or model; but it may be the hypothesis or model which helped us to build the theory: the hypothesis is only the scaffold. The real thing is the verified theory which should be viewed, and even presented, as *prior* to the hypothesis. The idea of a scaffold is already used by Bacon, who believed that Thales had arrived at his speculation by induction and afterwards removed the scaffold – the *observed facts* (*Novum Organum*, I, Aph. 125).

Herschel never says that the use of hypotheses is essential to the method of science; he does not say if it is ever part of the theory, or if, as a part of the theory, it is verifiable. Herschel is simply not clear; he gives two examples from the theories of caloric and of light, but the examples do not help much. True, one can distinguish between Fourier's facts and his theory without much difficulty, but I, for one, do not know how to distinguish between Fourier's theory and his model, especially since he persistently stood above the controversy between the caloric and the heat-as-motion schools. The caloric hypothesis was a model, to be sure; it is not clear whether or not Herschel accepted it as verified, nor how he would distinguish the caloric theory from the caloric hypothesis or model. In the passage above he says that the theory even supports the model, so that it may have been verified first. But the problem is how to present such a theory, how to relate it to the model. Lavoisier had spoken of caloric without committing himself to what it meant exactly. Only *after* having presented the theory did Lavoisier explain that caloric is the matter of heat. Perhaps Herschel approved of this.

After discussing the use of hypotheses, Herschel moves on to make an important contribution to the theory of testing hypotheses – the theory of *independent* tests, which, I think, is almost entirely his own invention – and a very important one. True, he claimed that Kepler's theory is in accordance with Newton's. But he also claimed that planets deviate from Kepler's orbits, a fact which enabled one to find independent tests of Newton's theory. His new and important idea, then, is this: a theory must be tested by other facts than those by which previous theories were tested; otherwise it is not a new theory. Now, to avoid confusion, let me note that we have nowadays two very closely connected criteria for the novelty of a theory: one is that the new theory must contradict existing theories (Popper), and the other is that a new test for it can be devised (Herschel). If a theory is empirical, i.e., testable, and if it contradicts previous theories, then it is certainly testable *independently* by *new tests* (the crucial tests). Possibly, however, we may find a theory which is testable by new facts but does not contradict any previously accepted view. If this possibility is denied, then the two criteria (Popper's and Herschel's) are coextensive. When one accepts this possibility (as Herschel does), one must conclude that Herschel's criterion is the broader of the two. The interesting fact, however, is that Herschel began a trend of postulating stringency of tests.

In this vein Herschel requires that we test a theory by "a great mass of observed facts" (p. 156). But this does not mean many repetitions. We repeat one test *many times*, making the observations more accurate (using statistical laws, p. 162) in order to compare them with more exact results of the theory. The moment we are satisfied that fact and theory accord we look for *other* facts, and this raises the problem of the criterion for *otherness* or novelty of facts. Herschel does not cope with this problem, but he does suggest that we somehow use background knowledge, or common sense, to solve it. In this he surely is right, and his claim that we use deviations from old theories as new tests (Keplerian irregularities to test Newtonianism) is strikingly new. The idea was later used by Whewell in his paper attacking Hegel's accusation of Newton as a plagiarist. It was later still used by Duhem to disprove the classical inductive theory of the ladder of axioms: Newton's theory does not rest on Kepler's, but rather modifies it. It is here, I think, that Herschel's idea led to serious and important developments; in any case, most of the methodological literature is still too glib concerning the problem of independent evidence and the difference between new evidence and a variant of old evidence.

8. THE HISTORY OF SCIENCE

Part III of Herschel's book is a history of the physical sciences. First comes a classification of the sciences. This includes a passage on light, where Newton's corpuscular hypothesis is described (p. 188). In defense of Newton's conduct Herschel says that the corpuscular hypothesis had explained all the then known phenomena, including Newton's own discoveries. He was confident that "had the properties of light remained confined to these, there would have been no occasion to have resorted to any other mode of conceiving it". This has been meanwhile refuted: Whewell has convincingly argued that Newton's theory of light had never been properly tested and confirmed; Mach has shown that Newton had overlooked the most important part of Grimaldi's discovery.

Huygens, Herschel says, had a rival hypothesis, but it seemed to be less capable of explaining diffraction. Other phenomena were discovered which could serve as a further trial of the explanatory power of these hypotheses; e.g., Grimaldi's diffraction. Moreover, Newton's rings were explained by Newton and not by Huygens. Here the history of science is presented for a while as the history of competing hypotheses: Herschel even implies that *there is a criterion of choice of a hypothesis which is not inductive*: the theory with a *higher explanatory power* is to be preferred. Although Newton's theory is false while Huygens' theory is true, our predecessors were right in adhering to the false theory because it had a higher explanatory power, where explanatory power is assumed (as usual) to be monotonic with the paucity of assumptions and the multitude of explained facts.

Herschel was historically mistaken. The explanatory power of Newton's theory of light was much poorer than it appears, because, as Whewell showed, for each new fact explained by Newton a new hypothesis was made. Herschel could have found this out had he asked himself if Newton's theory had ever been tested by *independent* tests. Instead, he merely based his argument that the explanatory power of Newton's theory was *greater* on the claim that it explained refraction, diffraction, and Newton's rings. Instead of examining whether Newton's explanations of these phenomena included new assumptions or not, Herschel was engaged in defending Newton against the charge that he had presented theories as facts. Herschel approached the problem in a roundabout manner, using Biot's hypothesis to illustrate Newton's hypothesis of fits of easy transmission and easy reflection:

> The simplest way in which the reader may conceive this hypothesis, is to regard every particle of light as a sort of little magnet revolving rapidly about its centre while it advances in its course, and thus alternatively presenting its attractive and repulsive pole, so that when it arrives at the surface of the body with its repulsive pole foremost it is repelled and reflected; and when the contrary, attracted and so enters the surface.[12] Newton, however, very cautiously avoided announcing his theory in this or any similar form, confining himself entirely to *general language*. In consequence, it has been confidently asserted by all his followers, that the doctrine . . . as laid by him, is substantially nothing more than a statement of facts. (p. 190)

So, first of all we have cleared Newton, although at the cost of an unjust smearing of all his followers and a distortion of Newton's text. And this is the lesson to draw from the mistaken belief, allegedly of Newton's followers, that the theory of fits is only a statement of fact:

Were it so, it is clear that any other theory which should offer a just account of the same phenomena must ultimately involve and coincide with that of Newton. But this . . . is not the case, and this instance ought to serve to make us extremely cautious how to employ, in stating physical laws derived from experiment, language which involves any thing in *the slightest degree theoretical*, if we would present the laws themselves in a form which *no future research shall modify or subvert*. (p. 190)

This is Bacon's great discovery: if we want our factual reports to be unassailable, we must employ a language which involves no theory whatsoever. At the period of transition from the corpuscular to the wave theory of light physicists were engaged in an interesting and difficult exercise: translating statements of general facts from the language of one theory to that of another. Duhem used the existence of this procedure (his example is, indeed, Newton's rings) as an argument against induction, claiming the general impossibility of divorcing any so-called factual statement from some theoretical language, though it could be divorced from any one theoretical language by translating it into another.

In concluding my examination of Herschel's historical examples, I should say, in fairness, that I chose the worst of his examples. Not that I think that his history of galvanism or magneto-electricity is sufficiently candid, but at least there he does not distort history so obviously as when he presents his own interpretations as hard historical facts. The major subject of his history was, I think, the revolution in optics, no matter how well he concealed this behind observations on the history of other fields of science.

9. THE PHILOSOPHY OF SUCCESS

There is a fundamental difference between eighteenth- and nineteenth-century views on induction. Eighteenth-century philosophers were enthusiasts. They collected information and made theories with the hope that soon something grand would come about, something similar to the creation of Newton's mechanics, only much more universal and significant. It was felt that the smallest contribution was welcome and important because it hastened the coming of the kingdom of Reason on earth. This was the philosophical, intellectual, moral, and political atmosphere of the age of Enlightenment.

By the 1820's the picture had changed. Lavoisier's revolution had been overthrown by Davy, and Newton's optics by Young and Fresnel. Confusion reigned. The difficulties inherent in Ampère's theory caused even despair in some quarters. There was a feeling that Rationalism was going a little too far, slightly beyond its own natural limitation. It is not true, of course, that everybody was confused, nor that there was no confusion before. But a glance at contemporary works dealing with the problems of atomism or electrodynamics or science and religion or the social organization of science (see especially Babbage's *The Decline of Science in England* [1830]) shows how much the atmosphere had changed.

But then there came a sharp turn. The nineteenth century will be known as the century of the philosophy of *success*. Science rapidly progressed, even though it had more problems, and more formidable ones. The prevailing feeling was that everything was improving fast, and that theoretical physics was approaching its ultimate stage. The eighteenth-century ideals shrank in a number of respects; in particular, due to the failures of the French Revolution, scientists qua scientists shied away from public affairs. But the old ideals seemed no longer so much ideal as reality, or the reality of the next day. The philosophy of induction was utopian since its beginning, but its features had altered. The seventeenth century was one of hope, the eighteenth, one of progressive work, of the advancement of learning and the improvement of the mind; the nineteenth century was the reaping of the dazzling harvest of success.

The philosophy of success is, of course, as old as the success of Newton's theory. It shows its first mark in the preface to the second edition of Newton's *Principia*, and it plays a very significant role in Laplace's philosophy. Still, Laplace's philosophy is more of a philosophy of improvement than of success. The shift of emphasis, one may argue, from improvement to success, is insignificant and gradual, and this indeed may be the case. Yet, in my view, the change was not so smooth.

I do not know how much Herschel's work was a part of this trend and how much he was one of its causes. I have too little evidence to support my own view that he was more of an originator than reflector of the philosophy of success. In either case, Herschel's philosophy of success deserves mention. Herschel's emphasis on success permeates his book. The idea that science is identical with scientific success, intellectual as well as material, is implicit throughout, and explicit in quite a number of places. Failure is mentioned only once: the success of science is predicated on the avoidance of possible failures, the results of working against the laws of nature. One failure which science could have helped to avoid is the disaster of an inventor of a submarine who sank with it at the very first trial (p. 45).

It is a measure of the difference between the atmosphere in which Herschel wrote his book, and the one in which these lines are written, that I find it hard to explain to my reader that Herschel knew that failure in scientific inquiry is unavoidable without sounding as if I am calling him a liar: it seems so strange to us today that a man of Herschel's stature should have felt obliged to mention only failure which science can prevent and pass silently over failure which is unavoidable. He felt obliged, I suppose, to fuse philosophy with some degree of propaganda. This is the highway to self-deception.

The propaganda is at places rather thick. Herschel's philosophy is "accustomed" to create science (p. 12); he is entirely disinterested and free from authority (p. 13); he is "deeply imbued with the best principles of sound philosophy" (p. 39); and he goes on verifying his theories and sagaciously generalizing them to the moral and physical benefit of all mankind. As a portrait this is not new in the least: it is Bacon's philosopher, with a small difference. The ideal is now the real.

In Herschel's methodology there exist no more problems than in Bacon's. He names only tasks – to observe, to find causes, to generalize – but no problems; and, like Bacon, he even offers arguments from success. The task of Herschel's philosophy of success is, however, not really to eliminate problems; rather it is to put them in their place – within the workshop – and leave no trace of them in the shop-window, where success alone is to be displayed. The greater the problems in the workshop, the greater was the display of success in the shop-window. This display was the method of concealing real issues, perhaps also of burying them, but not intentionally.

Herschel's attitude has one important consequence, probably unintended: it widens the cleavage between researcher and general public, between insider and outsider. Insiders, professional scientists or dedicated amateurs, existed throughout the history of science; in the eighteenth century, however, the pretense was that everyone was an amateur and even somewhat of a dilettante, there were allegedly no insiders, no mysteries, no esoteric teachings. The inductive style was instituted, as I have discussed elsewhere,[13] in order to encourage the amateur, and in a variety of ways. The inductive style of writing encouraged self-training in experiments, as well as the publication of even very minor discoveries; it also pushed aside the method of explicit argument and criticism as too frightening, not to say offensive, to the amateur. With the advent of nineteenth-century science, style changed with pace (Sir Humphry Davy says: I shall not start with a description of an electrochemical pile, since those who do not know it will not understand me anyhow). Thus, the chief purpose of employing the inductive style was being ignored even before Hershel arrived on the scene. More and more educated people were losing contact with natural science and were told instead that the scientist almighty is going to set everything right. The argumentative method was pushed even further out of sight and *the inductive style was now functioning not only as a method of concealing arguments, but also as a method of concealing problems*.

True, Faraday was just starting to violate the taboos of the inductive style, and towards the end of the century his openly critical style was gaining ground. For the greatest majority, however, the method of avoiding argument remained that of presenting the history of science in a doctored version as the history of inductive success, and of concealing all current problems by presenting present theories as if they were utterly unproblematic.

It was in 1860 that a paper by Newland concerning his hypothesis of the periodic table of the elements was rejected because the Royal Chemical Society "*made it a rule not to publish* papers of purely theoretical nature, since it was likely to lead to correspondence of a *controversial* character".[14]

Not only the unpublished controversial material of the period suffers from the philosophy of success, but the published material does so too. Distortion of this kind still makes us present most of nineteenth-century science as if it involved no controversy. The controversial character of Faraday's ideas has still not found adequate expression in the historical literature. Disagreements galore concerning theories of force, energy, and heat are still ignored almost regularly. Kirchhoff's claim that Balfour Stewart had no priority over himself concerning the radiation law is based on the fact that Stewart had made quite a few errors and offered an unsatisfactory proof. Rayleigh's famous counterclaim is based on the fact that Stewart was successful in his experiments. Historians accept one claim or the other and the matter is still waiting to be clarified. I had myself accepted a version of the story from a nineteenth-century source and found later how mistaken I had been.[15]

In Herschel's work I have found one interesting departure from the philosophy of success, or at least a statement in a different mood. When attacking irrationalism he mentions (pp. 6–7) that science is based on the *honesty* of the witness and on universal skepticism; truth is capable of standing up to *all* tests "and coming unchanged out of every possible form of *fair* discussion". If I am not mistaken this is the only reference Herschel makes to discussion. It seems to me quite traditional that when arguing with irrationalists, rationalists are more humble and critical – more rational – than when talking to other rationalists. This is so because critical rationalism suffices for the rejection of irrationalism. Most rationalists, however, at least feel that criticism must be supplemented with something more positive, something more than what is necessary to combat irrationalism. The negation of other views is not enough; we must show the strength of our views to those whom we wish to recruit. Bacon expressed this mood: "the greatest obstacle to the progress of the sciences, and the undertaking new tasks and provinces in the same, is found in the *despair of men and the supposition of impossibility*" (*Novum Organum*, I, Aph. 92). And therefore we must teach our pupils that science is demonstrable. Unfortunately or otherwise, the claim for demonstrability is baseless, and so it leads to despair, or at least to much dissatisfaction. Nevertheless, it is clear that Bacon had hit on a significant problem. Permanent ill-success *may* lead to despair, to barren skepticism, and to cynicism. But great success may even be worse, since it *may* spoil. The only solid hope there is, is that people work because they are *curious*, because they are ready to *try*, no matter what the outcome may be. We must base rationalism on other foundations than past success or even the hope of entering the promised land of success.

In conclusion, I may state that, in view of the great fame and wide influence of Herschel, I was somewhat puzzled, upon reading his *Preliminary Discourse*, to find so much in it which is so very uncritical. The explanation of this puzzle is possibly that Herschel was arguing against the surge of skepticism – from a position of strength[16] – claiming that the new skepticism may spoil all past achievement. His philosophy of success was to put down the threat from skepticism, accounting for its presentation in a mood rather than a statement.

Herschel was certainly not the first philosopher of success. Bacon's philosophy of hope is in a way its early predecessor, and thinkers from Mersenne to Laplace spoke increasingly about the importance of success. With Herschel failures became entirely uninteresting. There is one positive element in this transition from hope and feigned success to real success which may merit mention – at least for historians of science who care about the spirit of one age as contrasted with the spirit of another age. During the period of hope popular error was attacked with the hope that it would be soon eliminated; during the period of success error was tolerated as a permanent feature of public ignorance to be overlooked as much as possible.[17] A fervent quarrel, like that between the phlogistonists and the anti-phlogistonists or between the anti-phlogistonists and Davy, could hardly take place in the mid- or late-nineteenth century. Success to a certain measure may make us less apprehensive and thus less intolerant. Though Herschel did not excel in toleration, his followers did. Some knights-errant of the Enlightenment tried to raise indignation concerning prejudices even during the nineteenth century; most scientists, however, felt that positive popular lecturing and positive research would do. And so, perhaps social reformists who prefer the fervent battle against prejudice to token popular lectures may point out, not without some measure of justice, that success also makes us less sensitive. It is always a mixed blessing.

NOTES

[1] Sir John F. W. Herschel, *Preliminary Discourse, etc.* (London and Philadelphia, 1831; title page to English edition says 1830, but the page with portrait contains correct date). Facsimile by Johnson Reprint Corporation, with a new introduction by M. Partridge (London, 1967). Page numbers refer to the American edition. For the English edition page numbers read thus: 6→9, 12→15, 39→52, 60→79, 61→80, 63→83, 64→84, 73→96, 78→104, 79→105, 85→113, 108→149, 141→188, 143→190, 145→194, 147→196, 149→198, 153→204, 156→208, 162→215, 188→250, 190→253.

[2] *Phil. Mag.* **55** (1820), 417 ff. Forman, thus, is the discoverer of the chief error of Baconian philosophy, its fear of error. Note also that the reference to Bacon is echoed in Herschel's often quoted eulogy on Bacon, quoted here at the opening of Section 5.

[3] H. Bence-Jones, *The Life and Letters of Faraday*, in two volumes, *1* (London, 1870), 303–311, extracts from a lecture 'On the Forms of Matter', from which the above is an extraction. The full lecture is extant in the Royal Institution. I hope someone will soon publish Faraday's early works.

[4] See my 'Methodological Individualism', *Brit. J. Sociology* **11** (1960), and my *Towards an Historiography of Science, History and Theory*, Beiheft *2* (The Hague, 1963; facsimile reprint, Wesleyan Univ. Press, 1967). See also J. W. N. Watkins, 'Epistemology and Politics', *Proc. Arist. Soc.* (London, 1957).

[5] *Phil. Trans.* (1820), 45 n.

[6] *Encyclopedia Metropolitana* (London, 1845), *4*, 533. As to Fresnel's sensitivity, see, for example, his correspondence with Young concerning priority in Peacock's life of Young.

[7] See my 'An Unpublished Paper by the Young Faraday,' *Isis* **52** (1961); also 'The Confusion between Physics and Metaphysics in Standard Histories of Science,' in *Ithaca, 1962* (Paris, 1964), reprinted in my *Science in Flux, Boston Studies in the Philosophy of Science* **28**, 1975; also *Towards an Historiography of Science, op. cit.* (note 4), Section 5 and notes; also my *Faraday as a Natural Philosopher*, Chicago University Press, 1971.

[8] 'Who discovered Boyle's Law?', *Studies in the History and Philosophy of Science* **81** (1977), 189–250; also my unpublished doctoral dissertation, *The Function of Interpretations in Physics* (University of London, 1956).

[9] See my 'Unity and Diversity in Science', in R. S. Cohen and M. Wartofsky, eds., *Boston Studies in the Philosophy of Science* **4** (1969); reprinted in my *Science in Flux.*

[10] See my *Towards an Historiography of Science, op. cit.* (note 4), Sections 2–5 and my 'Sensationalism', *Mind* **75** (1966), 1 ff.; reprinted in my *Science in Flux.*

[11] See Goethe, *Gedenkausgabe der Werke, Briefe, und Gesprache* (Zurich, 1949), *9*, 653, para. 1222. See also my 'Unity and Diversity in Science', *op. cit.* (note 9), 376 or 457.

[12] This, by the way, was Biot's attempt to reconcile Newton's theory with the facts of polarization and double refraction, which, he showed, could be explained by assuming

that the spin of the photon (to use the modern idiom) assumes only definite discrete values. But this assumption, now employed in quantum theory, was then rightly rejected as too arbitrary.

[13] See note 8 above.

[14] J. A. R. Newland, *The Periodic Law* (London, 1884), 23; quoted by E. T. Whittaker, *A History of the Theories of Aether and Electricity, 2* (New York, 1960), 11.

[15] See my 'The Kirchhoff-Planck Radiation Law', *Science* (April 17, 1967). The part on Balfour Stewart (p. 32) is based on A. Cotton's excellent but historically inaccurate paper in *Astrophys. J.* 9 (1899), 237. See my note 20 there. More details are in my *Radiation Theory* which may one day find a publisher despite its unorthodox stance.

[16] See W. W. Bartley, 'Approaches to Science and Skepticism', *The Philosophical Forum* **1** (1969).

[17] See L. Pearce Williams, 'The Politics of Science in the French Revolution' in Marshall Clagett, ed., *Critical Problems in the History of Science, Proceedings of the Institute for the History of Science at the University of Wisconsin, September 1–11, 1957* (Madison, 1959).

Of course, the nineteenth century had more success in popular adult education than its predecessors. Boyle's *Seraphick Love*, and Spinoza's and Locke's equivalents, were all meant for the intellectual. Isaac Watts's *Logic* and *Improvement of the Mind* were much wider in influence, but nothing like Sam Smile's *Self-Help*. Yet the combination of the industrial revolution and the rise of socialism and allied radicalist movements was the crucial factor. Professors' lectures and the likes were but the trimmings. Doubtless, the rise of the British Association was an expression of the rise of the new class of technologists and experts, but it too was not as important as the movement of literacy classes for workers run by radicalists – as described, e.g., by C. P. Snow in his *The Two Cultures and The Scientific Revolution*, which the introduction above studies in detail.

CHAPTER 28

WHAT MAKES FOR A SCIENTIFIC GOLDEN AGE?

The expression "golden age", originally signifying a mythical period in history when our ancestors lived in plenty and harmony and bliss, has become a fairly technical term in the hands of historians, especially of culture, to designate not so much peace and plenty as cultural flourish. The Golden Age of Moslem Spain, as well as of Christian Spain, are the paradigms, but the age of Pericles, the high Renaissance in Italy in the sixteenth century and in the Low Countries in the seventeenth century, primarily in the works of historians of the plastic arts, are not less celebrated. What makes for a golden age?

The question is pressing on the generally accepted hypotheses that talent is not concentrated in some space-time region, in some parts of history. Talent is thus assumed to have been on the average no more nor less present in Periclean Athens than before or after, or east or west of it. Why, then, was such a brilliant cascade of a display of talent just there? It seems, then, that some conditions are very conducive for talent to develop and flourish, others not. And it is, obviously, of a great practical importance to pin-point these conditions: if we knew what made Athens great in ancient times, then we may try to make it great again now! Is there a recipe for a cultural flourish?

1. THE TRILEMMA: BACON VERSUS WHEWELL VERSUS POPPER

Can we induce the growth of culture in general, or of science in particular? In the first quarter of the seventeenth century Sir Francis Bacon said, yes, very much so. And he said how: throwing away prejudices and superstitions and paying heed to the smallest facts of nature and patiently collecting and classifying small facts will bring back the golden age of science, will revive ancient glory.

And the golden age of science came back. The Royal Society of London was founded in 1660 and its patron saint was Sir Francis Bacon and its rules and regulations reflected his teaching to a large extent. And then came Newton, who was president of that society, and the age of Enlightenment, and all that. In the eighteenth century, tells us Paul Hazard, the great modern historian of culture, the only name that could be joined to the name of Newton without blasphemy was that of Bacon. Many historians of science view the florishing of science as the result of the adherence to Bacon's teaching. The periods of smaller success in science, such as the age of phlogistonism, they saw as the result of deviations from the Baconian norm.

The story becomes, then, very much like the story in the ancient *Book of Judges*: When the Children of Israel did right in the eyes of the Lord, they sat securely each under his vine-tree and under his fig tree; but when they did evil in the eye of the Lord he sent them enemies to trouble them. The trouble with the *Book of Judges* is that it is circular: the very fact that people enjoy peace is its evidence that they were righteous and that they suffer that they were sinful. Jeremiah and Job presented evidence to the contrary. Otherwise, if evidence, independent evidence, were to support the *Book of Judges*, we might think differently. Indeed, some religious leaders still challenge us to make the experiment. I am all for that experiment, except that I first throw the religious leaders the counter-challenge of describing exactly what should count as the performance of the experiment, and what outcome it may have should count as a refutation of the view of the *Book of Judges*. For, it is my considered opinion that the view of the *Book of Judges* that justice prevails was already refuted and in one of the most interesting experiments in world history. And the children of Israel had hoped to achieve peace through religion and social justice, and they failed.

Similarly Bacon's view. It was applied to history. It seems that history has vindicated Bacon: he came, he announced that following his method people can recover the golden age of science and more, they did it, and the results were stupendous! Is this no proof? Perhaps. When I read in the *Book of Deuteronomy* that the Lord is going to scatter the chosen people to the four corners of the earth and then return them to their land I have the same suspicion. Scholars, however, are usually of the opinion that the suspicion is silly. A prediction that came true was usually written after the fact, as in *Isaiah*, Book II, so-called; or else, the prediction's success is sheer accident; otherwise it might be a self-fulfilling prophecy, an Oedipus effect. Take Jules Verne's forecasts. Scholars have this excuse for the success, the stupendous success, of his prophecies: they speak of reasonable expectations and of the demand caused by arousing expectations. Both these things may, perhaps, be said of Bacon's utopian dream too. How can we judge matters at all?

We can take seriously what Bacon has and science fiction and other fantasy do not have: a mechanism of the process described. As I have said, I challenge religious leaders to specify the experiment so that we can test the hypothesis that doing this or that brings peace. Science fiction and religious leaders alike refuse to specify. Bacon did, and by the description of the mechanisms of success and of failure.

Bacon repeatedly fell back on one question: how come the golden age of ancient science was replaced by mediaeval stagnation? To this question there was a standard answer, and one that happens to be central in almost all great cultures, and even in some primitive cultures: antiquity was a heroic age (hero means literally, ancient, Gilbert Murray reminds us), and later ages cannot possibly aspire to a higher grandeur than earlier ones. This myth, the myth of decay, was not universally accepted. In accord with Claude Lévi-Strauss's theory of duality of myths, the myth of decay is coupled with its opposite, the myth of the eternal return. And nowhere as in mediaeval Europe was the polarization stronger, the myth of decay was felt more painfully and so was the strong yearning and eager awaiting of the return, the second coming, the coming, the rebirth, the return of the Golden Age of Antiquity. But Bacon took a step further, though he was not the first, perhaps: he said, the myth of decay is false, since though the Greeks anteceded the Egyptians, they still were greater than the Egyptians.

What, then, led from the Golden Age of Antiquity to the mediaeval stagnation? In one word, Bacon's answer is, hypotheses. One makes a hypothesis, teaches it, makes it his school's dogma, and spreads both dogmatism and scholastic disputations. This, for example, is why Bacon was so harsh on Copernicus: Copernicus was creating a new school and he may succeed and replace Aristotele. What of it? Instead of the first Aristotelian Middle Ages we are now entering the second, Copernican Middle Ages. This is not good enough, he said. Great as Copernicus might have been, and at times Bacon was willing to concede the possibility, at times not, surely one Copernicus a golden age does not make!

Strangely, there is little disagreement even today with Bacon about dogmatism and scholastic disquisitions. The little that exists, very important and interesting as it is, is almost wholly due to Pierre Duhem, the creator of the field of the history of mediaeval science. I will leave him for now, since he denied the very existence of a golden age of science.

Supposing we have removed the impediments to the growth of science. Is then the growth of science secure? Yes, said Bacon. The natural disposition of humans to learn suffices. The trouble is that one lazily and impatiently makes hypotheses instead of doing the natural thing, of looking for facts of nature and collecting them patiently and refusing to make up one's mind prematurely.

This formula was endorsed by the Royal Society of London and was reflected in their book of rules. The culmination of the Baconian method was indeed, in Newton's achievement. *Hypotheses non fingo*, he said: I do not feign hypotheses.

This is how it came about that Newton was judged infallible: he had a method, the Baconian method, that forbids the announcing of any theory prior to its proof. Newton proved that man can, by his own powers, reach the divine, he was infallible.

This is also how Baconianism was refuted: when Newtonian optics was refuted it became clear that he had feigned hypotheses. Of course, just as so many people gloss over the unpleasant details of the experiment described in the *Book of Judges* so as to deny the refutation of its view, so did many nineteenth century thinkers. In particular, Sir John Herschel declared that Newton's optical theory was not really Newton's last word, but a kind of dogmatic misreading or over-enthusiastic reading of Newton's texts by careless and devoted disciples, meaning Laplace himself, no less! And Sir David Brewster declared Newton a follower of Thomas Young! And the two authors were sufficiently powerful and influencial and important and they nearly made it. And, to some extent they have: most historians of science to this day are still Baconian, and Newtonian to boot. But something else happened then, namely William Whewell developed a new philosophy.

Whewell was a Kantian in that he thought there can be no mind empty of all expectation, and he was a Baconian in that he thought theoretical science was truly empirical to a large extent. He synthesized two great philosophies into an original idea: the unprejudiced mind can produce new hypotheses and verify them. But there is no guarantee that new hypotheses be created, and much less a guarantee that they be verified: we must try and repeat trying. Hence, the growth of science is made possible with the removal of impediments, but this is no assurance of progress. Because Whewell denied the assurance of success his philosophy was criticised by John Stuart Mill and his successors. Only after the overconfidence of the Newtonians was shaken again by Einstein, was there a return to Whewell. And then even a more pessimistic view developed, by Sir Karl Popper. Not only is there no guarantee of success, he says: there is no knowing that the impediments are removed: we never know that we are free of all prejudice, or that there are no other impediments to the growth of knowledge.

All three views, each in its own way, seem to conflict with the historical facts, with the ups-and-downs, with the presence of blossoming space-time regions near arid ones. If we ever can explain this blossoming, then the explanation will, *ipso facto*, become a recipe for the making of a golden age.

Hence, either these are not at all explicable, and with it the little sense we hope to see in history becomes miraculous, or else they are explicable and the explanation will be a Baconian recipe, what Popper calls a science-making sausage-machine. The idea that there is a sausage-machine may appeal to some, but it means the end of a challenge and the end to creativity. And whatever is said here of the sciences may be said, with little alteration, of the arts as well! What, then, shall we assume? A mechanical view of culture (Bacon), a view of it as a near-miracle (Whewell), or decide that it is all inexplicable (Popper)?

When all options available are unacceptable, we may search the mistake they share. Indeed, all these three views share at least three mistakes. *First*, the idea that once we know what causes a golden age we can implement it. The conditions may be known but beyond our power of reconstruction. *Second*, all three views share the idea that science is either all sausage-making machine or not at all. In fact, some science but not all of it is fairly routine. *Third*, all three views assume that either we know what makes a golden age or we cannot explain the phenomena. Strangely, even Popper repeatedly makes this mistake which amounts to confusing explanation with true explanation. I will discuss the importance of this point later on, and show why this enables us to hope to create golden ages without thereby making all science a mere routine job. Let me begin, however, with the first mistake: let me discuss the question, how reproducible are conditions propitious for the growth of knowledge.

2. THE HISTORICAL DIMENSION

I wish first to discuss the possibility that a known factor causes a great spurt of a cultural activity, yet with no possibility of a repeat. Consider as an example the possibility that what caused the outburst of cultural, artistic and intellectual, activity in the Renaissance, was the sequence of an enormous thirst for anything cultural with no idea as to how to quench it, followed by the discovery of a way out of the predicament. This means that we can hardly expect to have a repeat performance.

There is no saying whether this theory is true. We know that mediaeval thinkers did feel an enormous yearning for the return of the golden age and a profound sense of impotence. The mediaeval thinkers not only were convinced that the ancients had knowledge and happiness; they were convinced that ancient knowledge was preserved and transmitted from generation to generation. Only, they felt, the later generations lost the ability to read the meaning of the message; the open book looked like a closed book because the code to the meaning had been lost to those not worthy of it. We also assume that this led to an enormous sense of impotence that was barely escapable. And this sense of impotence was overcome in the Renaissance and thus gave way to exuberance. We can even point at the event that was the break-through: Brunelleschi's construction of the cupola of the dome of the Florence cathedral in the first half of the fifteenth century.

Let me stress that I do not think the structure is a significant work of engineering. And as a work of architecture it is inferior to Brunelleschi's best. Vasari stressed the uniqueness of the construction as a huge and complex piece of architecture and engineering, and, most particular, organization. This indicates that he really did not think much of the finished product. Indeed, it is not an impressive piece. It embodies, nonetheless, a most impressive piece of history.

This time was the aftermath of the Black Death, when the land was impoverished, the cities almost emptied and the countryside roamed by flagellists, who accentuated all the weakness and confusions of the era, masochism, fanaticism, mysticism, anti-intellectualism. One result of the decay of the period was the impossibility to finish the Florence cathedral: cathedrals were previously filled with wooden structures on which domes were mounted and only at the end of the work could the domes support themselves. But there was not enough wood for the Florence cathedral after the plague. And then Brunelleschi came and deployed an ancient Roman method!

I do not know how much the period known as the Renaissance was really a renaissance: after all, the Renaissance people never went around wearing togas, never built temples to ancient deities, etc. Yet the attempt to rebuild the glory of the ancient Roman Empire produced Machiavelli, the first obviously modern and frankly secular thinker. Moreover, the Renaissance natural philosophers discovered that there was no uniformity in the ancient world. Copernicus discovered that Aristarchus disagreed with Aristotle, Buonamici discovered that Archimedes disagreed with Aristotle, and Galileo Galilei was a student of Buonamici and a disciple of Copernicus. Moreover, Galileo Galilei's father, Vincenzo Galilei, argued against the attempt to revive Greek music, and so at least Western music broke away from the Renaissance mold of reconstructing antiquity, with the works of Vincenzo Galilei, the Gabrielis, and others. Did Galileo follow his father in anti-Renaissance iconoclasm, or did he react to his father's reaction and try to reconstruct Greek science?

I do not know. The fact is, Galileo wrote a book on floating bodies in which he reconstructed the book of Archimedes by the same name, and fairly well, of course. What I want to stress is the fact that the kind of things Galileo did, which might have been terribly important for him are barely repeatable: no one today will try to reconstruct Archimedes. We have him in full. And trying to reconstruct is hardly physics these days, merely the history of physics. It was physics for Galileo and his contemporaries.

History is unrepeatable, historical settings are not reproducible in any laboratory. Yet attempts to make sense of history were made, and repeatedly so. The most famous philosopher of history was Georg Wilhelm Friedrich Hegel, whose philosophy has excited so many people because of its historical orientation, because it views history as the history of culture, because it saw culture as religion plus art, plus science, and because it is a theory of golden ages.

Hegel tried to see the broad outline of history while ignoring as many details as possible. The claim against him made by many historians and philosophers of history is that he violated so many details. I suppose this criticism is not to the point. The point is, when you ignore as many details as possible, what do you see as the most important? You see not details, but large concentrations of details. But concentrations of which kinds of details? Hegel noticed the history of golden ages of cultures and of military glory. And he made the conjecture that the two go together. This enables us to refute him on the large scale that he chose to study.

For, why should culture and power go together? God could order the world so as to divide culture and power; the Jewish myth says, Jacob had culture and his brother Esau power. But Hegel saw no justice and no division of labor in history. God, he thought, graced those he chose to grace and for the duration they were graced, with both military victory and cultural growth. (This is known today as the Matthew effect.) There is a simple and very commonsense explanation for this: the powerful nation can purchase art and science; its capital becomes the most exciting and stimulating attractive place; its wealth enables it to build its culture further.

Parkinson of the celebrated Parkinson's law disagrees: growing empires, he said, are too busy to build palaces; these belong to decaying empires. Perhaps. But whether decaying or growing, powerful countries do purchase culture. It is a fact that Michelangelo went to Rome and Leonardo to Paris, that Galileo went to Rome again and again, that Newton went to London. Yet how come both Leonardo and Michelangelo stem from Florence? Why were Florence and Padua more important than Rome and Paris? Why did Descartes leave Paris?

Are centers of learning always the strongholds of victors? We have Pericles' Athens to conform to Hegel's theory, but we have Alexandria to conflict with it perhaps. Hegel would, and did, consider Alexandria as part and parcel of Rome. Yet Genghis Khan, the greatest victor of them all, his left no cultural legacy to speak of, and Talmudic Judaism was built in the village of Yavneh as a response to the fall of Jerusalem. Hegel was just plainly mistaken, even in large scale points in history. He had no room in his theory for Jewish culture and so he said it did not exist, and so Jews are merely mechanical. Yet we can learn something from Hegel nonetheless, I think, and it is at least the fact that a historian can look at different kinds of history, see different large scale historical events, or generally, historical trends. For example, the rise of golden ages.

To show this, I will discuss for a short while Popper's historiography of science. Popper sees in history the history of diverse cultures, of diverse cultural traditions, and especially the scientific tradition. And he notices that there is only one scientific tradition proper, the Greek tradition. Popper thinks that science was created once and then recreated as a by-product of the Renaissance, and everywhere we have science today, whether in Japan or in South America it was an outgrowth of Renaissance Italy.

Popper's theory is, first and foremost, rooted in his heterodox view of science. Whereas most philosophers claimed that science is verified, and others that science is only probable, Popper says, science is refutable. This raises the question, are all theories scientific? For example, is the Biblical theory that justice prevails scientific? No doubt, for Popper (as well as for Bacon and for Whewell; but not for Poincaré and Duhem, however) *ad hoc* rescued theories are unscientific. This includes *The Book of Judges* as well as the very paradigm of science, Newtonian mechanics as rescued by fanatic Newtonians (Hugo Dingler) from the refutations it suffered in the hands of Einstein's followers. But the unrescued Newtonian mechanics, refuted and hence false though it is, surely is scientific. Can we say the same of the theory that justice prevails on earth without the excuses made for it by the *Book of Judges*, but as presented, say in the *Book of Job*?

I wish to press this question home, because Popper gives both a demarcation of science as a set of theories, and a demarcation of science as a tradition, and these two need not coincide. Indeed, one can hardly expect them to, seeing that traditions are seldom as clear-cut as refutability is. Of course, we should therefore not insist on utter agreement and use Popper's criterion of demarcation of theories only as a touchstone for the demarcation of the tradition. For example, though the *Book of Job* is critical, it need not be in the Greek tradition; but that book of *Ecclesiastes* in the Bible is so imbued with the critical spirit that already earlier writers saw it as a combination of Hebrew and Greek traditions. For another example, long after Newton admittedly won over Descartes, there was a strong disposition to rescue Descartes; efforts were repeatedly made in attempts to rescue him. Great men of science such as Euler were Cartesians of sorts. Even Oswald Külpe, the teacher of Popper's teacher and the founder of the Würzburg school of psychology to which Popper belonged, said around the turn of the century, that Cartesianism was still alive and the contest between it and Newtonianism is not concluded as yet. Since Descartes expressed the idea, already voiced by Galileo, known as mechanism, it is very customary to view Western science as an offshoot of the mechanical philosophy. Recently a very confused and superficial historian of science, R. Dijksterhuis, his won fame for his volume, *The Mechanization of the World Picture*, in which he does violence to the facts and declares Newtonian mechanics mechanistic, thus regaining by default the view of science as mechanism. Popper, by contrast, will view Euler as a scientist par excellence because he did not follow his bent to the extent of rescuing Descartes *ad hoc* but did recognize the superiority of Newton over Descartes. But this Popperian move, enlightening as it is, and enriching our view of the tradition, much depends on the claim that refutations, even criticisms, are rare outside the scientific tradition. Thus, the rescue operation which rescues the view that justice prevails, should be characteristic of the Hebrew tradition, whereas the reluctance to rescue Cartesian mechanism, should be characteristic of the scientific tradition. The question remains, is Popper's view not too *ad hoc*?

Popper's view of the scientific tradition as Greek should explain why in other traditions criticizable and refutable theories were seldom presented, criticisms and refutations seldom offered, and seldomer accepted without further ado. There is a lot of evidence conforming to that view, such as the history of Chinese astronomy, much of which turns around the attempt of Chinese astronomers to cover up their ignorance and not be ridiculed for the shortcomings of their forecasts of eclipses, as Nathan Sivin has showed.

Popper's theory also makes use of the fact that Greek writings were never lost, and transferred from Greek sources to mediaeval Italian traditions in a roundabout way, via Syriac, Hebrew, and Arabic translations. Nor were these writings sufficient, since writings need not transfer traditions. Only when the Italian Renaissance thinkers tried to revive Greek culture, the Greek way of life, says Popper, did they also, and not knowing what they were doing, thereby revive the Greek critical tradition and thus also Greek science.

Popper thus explains why the Renaissance turned via philosophy, through art, into science. And perhaps this is true.

Perhaps not. We know too little about the earliest phase of the Renaissance. No doubt, the high Italian Renaissance was Platonist or neo-Platonist; the leading intellectuals of the high Renaissance combined philosophy and art in one tradition; and in the ideal of the universal man that they at times realized. Yet, it is hard to say the same of the early Renaissance innovators. It is even hard to say with whom the Renaissance begins, with Cimabue or Brunelleschi. Cimabue, says Vasari, learned from Byzantine artists. The Byzantine influence on the Renaissance is not very philosophical, scientific, or critical, and it is not clear what role it has played and how large. Brunelleschi is the one who invented perspective and archeology and also revived ancient building styles and methods in the true Renaissance spirit of revival and of the ideal of art and science combined. Yet we know too little about him. He worked with a handful of individuals who dominate our view of the first half of fifteenth century Italian art, his opponent Ghiberti, his disciples Nanni and Masaccio who died very young, and his friend and collaborator Donatello, in particular. We do not know even how literate these people were. Filarete tells us, just after the mid-century, in his life of Brunelleschi, that his hero learned to read and write even though in his youth only medics and clerics learned this art, because his parents wanted him to have an educated profession, and only because he insisted was he apprenticed to a silversmith. This story shows clearly, that Filarete's readers already lived in a new era in which it was already expected of an artist to be literate. Ghiberti probably was not lettered, as Krautheimer has argued, and perhaps Donatello learned literacy later in life whereas the two others died too young for that.

I do not know, therefore, how much philosophy has played a decisive role in the history of the rise of the Renaissance. It is fairly well known that this question has seldom been asked, and was answered by a bunch of Marxist scholars who have claimed that the Renaissance was the culture of the new class of artisans. As evidence they bring the fact that the Renaissance artists all had workshops and apprentices, and the fact that the science of mechanics was of a central interest in the Renaissance, from the start to the days of Galileo, and that Galileo himself learned about the limit of a suction pump's maximum height from an artisan.

The Marxist historiography is an example of how an attitude influences both our views on traditions, and our selections of details to construct our large-scale views of history. The central idea of Marx was of culture as a superstructure with economics as its foundation. It is too vague and even half rejected by Marxist historians; even the politics of the Renaissance the Marxists get wrong because they see a division and a conflict between the landed aristocracy and the commercial-manufacturing rising city-dwelling bourgeois class. It is the manufacturing that the bourgeois controlled that makes the Marxists speak of the artisans as the *avant garde* of Renaissance culture. But this is just not true. The *avant garde* stayed in courts, even when they were originally artisans who came from workshops. But even the politics of the artisans the Marxists get wrong: the real class struggle was often between guilds. And it was the democracy of the guilds that mattered, the democracy that made the famous competition between Ghiberti and Brunelleschi, and that made the guild vote for Nanni as the executor of the sculpture of the dome of Florence.

As to the story of artisans teaching Galileo about waterpumps, it is sheer fable, based on the misreading of Galileo's text. He never said he learned anything from an artisan. In his *Dialogue on the Two New Sciences* he makes Sagredo the amateur say that – thereby allowing Galileo to observe that artisans know the limit of the height a suction pump can work. As to science in workshops, it is also a myth. Workshops existed in the Middle Ages as well as in the Renaissance, yet in the Middle Ages they had no connection with learning, much less with Archimedes. How much learning was in workshops such as that of Michelangelo we do not know. Galileo's interest in Archimedes, anyway, was not unusual for a scholar; as Clagett and Moody have shown, this was general. It was Galileo's contrasting him with Aristotle, that was unusual.

3. THE EXPERIMENTAL DIMENSION

I have mentioned the fact that Hegel introduced a methodology of historical trends. His ideas are very popular amongst historians of all sorts. Since his trends were largely political, political historians have little trouble to follow him. Cultural historians have a bit more trouble. It is still customary to call the music contemporary with Baroque architecture, Baroque music; and even the term Baroque philosophy has currency; all that remains now is to speak of Baroque science. The reason many historians refrain from this is that they view science in the ahistorical way Bacon and Descartes did, thinking that science is above the vagaries of periods.

I think all this is clearly rather silly. There is an inner logic to every art and every science. The inner logic does not uniquely determine its trend, but is there all the same. What art or what science is taken up vigorously may depend on both inner and outer conditions, and the choice of one of a few given possible directions may also be given by outer factors, but all arts and all sciences have both inner and outer logics. Nevertheless, we can see trends, both in the arts and in the sciences; or, if you do not like trends, you may speak of traditions.

Since the seventeenth and eighteenth century classical rationalists were opposed to traditions and demanded from science utter universality, admitting traditions and fashions into science sounds irrationalist, and today with Michael Polanyi and Thomas S. Kuhn it is. Since trends are generally allowed in art, Polany supported his irrationalism by declaring scientific research to be an art, learned in workshops by apprentices from masters, master-researchers, who exercise their authority and teach the tradition.

Being irrationalists Polanyi and Kuhn do not have to explain the rise of science or of a given scientific trend. Kuhn himself vacillates between Polanyi's claim that the leaders of science decide, after consulting their intuitions, whether to change the trend, and the more conventional instrumentalist view that says, the more cumbersome the traditional system is, the more the demand for a change is felt.

Popper has to look at things differently. He does not deny the existence of an irrational component in research and even stresses it; but he viewed the trend as rational and rationality as the search for new explanations and the criticism of existing ones. But how can Popper at all explain a trend? There is no place for it in his theory of explanation. His theory of rational explanation is this: the explicandum is human action, and it is explained as the action optimally conducive to a given end that is ascribed to the actor in circumstances in which the actor supposedly finds himself. Looking at the whole given society, we may find that as long as the physical conditions of that society do not change, and as long as its interaction with other societies is stable, their actions perpetuate their overall situation. For example, their marital and sexual habits keep the population constant, or harmonize with an emigration policy that together keep the population constant. In general the stability of a society is explained when one sees a cycle of explanations closed: condition A and aim α cause condition B and aim β which cause . . . which cause conditions A with aim α. The development of a trend likewise must be read off from the cycles of explanations. In other words, neither stability nor trends are explicanda! Of course, we can make them explicanda, by viewing stabilization as a successful action, by viewing growth as a result of successive incentives. But even then the trend will only be seen as trends, of stability or of growth, only when looking at series of explicanda, not as an explicandum!

Popper himself has noticed that. He allows us to read trends off series of historical explanations the way I have described, but since they are not explicanda, we may read them off quite arbitrarily, and even seemingly contradictorily: we can write a history of human cruelty and of human kindness, he says. Here he is insensitive to his own problem: trends exist, like the coming of golden ages, that he cannot place as explicanda anywhere since they may be the stable points of turbulent lengthy historical periods! Popper will say, but then we do not want any explanation of golden ages, since this will make its cultural product merely mechanical. But here he is in error. First, the question is, can he explain even fairly mechanical trends, such as the growth of the economic and military power of a country? Second, the explanation does not make the struggle of the past mechanical, only its reproduction, and we all consider the reproduction of the past, especially of past science but also of past art, rather inferior. Third, our explanation need not be true; the question is, can we at all explain trends?

To this Popper may respond, why bother with an explanation that cannot possibly be true? The answer remains, it may be nearer to the truth than its predecessor and so constitute some gain; moreover to the extent that it will be a success, it will render a part of art or science fairly mechanical, but we know that this is, indeed, the case!

Popper's theory of science is in part similar to Whewell's: like Whewell he says we must be lucky and imagine a good theory. Unlike Whewell, he does not say we must be so very lucky as to guess the truth: any good guess may be an advance over past guesses prior to empirical test: and so it would *a priori* constitute progress! And so, in a sense, Popper makes science, and art, less of a miracle than his predecessors. Yet, unlike them, he does not profess to know under what conditions art or science may flourish. For example, we never know our prejudices, he says, and these may limit us hopelessly.

In line with this, and contrary to Popper, I would say that we may profitably conjecture what are the conditions conducive to the growth of the arts and sciences. Popper's theory of rational action makes us expect a repetition of a successful act as long as aims and circumstances remain in the same. He expects us to be rational and try to replace an idea once it has been effectively criticized. But the canons of both rational action and rational thought are themselves conjectural and criticizable, and so in a historical reconstruction we can ascribe false canons to historical persona so as to explain their conduct rationally; we can also criticize past canons! For example, we can understand Galileo's and Descartes' adherence to the mechanical canons, just as much as Newton's deviation from them, and also those Newtonians who found the return to mechanism impossible. And we can likewise understand Baconian or Cartesian methodologies! As theories of rationality that were progressive yet false!

But still, the main point made by Popper stands. We may value growth and encourage it. We may do so even when it harms stable elements in our settings, or we may do so within limits. Popper shows that innovations, by their very nature, violate accepted views, maxims, standards, norms. He therefore advises us to encourage anyone who criticizes the accepted ways; never oppose an innovator by telling him to avoid violating the established views, norms, etc.

Workshops, then, are not what Polanyi said they are: masters do not always terrorize apprentices as he and Kuhn envisaged. They did so in the Middle Ages, they do so in what Kuhn sees in the Harvard University Department of Physics, and in other mediaeval relics. They did not do so in the leading Renaissance workshops as ample evidence indicates, including many anecdotes of Vasari: here both freedom to innovate and artistic growth prevailed. Not so the Renaissance universities, that were authoritarian mediaeval bodies. It thus is no accident that Copernicus, Brahe, Kepler, Gilbert, Harvey, Descartes, Mersenne, etc., etc. were no academics, but courtiers of one sort or another, that Galileo moved from university to court, as did even the great Newton. But England after the Restauration was different: it had barely a court, yet it had research sciences outside the universities. Nor did research turn to universities until their secularization as the result of the French Revolution. But the brief golden age of the universities in the last century is another story.

To return to the Renaissance, especially of science. Some say modern science started afresh in the very late Renaissance *à la* Bacon and Descartes. – There was no real Renaissance of science, then. Some say it was the continuation of the academic tradition, the Aristotelian-Ptolemaic tradition with small reforms one following another. This is Duhem's view. He denied the existence of a scientific revolution and of any golden age of science. Facts speak against him. Clearly in the Renaissance, when Aristotle dominated the university, neo-Platonism, mysticism and Cabbalism, were the fashion outside the universities, and became the spirit of the Renaissance. The world Cabbalah was replaced by the word Pythagoreanism and both Kepler and Galileo were Pythagorean. Much before that the spirit of the Renaissance was expressed in a young man's expression of exuberance – Giovanni Pico della Mirandola and his *Oratio on the Dignity of Man*, where he expresses the faith in man, in the Cabbalah, in technology that can be created and implemented – natural magic, that is: mechanics, combinatorics, numerology.

Most writers are obliged to view the outburst of Pythagoreanism and its prevalence among the *avant garde* of science as an external factor conducive to their search for explanations and for their development of empirical tests. Yet in this way one artificially imposes on the Renaissance thinkers later canons of rationality in an ahistorical fashion and obscures the historical situation. As long as it was rational to be a neo-Platonist and as long as that was helpful to development, it was a major factor in the development of the trend, and the trend has to be so explained. There is no way to explain the creation of such exciting metaphysical ideas, but clearly their public acceptance is a major factor in the trends. The same can be said about the growth of the arts.

This, then, is my recipe for a golden age: develop some exciting general ideas, paradigms, models, methods, techniques, and create workshops where ideas are freely exchanged and criticized and where innovation is encouraged with no reservation. No doubt, easier said than done.

CHAPTER 29

MAX WEBER'S SCIENTIFIC RELIGION

Sociology dates back to the dawn of science; yet contemporary sociology, whose founding fathers were Émile Durkheim, Max Weber and Georg Simmel, goes back only to the early twentieth century. The sociology of science is twenty, or at most thirty years old, though it was foreshadowed by Robert K. Merton's doctoral dissertation of the thirties, in which he develops Weber's view of science as Protestant and puritanical. Indeed, the oldest text repeatedly referred to in the literature of the sociology of science is Merton's text on social science in general, of 1952. Philosophically, the empirical study of science ought to accord with our views of the rules of scientific method, since, just as the sociology of religion or of law or of prostitution offers the rules exercised by the practitioners of these ancient activities, so must the sociology of science do. This, indeed, is the minimal philosophic requirement; it seems that though the attempt by the empirical method to legitimize itself empirically begs the question, its failure to do so is suspect, to say the very least. It is therefore not surprising that a methodology backed by empirical sociology is taken more seriously than any other methodology. What is puzzling is that the fathers of modern sociology, though they laid stress on method, over which they sharply disagreed, contributed almost nothing to the sociology of science. Possibly they felt that their methodologies would be seriously threatened by sociological examination.

The nearest we come to have a work on the subject is Max Weber's *Science as a Vocation*; its prominence is thus understandable. Historically, it is very important indeed, but not so intellectually, as I shall venture to show. It originally took the form of a lecture, delivered soon after World War I and soon afterwards published, just before Weber's death. It is nearly thirty pages long, and thus obviously an extended version of the spoken lecture. It is not easy to follow even when closely studied, let alone when heard. I offer here a highly selective summary of it, intertwined with my comments. To avoid surprise, I wish to present my conclusion now. The lecture is an admonition to work hard and an irrationalist retreat to commitment: we are committed to science for want of any better commitment. Put together these two points make little sense: we may well be committed to science in a manner permitting us to be lazy and hedonistic about it, rather than in a manner exacting from us a way of life of asceticism and dedication. To this objection Weber replies in his very closing sentence: for him science is an obsession. Now there can be no objection to the hard work of the obsessed, just so long as others are allowed to remain uninfected. In the following summary, then, there will be no reference to the fact that Weber shows little patience for those who do not share his obsession.

1. A PICTURE OF THE SCIENTIST AS A MANIAC

The first fifth of the lecture is devoted to science as an occupation. Its point is to prove that devotion to science is the only rational motive for becoming a scientist, since a beginner, even a bright beginner, has very little chance of becoming a professional. For, Weber assumes that a man of science is a university professor and he stresses that while candidates are many, only a few are chosen and not the best – i.e. he also assumes that the university system will not expand (whether absolutely or only relatively) and that no improvement on the appointment system is possible. Assuming, then, that a young aspirant knows that his chances of becoming a professional scientist are slim, even if he is good, he can only choose science as a profession because it is for him a calling, a vocation.

This, of course, does not cover all aspects of the situation: why must a scientist be professional? Because, needless to say, as a vocation it will not be satisfying unless he is fully occupied with it. This is a vicious circle. Hence, we can ignore science as a profession altogether and ask, why is science better as a vocation than as a leisure activity?

Weber has an immediate answer to this question: it is simply a matter of fact that these days, proper science is given, not to dilettantes, but to dedicated, hard-working specialists. This is not to say that it is the external constraint that forces Weber against his will. Bertrand Russell, for example, admits that the growth of science imposes specialization on its practitioners, but finds this regrettable and insists that at least the fruit of their labor be made open to the wide public. Weber's approach is entirely different. He regards specialization as highly demanding and approves of it because it is both demanding and rewarding. This, then, is the point, and not the economic constraints which are irrelevant. Also, as we shall soon see, the gamble taken by the specialist scientist is far more daring than the gamble taken by the candidate for a university position.

To the intrinsic value of specialized scientific research, then. In Weber's words, "the individual can acquire the sure consciousness of achieving something truly perfect in the field of science only in case he is a strict specialist." This is, of course, problematic, quite apart from the fact that some of us – myself, at the very least – may reject perfection bought at such a high price; or at any price, come to think of it: of what good is perfection? Is perfection always required? Weber himself clearly indicates that he has no claim to steady perfection: some of his sociological studies of borderline areas are merely preparing the ground for future specialists. Apart from these, he is perfect. For, in the same paragraph in which he proposes temporary resignation to imperfection, he also admits and aspires to "good and definitive accomplishment"; he says that the chief asset of science is a "personal experience" which is a "strange intoxication" and which, although ridiculed by outsiders, makes one an insider within the true scientific vocation. The description of this experience for which all sacrifice is justified makes me shudder:

> Whoever lacks the capacity to put on blinders, so to speak, and to come up to the idea that the fate of his soul depends upon whether or not he makes the correct conjecture concerning this passage of the manuscript, may as well stay away from science

– the manuscript being the Book of Nature, I presume. It all depends, then, on your realizing the fact that it all depends on "whether or not you succeed in making this conjecture" i.e. the correct conjecture. You must be right, know that you are right, and know that otherwise you lose your soul.

Science as a vocation is Faust's supreme gamble – with one's own soul as the stake. In addition, the game must be played with keen enthusiasm which is a necessary but not sufficient prerequisite for the "'inspiration' which is decisive": one also needs "a correct idea, if one is to accomplish anything worthwhile. And such intuition cannot be forced". Calculation, too, is essential but not enough. The correct idea is "normally . . . prepared on the soil of very hard work". Even in the exceptional case in which the dilettante beats the specialist, it is the latter who appreciates the idea and works it out. Hence, hard work is necessary but is no guarantee here, any more than in other avenues of human activity: imagination is everywhere equally important and thus equally risk-generating.

Imagination or inspiration is a gift. Gift is often viewed as a matter of personality and personality as the outcome of life experience; hence, there may be an erroneous search for this very life experience. Yet, in science, as in art and politics, this is given to one "who is devoted *solely* to the work at hand"; *solely*. The difference is that while science is progressive, art is not. Progress in art is in techniques, not in beauty. Comment: this idea has gained recent popularity due to Sir Karl Popper and his disciple Sir Ernst Gombrich, the aesthetician and art historian. The idea relates to ends: science progresses relatively to its own one end which is the truth; for, its products are at times of unsurpassed beauty, simplicity, or other qualities. Does art have one and the same universal end? Whether the idea mentioned here is true depends on the aim or aims of art. If the aim of art is beauty, and one work which is of unsurpassable beauty does not block the creation of other works of art, then, obviously, there is no progress in art. But then the question arises as to why successive artists create at all. This proves the Weber-Popper-Gombrich idea inadequate. Moreover, for Popper, the truth is one towards which science can progress through the reduction of its errors. In the same way, perhaps, art can progress towards the one beauty by overcoming its aesthetic limitations. For Weber, who demands perfection from a scientific idea, the situation is worse: the difference between art and science regarding progress is not at all clear.

Weber may have noticed this difficulty. If so, he should have raised it in reference to his demand for perfection. He does the very opposite: he moves from his comparison of art and science straight to the claim that all scientific works are obsolescent, even as "the very *meaning* of scientific work". How, then, can one and the same work be both perfect and obsolescent? No reply. Perhaps I am in error here. I cannot judge whether he asks this same question or not. Anyway, he now asks why do scientists perform a never-ending task? In part, he admits, because scientific progress, however partial, has pragmatic results. This is true but irrelevant to the fact that science is a vocation. The true answer requires some preparation. Let us leave the question now, then, and follow Weber's preparation.

2. THE PRESUPPOSITIONS OF SCIENCE

Scientific progress is part of a broader context of progress of "increasing intellectualization and rationalizaiton" which does not mean that each and every one of us knows more about the parts of our environment, but that "one could learn it at any time". This possiblility, although only in principle, entails a very broad consequence (well beyond the realm of science): "This means that the world is disenchanted". Every fact is, in principle, subject to calculation.

It is very hard to comment on this. What should one say when such a great thinker on such a great occasion makes such an obvious error? Most things are not given to calculation, and even if they were, then the enchantment would not vanish but merely transfer from the amulet to the abacus. But this may be contested. What is not contestable is that the logic of Weber's argument is very obviously faulty, even quite hopelessly so. Indeed, the situation is very frustrating, since the inner logic of the paper gets worse. Weber first says that scientific results are perfect and then asks, why strive at an unattainable perfection. He then moves from the perfection of science in a narrow sense – meaning truth – to its perfection in a broad setting – meaning disenchantment – and again he speaks of it at times as totally achieved and at times as unattainable. And, after all, since disenchantment is the corollary of the attainment of science, things cannot improve, but only worsen; because recognising that disenchantment is an endless process, as at times Weber does, means, of course, that we shall never be fully rid of magic. Rather than noticing this, he asks as to the good of disenchantment beyond its practical uses. Yes, it beats death, since no matter how long you live, if the process in which you partake is endless, your death is meaningless as the process goes on without you. This is clearly so for those utterly devoted to progress, those who choose science as a vocation. This seems to be the heart of the matter: gamble your soul and hope to win immortality. But let us continue with Weber's lecture.

In antiquity, he says, the idea of truth appeared, and Plato taught that one knew nothing or that each item of knowledge is final. During the Renaissance "rational experiment" was discovered. At that time, science and art were the same as truth and nature. But today, science is divorced from both nature and art. During the Renaissance, things stood differently, and Protestantism and Puritanism ("indirectly") influenced science and made its self-image as the image of "the path to God". These ideas have been (erroneously) pushed aside under the pretext that science is free of all presuppositions. But this is a false pretext: we presuppose that science and its results are important. This presupposition cannot be proven; it is one "which we must reject or accept according to our ultimate position towards life", i.e. it depends on our commitment and way of life. This idea is now gaining increased popularity in the form given it in the philosophy of Michael Polanyi; except that whereas Weber begins with specialization and ends in finding the meaning of science in commitment Polanyi preaches the view that commitment and specialization come together, that becoming a specialist is a matter of training and that the same training that imparts specialized proficiency also imparts the value system and the life-style of the specialist.

Weber mentions, to continue, specific sciences and their presuppositions: the supposition in each field is that there exists order which is worthy of discovery. The only intricate case is political science in which a political party stand is not to be advocated, yet study of which should not be deemed worthless. This point is, of course, one major contribution made by Weber to our political life in general, and need not be further defended. In the lecture at hand, this point is a side issue which Weber had to clarify yet again, but which we may now take as not particularly problematic. The problematic is not whether we should avoid prejudices not shared by all our colleagues, but rather whether utter objectivity and truth can be attained, as Weber explicitly demands, although science has its presuppositions and is obsolescent. He seems to demand that science be value-free, yet value-laden; meaningful, yet disenchanted. It looks hopeless from the start. As it turns out, it is even more hopeless than at first appearance, and Weber's lecture loses its structure in increasingly rapid steps.

After a few asides, Weber declares that he is returning to the subject. A student who wants more than facts and figures from his professor, "experience" or something of this nature, wrongly expects the teacher to be a leader. The American student who purchases knowledge from his teacher so as to pass an exam in order to get a job and who desires no *Weltanschauung* and no code of behaviour is vulgar, but there is truth in his stance which saves him from the slavishness of the German student. Academic excellence is no leadership, least of all in politics, a professor may advise his student outside the classroom; inside, where his opponents cannot answer, he should not make propaganda for his own personal opinions.

This, again, raises the question as to whether science is of value. What does it have to offer outside of the classroom? First of all, technology, which is systematically (and rightly) ignored in this lecture. Second, the habit of clear and systematic thinking. This is also a practical affair which may be ignored. What else? With the aid of clear and systematic thinking we find our own personal commitment to our own values. Whatever these may be, one thus finds "*the ultimate meaning of one's own conduct*".

Here, again, the very dichotomy between facts and values which makes us despair of finding value in science, is used scientifically to find value in the world of fact: we do have values; we *are* committed. Here Weber's irrationalism shines. Today, that is after having learned existentialism and the retreat to commitment that it amounts to, after having seen its explicit implementation by Michael Polanyi and Thomas S. Kuhn and after endless debates, we should not find Weber as express or as clear an irrationalist in his own time as we may find him now, in retrospect.

My reader may be uncomfortable; at least I hope that he is. We have asked clear and simple questions concerning dedication to science, and the perfection it offers; concerning the endless progressivity of science, and the question as to why bother about the unattainable. It now turns out that when science is applied to the devoted scientist, it shows that he cares. We have here two different questions: why does he care? and, should he care? The first is biographical and interests biographers and cultural historians. It is the second that is at stake since Weber is now in a recruitment propaganda lecture in which he defends science as devoid of propaganda in action, yet of enormous propaganda value anyway. Which value? From where does it come? From the converted listener, of course. But can the unconverted audience be rationally converted? No. Does Weber admit that much? I cannot say. If his lecture is science and science is value-free, then, of course, he knows it. But if his lecture is a sermon, the question is, can a sermon be rational? If it founds one value on another (in given circumstances), then, yes of course, this is rational. But can ultimate values be rational? No. Does Weber know this? Does he say this? I cannot say. I am as uncomfortable as I hope my reader is. Indeed, a little after the appeal to a final commitment, Weber advocates science to those not committed, in accordance with the slogan 'know thy enemy'. This is but a sleight of hand: to be dedicated to the cause and to know it, as the enemy or not, are two different things; indeed, this is the thesis of the lecture!

No science is absolutely free from presuppositions, and no science can prove its fundamental value to the man who rejects these presuppositions. Every theology, however, adds a few specific presuppositions for its work and thus for the justification of its existence . . . Every theology . . . presupposes that the world must have a *meaning*, and the question is how to interpret this meaning so that it is intellectually conceivable.

This is a revolution in theology, since thereby theology becomes universal rather than particular. Weber makes it quite clear that he is aware of this fact. I agree, however, that if we wish to combine science and religion, we must eschew or play down particular theology. The great impact that the philosophy of Martin Buber has on Protestant theological schools and that the philosophy of Michael Polanyi has on Catholic universities, especially on those there who specialise in the philosophy of science, is evidence both for Weber's immense influence and for his irrationalism.

3. CONCLUSION: WEBER'S OUTLOOK

The preparatory material has now been concluded, and we may now, at the very end of the lecture, return to the question that prompted it. Let me remind my readers that the question, though Weber glosses over it, is why should one commit oneself to science when, since the task is endless, each contribution to science is obsolescent?

Our period "is characterised by rationalization and intellectualization, and, above all, by the 'disenchantment of the world'" in which there is no room for either monumental art, sweeping new religions or any "academic prophecies". It is pointless to sacrifice one's intellect and return to a church, though this is better than "academic prophecy". Thus, he concludes, defending intellectual integrity, we must admit that there is nothing left for us but to continue in our search, a search that is pointless as sheer yearning and meaningful only as investment of work and thereby meet "the 'demands of the day' " in both our human relations, and our vocation which is our obsession as well, perhaps.

As a final comment on this conclusion, let me observe that here Nietzsche's influence is most patent. It is the same as that which influenced Niekisch and other leading disciples of Nietzsche, except that Weber hoped that the teaching of the master had forestalled the risk, whereas Niekisch declared Nazism the realization of the master's darkest fears – the growth of a new monumental art, secular religion, and academic prophecies combined. It is clear that Weber underestimated the dangers. Can this overoptimism be the corollary of the overoptimism of his alternative choice? Or should we disregard the overoptimism as part of the style of the professor at that time and place, and see beyond it the individual in search of meaning, hoping to find it in the life of reason? I cannot say.

CHAPTER 30

ON PURSUING THE UNATTAINABLE

The aim of this chapter is to criticize the view that it is never rational to attempt the impossible; which is not, however, to advocate all impossible aims. The ideal of positivism, which positivists deem obviously attainable, namely the unity of science in rationality and the rational unity of mankind, is here viewed as very worthwhile, but quite possibly impossible and certainly not obviously possible. Yet, to repeat, not all impossible aims or unattainable goals are reasonable to pursue.

1. REDUCING REMOTE GOALS TO NEAR

There is a simple and straightforward sense in which pursuing the unattainable is palpably irrational: it is the pursuit of what we know to be unattainable while ignoring this relevant knowledge. However, this does allow for rational or reasonable error: when speaking of the unattainable we do not mean what we take to be attainable by some error or another. It is quite obviously rational to attempt to attain an aim on some reasonable assurance, and this remains so even if later on it turns out to be unattainable. After the destruction of a work of art, but before the news of the destruction is broadcast, the search for it may be reasonable even though it is unattainable. *A fortiori*, it may be rational to try to attain the almost impossible; for example, if it is one's only chance for happiness or if the stakes are sufficiently high: the young hero may rationally try to marry the princess, even though all his associates know his venture to be hopeless – but only insofar as he still retains hopes, however faint, and insofar as he feels that the stakes are high enough for him to make the odds fair enough. Otherwise, if the error on which the pursuit is easily detectable, or if chances are grossly miscalculated, or if the cost of possible error is wilfully ignored, if the pursuer refuses to consider the possibility that he may err and the cost of his error, then we may rightly view him as rather irrational.

There is the claim of the psychologist to muse about, which is that some person will permit himself to fall in love only hopelessly: only when assured of the hopelessness of attaining this end will he dare pursue it. Moreover, the psychologist will assume that in such cases, the hopeless pursuer will refuse to consider the possibility that his pursuit is hopeless, and he will argue obsessively to prove to himself that he has high hopes. Let us assume all this to be true. Is the end, hopelessly pursued, the unattainable end, simply the one which the pursuer is really after? Not in the least; his real end is to avoid attaining it, says Freud. If so, why not simply avoid pursuing it? There may be different answers to this. For example, the flesh is conditioned to chose an end that the spirit wishes to avoid attaining. It is relieving a pressure to let the flesh, in its ignorance, pursue what only the spirit knows to be better unattained. Then the flesh is rational and pursuing the princess, while the spirit is rational and pursuing avoidance of love plus relief of the pressure of the flesh. Freud's theory of the id is a variant of this. Usually Freud suggests that the id and ego stand for flesh and spirit; however, the ego is conscious and often the flesh is taken account of by the conscious and the fear of attaining this aim by the subconscious (as in the case of the groom who 'forgot' to go to his own wedding).

The Freudian theory of conflict will present matters even more subtly. Conflicting aims may remain unresolved, Freud indicates, yet their pressure may be relieved by playing against the odds; a person in conflict may wish to retain his conflict, or more precisely, is in conflict as to whether he wishes to retain his conflict. Here, then, relief of pressure from conflict while retaining the conflict may, indeed, be the true end of one's action. Whatever is our opinion about the truth or falsity of Freud's view, we can learn from it that possibly it is within one's aim to pursue another aim, and yet these two need not be in a simple hierarchic order; there is a difference between hopeless love aimed at the avoidance of consummation and courtship aimed at consummation. Regardless of any given facts, this may help us develop a more sophisticated approach to all goal-directed behavior: the real goal, the final end, may conflict (truly or seemingly) with the partial goal, with the more immediate end.

Let us take a more extended, long range, abstract end – any promised land, private or tribal, perhaps even religious, scientific, or aesthetic – any end beyond achievement within one lifetime. What is the rationality of an individual's pursuing it, even though it is admittedly unattainable for him and perhaps even unattainable for his descendants, or for the tribe as a whole, or for the whole of mankind? Let us first discuss the case where the end is attainable for the species but not for the individual, and later the case where it is not even attainable for the species.

In the first case, where, say, the final goal is not given to one's own self, but is given to one's descendants, the rationality of the pursuit is not really problematic. Any act for the sake of posterity, be it one's children, or one's future honorable mention in the future textbook of science, etc., can be viewed as a special, slightly more complicated case of a long-term project which lies within one's own reach. The rationality of working for one's posterity can be reduced, with little effort, from that relative to the goal attainable only to one's successors to that relative to some goal which one may well achieve. We can say a father works not for the good of his children; his true end is the knowledge of, even mere hope for, his children's good opinion of him while he lies on his deathbed, say. And, clearly, in order to gain their approval, he has to pursue their ends, which are, indeed, not within his reach, yet well within theirs.

Here we have a curious case, where the partial end is farther ahead than the true and final end. Usually the partial or subordinate goal is more immediate than the final or primary goal; the attainment of the partial goal is means for the final or true goal; as means, as a link in a causal chain, it is achieved first. One may offer a criticism of any reduction of an end to a part of a more immediate end by the suggestion that it is subjectivist, since self-deception will satisfy the reduced end just as much as real work for one's son's future interests. This criticism is clearly acceptable on occasion, but it may, generally, be not harmful to the reduction proposed. Some people can deceive themselves, and when pressed hard enough they prefer the self-deception that they labor for their children's future over actions to such an effect. Other people are less capable of self-deception and so have no option. They may be unable to deceive themselves, whether from habit against it, or from having frank and free exchanges with their children. In either of these cases they can only feel having done something for their children's future when they have reasonable grounds for such a feeling – by standards of reasonableness acceptable in their community, particularly to those people whom they except to be present when they might die. Of course, taking the moment of one's death as all-important is here a mere simplification. The real point being that some philosophers wish to reduce all aims regarding posterity to aims regarding one's own lifetime.

The debate, however, has here deteriorated from a general complaint to a particular instance of it, which may not be good enough. The general complaint was that the partial end, when allowed to be later than the final end, has no causal link akin to the partial end whose actual attainment is a causal step towards the final end. Thus, for example, if the immediate price of a commodity, in money terms, or more abstractly, as the exchange of concern for appreciation – if the immediate price depends on future expectation, we have a funny infinite regress! This regress, however, may be admitted as rather innocuous. It becomes important, perhaps, when we wish to improve our expectations. But when explaining a father's concern for his son's remote future, all we have to notice is that he and his son have some expectations in common, and on the basis of these they trade concern for remote future as against a promise of appreciation in the less remote future. And the difficulty of a detailed and correct reduction, then, may well be rooted in our own ignorance of the exact terms of contract between father and son, as well as the exact area of common expectation, the influence of disagreements, of misunderstandings and self-deceptions, and similar complications – all of which are immaterial to the principle of reduction we have discussed here.

The reduction of the rationality of working towards one's children's future to the rationality of working towards one's own future, say one's own future peace of mind – particularly in one's old age – is one which we need not insist on. Some philosophers, following Spinoza's wake, insist on it. They reduce all unselfish motives to selfish ones, and by implication all motives concerning posterity to motives concerning a more immediate future. Now, quite obviously, the demand to reduce all motives to selfish motives can always be met – on the condition that selfishness is defined broadly enough, of course. The reduction, thus, is not too satisfactory: it may be made easy by assuming the following ploy: being morally conditioned, one can only attain happiness, peace of mind, or any other selfish goal, when acting morally – even if this means trying to attain some unattainable ends. The result is that any reduction is handy, from any unselfish goal to a selfish one, including an unattainable goal to an attainable one. This Spinozist ploy was repeatedly and systematically used by Freud, in the guise of the theory of the super-ego, its formation and content, and the guilt pressure that it effects.

Briefly, according to Freud, any goal, selfish or absolutely altruistic, reasonable or utterly mad, may on occasion be stored in one's super-ego during childhood (before one can either examine or protest) and thus be operative in the sense that a man may have to obey his super-ego without questioning it and without endorsing it – simply in order to attain no other end but an immediate relief of a painful sense of guilt. The immediate end – peace of mind – is here the final end, and the end of the super-ego is served as a means for it. This is how Freud achieved a causal explanation of purposeful action: by seeing the sense of guilt as a means of pressure, as a cause, a motive force, or a propellant. We need not view the theory as strictly causal, however; we may view the relief of pressure (of one's sense of guilt or of one's bladder) as goal-directed, as directed towards relief. We may agree, then, that Freud reduces with ease – too great an ease – all remote ends to an immediate one, namely that of relieving one's sense of guilt.

2. THE POSITIVISTIC UTOPIA OF RATIONALITY

It seems, then, that all remote goals are now somehow taken care of, and with great ease; the goals attainable to oneself in the remote future as well as the goal unattainable to the species on principle, such as Heaven on Earth, some other Utopian dream, or the attainment of full rationality, or of knowledge of the whole truth and nothing but the truth, which is the discovery of the secret of the universe. There is no difficulty in reducing the unattainable goal to attainable ones in the manners described above. The attainable goal may be approaching the unattainable goal, or facilitating the approach to it; the end need not be the final one, but the coming to a point as near to it as possible. Also the goal may be the more immediate one than that of finally achieving the highest end possible: we may declare the goal to be the immediate pleasure of the search rather than the find, as in the case of going for the princess, described above. One may even declare the goal to be, as a matter of fact, the immediate pleasure derived from the search. This really is the easiest reduction: the benefit of the exercise might be proportional to the effort invested in the search; one may thus end up recommending the maximum investment with assurances against success! For this end one may require the pursuit of an impossible end plus the maximum desirability to approximate it as much as possible; one might proclaim that the greatest effort made by the greatest numbers may lead humanity only infinitesimally nearer to the unattainable goal, yet we should all put all efforts unsparingly to attain this slight improvement. Particularly when one is ambivalent about one's goals, one tends to favor such a philosophy. Thus, one serves a few and even conflicting ends when one aims hard for the impossible.

This is not to say that the only cause for the endorsement of unattainable goals is psychological conflict. Indeed, some such endorsement is advocated here, though psychological conflict is not. It is possible, however, to suggest that many thinkers, particularly in the nineteenth century, preferred remote ends to immediate ones, in typical pre-Freudian self-doubt. When John Stuart Mill asked himself, we remember, what if his end will be attained in his lifetime, he suffered a severe nervous breakdown. All this, however, is no argument for or against unattainable goals. It is interesting to notice that strong emotional drives may also stand behind the opposite view, which condemns all unattainable goals as chimerical.

As experience shows, philosophers, especially as graduate students, show sometimes enormous hostility to the idea of going for unattainable goals. The reason they often give (I am reporting from my own limited experience) is not so much that it is irrational to go for an unattainable goal – this they keep in reserve and use only when very hard-pressed – but that if the goal is admittedly forever unattainable it then simply does not exist, and so talking of pursuing it is plainly meaningless! This argument is not from the theory of rationality but from epistemology. If one analyzes it carefully enough, one finds it to rest on the positivist verification principles of meaning: what we cannot verify we cannot understand. That this principle has a terrific emotional import is obvious, and it may well merit some examination, however superficial.

The verification principle is so vague that philosophers could never quite get its meaning straight – if it has a straight meaning at all. (In my view it has only emotive import, but no cognitive meaning.) It is all right to say, what *I* do not verify *I* do not feel I fully understand. Faraday, the greatest fantasist of physics, said he could never feel he fully understood a description of an experiment unless he could call it his own – that is to say, unless he performed it with his own hands and saw its results with his own eyes. Faraday was no positivist; he only expressed a very strong positivist feeling, and he strictly confined this feeling to matters regarding experiment alone. No positivist, not even the most extravagant one, ever asserted, what *I* cannot verify, *I* cannot possibly accept. One reason why Descartes' philosophy sounds so frightening – Kierkegaard regarded his going through with his doubt heroic because it was so frightening – is that he began with this dictum; but, of course, once he established the veracity of God and of the Natural Light, things looked much less frightening. Perhaps the main reason why no philosopher, not even Berkeley, could fully consider the case of solipsism is that it really is frightening to believe no one around, to trust no one's testimony except one's own. (Indeed, it is quite mad to doubt that one is born to a human female, regardless of one's source of information concerning the facts of life or the reliability or otherwise of that source: do you remember who told you that you came from a womb and does it matter to you? Will you ever calculate the probability that indeed you came from a woman's womb?) Once verification means strictly the acceptance of only the evidence of one's own senses, then verifiability becomes too constraining a principle – too constraining even for any positivist to contemplate it as a serious option.

On the contrary, the positivists always stressed that verification was required only in principle, not in actual fact. Whatever this meant, it allowed for the ingenuous acceptance of testimonies of other scientists – without first calculating on the basis of evidence the degree of their veracity. In the heyday of positivism, when the moon was just coming to within man's reach, the example 'there are craters on the far side of the moon' illustrated what positivists meant by a statement verifiable in principle but not in fact. This really expresses the positivist ethos: When Man – *any* man – reaches the back side of the moon, then we – *all* of us – will know whether there are craters there. When a positivist speaks of Man or of people or of us, when a positivist uses the first person plural, he really means the scientific community. And he means to exclude from this, first and foremost, the obscurantists – who do not count since they have abdicated their rationality. Further, amongst the obscurantists he includes the speculative metaphysician. The ethos of the positivist is one which contrasts science with metaphysics by stressing that science is bold but only to the limits of the practicable: reach for the moon, not for the outer galaxies; certainly not for the outer reaches of the mind, where experiment possibly cannot follow. Thus, in my opinion, the emotional import of the positivistic verification principle is clear, and at least quite unobjectionable, if not also laudable, in spite of its branding the metaphysician as an obscurantist. When a metaphysician turns out to favor science and, more so, when his speculative system has a benign effect on science, then the positivist – at least Carnap and Waissmann – has no qualm in annexing the metaphysician into the community of science: he simply stresses that he requires verifiability only in principle. What, then, happens to the cognitive import of the verification principle?

Statements about the back of the moon, we said, are in principle verifiable; even statements about temperatures in the heart of the sun may be so declared. In spite of all criticism of logical nature, we can say that these were meant to be included. But what will positivists say about fantasies regarding outer galaxies, wild science fiction stories bordering on the metaphysical? Where and how will they draw the line between where experiment cannot follow as yet, but may one day, and where it cannot ever follow at all? This question was studied by Kant. His studies are now obsolete. There is none to replace them. It is fairly common sense to admit that this question is beyond our reach: Kant could not suspect, but we can argue, that the limits of reason are beyond the limits of reason: we cannot find out what one day we may be able, and what we shall never be able, to find out. The answer to the question, posed by Kant, what goal is attainable, what is not, is, itself, quite unattainable by Kant's own standards and even by more lax ones. The very examples with which the positivist ethos is illustrated, indicate that positivists want common sense more than an abstract formula, and by common sense they wish to put before science challenges not too easy to ignore and not too hard to undertake; surely this is too dependent on circumstances!

The young positivist who rejects vehemently the idea of the rationality of pursuing some unattainable goals, is quite rational, and on two counts. On one count, he sees no need to start with the assumption that a goal – Utopia, the ultimate truth – is unattainable; and when early in his career he is told that they are beyond reach he sees no reason for believing this. On the other count he fears that the unattainability of the goal, if openly admitted, might be too discouraging. Yet one may reach a point where the faith in the attainability of a goal is but a mirage, retained by a voluntary act of self-deception, by the fear that the recognition of the unattainability of a goal may bring a halt to the search. But it may be, as I have argued, perfectly rational to go on searching the seemingly unattainable. The positivist's fear which leads him to self-deception is ungrounded.

The positivist Utopia is well worth retaining: the unity of mankind in rationality: the unity of science in humanity. For all we know it may even be attainable. Suppose we stress that the attainability of the goal need not be immediate, or even demonstrable here or now, but we may, in principle, hope that one day it will be. In that case the great fear that the goal is meaningless need not be so paralyzing. The very suggestion to the contrary – the claim that the positivist Utopia is obviously attainable – if at all seriously entertained, is self-defeating. Regardless of whether one thinks the aims of science are attainable or not, one can agree to the following central point. It is all right to leave the question open; but answering it in the affirmative as a matter of course renders the Utopia not very exciting; whatever we (nearly) have, as a matter of course, since we are so obviously poor, is hardly worth having. Whatever we obviously can find is not something very exciting to look for. But the positivist Utopia – the rational unity of mankind and of science – is the possibly attainable, possibly unattainable goal, well worth pursuing.

3. THE FRUSTRATION OF SEEKING THE UNATTAINABLE

One thing, then, a skeptically minded student may learn from positivism. The contrast should not be made between the attainable and the unattainable goal, and not (as the positivists suggest) between the obviously attainable and the rest, but rather between the obviously unattainable and the rest. This is but the application of skepticism to itself, quite central to the ancient skeptic doctrine of ataraxia. Yet it is not so obvious that we should apply it to the case of goals, and the positivist insistence of the obviousness of the verification principle may be a proper stimulus here.

There is, indeed, a slight bonus here. The doctrine of doubt, when applied to itself, hardly leads to new insights (though it may relieve some pressures). Yet when applying it to goals, it becomes more interesting. At first, superficially, one may hastily conclude from doubt that at least one important goal is clearly unattainable. Since the aim of science is knowledge, and since the possibility of knowledge is seriously doubted, we may conclude (with Socrates at the end of the *Symposium*) that the end of science, knowledge, is quite unattainable. Yet, on a second thought we may view things differently. Strictly speaking, the positivist, when wishing to insure the attainability – or even the mere existence – of a goal, speaks not of any objectively attainable goals, but only of some of these, namely of those which are certainly attainable. This raises the question, are all *objectively* attainable goals *demonstrably* obtainable? This hinges on our question, what does the positivist mean when he says he requires verifiability only in principle? I do not think anyone has worked out this question. (Even if you declare that all true laws of nature are discoverable, you need not say all attainable goals are certainly attainable – unless this is a law of nature!) This is why I deny that the verification principle has cognitive meaning. Hence I also deny that the principle has ever been refuted, of course. (Only very simple variants of it were refuted with the phrase 'in principle' sufficiently clearly, but quite naively, construed for the sake of the argument.) And so, one chief reason for pursuing the seemingly unattainable may be, that we do not know whether in principle it is attainable or not. We can even say, we do not know what our goals are – whether individually (remember Freud) or collectively (remember the positivists). The princess may, and sometimes does, return love to the daring insolent unworthy suitor just because he is just this; with a happy or an unhappy ending possible. The unity of mankind in rationality and of rationality in science may be attainable in one sense of rationality or another not yet sufficiently explored, or even in a known sense of rationality yet by hitherto undreamt means. We do not know whether this is so, and the doubt itself may be a strong motive for the search. The word 'skeptic' means searcher, and it is a pity that the old skeptics stressed the value of peace of mind (ataraxia) rather than search, though they did say that peace of mind came, as an afterthought, as a result of search, and though, no doubt, many have found peace in the search. To conclude, if we doubt even our claims about our goals, our study becomes more interesting.

The question, I feel, still remains: why do so many agree to aim beyond reach only when they are ambivalent, and why do so many find the idea that their aim is utterly beyond reach so disturbing? It is not so counter-intuitive, after all, to think (as I tend to do) that it is a disaster to have no secret of the universe to hanker after, namely to think that hopefully the end of science will never be reached. After all, Lessing already expressed the sentiment, among other classical writers, in a manner which did win popular acclaim.

What both the ambivalent seeker of the impossible and the confident seeker of the possible share, is the view that seeking the impossible is highly frustrating. This view, as a personal expression of individual tastes, is not objectionable – merely regrettable. But all too often it is presented and even pressed hard as a principle of objective validity, psychological, methodological, even logical. This, obviously, has been empirically refuted. Though endless search can be frustrating, and though peace of mind can be achieved with little or no search, we do have empirical testimony of lives devoted fruitfully and happily to tasks not yet attained and perhaps unattainable. Kepler's search for the harmony of the spheres and Einstein's search for a unified field theory are such instances, and by no means the only ones.

Things might fall better into place, perhaps, if we rectify the reduction of all motive to selfish motive. No doubt this reduction is one which has a laudable and an objectionable aspect. Consider a true martyr, i.e. one who loves his life and would not see any merit in sacrificing it, yet who sacrifices it for the sake of preventing some worse catastrophe, such as the violation of the principles of humanity (in accord with his opinions, whatever these may be). One may say, and it has been said, that the martyr is giving up a long life of shame in preference to a moment of life with dignity, and that really in his act of martyrdom he only expresses his high preference for dignity. No doubt, there is a lot to this, yet we know that the martyr may not be thinking in this manner no matter how deliberate his action is. Take, more specifically, the Freudian theory, according to which the martyr is avoiding a supreme stress created by a sense of guilt which would crush him if he avoided the act of martyrdom. This theory is more comprehensible, yet there are more and clearer counter-examples to it.

The merit of all reduction of motive to selfish motive, however, is in its staunch individualism, in its incorporation of a blanket refusal to consider any action of the race, the tribe, or the General Will, through the individual. In brief, the strong part of the reduction is its incorporating the autonomy of morals, of the attribution of responsibility to the individual alone. But it does overshoot its mark, not only by denying altruistic act or act for the general interest, it even denies that one can construe the general interest and act in accord with it. We know for a fact that people, small and big, do have views concerning the general interest and act at times in accord with their views about it. They may sincerely act in accord with what (in their opinion) is no less than the interest of the whole human race. Every time we applaud a person who puts world peace before any other interest we acknowledge – rightly or wrongly – that he falls into this category.

4. THE SCIENTIFIC MILLENNIUM

Can we, then, not act also in accord with an end which happens to be not achievable even by the human race? Surely this does happen; sometimes we do it knowlingly, sometimes in ignorance. Inasmuch as science is taken for religion, it is no doubt a shallow religion; yet this happens regularly. And science as a religion-substitute contains its hagiography and folklore and mythology, these including some science fiction. And repeatedly science fiction, both in its less pretentious and in its more pretentious instances, speaks of a more general purpose, to which humanity as a whole is but a small contributor. We need not assent to this view; we merely must admit its legitimacy even while we wince at it. This, however reluctant, is an admission that in principle one may crazily devote one's action to what one deems that one dimly perceives as the end of a huge process to which we may be but a small party.

There is no doubt of the religious dimension, and even piety, of all this. The better science-fiction writers make no bones about it. What is unpleasant about it is its inherent pretentiousness, however, not its religiosity. Some science-fiction writers, philosophers, and religious leaders of various schools have declared (and others denied), that ofttimes the effort is laudable, and is imprinted on the cosmic book of history, quite regardless of its outcome. The positivists who found all this outrageous, were particularly outraged by the infusion of the religio-metaphysical into the domain of science.

Thus, it all falls into place. The positivists insisted on the unity of science as a fact, and on its dictating sobriety to us, not so much in order to declare that the millennium has arrived, as in order to prevent talk about the millennium altogether. This was to no avail. Once we hanker after the millennium, we can only claim that it is here, or that it is remote, or that it is the unattainable goal. The positivists did not mean to tell us that it is here; but to the hankering – not as a mere feeling which, like any feeling, of course, they allowed – they refused to allow the cognitively recognizable expression. They simply did not succeed, and the feeling, the hankering, the religious dimension of science, kept finding diverse expressions in our culture. And so, the positivist thesis of the unity of science itself became one expression of that hankering, plus the implicit and unintended suggestion that the paradise wished for is around the corner. In the darkest days for humanity, preceding World War II, this was the message – clearly the unintended message, of course. It was meant to preach sobriety, but it preached the millennium of sobriety.

This conclusion may be answered with ease. Suppose that positivism preaches sobriety and hence (in some sense of 'hence') the avoidance of a discussion of the scientific millennium (which is the uncovering of the secret of the universe); suppose that raising the question nonetheless does make positivism millenarian; why should this trouble the positivist who preaches the avoidance of the question? The trouble, however, lies in the positivists' change of this philosophy of science. Already Pierre Duhem, the last great positivist of the period preceding that of Wittgenstein and the Vienna Circle, was caught in this same difficulty. He declared that we can speak the truth and nothing but the truth, yet that we cannot speak the whole truth (which is the secret of the universe). Now, as long as we can have a class of absolutely true statements and increase its content in time, we may say that the question, does the secret of the universe exist? is meaningless: we can say that the answer to the question is in principle unverifiable, unconfirmable, untestable, etc., and we can try to dismiss the question altogether. But suppose that science does not progress as the increase of content of the class of all known true statements.

Suppose, science is a series of different pictures of the world, such as the Newtonian and the Einsteinian picture in physics, or the Darwinian and neo-Darwinian in biology, etc. And suppose we agree that there is some preferability here, which is not oblivious of the value of truth. Suppose, that is, that there are some scientific world-pictures which are nearer to the truth than others; then we have admitted, willingly, that there is a secret of the universe which may or may not be attainable to science.

Duhem himself was aware of all this. He usually declared that the preferability of one world-picture over another is judged with the aid of criteria other than the truth – such as utility and beauty. He could not, however, altogether relinquish the idea that the preferability also relates to the approximation to the truth. Louis de Broglie, the famous quantum physicist, remarked in his introduction to Duhem's *The Aim and Structure of Physical Theory* that this amounts to a declaration of bankruptcy.

Duhem also realized that having a succession of world-views destroys the idea that at least reports of observation are final and unalterable. Yet he insisted that we do have a class of verified statements of observation which increases in time. He somehow felt that though put in scientific language an observation report is made open to revision (with the revision of the language of science), when put in ordinary language a statement is vague enough to remain unchallengeable. This is a hypothesis concerning the unrevisability of ordinary language which is false, and whose reputation has led the positivists of post World-War II to a new theory of the reform of ordinary language through the process of explication of concepts so-called. With this admission that both ordinary language and scientific systems are revisable, there is nowhere to turn, it seems; we have criteria of improvement, and these may point the way towards the very best, and the very best may, indeed, be the uncovering of the secret of the universe. All we need to solve so many traditional philosophical problems is, I think, to admit that the secret of the universe is quite likely the unattainable goal of all research.

CHAPTER 31

FAITH HAS NOTHING TO DO WITH RATIONALITY

The problem of rationality is traditionally viewed as the most fundamental problem of both abstract philosophy and the philosophy of life. Abstractly, what is the foundation of my intellectual system? and concretely, on what should I base my way of life?

1. THE BASIS OF RATIONALITY

The abstract problem is the problem of first principle, of foundations, of the Archimedean point. The more unified a philosopher's *Weltanschauung* becomes, the more one can make it hinge on one fundamental axiom or postulate. Like Archimedes, he seems to plead, grant me this small point and I shall lift my whole system to the level of demonstrated truth. Yet, the more we tend to accept his contention that he can do so, the more reluctant we are to grant his wish.

Some people, I am reporting an observed fact, are so impressed by the fact that one principle seems to make sense of everything they experience, that they cannot but give their assent to that principle. And, when they encounter an opponent they try to convey to him the experience, assuming that once he sees the point he will share the experience and then the sense of marvel and then the profound sense of conviction. At times this happens and the circle of believers gains a new convert. At times, however, the opponent has undergone a few cases of such conversion, and has learned that they are not unique, that there is more than one Archimedean point. What then? He may say, one Archimedean point is the better, and I choose that one. He may say, there is no better and no worse Archimedean point. Indeed, he can prove that: if there were, then the criterion of choice between competing candidates for the status of Archimedean point is the true candidate. If so, then he may stay at the point where he happens to be, out of indifference. Or, he may choose the most convenient. In either case, indifference or convenience is the true, and rather disappointing Archimedean point. Or he may be disappointed and try not to think, or say thinking is all no good.

Non-thinking is leaving the game. Some thinkers say, you cannot leave the game: it is the game of life. And there is no Rational Criterion for choice of

Archimedean points. The criterion of choice is extra-rational. Or the choice is arbitrary. Or the choice is made by sheer accident. These thinkers are called irrationalists. The word is misleading since, because the prefix means non, irrationalists may be misunderstood to be non-rationalists. Yet they are (a) anti-rationalists about choice of principles and (b) rationalists about all else.

In the philosophy of life for most people the question either does not obtain or arises when they encounter the real freedom of choice between different communities to join, as Martin Buber and Michael Polanyi have said emphatically. I can join the community of scientists, or the international proletariat or the community of the practitioners of the faith of my father, or even of the faith of your father though I would have to know a bit more about it. I can join, perhaps, a European monastery, but not an East Asian one. Each of the possible ways of life is commended by its own system of thought and each tells us it has an Archimedean point – The Trinity, the Ten Commandments, scientific socialism, or science. Why should I join any of them? Should I perhaps join none of them? Or join them each for a year or five years on a rotation basis? What am I to do about the choice of a way of life?

The problem is the practical problem of rationality. It is a characteristic twentieth century problem. In the Age of Reason, the Enlightenment, *the Aufklärung*, the question was known but not discussed. Rationality, it was taken for granted, is best exhibited in science. If the person engaged in the practical problem of rationality wants to be rational let him, then, join the community of practising rational thinkers, the philosophers, namely the commonwealth of learning, the scientists or the intelligentsia.

Yet the Enlightenment had its own faith about faith and reason. And I should expound it and expose it as a dangerous myth, though I confess I am a great admirer of the Enlightenment. Indeed I am for the revival of Enlightenment, for the Re-Enlightenment movement. But we cannot accept all the mistakes of our noble predecessors and especially not their faith concerning rational faith.

Briefly, the theory of rationality has four components, made by two divisions: thought and action, and private and public. So we have private thought that should be rational, and public thought, and private action and public action. Yet the Enlightenment made a very interesting but mistaken unification of the four. They thought if one acts in accord with one's own beliefs, as one must in order to be autonomous, to be one's own master, then one acts in accord with one's own beliefs. Hence, the autonomous rational thinker is an autonomous rational actor, and since rationality must be autonomous, rational thinkers must be rational actors. So it is enough to center general considerations on thought and I leave actions for specific considerations alone. As to public thinking, if the public consists of autonomous individuals, then the public thought will be no less rational than that of the least rational member of the public. Hence, the individual rational thinker will do.

2. THE REASON FOR RATIONALITY

Why, then, should the individual be a rational thinker?

The Enlightenment accepted the answer to this question from Sir Francis Bacon and René Descartes – the fathers of the movement. The answer is Bacon's doctrine of prejudice.

Briefly, Bacon noticed that it is rather easy to choose an Archimedean point, that the choice of that point largely decides what one perceives and how one perceives – the choice acts as a blinker and as a sieve. Facts that do not fit stay unobserved or get observed in distortion. And so different communities see different facts, different worlds, and talk different languages. As the mediaeval Aristotelians put it, *contra principes non est disputandum*; or, as they say it today, you cannot argue with a Commie. But, said Bacon, if we are careful not to believe in anything, then one removes the obstacles to knowledge and can see facts as they are and learn the truth. And in truth there is no division to schools of thought, to conflicting principles, etc., etc.

This theory is plainly false and refuted by the fact that we always had schools of thought – even the Enlightenment had its schools – and it is not true that people who hold false principles cannot be men of science. One of the greatest thinkers of the Enlightenment, for example, was Joseph Roger Boscovich, who was a Jesuit and hence an Aristotelian of sorts, yet he developed a system compromising between the system of thought of Newton and of Leibniz and so made quite an important contribution to the growth of science.

The Baconians have assumed that we can give up all opinion, consider all arguments for and against each possible theory, decide which comes up tops, and then rationally decide to believe that one.

We know that facts are different. We know how complex is the development of an individual's system of belief, and seldom does it follow the scheme just mentioned. Of course, we know what may influence some beliefs we may hold. For example, one who is hesitant about one's proposals to reform an ailing economy can go and ask friends to criticize them, and then he can ask opponents to do so, and then decide to publish the proposals or to call a press conference. Does this mean that one has decided to believe in the advisability of one's proposals before one has decided to call for a press-conference? Does it mean that if he decides to give up the proposals, he did so on the basis of facts? Certainly not. We know how complex even the simplest decision procedures can be.

Sir Karl Popper often said, we must decide what kind of fact may force us to give up our views, but I know of very few people who can do so, and even they often find *post hoc* that certain surprising criticisms surprisingly made them give up their views, that others ought to but did not, etc. Calling all this irrational will only make everybody irrational.

The fact that we do not think and act in the same way is all too well known. For the classical rationalists this is just impossible. Just as mediaeval philosophy said man is irrational since he sins in spite of his knowledge of the eternal damnation that is the wage of sin, so Enlightenment philosophy said, man cannot act irrationally unless he does not think, unless he is confused, etc. Yet often enough clear thinkers acted not in accord with their own thought. David Hume is an example. He admitted it was easy to be a hypocrite, i.e. act in accord with public opinion!

Whence irrational public opinion? Either it is due to individual people's lack of rationality, and so it is temporary. Or else due to custom. But then custom is either rational or irrational and we may well rationally give up the irrational part of our customs. This is not the whole story. Even Descartes noticed the fact of the influence of history. If there is a mountain path that is not the best, the individual mountain traveller is better off taking it than breaking a new path. But then, it was said, a community of rational men are better off ignoring the old and not so good path and building the best path possible.

This is utopianism and radicalism, of the kind that has failed in both the French and the Russian Revolutions. Now, once the utopian rational society is not around the corner, then, even by classical standards a rational person has to take account of existing customs and traditions, as well as of individuals who do not share his rationality or ideal of rationality. And so, rational thought can no longer be identified for the private and the public cases; and rational thought cannot be identified with rational action; and so, the idea of rational thought stands alone and it has only one rationale – that of Bacon, i.e. he who is not fully rational by Bacon's light is not qualified to contribute to the stock of human knowledge – which idea is quite amply refuted: no scientist we know of qualifies by Bacon's standards.

The theory of rational action that seems today most viable is that of Abraham Wald, though one must admit that it is as yet very fragmentary and yields definite answers only for marginal and extremely simple cases. Quite briefly, Wald considers the list of possible hypotheses to act on, the reward and the punishment of the expected outcome of acting on each given hypothesis, once on the assumption that it is true, once on the assumption that it is false. Moreover, in the same way, one may calculate the cost on the test of a hypothesis and the expected reward coming out of the test, once on the hypothesis that the test goes this way, once that it goes the other way.

All this will not do, except to show how many considerations are involved. On needs, in addition, a policy. When considerations show one choice to entail only gain, and others to entail only losses, then decision has already been made. Alternatively, in cases where risk is very small but possible gain is very large, decision is easy. And so on. But most decision problems are viewed by decision theoreticians as still open.

Nevertheless, I think this theory accords with a theory of rational action that I would like to offer for consideration. According to this opinion, rational action is not uniquely determined and is any action that is possible to deem responsible. And we deem an action responsible if we can explain one thing about it. Suppose an action turns out to be a mistake, and suppose we can ask the actor how come he made the mistake. If he can show that he took a calculated risk, then he is responsible. A calculated risk is just that: the calculated cost of having made a mistaken decision.

As to rational decision in the public domain, the public institutes means of assuring the exclusion of irresponsible public decision, such as parliamentary debate, tests of new products along agreed lines, etc. These are not sufficient and the public attempts to improve them regularly. Rachel Carson's plea is a prominent example as is Bertrand Russell's. Reform is largely about such matters.

As to rational thought, it is simply the dialectics of given questions: what alternative answers we can imagine, what arguments pro and con each of them can we generate.

3. THE FAITH IN RATIONALITY

All this relates to faith in many ways, most of them as yet unstudied. But we can say, empirically, that each of the steps of rationality here described can, and at times does, come from people, individuals, groups, or institutions, with no endorsement of any alternative. When an alternative is endorsed, it is not clear what the act of endorsement means. Pertaining to action, it means a recommendation of practical choice. Pertaining to thought, it is not easy to say. More and more philosophers of science now endorse Sylvain Bromberger's proposal to view the scientist's endorsement as a decision to act – to develop his research along given lines. This means that faith in a given scientific hypothesis is a matter of action, often the action of trying to test the hypothesis, i.e. trying to disprove it. Religious faith, whether in a Christian doctrine or in Scientific Humanism, means something else. What?

The positivist theory of faith is best represented by diverse schools of thought that look so very different from each other, including Durkheim and Malinowski, Buber and Polanyi, and others. It says, a choice of a faith is a choice of a way of life. For my part I find it appalling that this doctrine circumvents the whole question of truth. If I thought Christian doctrine is true I would of course choose the Christian way of life. To tell me that I can choose the way of life even if I cannot say whether the doctrine is true is to tell me something that does not help. In other words, faith is, *inter alia*, an opinion about the world of facts. But what singles out an opinion that it becomes a faith? Positivists say, practising a system means believing it. This is nonsense, I regret to notice, since we all believe in peace and practice war. Yet to say that I declare true what I believe in, what I have faith in, is far from being enough: I declare true what I have faith in because it looks to me true. *Why it looks to me true I do not know* and in fact there are known cases for which the reasons differ. There is *no general answer* to the question why do people think that what they believe in is true, why it looks true to them. It just does.

The theory of rational thought says, ah, but there is a general answer to the question, why do people believe rationally in what they do? This is empirically refutable by many known facts about people who are generally considered rational. Make a list, says Shaw, of people you consider rational and ask if they decided in the same way about their different beliefs. And rational belief is a matter of a decision by definition: a person must rationally decide to believe what stands up to the criterion of rationality. To this Robert Boyle and Benedict Spinoza said, you cannot decide to believe in any idea. You can decide to believe your business partner even though you suspect or know that he is lying, in the sense that you can pretend. But you cannot possibly decide to have the facts look to you other than they do. C. S. Peirce said in a famous letter to William James the same thing – and regarding his own famous essay on the fixation of beliefs: we cannot decide what we should believe in, what not. One day, says George Orwell in his *Clergyman's Daughter*, you find your faith gone and there is nothing you can do about it.

Well, not quite. Karl Popper demands from a person who holds a view, we remember, to specify under what conditions he will change his mind. This looks like a matter of a decision, but I must stress as an empirical fact that it is not. Nor can we always specify. You can ask Popper to specify under what conditions he will change his opinion about rationality. He cannot answer, and he says so very frankly. In Chapter 24 of his *Open Society* he discusses irrationalism. Irrationalists declare the endorsement of an Archimedean point to be an act of faith. They characterize it by considering it as given, unquestioned, dogmatically upheld.

This is stupid, since we never know whether a central opinion that seems to us true today will seem to us true tomorrow. But Popper agrees with the irrationalists and only wishes to minimize the dogma that is the Archimedean point. And the minimal dogma, he says, is faith in criticism. When I was a student this seemed to me true and just wonderful. I still think it is wonderful but I now can see that it is not true.

A popular alternative to Popper's theory of rational thought as openness to criticism is the theory of constructive criticism, endorsed by Catholics and Communists and many thinkers such as George Santayana, Harold Jeffreys, T. S. Kuhn and Paul Feyerabend. It says, when I have a better view than you do – by diverse pragmatic criteria of the better – you will switch to my view.

I do not deny that this happens, but I can report cases where criticism caused people to give up an opinion before having an alternative, and even stranger cases. Indeed the seeker for a new alternative is usually a person not happy with any alternative he knows. And Galileo and Kepler both endorsed Copernicanism and then looked for rational reasons for their endorsement.

We do have faith, beliefs, in many propositions. We know that much of what we believe is false, but we cannot see which of our beliefs is false. And so, a sane person is usually ready to check his opinions. This holds about abstract as well as about concrete objects of belief. The great psychologist James Gibson says, a part of a normal person perception pattern is testing what he sees. Giving up that part one sees things that others do not see, and when they examine the facts they may say that he hallucinates. And so Bacon's theory of perception is false, since it assumes that unless one is careful one prefers not to test one's perception. The same holds for theory. The more perceptive mind is the mind that enjoys considering and testing more theories. At times one just sees the superiority of one existing alternative over all the rest, not knowing why. At times one sees defects in all of them and has a question without an answer and a quest for truth. Some enjoy the suspension of judgement, some hate it and wish to have a fixed faith. They may have their wish granted or not. The question, then, what governs people's faith, is an interesting empirical question. But there are enough observed facts to show that we can have faith with little rationality, and we can be rational while showing little faith. Hence, faith has nothing to do with rationality.

One last paragraph. Boyle, who said, I cannot choose what to believe in, was a skeptic. He advocated a philosophy, nonetheless, the mechanical hypothesis, on the ground of its usefulness in research: it accomodates all experience. This is question-begging and turns out to be false. Yet it is of a significance that I cannot discuss here. Spinoza, who shared Boyle's critique, said, truths manifest themselves: The choice of a proper belief imposes itself on us, and if we are utterly unprejudiced, the choice must be final. I half agree with Spinoza: the choice does impose itself on us and we cannot control it, except the way we can control falling in love – by indirect means. But this seems to me obviously to show that no choice is final and we can choose to put our choice to further tests in the hope of improvement. Rationality, then, entails not taking our faith too seriously.

CHAPTER 32

RATIONALITY AND THE *TU QUOQUE* ARGUMENT

The *tu quoque* argument is the argument that since in the end rationalism rests on an irrational choice of and commitment to rationality, rationalism is as irrational as any other commitment. Popper's and Polanyi's philosophies of science both accept the argument, and have on that account many similarities; yet Popper manages to remain a rationalist whereas Polanyi decided for an irrationalist version of rationalism. This is more marked in works of their respective followers, W. W. Bartley III and Thomas S. Kuhn. Bartley declares the rationalist's very openness to criticism open to criticism, in the hope of rendering Popper's critical rationalism quite comprehensive. Kuhn makes rationality depend on the existence of an accepted model for scientific research (paradigm), thus rendering Polanyi's view of the authority of scientific leadership a *sine qua non* for scientific progress. The question raised here is, in what sense is a rationalist committed to his rationality, or an irrationalist to his specific axiom? The tradition views only the life-long commitment as real. Viewing rationality as experimental open-mindedness, we may consider a rationalist unable to retreat into any life-long commitment – even commitment to science. In this way the logic of the *tu quoque* argument is made irrelevant: anyone able to face the choice between rationality and commitment is already beyond such a choice; it is one thing to be still naïve and another – and paradoxical – thing to return to one's naïveté.

1. THE FAITH IN REASON AGAIN

(1) The idea of rationality, of thinking and living by the light of reason, has fascinated some of the greatest minds in history. In the West, excessive claims for reason led around the turn of the century to severe disappointment and to the rise of all sorts of irrationalism. This is quite understandable. Certain irrationalists have declared that the disappointment is inherent in the situation, that reason, excellent as it is, is severely limited. The limitations of reason, they said, are acknowledged in the very recognition of the need to supplement reason – by a leap of faith. Now, the view that reason is limited is ancient, an essential part of latter-day skepticism in the Hellenistic world. Perhaps the leap of faith was no news to the ancient Christians either. What

is new these days, then, is largely the clarification and sharpening of positions. Also, this led to the *tu quoque* argument: *tu quoque*, says the irrationalist to the rationalist; you likewise have made a leap of faith, in your very jump to commit yourself to the life of reason: for prior to your becoming a rationalist you had no reason to become one. Hence your central choice was as irrational, as arbitrary, as mine.

(2) The two great philosophers of science of our own day and age who have responded to the *tu quoque* argument are Sir Karl Popper and Michael Polanyi. Both have openly accepted the *tu quoque* argument as valid and have explored its implications. (Popper, Ch. 24, *The Open Society*; Polanyi, *The Logic of Liberty* and 'Critique of Doubt' in his *Personal Knowledge*.) Oddly enough, Popper is a rationalist whereas Polanyi is an irrationalist. It looks easy to pinpoint the rationalism of the one – who is anti-authoritarian – and the irrationalism of the other – who commends the authority of the scientific leadership yet as something far from absolute (as reason is severely limited). It seems easy to argue from their texts that the one advocates as much doubt and criticism as possible, yet admitting that total doubt is impossible, whereas the other advocates that doubt be concentrated and limited since total doubt is impossible and effective doubt must be well aimed. The harder one looks at the difference between Popper and Polanyi, as Imre Lakatos has noted, the more they tend to vanish, and the harder it is to see the difference between their different commitments to reason. This surely is an intriguing observation (and explains Lakatos's own shift from Popper to Polanyi).

Regarding rationality, Popper's chief disciple is William W. Bartley III in his *The Retreat to Commitment*, and Polanyi's is Thomas S. Kuhn in his *The Structure of Scientific Revolutions*. Textual hermeneutics become harder here, since Bartley makes excessive acknowledgements in his book and Kuhn in his tends rather the other way. (They both made corrections on other occasions, let me hasten to add.) Yet, for what it is worth, I shall take their views as mere elaborations on earlier ones. I should say, however, that though the *tu quoque* argument is not new, and although it was taken seriously by a few philosophers of science, it was Bartley's *Retreat to Commitment* which gave it the new status it enjoys today.

(3) Popper's concern is to minimize the constraint on reason due to the *tu quoque* argument, so as to let reason have maximal scope. This very desire to be rational is, of course, by the *tu quoque* argument quite irrational. Popper admits that much. He tries to admit no more. The hope that comes with rationality is irrational, but apart from it we lay any assumption we make wide open to criticism, as wide as possible, indeed, and this makes all

our views rational enough. To this Bartley adds, we can lay our hope in rationality, too, wide open to criticism and with this move we leave no vestige of a concession to irrationalism. The *tu quoque* argument is thus fully taken care of.

Polanyi's admission, is, like Popper's, that he has made a leap of faith when he endorsed science. (Notice that Popper says he leaps to rationality; Polanyi says to science.) In science, he adds, there is a social structure and a leadership, both tentative and open to criticism and improvement; even to revolutions. Kuhn is more specific: the organization of science is aimed at the proper coordination of scientific training and research and this is both essential to science and not too well justifiable rationally since the outcome of research is not predictable. And so, at times when research is sluggish it is time for a revolution; the revolution can be led by an existing leadership or by Young Turks who become the new leaders after the successful overthrow. This, Polanyi and Kuhn suggest, does not do away with the authority of the scientific leadership, but it legitimizes the authority within a democratic society.

(4) Yet the prior question is, can we at all choose reason or unreason? Is it really up to us to make such a momentous decision? If the answer is no, then the debate between all parties involved is purely academic and therefore futile and therefore self-defeating. Not all futile debates are self-defeating, of course; only debates meant to lead to a course of action. And the debate about rationality is indeed about the act of commitment, whether to rationality or to some other creed. This is so much so that one may wonder if it is not the case that, since the debate is practical – conducive to an act – and therefore fundamental, these two facts do not make all participants in it pragmatists of one sort or another. To support this one might notice that already in his *Logik der Forschung* of 1935 Popper defended the scientific venture on the ground of its fruitfulness.

Fruitfulness is, as Lakatos notes, an ambiguous concept. This, perhaps, does not matter, since pragmatists may indeed differ in their view of what is desirable; which obviously only adds power to the *tu quoque* argument. Popper makes it amply clear that the fruitfulness of science which impresses him is not material but intellectual. This will not matter at all except in the following one case. If science is so fascinating that I simply cannot tear myself away from it, then surely we may say that for me the problem of choice does not exist to begin with, that my addiction to science is a matter of neither a rational nor an irrational choice – it is no matter of choice at all. This case, however, has not yet been discussed in the literature. Many philosophers, including Heine, Marx, and Popper, spoke of the moral obligation to accept rationality. This, of course, is an eminently respectable sense of 'fruitfulness'; it still makes its followers pragmatists enough to face the *tu quoque* argument: without reason we cannot say that reason alleviates misery, and with reason we have already prejudiced ourselves in reason's favor.

The question, should I commit myself to the life of reason? is, to repeat, circular in the way the irrationalists declare: if you use reason to answer it you prejudice the issue; if not, your answer is not rational but arbitrary. Let us, nonetheless, examine the meaning of the question. For a pragmatist the question is hardest, since it is a question of fact: does the use of reason in fact further certain given practical ends? We use social science to answer such questions, but this betrays rationalism of even a fairly extremist type (not all rationalists are committed to 'hard' social science). Moreover, why one practical end rather than the other? But anyway, pragmatism is only a specific kind of moral philosophy, and quite generally, the choice of a moral philosophy may or may not involve rational thinking and, in return, any given moral philosophy may advocate the use of reason, demand to limit it, or to oppose it, or may be indifferent to it. This, then, is the vicious circle involved in rationality.

(5) The situation is disconcerting because when we have a vicious circle like that we do not agree as to what to do about it. The major error behind the *tu quoque* irrationalist argument is the idea that there is nothing a rationalist can do in the face of any circularity but admit error. This is just not true, either historically or in point of philosophy. The reason circularity has been mistaken for a deadly criticism is very obvious and has been repeatedly explained by both Bartley and Popper. (On this Popper has forcefully acknowledged his debt to Bartley in a radio talk, 1962 and 1964 at the University of Illinois, which I do not think he intends to publish.) When rationality is identified with justification, and when the skeptic's critique of justification – one witness requires another witness to testify to his veracity – is fully savored, then both infinite regress (infinitely many witnesses each testifying to the veracity of his predecessor) and circularity (two witnesses each defending the other's veracity) do undermine justification enough to refute it by its own very lights. Otherwise, neither circularity nor infinite regress is in itself (or as such) necessarily damaging.

Supposing this to be so, the question is, is the circularity of rationalism as bad as the leap of faith of the irrationalist? Is either bad? Here I feel both Popper and Bartley have skipped a step in the debate, and here Polanyi and

Kuhn have the upper hand by merely declaring the leap of faith good – *their* leap of faith of course.

Is, then, the circularity of rationality bad, or rather the irrationalists leap of faith, or for that matter the rationalst's?

This way of putting the question seems to me quite provocative. Bartley says, we owe it to ourselves to answer the *tu quoque* argument. But for my part I feel it is much better to backtrack a little. The *tu quoque* is a reply to our criticism, a riposte to our attack. It is possible, of course, that our attack was overzealous. Of course the question, should we at all attack irrationalism? is only a pragmatic question; I cannot possibly see any expected intellectual benefit from attacking people who have given up, or voluntarily limited, the use of their intellect. But morally, they may be a nuisance, both as power politicians and as educators, whether in military academies or in public schools. And the best way to fight them may – or may not – be to show the weakness of their irrationalism.

2. RATIONALITY AND MORALITY

(6) Thus far the debate was based on the assumption – a question-begging one! – that the irrationalists are immoral. I do not mean that they may endorse any old immorality in their leap of faith – some do and some don't and this is neither here nor there in the present context. It is not that any old immorality is of no concern in the present context; rather, the question, does a faith one leaps into have this moral character or that? – this very question requires some application of some standards of morality and of rationality. Hence, either we choose an arbitrary standard (leap into the faith in a standard), or we attempt to use the standard of rationality (provided rationality does incorporate a moral standard; and, indeed, all parties in the present discussion agree that it does). And so, the deepest question, both moral and rational, is not so much whether the faith one leaps into happens to be moral or immoral, but, and much more so, whether the very jump is moral or immoral.

And so, we may declare, for the rationalist the greatest immorality of the irrationalist is his very act of leaping into any faith whatsoever, except the faith in reason itself. Any betrayal of reason, says Popper (here he follows Russell and an earlier tradition), is immoral and leads to evil deeds, even if it is the preference of love over reason.

Strangely, an abstract argument gains here great plausibility from concrete illustrations, such as religious wars. Yet we ought, in all rationality, to be able to conduct the debate on the abstract level alone, and ask, abstractly, what is the evil of a leap into a faith (other than the faith in reason) as a leap into a faith?

Suppose we accept a version of critical rationalism, or of non-justificationist rationalism. That is to say, we accept any doctrine which (a) admits the inability to support or justify any belief, and (b) allows any faith on the condition that it be held tentatively, i.e. be left open to criticism and when effectively criticized be relinquished. First and foremost, this means that versions of critical rationalism are admitted which permit any leap of faith provided the faith is held tentatively. This looks much too permissive, especially when we say that ever so many faiths and doctrines of all sorts exist which seem impervious to all possible criticisms, from certain religious doctrines to stories about goblins to fantasies generated by paranoiacs of all sorts. Is it really rational to hold *any* faith if it is tentative? Is it really not objectionable ever?

(7) The question seems to me to be still wide open. The old rationalists and the new positivists, being justificationists, could hardly be expected to examine it, and non-justificationists are just now coming to cope with it. Interestingly enough, it seems to me that Popper's theory cannot give us an unequivocal answer. In his *Open Society* (1945) he suggests, it seems that as we should minimize the leap of faith, we should only accept tentativity in the above sense (openness to criticism) on faith and hold tentatively only such views as are open to criticism. But this cannot be. In his *Logik der Forschung* (1935) Popper deviates from this requirement in one direction as a private citizen and in the other as a member of the commonwealth of learning. In his first capacity he admits to the holding of an uncriticizable faith in the existence of an external reality, so-called. In his other capacity he recommends quite explicitly that among all existing alternative views we should hold the one most highly testable, which, in our context should read, the most highly criticizable. This is not an inconsistency, of course. Yet it leaves us with ample interpretations. Perhaps Popper says, publicly we should hold only the most criticizable theories, perhaps any criticizable theories; and perhaps he thinks private uncriticizable theories are private weaknesses, perhaps things too pale to be noticed, perhaps a matter of sacred privacy – I really do not know.

Whereas Popper comes to minimize the content of the leap of faith, Bartley cares only about the tentativity of the outcome. Hence, he says, as long as we hold even tentativity itself tentatively we have no trace of irrationality left. Of course, once we accept Bartley's proposal, then not only rationality, but realism, and goblins may be allowed as well – tentatively, of course. It is an interesting fact, and less amazing than I thought at first, that Popper once dismissed Bartley's proposal as rather silly and once endorsed it as rather obviously commendable.

It is, indeed, not easy to see what exactly is involved. In a concrete predicament, every rationalist, when effectively criticized, will change his view, including the justificationist who holds his view as final and unassailable. Indeed, the pre-Einsteinian period was full of Newtonians committed to their views, they said 'forever', yet who changed their views when effectively criticized. And Ludwig Wittgenstein who published his early ideas on the assumption that they were final and unassailable even ridiculed them in his later works. It is not a concrete situation, then, that Popper and Bartley differ about, but the abstract reading of it. I confess I cannot myself put my finger on it, especially since the idea of criticizability is so vague. I find only one thing clear and common to both Popper and Bartley: they are reluctant to admit any leap of faith, they admit that some leap is unavoidable, and they try to minimize it, either in content (Popper) or in commitment (Bartley).

What, then, is so immoral and objectionable in the irrationalist's advocacy of a proper leap of faith? Only the mode of the recommended leap, the fact that they willingly recommend a leap of faith, that they recommend that we take as above or beyond criticism at least one major dogma, religious, social, or political, and perhaps a few ancillary ones as well; but chiefly their immorality is, for the critical rationalist, rooted in their recommendation that the leap of faith should be final – that the act of leaping should be one act that should suffice for a whole lifetime. It is here that opinions differ, then: for the unreasonable suggestion is of one leap of faith in a lifetime whereas the more reasonable suggestion, since we are fallible, is never to burn our bridges and always permit alternative (major) leaps if need be.

(8) This does not do away, however, with the *tu quoque* argument. Rather it makes it quite different, and perhaps quite unanswerable.

The man who commits himself once in a lifetime may be mistaken; it is, perhaps, wiser to commit oneself to different doctrines at different times and perhaps finally to return to one or another, but even then not necessarily for a lifetime, since a new and promising alternative may turn up later on. This last statement, however, itself becomes a lifetime commitment.

There is no pun here, nor a question of levels of commitment or any other old nicety. It is a matter of brute fact that some possible ways of life are open to one, some closed; and by the very decision to become a roaming fox of a rationalist a person gives up forever the very option of being a hedgehog who stays put, as perhaps his father or brother or sister do. The rationalist relinquishes the option of clinging to the *status quo* which so many prefer.

It often happens that a person is born into a fairly stable way of life, never questions his being destined to his station, and then all of a sudden he learns – as far as my experience goes, always from others – that he can question, that he can choose. I have met such persons, sometimes a student in university courses, undergraduate and graduate, at home and abroad, east and west. They are all fairly quick to present the argument as I have put it in the previous paragraph.

(9) Now brute facts of life are not subject to debate, and no one can deny that the life of experiment is incompatible with the life of stability. What is there to do about this fundamental and irreversible choice of a life of experimentation? Is this not the leap of faith and the irreversible commitment to rationality?

We have now put a very heavy burden on irreversibility. First it was the irrationalist who advocated irreversible commitment. Now, it turns out, it is the rationalist's way of life which involves a commitment as irreversible as the loss of naïveté! This, indeed, is a *tu quoque*.

But here a surprise awaits us. If the rationalist's irreversibility means a factually irretrievable loss, what then is the meaning of the irrationalist's irreversibility? In what sense is the irrationalist's commitment irreversible? Have we not seen people declare an irreversible commitment and then reverse it? And if so, is not the irrationalist demand for irreversibility mere talk?

(10) In diverse committed circles, religious, philosophical, ideological, political – I here make an empirical observation – there is an answer to that: once committed, they say, always committed. What they mean is, of course, that the very failure of a person to stay committed proves that he, at least, never entered – perhaps never fully entered – the life of commitment.

Qua rationalist one might pounce on that answer as *ad hoc*, as an excuse, as a defense of a refuted dogma. And for any argument between two rationalists such a move may – I am not convinced even of that – be quite in order. But in a debate between a rationalist and an irrationalist surely that is a wrong move, question-begging, side-tracking, asking for trouble.

No; what is meant by the life of commitment is surely the recapture of naïveté! Can naïvaté be recaptured? Here I wish to make my own partisan rationalistic appraisal: when in process of just losing one's naïveté one does have a period of grace, when one is still not sufficiently adept at rational debate and experiment, when one's slight deviation into the rational way of life is still accountable in terms of the tradition from which one comes, in a manner which will still seem fair enough. But the period of grace is brief. All examples beyond it are of self-deception and self-torment.

There is, in practice, a more tangible part to this torment than mere self-deception. As gangster movies of the thirties made amply obvious, when one chooses the life of crime one is not accepted into criminal society before one is seriously committed – usually by becoming wanted by the police for some very serious and not easily shakeable charges. In universities, administrators often look for faculty members' commitment to their own universities, a commitment expressed by the purchase of a house, by joining some country club, and by public statements. The social aspect of all commitment, ideological, religious, political, is similar. Underground movement leaders love their novices to be caught distributing leaflets.

Here rationality turns out to be a certain degree of character, of independence. It is easy for a scholar to endorse a view and then reject it, especially when he belongs to the society of rational scholars. It is much harder, I observe (I confess I was never able to understand why), to free oneself of commitments to institutions which guard one's past convictions.

Even here, of course, the *tu quoque* argument is very powerful even against the most rational scholar: it is well known that there is an entry fee, and a fairly high one, to the community of scholars. Even for an acknowledged scholar it is very costly to enter a sub-community of scholars to which he has not yet been admitted. Again, let me observe, it is very hard for a scholar from one field to publish a paper in a journal devoted to another field even if it is admittedly an eminently suitable paper.

3. RATIONALITY AND SOCIETY

(11) In this light the philosophy of Polanyi and of Kuhn turns out to be fascinating as the sociology of those caught in the middle, of those who have lost their naïveté but who are not strong enough for the stormy adventures which the rational life may offer, of experimentation on a large scale. What Polanyi and Kuhn describe as normal science is an in-between tradition, a life of commitment to scientific research, to scientific doctrines, and to the readiness to make endless *ad hoc* adjustments. Indeed, perhaps the most important technical point which distinguishes the rationalists from those who are pro-science yet not quite rationalists, is the question of *ad hoc* hypotheses: the rationalist has an aversion to them even when, as at times he does, he employs them for want of anything better; the irrationalists tell us, on the contrary, that *ad hoc* adjustments are standard in science. They add,

however, that when there are too many of these for comfort, a revolution may be under way.

And so, for Polanyi and Kuhn the commitment to science is a matter of a lifetime, and the commitment to today's doctrine is less final as there are, alas, unsettling scientific revolutions from time to time!

Why is the commitment to science a lifetime one? Because its victim is too critical to return to the life of a stable tradition – perhaps there just isn't a suitable one in his vicinity – yet not brave enough to become truly rational and boldly experimental.

But this last point is purely psychological. What happens to those who in fact drop out of science and return to the fold of the Church or Party? What happens to those who stay within science and become rebels? Polanyi and Kuhn will not say. Of the former, evidently, they will agree with me when I say, either he was never competent at being critical or he suffers torment in repressing his competence. Of the latter, they will say, I suppose, he aspires for the leadership of the revolution, he is a Young Turk and a potential leader. If the Young Turk succeeds he becomes a leader; otherwise, says Polanyi, he simply loses: there is no guarantee for success anyway.

(12) By contrast, for Popper and for Bartley the man caught between the life of naïveté and the life of experimentation, falls between stools. For them science or rationality is experimentation, and whether the brave experimenter is a leader or not signifies little.

I find this view more satisfying, morally, intellectually, temperamentally; but, above all, I find it true that the rank-and-file scientist is a sad case, a person who has neither stability nor adventure and so misses everything.

Perhaps here is the place to say a little more about incertitude or lack of guarantee. The normal scientist is fairly secure; he ventures little and can expect little surprise. The ambitious scientist, says Kuhn, spends sleepless nights; says Polanyi, he gambles, he covets bitter controversy, he may lose his leadership, his position, etc.

The psychology of the sleepless night is a virgin field. Kuhn does not say so, but I understand him as saying that the normal scientist is spared the agony of the sleepless night. This is still the old-fashioned and misanthropic work ethic which exhorts us to strive to overcome our natural dispositions and suffer and work; for the scholar the work ethic recommends burning the midnight oil and brushing the sleep from one's brow. I will not go into all this here. I would only say that the work ethic deprives the scholar both of sleep and of the pleasure of work; that the genuine searcher is too excited to go to bed and he finds the experiences of the sleepless night rewarding enough. This, I think, is Spinoza's view, and the strongest defense of rationality.

(13) Now what kind of argument is that, and what rationality is it bound by, and how strong is the bond? I tend to agree here with the view asserted in passing by Robert Boyle, three centuries ago, in his light *Occasional Reflections*. I do not know what makes me think as I do, he said, and if a prince would try to make me think otherwise, I would not know how I could obey him. (This view was in a way endorsed by Spinoza.)

Rationality is really a part of our way of life, and that goes alike for the rationalist and the irrationalist in our midst. What we can decide is whether we like it or not, whether we wish to drive it to its limit and expand it or take it as it comes and be content or even try to avoid and repress it and then succeed to some measure.

And so the question, in my opinion, is not absolute, but relative: Do we wish to become more rational or less? And once we accept it this way we are, when we ask it, already rational to some degree. We may be rational to a sufficient degree, I fancy, to make the desire to increase our rationality trivially preferable. But we may go the other way, as Polanyi and Kuhn claim, quite legitimately.

(14) The question, then, should we be rational? is to my mind a rather artificial one. The very ability to ask it tells us we are. Admittedly, when we are just starting to exercise our ability to be rational we can still revert to naïveté. Admittedly we may try hard to suppress criticism and skepticism and say, *credo, credo, credo.* What this amounts to is, indeed, the repression of a thought process. It is hard to see what good this brings. Admittedly we may try to control the thought process and think timidly and carefully, like Kuhn's normal scientists. To the extent that, and in cases where, this is reasonable, it would be rational to discipline our thought. And if in doubt we may try at times to think timidly, at times wildly, or have a division of labor based on temperament. These issues are still open.

(15) With this, I suggest, the *tu quoque* argument vanishes.

Or rather, to be precise, it is transformed. For those of the *avant-garde* who wish to push rationality farthest, the reasonable thing is to attempt to present theories of rationality and to try and improve upon them rationally. Here we are back with both circularity and infinite regress. This makes progress difficult, but since rationality is no longer justification there is no harm in that, only problems piled upon problems. For some, that sounds tiresome and tedious, for others not. That seems to be the situation now with rationality.

To conclude, whereas the rationalists do have certain presuppositions, as Popper observes, the *tu quoque* argument that these presuppositions are therefore irrational need not be accepted, as Bartley observes. The reason is that these presuppositions are not easy to shake. I think this, the view presented here, is of J. W. N. Watkins, and I am glad to join him in it; yet I do not think he would wish to present this view as opposed to Popper, only to Bartley. So I take the blame myself for contrasting this view with all other extant views, and to present it thus. Both rationalists and irrationalists are prone to assert that one cannot reasonably deny some claim – that we need certainty, that we need God, that God exists, that 2 + 2 = 4, that I exist, that my arm exists, that I have a terrible headache or toothache now, that I see a brown desk. (My own choice is that I am a man born of a woman, as the Bible puts it.) All proposed corollaries to these claims, even in case the claims are true, are objectionable. The correct corollary is that it is very difficult to criticize certain claims at certain times, but that we are in a tradition of criticism that at times makes progress and so raises the degree of our rationality. Attempts at criticism are highly valued when successful, and should be valued also when not. This demand, I think, will also improve our rationality.

CHAPTER 33

TECHNOCRACY AND SCIENTIFIC PROGRESS

1. ROBINSON CRUSOE TODAY

When Newton's contemporary, the English author Daniel Defoe, heard a sailor's tale about having been shipwrecked and marooned, he was seized with great excitement and ran home to start writing his classic *Robinson Crusoe*. It is not that a marooned sailor is such a rare bird, that the very meeting of one fires the imagination. Nor is there in the story Defoe wrote any detailed knowledge that requires an interview with an experienced informant to find it out. On the contrary, inasmuch as there is detailed information in Defoe's book, it has to do with the inner life of his hero; most likely this reflects the mentality of the author much more than that of the informant. For, what we hear about the ecology of Robinson Crusoe's desert island, and about the desired knowledge concerning survival on it, is less than minimal. If anything, the author takes literary license and allows his hero to salvage from the wrecked ship a few essential tools that he could not possibly manufacture, a few seeds which he certainly could make do without, and a Bible. The story is most impressive in its austerity and bareness. When things start moving at all and Man Friday appears, the story is drastically altered. Up till there it is a parable of the self-sufficient individual; from then on it is the parable of the role of the self-sufficient individual as lord and master.

The self-sufficient individual, Robinson Crusoe, is a symbol of the age. Most introductory economic theory texts, for example, begin by presenting a Crusonian economy. A Crusonian economy is marketless, yet economic theory concerns the market. This is why Robinson Crusoe exits from the introductory economic textbook for good on the first or second page, never to return to it; in the whole economic literature he makes extremely rare visits and always very brief ones. It is not his absence that is puzzling, however, but his very presence, even for the briefest of moments, not to mention it being in the opening scene. And, indeed, the secret of the situation is that Crusoe never really leaves the economic textbook; rather, instead of acquiring a man servant he meets others like him. For, Crusoe has two characteristics: autonomy and loneliness. His loneliness, in another novel, might force him to become autonomous; indeed that other novel has been written by Jules

Verne: *School for Crusoes*. Defoe's novel portrays his loneliness only as a means to illustrate his autonomy and his autarchy – his independence and his self-sufficiency. In the economic textbook, the individual's autonomy is introduced with Crusoe; his loneliness then leaves the scene to permit the opening of a market; but his autonomy remains. Market economy, we can say, is a system of Robinson Crusoes transacting sales and purchases of goods and services; it is what Karl Popper calls an abstract society. When the abstract society becomes real, that is to say, when real society loses its everyday characteristics that make the individual a person, and leaves him a nameless social security number, then we have not only a market economy but also a society of alienated individuals, the likes of which we meet in the novels of Kleist, Kafka, Sartre, and Camus.

Robinson Crusoe is self-sufficient or autonomous, and as a matter of course. One can say he hardly notices this fact. He suffers hardship, both physical and mental. His physical hardship is described clearly enough but with no depth and no exceptional perception or familiarity. His loneliness is described vividly: the author had presumably had firsthand experience in loneliness and was marvellously perceptive about it. Yet the treasure he discovered was lost: the loneliness he described was that of a man marooned, and this is as expected – indeed one cannot but expect loneliness on a desert island. Indeed, the power of the book is just in its lack of any surprise, in its presentation of Crusoe as a most familiar figure.

Only recently, and in the light of newer developments, was it observed that Crusoe was not at all familiar; that he entered the family circle as if he was a member from birth, even though he really was a stranger – a total stranger, perhaps; an ideal. The philosopher Ernest Gellner has recently viewed him as the symbol of the philosophy of the Enlightenment. René Descartes, the father of modern philosophy, undertook a journey no less frightening, and entered a loneliness no less engulfing, than Robinson Crusoe; and this he did while sitting in an armchair in front of his hearth one snowy night while a soldiering in a foreign land. Descartes, like Crusoe, faked a bit. He did not take household implements with him, since the journey he undertook was a journey of the mind. But, like Crusoe, he did take the Bible with him. When he entered his period of utter doubt, including the doubt in the existence of his fellow men and even of his own body, he was doubtless terrified. The great irrationalist anti-Cartesian Søren Kierkegaard admired him for his guts. Yet before he undertook the voyage he made a vow that if he came back intact he would go on a pilgrimage to a shrine of the Holy Virgin. And, for all we know, he did come back intact. But philosophy did not; nor did the social sciences. They stayed Crusonian. The major division between the modern philosophers, between the intellectualists and the empiricists, was not concerning Crusonianism. All the leading philosophers of the Age of Reason, Locke and Berkeley and Hume, just as much as Descartes and Spinoza, and as Leibniz, they all were Crusonian. And this Crusonianism permitted taking seriously the stupid and paranoiac doctrine, solipsism, which says that only I exist and you and they are all figments of my fertile imagination. And since the Crusonian ethos allowed for solipsism, this doctrine was advocated by this or that philosopher, for example by the famous Johann Gottlieb Fichte. As Heinrich Heine describes it, Frau Professor Fichte invited the wives of her husband's colleagues for the usual afternoon tea, meaning tea and cakes and crumbs of university gossip. Is it true, Frau Professor Fichte, they said, is it true that your husband does not believe that anyone but he exists? Does he even deny that you, dear Frau Professor, exist? And do you permit him to deny that you exist?

This terrific and ruthless joke of Heine sounds good enough because it takes us away from the clouds of Crusonian metaphysics and lands us back on the ground of everyday life. But this is a superficial reading of it. Crusonian loneliness is not so metaphysical; Heine had experienced it often enough; and many writers from Kleist before him to Kafka and Camus after him have described it in details. The joke is so cruel because it tells us that Fichte was not autonomous, he was his wife's Man Friday, and so, lonely or not, he had no right to Crusonian metaphysics. The same holds for others, for many others. Bertrand Russell tells us that D.H. Lawrence was obsessed with sex for fear of solipsism: sex secures the existence of at least one other person besides one's own self: one's sexual playmate. But this is not true: Lawrence was not an autonomous Crusoe but a Man Friday to his wife Frieda. Anyone who has thrown a philosophic glance at his *Women in Love* knows that. And, indeed, D. H. Lawrence was tired of the autonomy of the individual and preached the herd philosophy of Fichte and Hegel and other romantic reactionary irrationalists.

The advocates of herd philosophy hate the classical economic theory not because it fails to deliver the goods, and I think it does fail to deliver the goods; and they hate it not because the individual human beings it describes are abstract lonely Crusonians, and I think the description is indeed unreal; they hate classical market economic theory because it preaches moral autonomism, and in the most powerful way there is: by taking it for granted. Economic theory takes the autonomy of the individual as much for granted as Daniel Defoe did – in all of his many works, big and small, concerning moral and immoral people.

Of course, classical economics assumes that, unlike Crusoe, humans do interact. Since they are allowed to stay isolated in all senses except in the economic sense, what makes them interact is one and only one factor: the economic factor that is the prime cause of all exchange. It is, as Adam Smith has taught, the division of labor. Take two Crusoes, allow them to specialize, and if they do they must trade; allow them to specialize and trade and nothing else, and they are two Crusoes economically better off than without trade. The principles of classical economic theory are just stated in this last sentence. Hence, the romantic enemies of the Enlightenment movement attacked the division of labor. They replaced moral autonomy with economic autonomy. Whereas the Enlightenment philosophers broke Crusoe's economic isolation but forgot all about his moral isolation, the Romantic philosophers, complained about the alienation or estrangement or loneliness of the modern man; and they called people to return to the community in matters moral, and to autonomy in matters economic.

This is schematic. History, as usual, shows that people who hold fantastic sehemas adjust them to accomodate facts from the real world. Thus, Adam Smith had a great disciple called Jeremy Bentham who declared that Crusoe can purchase friendship. Bentham thought he was the Newton of the human sciences because he discovered that humans gravitate or attract each other. In other words, he notices that there was no human attraction in the philosophy of Adam Smith. Yet the attraction he discovered, the need for company that can be purchased, is uncomfortably near to its caricature, Dale Carnegie's *How to Win Friends*. Similarly, the peasant who is morally dependent on his community but economically autonomous is not quite fully economically autonomous; and he may have a son who is morally autonomous, though very seldom. We can overlook these details here.

It is easy to also overlook the fact that economic theory was the prime model for all social theory of the Enlightenment, that the Enlightenment has presented people as autonomous beings marooned on this desert island Earth having to make do as best they can. But such oversight will prevent us from noticing the force of the romantic reaction which rightly complained that the image of man projected by the Enlightenment philosophy was that of man doomed to alienation, and which took history rather than economics as the prime model for a social theory. The Enlightenment thinkers were not so foolish as to ignore the possibility of the failure to implement their grand scheme of a rational society, due to a dangerous slip; but they thought a little care will go a long way to prevent calamity. Smith was aware, for example, of the critique of extreme division of labor. He said, too much of it destroys the worker; and he said, thinkers as thinkers contributing to the economy, must be exempt from it. But the whole matter seemed to him to deserve no more than a couple of cautionary remarks. Had he seen Renoir's *A nous la liberté* and Chaplin's *Modern Times*, he would no doubt sympathize; but also he would be puzzled that things were allowed to get so much out of hand.

2. THE RISE OF TECHNOCRACY

How did things go so much out of hand? First there was the dogmatic insistence that things cannot go out of hand and hence action should not be taken to remedy the obvious defects of the system. John Meynard Keynes enlarged on this point in his *General Theory* and I will not go into it, since I wish to avoid debating economic matters. But what added to the severity of the situation, perhaps as the major factor, was not the market mechanism, nor the dogmatic over-reliance on it, but the vice of technocracy. We get echoes of this in Aldous Huxley's novel of the same era. In his *Brave New World* we have a charming female lab assistant who works in a test tube baby factory. Her work is very sophisticated but also so specialized that outside her exact location her whole training is of no use whatsoever. She finds herself in a situation not much different from that of the heroes of Jules Verne's *Mysterious Island*. But they are Crusoes and use elementary trigonometry and elementary geometrical optics and other bits and pieces of physical science as means of survival; she is the exact opposite of Crusoe: she is utterly helpless. But at least she is no scientific researcher, only a scientific industrial employee. The scientific researcher appears in other novels of Huxley, such as *After Many a Summer Dies the Swan*; he appears as crude, proud of his being inarticulate, and even narrow-minded and unscrupulous. In other words, research became an industry, and the researcher lost his autonomy and became someone else's assistant. At least in the novel cited the difference between the autonomous researcher and his intellectually able but morally poor assistant, is made amply clear. In Werner Braun's autobiographic movie, the specialized assistant has grown to be the researcher devoid of all moral or political sentiment.

Does specialization demand the loss of autonomy? The loss of moral and political interest? What is the rule to distinguish good from bad specialization? What is the permissible limit?

I will not answer these questions. They are put in a Crusonian framework and so they beg the question. To assume that people are morally and economically autonomous, that they must specialize so as to retain economic autonomy, but should not go so far as to lose their moral autonomy, this sounds noble but is too defeatist a position and too unrealistic a hope.

The problem is in part educational, since we can educate the young to increase or decrease their moral autonomy and economic resourcefulness. And we must notice that an ambitious research scientist today is likely to be an educator, and if he knows what he wants he can do the routine part of his educational job fairly perfunctorily while screening from the scores of students he meets every semester no more than one or two, train them as his assistants while getting for them adequate grants and while his wife invites them and their mates for dinners at regular intervals, and in a few years he can develop a powerful research team, direct them with a sure hand, and make a place for himself and them in his specialized community, perhaps even win a Nobel prize and share with them its emotional and economic fringe benefits. Usually the achievement of such a researcher, if he has one, is a benefit to mankind; yet the cost of this achievement is such that I do not hesitate to call the Swedish Academy the unwilling modern Moloch which absent-mindedly devours innocent youths in many universities the globe over: for the training they receive is for the absence of both moral and economic autonomy which they trade for careers.

The problem, then, is in part educational, especially since the Nobel Prize winning slave-driver is often an educator taken as an educational model. Yet even in the educational part of the problem it also pertains to the organization of the research side of the university, especially the part of research for which the Nobel Prize is attainable. It thus affects profoundly the university structure as a whole, including the fact that some universities are structured to grant higher degrees, some not.

The problem, then, is educational and relates to the structure of the university. Since many a high school excells in sending so and so many of its graduates per year to institutions of higher learning, especially famous ones, the trouble seeps from the Swedish Academy down to the small rural high school whose headmaster happens to be a bit over ambitious. But I will not stay with this topic. Rather, let me broaden the scope: the structure of a university interacts with the broad structure of the society to which it belongs, particularly in catering for the spiritual and ideological needs of that society. It is neither an accident nor a secret that academics are not great supporters of democracy these days, since they are too often both elitists and technocrats. Technocracy, I think, is a major cause of specialization and to the point of loss of all autonomy. Let me elaborate.

What we want is efficiency. This is a fact. We may condemn this as taking means for ends and we may condemn this as materialism. In fact it is Crusonianism devoid of its autonomism, and thus pointless or degenerate: it is Crusonian to want from each other only specific goods and services and in a manner as matter-of-fact as possible, since it narrows our commerce to a minimum and permits us to have the rest of our time and concern for ourselves. It is, however, degenerate and even pointless in that when we have specialized we also lost our autonomy. A Jew who devotes as much of his time and energy as possible to the study of the Law and living in its lights, must also work to make a living, but ideally he should spend as little time on work as possible, and therefore do so as efficiently as possible so as to earn his minimum required wage while losing as little time as possible. Jews were a Crusonian community among the nations; as long as they took their own affairs seriously they took other business efficiently. Once one loses one's ends, efficiency becomes pointless indeed, a mere habit or a substitute end, or at best a stop-gap until we find a new and a better goal than the one we lost – whether is Jews, or as Christians, or as members of the Enlightenment.

And so we do want efficiency, but not efficency alone or as a supreme end. And in order to attain efficiency, we want the people we engage in a specific task to be expert in it. So we do want expertise. The inference, by the way, is invalid: we can only want expertise when the task is known and well defined. If we want efficiency as a means, then the end limits the demand for expertise. And if we want efficiency as a stop-gap and translate efficiency as all round expertise, then we want efficiency in a known manner concerning known things, and we unwittingly exclude search. The stop-gap then may become a chronic condition.

And so we do want efficiency and hence expertise though only up to a point. Now the diversity of tasks in our society makes it unreasonable to require that anyone should acquire expertises in many tasks. Hence we do want specialization. And, of course, taking this all the way amounts to the same mistake, because both the task and the way of performing it thus are in a danger of becoming fixed even if they should conflict with the aims of society. But, finally, we want the specialist to make his specialty a life-task, so as to insure that he is devoted to it, and so we want him professional. He is thus made to lose his autonomy.

This is technocracy. Technocracy is not government by nuclear engineers or by brain surgeons. Whereas aristocracy is rule by noblemen, clericalism or theocracy is rule by priests, plutocracy is rule by the rich, etc., democracy is not rule by the people (literally the poor) but rule controlled by the people and catering to their wishes; technocracy is likewise government by expert political scientists who see to it that the leadership of society on all its walks, segments, sectors, and aspects, is by professional specialized expert technicians. Thus, technocracy is the realization of the abstract society at its worst, corroborating all the criticism that the Romantic Reaction launched against the Enlightenment.

Yet technocracy survives. Why? There is the claim of Martin Heidegger and his disciple Herbert Marcuse: technocracy offers the comfort of the body and of the dulled life of the spirit. This is traditional Medieval Christian preaching in modern circumstances. I think it is only partly true, and only as the last resort: people fall for it out of despair, not knowing how to use their lives to better ends. Certainly Heidegger's Nazi convictions and Marcuse's communistic convictions sound less convincing than their admonitions to their fans who endorse the admonitions but do not try to effect a change. There is the inability to change, then, and merely for want of new directions. But, one might irritatedly respond, when there is no aim, any aim should be preferred to aimlessness. This is debatable. It was Albert Camus who saw his *Stranger* as an optimistic novel and his hero as a Christ figure, though of a negative sort. Camus analyzed his novel this way because its hero, he said, refused to endorse any end that did not capture him and he bore the cross of aimlessness and loneliness in the burning conviction that he must have all channels open for the possible new message to come – if and when. But this is not all. For, technocracy, even at its worst, is not quite aimless: it aims at the growth of science. This, perhaps, is its best defense.

3. SCIENCE IS NEITHER CRUSONIAN NOR TECHNOCRATIC

The Crusonian philosophy, we remember, is the philosophy of the Enlightenment; it rests on the conviction – the Baconian-Cartesian conviction, incidentally – that only an independent unbiased mind can be engaged in proper scientific research. This is what the leading contemporary student of scientific method, Sir Karl Popper, labelled as Crusonian science: Crusonian science is scientific research conducted by Robinson Crusoe on his desert island. No doubt, Bacon also insisted on the demand for collaborative science. He complained that he had neither free time nor assistants to conduct research. He exhorted anyone who had any free time, however little, to make experiments, however minute, because every little bit helps. In his Utopia, the *New Atlantis*, there is a secular college whose members are engaged primarily in research – this in the early seventeenth century, two or more likely three, centuries before it happened in real life – and in accord with a detailed division of labor with which the small book abruptly ends. That is the first place specialization is described, as far as I know. Yet, we must remember, the division of labor is a mere technicality. In principle, research can be divided only because it is indeed Crusonian! Popper, to drive the point home, not believing in the possibility of Crusonian science, denies the possibility of division of labor in science. Since so much energy is spent these days in order to avoid duplication of, perhaps also to effect some integration between, diverse researchers – from writers of Ph.D. proposals to smashers of nuclei – this is a matter of much practical import, and, to use William James's criterion of significance, it involves millions of dollars of annual budgets.

As ever, the schemes are modified when applied to real life. Popper will admit that for two individual doctorants to decipher the same code and read the same message, for two students to discover the same proof for the same mathematical theorem, is a bit wasteful. Of course, he will say, why not let two people seek the same object? The answer is, one of them may fail to receive his doctorate! This does not hold for atom smashers who burn public money on their race to Stockholm, and so here the Crusonian scientists accept Popper's view and allow duplicity as means of competition and of double insurance. Nevertheless, the detailed qualifications aside, it is clear that the chief argument in favor of technocracy is that it is enormously conducive to scientific research and thus to progress. Provided, that is, that science is Crusonian. Is it?

Let me emphasize that viewing the economy as that of a division of labor, is accepting specialization in production and in services. The question there is, how much? Viewing science as extra-economic activity forces us to view it differently and even ask whether it leaves any room for specialization at all. Indeed, I will be quixotic enough to say, no: science wants no specialization at all. Let me elaborate these two points, concerning the market and concerning scholarship, and bring this chapter to a close.

The economic question is traditionally put in terms of economic efficiency, as we have seen. The excuse was that the individual who lives in an efficient market has a wider choice and decides, for example, to work little and study much. This idea sounds very liberal and appealing but its implementation is neither: most workers invest most of their time and energy on the job, and efficiency considerations applied to them get things wrong – even from the viewpoint of efficiency. It turns out that the very improvement of the quality of working life improves even productivity, and that this requires less specialization since it is best done by autonomous work groups which supervise their own product quality, thus eliminating the specialized profession of a supervisor, and who work in rotation so as to specialize in more than one task. Of course, there is also the problem of leisure activity of workers, since the excuse for efficiency is the demand for leisure, and this requires some training – training for moral autonomy indeed; education. I will say no more on these two matters, the matter of the improvement of the quality of working life, job improvement, etc., and the matter of the acquisition of more leisure, more taste for leisure, and more leisure activity. There is a rich literature devoted to each of these two topics.

Yet I cannot overlook one item here, and it is the fact that universities have unintentionally turned from places of leisure activity to places of professional training. I do not in the least object to professional training, in college or anywhere else; on the contrary, it is means for economic autonomy, and so highly commendable. Yet this is no excuse for the ousting of leisure from college which professionalization has effected. Let me elaborate.

As Aristotle has observed, learning requires leisure. Adam Smith concurred. Leisurely people are parasitic, unless they are researchers whose leisure is destroyed. It is impossible to avoid some parasitism anyway, and if within a parasitic class some research occurs, it is economically beneficial. But even otherwise parasitism is unavoidable. Monasteries were meant to include social outcasts of all sorts, also parasites. The same holds for the universities, except that the inmates in universities had to show some literacy, and that marginal tasks were left for both monasteries and universities to perform, some of which were educational. The attempt to have Protestant societies devoid of all parasitism created tensions that expressed themselves in vicious witchhunts which risked the very foundations of society. In the scientific revolution it was taken for granted that universities will do nothing to advance science, and amateur societies were developed to that end. The ideal then was the polymath self-educated researchers living on private means or working only part-time. The secular university was born to the American and French Revolutions, both of which belonged to the Enlightenment. The secular university contributed to research chiefly in a sort of national competition and in the name of advanced technology, yet on the whole its structure was traditional. Things moved slowly and the demand to raise standards was regularly translated as the demand to specialize since, it was said, there is much to learn in each field. All this was hardly planned. Reactionaries of all sorts wanted the university to be a place for leisure or for educating the leisure classes, and the progressivists wanted more technology. Some reactionaries demanded that philosophy and history receive honorable shares in specialized universities. This was granted them so as to get their blessing for specialization and so specialization in the arts was born. The last cry here is Sir Charles, or Lord Snow, *The Two Cultures and the Scientific Revolution*.

The result is that one can hardly go to college for fun any longer. The absorption power of the educational side of college, especially the large urban prestigious college, is geared to the labor market. The leisurely reputation of the Ivy League is now completely replaced by the reputation for professionalism. We have too many job applicants in profession x, let department x in all colleges close its doors to candidates, says the American professional association. Of course, this is not necessarily so, but the designers of education on the national scale, especially professional associations, take it to be realistic. If it is realistic, however, then we should offer both professional and non-professional courses in every department and enable students to choose. To this the objection comes from professors fiercely competing for Nobel Prizes, who have learned to slave drive their students as they were driven by their own professors. Fortunately, however, not all schools are as highly competitive. It is still possible to seek education, or even a combination of vocational training and liberal education, in many universities and colleges of different levels of quality.

It remains for me, as my last point, to speak of specialization in research, or perhaps in research and in teaching combined. This is, at last, the meat of the topic of this chapter. So, for the hard-nosed traditional advocate of specialization, my presentation begins now. My first point to him is that he is deceived. Research and the training for research are a minute part of the social life of even the most advanced society. Even if one takes artificially the most advanced university, one finds there a microcosmos that produces many things quite apart from research plus the education for research; but, more particularly, education not for research, both vocational and other, are commoner even in the most prestigious research university. The claim that all university education is for research is the necessary and sufficient defense for specialized education. And the claim that all university education is specialized is either a gross error or the policy of a Nobel Prize hunter who is willing to sacrifice the majority of non-specialists.

This, indeed, is the norm. A violin teacher who meets hundreds of students in his career remembers with pride the one who has made it big and justifies thereby his having administered the same bitter medicine to all of them indiscriminately, making many of them music haters for life. This must stop. We have better filters to spot the top student and need not torment the rest. As to the top student, he easily studies all sorts of things and so is seldom a specialist. Nobel Prize hunters often use terror to make their blue-eyed slaves desist from seeking intellectual pleasures for their own sake. Terror is, of course, never acceptable.

Yet the real researcher may indeed specialize. As I have written a book on Michael Faraday, I will use him as an example. He was self-educated. He was a fellow of the Royal Society, of the British Academy, of the British Association for the Advancement of Science. At least one conversation he had with Turner is recorded – on account of its quaint interdisciplinary character, I suppose – concerning the color of the sky. He was a lay preacher, a lay musician, a man of broad interests. He had a free box in the opera house whenever he had the time to go there. He did not have the time. He worked ever harder on an ever narrower field. He became a true specialist. Anyone who wishes to condemn him is but a moralizer and a fool. Yet Faraday specialized by choice, and the fruit of his labor was to become common knowledge as soon as possible. Indeed, he was himself a superb popularizer.

We have to allow knowledge to disseminate in as wide a public as possible, and we have to allow researchers to specialize at their own choice. And so we need experts in dissemination of new knowledge, expert popularizers, that is. Also we need to revamp the education system to make more people able to assimilate more knowledge, and, most importantly, we must alter the incentive system for research. The incentives are Crusonian; science is not. If we try to push science to its traditional Crusonian mold it will be fragmented, specialized, Crusonian in the extreme. And there is no science with no unification. Hence, we must rethink matters on a large scale and implement the new results in all institutions of research and learning.

4. SCIENCE AS A GUILD AND AS A REPUBLIC

I conclude with the statement and criticism of the philosophy of technocracy. It is a recent philosophy, invented by Michael Polanyi and expounded in his magnum opus, *Personal Knowledge* of 1958. It was popularized by Thomas S. Kuhn, perhaps the most popular philosopher in the United States at present. Polanyi fought for the autonomy of science, yet he thought scientists cannot be autonomous: they must belong to a rather hierarchical community of experts. The expert cannot explain his convictions to the outsider, and therefore he need not. He is accountable only to his guild. The young novice can learn the tricks of the trade, but only by becoming an apprentice. And great masters have good workshops which are more likely to produce the next generation of leaders. Hence a novice cannot know what he is choosing:

choice must precede practice and practice must precede knowledge. Also, practice decides what degree of specialization is advisable or permissible.

So much for the outline of Polanyi's theory, which calls for a brief comment. If this theory is correct, then we cannot but leave our topic of specialization to the experts – and experts in each field should decide how much specialization they require. Possibly, however, Polanyi's theory is false. If it is false, then experts are criticizable by outsiders. In particular, experts in a field, who must also act as educators, may be criticized by experts in education and in philosophy. If this be the case, then Polanyi's doctrine is inconsistent, since it supports the experts on both sides of any disagreement. To defend it from inconsistency, then, we will have to divide the empire very clearly and avoid all tresspassing and all border disputes. This task must be given to expert administrators, and their failure would render Polanyi's theory inconsistent. Hence Polanyi's demand for the autonomy of the sciences must be limited by expert intellectual administrators – college presidents, editors, administrators of learned societies, government liaison agents.

I take it we have here a clear-cut contrast, then. Not between specialization and well-roundedness, as many writers claim, but between the authority of specialized experts and citizens' autonomy. Autonomy is, doubtless, harder to maintain than succumb to experts' authority. But reduce and it is worth the effort. We can create specialized fields of criticism of diverse fields, of popularization, of education, and of job counseling, job rotation, job reorientation, and so on. These are mere suggestions; they deserve exploration. Nevertheless, I think I have presented a sufficient argument to shake the common faith in the best theory of expertise – that of Polanyi and Kuhn. I have described the sifting process of education by which specialized students are selected to join teams of great masters, and I have criticized this process administratively, educationally, and morally. I have said that this process is both highly wanting in efficiency and harmful for the rejected many who deserve a better curriculum than that for the chosen student; I have said that this curriculum is too narrow even for the chosen expert, since it is too specifically geared to his teacher's project, thus leaving the student both unequipped and untrained for the task of designing his own project; I have said that the education for moral autonomy is not only absent but violated in the curriculum as envisaged by Polanyi and Kuhn. They think some leaders will be at the helm come what may; yet we know that without proper democratic education of any community, its leadership will pass not to the hands of the able and the public-spirited but of the power hungry intrigue merchant. Indeed, the medieval guilds are more intrigue-ridden than democratic societies, and it is the medieval remnants of the structure of the universities in the modern world that makes them notoriously more intrigue ridden than is to be expected. Hence, the claims of professionals for expertise, and the elitist philosophy of Polanyi and Kuhn that supports them, are anti-democratic and medieval.

Rejecting experts' claims for exclusive rights to judgments, then, is imperative, but is easier said than done; indeed, the easiest is to leave the expert to judge the expert, so that the citizen's democratic duty and responsibility is reduced and even made minimal. But to call the citizen to his duty is only possible when we know what alternatives to undertake, and these are both difficult to find and necessary to repeatedly improve. The only permanent thing with which to counter the situation is to create in all advanced democracies special organizations, from government agencies, through departments for adult education in cities and universities, to clubs of concerned citizens, whose specific tasks should be to fight elitism and devise means to neutralize any political advantage the elitist has against the general public, and the neutralization can only be effected by means of educating the public at large to the point that they can understand the experts enough to prevent their standing above the normal means of democratic control. This kind of activity should achieve both democracy and public education. It illustrates yet again the famous thesis that democracy and high level of public education go hand in hand. The claims of technocrats, thus, to be offering better if specialized education, are false: education must remain democratic and so the research that is specialized, if it brings knowledge, must bring the knowledge that can be disseminated publicly to benefit anyone who would wish to benefit. The unity of society and of science, thus, are as important for society as for science. This is why science and democracy went thus far together and this is why the claim for elitism in the name of science is an error. We must specialize, but our education should allow individuals to choose responsibility whether and how to specialize, and our society should remain democratic, not turn technocratic. These two limitations are the logical limitations on the degree of specialization science permits. Yet this is not to say that since these limitations are logically necessary, we need not bother about them. For, when the logically necessary conditions for science are violated, logic is not thereby violated, but merely science ceases to be the progressive enlightened activity we are used to. Hence, in the last resort technocracy that speaks in the name of science is but a threat to science, and the education for moral autonomy and for democracy that limit specialization and so put a ceiling on scientific growth are as essential for the very existence of scientific growth as ever.

CHAPTER 34

STANDARDS TO LIVE BY

1. PRELIMINARY DIFFICULTIES

To begin with, let me present a preliminary problem, of the kind that practicing psychiatrists or psychotherapists are trained to handle. It is the problem of how to overcome defensiveness in order to reach the level of anxiety so as to get rid of the anxiety in order to handle the problem at hand. Except that whereas the problems psychiatrists are trained to handle are specific and concrete, and relate to symbolic and emotional disturbances, and the problem I wish to handle is general and abstract, and relates to intellectual difficulties.

The preliminary problem I have mentioned is not at all specific to psychotherapy or psychiatry. In the simplest form it appears in surgery where it is solved by anaesthesia: the anaesthetist relaxes the patients' muscles and neutralizes their nerves in order to let the surgeon do his job. Yet this model of surgery is extremely inadequate for the purposes of anyone who needs the cooperation of his patient. The surgeon prefers his patient passive, and if the result is that the patient cannot even breath or pump his own blood, then the surgeon may very well do it for him with a heart-lung machine. This will not do for anything above the sheer physical level of activity. When cooperation is required, then defensiveness must be overcome in a cooperative fashion too.

And so my preliminary problem is more like the ones psychotherapists or psychiatrists, than the ones anaesthetists, are trained to overcome: how can we enlist the goodwill of defensive and anxiety-ridden people in the service of the required cooperation that may result in improvement? Yet, to repeat, my problem is more abstract and general, more intellectual than the ones supposedly met in daily work in psychiatry or psychotherapy. Let me explain.

My self-appointed task is to overcome the difficulties that prevent the normal development of the careers of even intelligent and eager young people who do not follow the accepted ways of developing a career. It is a fact that such people are severely handicapped, and by problems that no clinic, psychological or otherwise, is equipped to handle. Not only are these young people unable to follow the trodden path because of defensiveness and anxiety, but also those who unjustly penalize them and treat them either like criminals or like ignoramuses, those too are burdened by defensiveness and anxiety. It is my view that most intellectuals, university people, physicians, engineers, free professionals on a broad spectrum, are all too often burdened with so much defensiveness and anxiety and guilt feelings, that they cannot help those beginners who are the obvious candidates for the junior positions in their charge but who do not quite qualify since they do not quite fit the image, do not quite do the expected, etc. We have here a typical Freudian vicious cycle – anxiety causing a trouble that in its turn causes the anxiety to deepen, and thus the situation worsens systematically – except that the Freudian classical defense mechanism cycle is of concrete individuals and the problems they personally have with their symbolic-emotional mechanisms that prevent them from properly adjusting to their own environment, yet in the case I am describing the problem is not mainly with the symbolic-emotional mechanism and not with the individual's adjustment to his environment, but rather the problem is with the sociological-intellectual mechanism and with the individual in charge being himself a part of the environment that is hard to adjust to. Indeed, as I say, the individual in charge interests me less as a patient, as a sufferer, than as an obstacle to other patients, to other sufferers, the young aspirants. The other sufferers, the young aspirants, will do well to learn how to adjust to their dogmatic and neurotic seniors without thereby stopping doing the new things they want to do. The situation will also greatly improve if, as a preliminary to this, the seniors in question could become less defensive and less anxious, so that we may help them become less dogmatic. And, as I say, the preliminary to that preliminary, is to enlist the goodwill of the patient, the dogmatic senior, who, as Freud has noticed about the candidate for psychoanalysis, is so far from recognizing that he needs treatment, that the very realization that he needs treatment is what he needs most, and as a major part of the treatment. In other words, Freud's neurotic patient who destroys himself and my dogmatic senior intellectual who destroys the careers of intelligent eager young aspirants, both are dogmatically unable to see the trouble, as both are defensive and anxiety-ridden so that the very suggestion that something is wrong with them turns them completely off. Communication breaks down where it is needed most.

I do not think it reasonable to expect senior intellectuals and free professionals to submit to any psychotherapeutic treatment, though I think they need it and we would all benefit if they had it. I do not even think we could offer them the treatment if they asked for it, because of technical and social difficulties. In particular, once we could enlist their cooperation and eliminate their defensiveness and anxiety, the treatment they would need, I say, is not the symbolic-emotional one which psychotherapists of all sorts are trained to offer, but an abstract intellectual one, which they are neither familiar with nor able to handle.

So let me now leave my preliminary problem which is of the kind psychotherapists are trained to overcome and I am not, and arrive at long last to the treatment I would like to offer to one who is cooperative and relaxed for a while. I would like to tell that person, what the source of his anxiety and defensiveness is. I would like to tell him that, being a professional intellectual, he finds it imperative to have at his disposal intellectual weapons that he does not possess, that he thus feels a great sense of inadequacy and fear, not to mention plain guilt. I would also like to tell him that there is no reason for all this. One of the most moving passages in Freud's writings I find, is his profound, humane and terrific observation that everybody thinks that he, and he alone, is ridden by guilt since everybody else just does the expected. Freud noticed, at that moment, when he wrote his *The Id and the Ego*, that is, what an enormous relief it is simply to recognize the fact that the moral standard we all accept is much too high. I want to say the same about the intellectual standards we all accept, and the harm it does to the individual and society alike to accept such high standards. But let me first speak of the damage of having too high moral standards, since Freud himself was mistaken in not proposing to remedy the defect he discovered by reducing the moral standards of the community and that he was likewise mistaken in endorsing intellectual standards he found in the scientific community that were much too high.

2. THE ILLS OF EXCESSIVE STANDARDS

The current view of standards is moralistic and pedantic. Everyone knows that the standards are going to be violated regularly, and that this endangers our very human existence. Yet the moralist and the pedant conclude from this that there is nothing to do but defend as high standards as possible, as vigilantly as possible, and the more so the more often the standards are violated. This moralistic and pedantic attitude seems to have been endorsed by Freud, for example. And it is not that he did not notice that this situation creates a neurotic vicious cycle of an unresolved tension between standards of conduct and actual conduct. On the contrary, this is why he said, humanity and neurosis are inseparable: humanity is the acceptance of standards, and this acceptance causes neurosis. This, of course, makes it questionable whether complete and successful psychotherapy is possible or even desirable. Yet, at other times Freud insisted that a complete and successful psychotherapy is both possible and even imperative, that is, in his own case he laid great emphasis on his own complete freedom from all neuroses. But then, perhaps he simply meant to moralize and demand from his followers the impossible complete cure – I do not know. What concerns me here, in any case, is his acceptance of high intellectual standards, which, of course, is also a form of moralizing and of pedantry. And perhaps he demanded complete cure just because he accepted too high a standard of intellectual activity, which required this of him. For, on the one hand he was the one to make that great discovery that no matter how emotional and symbolic a neurosis is, it also acts as an intellectual blind spot. Indeed, he said, when an intelligent person makes a silly mistake one may suspect it to be caused by a neurosis. On the other hand he demanded that a scientist have no blind spot. And demanding from one to have no blind spot while seeing neuroses as being also blind spots amounts to requiring from a scientist to have no neurosis. Briefly, then, Freud's demand for moral and intellectual high standards intertwine and amount to the imposition of neuroses of a high pitch while demanding utter freedom from all neuroses.

I suggest to view all standards that are unattainable and tension-creating as undesirable. I recommend that we make out standards as realistic as possible, i.e. just comfortably above current usage. This would enable people to relax, be undefensive, learn to raise the level of their conduct to the standard, and permit the raising of the standard again by just a little so as to cause further improvement with no excessive tension. I suggest this quite generally, for all anxiety causing tensions between current usage and standards. But rather than discuss examples from other fields where the regular and frequent violations are common, such as traffic or drugs, let me discuss now the heart of my present discussion, namely the intellectual standard endorsed by Freud and by most intellectuals I know of, namely the standard of proof or of justification of one's opinions. It is intuitively and historically clear that Freud demanded from himself to prove all that he said and to be ahead of everybody in the field, so that if anyone ever dared contradict him, the very contradiction proved Freud's proof not good enough and thereby proved Freud himself to be scientifically at fault. My own teacher, Sir Karl Popper, himself holds this attitude, and considers all my criticisms of him as personal slights on his moral character. And he has said it to me in so many words. It is clear that neither Freud nor Popper could take criticisms impersonally because of the pedantry or the moralism which I am describing, and which in Freud's case is rooted in the acceptance of the standard of science as being above criticism and in the case of Popper as the result of the demand for a pedantic effort to do the impossible, despite its creation of a neurotic tension that in principle cannot be relieved. Popper is in favor of this neurosis, and explicitly so. I am all against it.

To take a concrete example of the damage that pedantry causes, let us take Freud's attitude to women. There is no possibility to discuss here all the silly things that Freud said against sex and against women. Suffice it to say that it is well known that he took women's inferiority for granted and thought that their maturity can only be attained by their recognition of their inferiority. I am not scandalised by Freud's attitude to women; I am just opposed to it uncompromisingly. (Even if his theory of penis-envy were true, I should not advise women to learn to accept their inferiority but rather suggest to them that it is a mere projection to blame one's genitals for some sense of inadequacy or another, just as Freud would say to a man blaming the size of his penis for his troubles.) What I want to observe now, however, is the damage that high scientific standards cause today when Freud's view of women is at issue. I am speaking of living people, not of the great thinker himself. I think that maturity – intellectual maturity in this case – requires the combined recognition of two facts, one being the greatness of Freud and the other the stupidity of his views of women. The intellectually immature, including, I say, those who hold intellectual standards that are by far too high, like Freud and like Popper who has called Freud a pseudo-scientist, the intellectually immature are forced to either show contempt to Freud for his silly views or to declare his silly views not as silly as they look. Perhaps they are not as silly as they look because of some alleged important ideas they contain; perhaps because of the allegedly low general position of women in that period in which Freud wrote; or perhaps for some other reason, known or unknown. All this will not do. We have to agree now that Freud was an important thinker who has made immense contributions to human knowledge, yet one who had no proofs, and was silly to think that his theories had been proven by experience, and was even sillier to think that he was utterly and completely with no neurosis and hence with no prejudice, and that what he said about women is irredeemably silly. We are all, all prejudiced and silly, and if someone manages to rise for a moment above the level of common prejudice, as Freud has done regarding the prevalence of sexuality, then he has already secured our gratitude, and quite unconditionally so.

But the damage of the high standards of scientific proof is wider in the field of psychoanalysis proper. The theory of psychoanalysis still remains anchored in the idea that Freud was fundamentally right – evidently because he was a scientist, i.e. because he had proved his ideas. In practice things are a bit different. It is not clear what happens in psychoanalytic practice, especially since this takes place in many countries and in many ways and there is evidently no survey of all that goes under the heading of psychoanalysis and it would be very difficult anyway for such a survey to differentiate the national and religious and regional and other differences in psychoanalytic practice from differences that are due to different readings of the theory or due to different developments of it, etc. Nevertheless, I would venture as a conjecture that the main damage that the high standard of demanding scientific proof where there is none, of course, is that it creates orthodoxies – in the plural, mind you – and that orthodoxies act as smoke-screens that obscure simple facts. And I want to offer a conjecture as to what is the simple fact about all psychotherapies, since I think that by now the main difference between the different schools of psychotherapy is a matter of orthodoxy which is only a negative factor.

3. ROOM FOR MAKING STANDARDS REASONABLE

All psychotherapy attempts, with varying degrees of success, to enlist the patient's cooperation by relaxing him, reducing his defensiveness and anxiety that prevent him from discussing his troubles intelligently, and presenting the difficulties he suffers from, not as he is familiar with them, but as they look to the onlooker from the outside. Once this high stage is reached, most of the work is done, and the therapist and his patient can discuss intelligently the problem-situation of the patient, let the patient make his own decision, and perhaps help him learn how to execute his decisions. Of course, this is the ideal. Some psychotherapists allow their patients to bully them to make the decisions for the patients, some therapists do other things wrongly, and even when they do things right they may fail: there is no guarantee for success, especially since there is no high road to success.

So much for the patient. Of course, a major factor inhibiting the success in reducing the high standards of the patient is the high standard of the therapist. Even therapists aware of the patients' high standards as a major source of anxiety are unable to help much because they apply high standards to themselves and to their techniques and ideas in general. As an example, they forbid themselves to fail or to make mistakes. But the mistakes and the failures are less surprising and less interesting. It is much nearer to the truth of the situation, and much less anxiety causing, to see in the history of psychotherapy since Freud, whether Freudian or anti-Freudian, a history of a few breakthroughs and of some stunning successes, less frequent than Freud claimed, but no less impressive; to disregard much, perhaps most, of what Freud said, as understandable errors; and to sum up what we all – Freudians and anti-Freudians alike – have learned from him, in theory and in practice. The recognition of this, it seems, is essential to the acceptance, and so to the advocacy, of more reasonable and more humane standards, and so to the opening up (to therapist and patient) of the field to more critical amicable discussions, to what Popper calls the friendly-hostile cooperation that is the heart and soul of scientific method, and to learn to promise to patients as little as possible, and to learn to share with patients doubts and difficulties. This will objectively make less room for anxiety and so improve therapist-patient relations from the very start. Also, this will objectively make the patient both active and cooperative and thus behave in a healthy manner and thus train himself for a more healthy way of life.

The picture which emerges from this is rather odd. The regular psychotherapist recognizes the fact that there are norms for the way of life of his patient, assumes that the norms are fairly reasonable in the sense that the patient has reasonable chances to do well by them and be fairly rewarded without too much effort, that the patient has a wrong estimate of the standards, that he therefore tries harder, yet that he is self-defeating. And at the same time the psychotherapist has the same defect he tries to rectify in the patient as a person practising a scientific discipline, and more so if he is a research scientist. For, the research scientists, to repeat, are suffering from the same malady described here.

This raises a host of questions. The most obvious one is, how can all scientists hold an excessive standard? Is it not the case that the standard accepted by most members of a community is the standard of that community?

This intriguing question seems to demand an affirmative answer by definition. Nevertheless, almost simultaneously it received a negative one in methodology from Pierre Duhem, in psychology from Freud, and in sociology and social anthropology from Émile Durkheim and Bronislaw Malinowski. The Freudian answer was already given: everyone thinks he is the only one to masturbate. Here, incidentally, statistical information, like the famous Kinsey Report, has done much to alleviate suffering, whether by legalizing widespread behavior patterns like homosexuality or by making violators of the standard know that its violation is very widespread. Duhem's discovery was that there are different views amongst scientists as to their practice, so that they cannot all be trusted as a matter of course. Durkheim declared the role of many customs to be so very different from their customary declared role, that he viewed the declaration as mere part of the custom. Malinowski said, it takes proper scientific research to find out when declared and practised standards agree.

Thus, when we ask, can the standard be universally misconceived? We may answer, in a sense, no: we see here standard misconception, and thus a standard of misconceiving, of the standard of a given practice. If so, can we declare all members of the community neurotics in need of psychotherapy?

The great contribution of Freud, Sigmund and to a lesser extent Anna, is the split of the question into two. First, we are all neurotics in the sense that we all misestimate the standards, and refuse to admit the fact by creating mental blocks about them. Second, most of us manage but some of us go too far and need treatment.

So far the social dimension of the situation was ignored in the present discussion, or rather underplayed. The fundamental assumption of all psychotherapy concerns the social setting within which the accent may be put on psychology. The setting, to repeat, is that the patient can do well in his life situation were he not in some psychological trouble. This can be clearly seen from the fact that psychotherapy is barely conceivable in times of war or famine or even mere physical hardship of the kind to be expected in a pioneering setting. And here we have made the standard mistake of polarizing matters: at one extreme we work on our external conditions, at the other on our psyche. But there is ample room to improve the norms so as to produce citizens less in need of psychotherapy.

4. THE REFORM OF STANDARDS

Unfortunately, if we accept that we need not polarize between society and individual and discuss fairly normal society which, nonetheless, puts too high demands of its individuals, then we hit another polarization. There is the view of a society as acceptable, as a viable organic entity, not to tamper with too much, which is known as conservatism, and the view of society as a mere collection of rational beings, each rationally pursuing his separate end and all together organizing themselves as reasonably as can be expected, which is known as radicalism. The view advocated here is unpolarized and recommends social reforms.

The radical view of society is the application of radicalism in scientific method to problems of social life. It is therefore hardly surprising that the commonwealth of learning, when it was socially established in the seventeenth century, was radicalist about itself. Scientists were not aware of the social setting that made science possible, of the integration of science in society, of the interaction between the scientific pursuit of the individual and the social institutions and norms of the social organization of science. Truth to tell, despite the falsehood of radicalism, both as a view of scientific method and as a social philosophy, things worked admirably well, perhaps because of the minimal social organization of science; research flourished and did not get into serious snarls until it encountered enormous success and until it became very well established.

The conservative view of society has been expounded by Edmund Burke soon after the French Revolution, and is still expounded mainly by political thinkers like Michael Oakeshott in England and Samuel Huntington in the United States. Yet it is also a social philosophy, mainly associated with the names of Tonnies and Durkheim, a religious philosophy of course, too, and a view of scientific method, best expounded by Pierre Duhem, and later by Michael Polanyi. It is today associated with the name of Thomas S. Kuhn. The Polanyi-Kuhn theory declares the standards of society as ones to learn by living in the society which practices them, that the articulation of the standards falls short of the truth. This is not different from what Michael Oakeshott says, except that he refers to political and cultural life in general and they refer to scientific research. The conservative view puts emphasis on continuity and on preserving the way of life by avoiding discontinuity.

For the same reason that radicalism works in the society of science fairly well, and for the reason that conservatism often works well in any successful society, conservatism applies to the sociology of science with ease and to a very large extent. The curious result is that here we have a domain where two extremely polarized views work. The flavor of having a way of life radical and conservative at once is very intriguing. In political philosophy it led to a new school of thought called radical conservatism or the radical right, which makes sense only in a country whose tradition is radicalist, namely, the United States, especially where the pioneering tradition is still alive. Its expression in the sociology of science has a locus classicus, the science fiction novel *The Black Cloud* of the British astronomer Fred Hoyle.

It seems clear that the radicalist view of society is liberal but has little room for democracy and so tolerates liberal meritocracy. The same can be said of Western conservatism, perhaps of all enlightened conservatism. To the extent that both theories apply to science, we can expect little democratic process in the learned world. And, indeed, much as the universities and the learned societies and the government and industrial scientific agencies are given to democratic control – both of society at large and within some of these institutions – there is no democratic institution governing science at large, nor can there be any, nor would any one really wish to see any. It seems clear that both radicalism and conservatism have an alternative to democratic control, and it is free competition. Radicalists have presented the market mechanism as the selector of the better practices, and conservatives have presented as the selector the struggle for survival and the survival of the fittest society. In either case, of course, meritocracy may well be preferred to democracy, except that the survival of Western democracy and of democratically controlled organizations calls for a moment of thought.

The reform of social standards, in society at large or in sections of it, is often done empirically, namely one tries it partially to see if it works. This is what Karl Popper calls piecemeal social engineering. Without discussing the question, is all commendable social engineering piecemeal? we can agree that in science piecemeal engineering will do amply well, since a reform that works well is bound to spread by the forces of competition that work so strongly in science, particularly due to the excessively high standards preached by its leadership.

Here, then, the refutation of conservatism is most forceful and general: it applies to reforms of standards within the community of science with a vengeance, because conservatives like Polanyi and Kuhn must, for logical reasons, endorse the scientific results of sociology of science, and because the reform of standards in science is not limited to a democratic process run by legislatures, where reform is much simpler, of course.

The psychological ill effects of excessive standards is, of course, well established. The conservatives, however, may well declare it regrettable but inevitable. Polanyi and Kuhn do so explicitly. They are not insensitive to the suffering caused by the scientific standards on aspirants who wish to join or on members who do not receive their fare share of the rewards. Yet they do not see how this can be ameliorated. It is not that they see no hope for improvement, but that they recommend no deliberate reform.

On this they are in error. History contains many cases of reform, and if sociologists of science recommend reform, then the Polanyi-Kuhn theory becomes inconsistent in recommending both the acceptance and the rejection of the reform.

The reform recommended here, on the basis of findings from psychology, learning theory, cybernetics and information theory, and Popperian methodology, is to introduce the concept of valiant failure, including that of a bold and interesting mistake, into our daily assessment of routine activities in science, coupled with the understanding that excessive tensions and excessively high standards are detrimental to the cause. The obstacle on the way to the implementation of all this are mental blocks of the scientific leadership that are institutionalized. We therefore need institutions competing with them, such as more liberal scientific periodicals where important results can be published.

BIBLIOGRAPHY OF JOSEPH AGASSI

1957

'Duhem versus Galileo' (*Aim and Structure of Physical Theory: Dialogue on the Two World Systems*), *Brit. J. Phil. Sci.* 8 (1957), 237–248.

1958

'A Hegelian View of Complementarity' (Cassirer, *Determinism and Indeterminism*), *Brit. J. Phil. Sci.* 9 (1958), 57–63.

'Koyré on the History of Cosmology' (*From the Closed World*), *Brit. J. Phil. Sci.* 9 (1958), 234–245.

'Commonsense Social Theory' (Ludwig von Mises, *Theory and History*), *Times Literary Supplement* (16 May 1958).

1959

'How are Facts Discovered?', *Impulse* **10** (1959), 1–3. (Reprinted, in *Science in Flux*, 1975.)

'Epistemology as an Aid to Science', *Brit. J. Phil. Sci.* **10** (1959), 135–146. (Reprinted in *Science and Society*, 1981.)

'Corroboration versus Induction', *Brit. J. Phil. Sci.* 9 (1959), 311–17.

'Jacob Katz on Jewish Social History' (*Tradition and Crisis*), *Jewish J. Sociology* **1** (1959), 261–5.

'The Stoic Background to Science' (Sambursky), *New Scientist* 9 (1959), 811.

'The Philosophy of Science' (Kemeny, *A Philosopher Looks at Science*), *New Scientist* 5 (1959), 888.

Review of H. Feigl *et al.*, *Minnesota Studies* **2**; *Mind* 68 (1959), 275–77.

'Wittgenstein the Elusive', two letters to the editor of *Times Literary Supplement* (22 and 29 May 1959).

K. Klappholz and J. A., 'Methodological Prescriptions in Economics', *Economica* **26** (1959), 60–74. (Spanish transl. in *Revista de Economica* 78–79 (1963), 259–77; Reprinted in D. R. Kamerschen (ed.), *Readings in Microeconomics*, World Publn. Co., Cleveland and New York, 1967, 60–74.)

1960

'Methodological Individualism', *Brit. J. Sociology* **11** (1960), 244–70. (Reprinted in John O'Neill (ed.), *Modes of Individualism and Collectivism*, Heinemann, London, 1973, 185–212.)

Review of J. K. Feibelman, *Inside the Great Mirror, Brit. J. Phil. Sci.* **11** (1960), 83–4.

Review of P. W. Bridgman, *The Way Things Are, Philosophy* **35** (1960), 374–5.
K. Klappholz and J. A., 'A Rejoinder', *Economica* **27** (1960), 160–1.

1961

'An Unpublished Paper by the Young Faraday', *Isis* **52** (1961), 87–90.
'Does Hong Kong Need Economic Reform?', *Far Eastern Economic Review* **29** (1961), 667–8.
'The Role of Corroboration in Popper's Methodology', *Australasian J. Philos.* **39** (1961), 82–91. (Reprinted in *Science in Flux*, 1975.)
Review of R. Kaufmann, *Methodology of Social Sciences, Econometrica* **29** (1961), 100–1.

1963

Towards an Historiography of Science, Beiheft 2, to *History and Theory*, 1963 (118 pp.). (Facsimile reprint, Wesleyan University Press, Middletown, 1967; Italian transl., *La filosofia dell'uomo libero, verso una storiografia della scienza*, Armando, Rome, 1978.)
'Between Micro and Macro', *Brit. J. Phil. Sci.* **14** (1963), 26–31.
'Empiricism versus Inductivism', *Phil. Stud.* **14** (1963), 85–6. (Reprinted in *Science in Flux*, 1975.)
Review of E. Nagel *et al.* (eds.), *Logic, Methodology and Philosophy of Science, Isis* **54** (1963), 405–7.

1964

'The Nature of Scientific Problems and Their Roots in Metaphysics', in Mario Bunge (ed.), *The Critical Approach* (Free Press, New York, 1964), 189–211. (Reprinted in *Science in Flux*, 1975.)
'The Confusion Between Physics and Metaphysics in Standard Histories of Science', in H. Guerlac (ed.), *Ithaca, 1962* (Paris, 1964), pp. 231–50. (Polish transl. in *Zycie i Mysl* **16**, 1966, Italian transl. in *Epistemologia etc.*, 1978; reprinted in *Science in Flux*, 1975.)
'Analogies as Generalizations', *Phil. Sci.* **31** (1964), 351–6.
'Variations on the Liar's Paradox', *Studia Logica* **15** (1964), 237–8.

1965

Review of K. R. Popper, *Conjectures and Refutations, Jewish J. Sociology* 7 (1965), 144–6.

1966

'Sensationalism', *Mind* 75 (1966), 1–24. (Reprinted in *Science in Flux*, 1975.)

'The Confusion Between Science and Technology in Standard Philosophies of Science', *Technology and Culture* 7 (1966), 348–66. (Reprinted in F. Rapp, *Contributions to the Philosophy of Technology*, Reidel, Dordrecht and Boston, 1974, pp. 40–59; Reprinted in *Science in Flux*, 1975; Italian transl. in *Epistemologia* etc., 1978.)

'Revolutions in Science, Occasional or Permanent?', *Organon* 3 (1966), 47–61. (Reprinted in *Science and Society*, 1981.)

'The Mystery of the Ravens', *Phil. Sci.* **33** (1966), 395–402.

'Starting on the Wrong Foot' (in Polish: Against the Science of Science), *Zycie i Mysl* **19** (1966), 49–51.

Review of E. Harris, *Foundations of Metaphysics in Science, Science* **154** (1966), 1047.

Review of R. Kahn (ed.), *Studies in Explanation, Phil. Forum* **23** (1965–6), 49–52.

Review of T. S. Kuhn, *The Structure of Scientific Revolutions, J. Hist. Philos.* **4** (1966), 351–4.

1967

'The Kirchhoff-Planck Radiation Law', *Science* **156** (1967), 61–7.

'The Uniqueness of the Idealism of Parmenides' (in Hebrew), in S. Perlman and B. Shimron, *Doron* (Tel Aviv, 1967), pp. 61–7.

'Planning for Success: A Reply to Professor Wisdom', *Technology and Culture* 8 (1967), 78–81. (Reprinted in F. Rapp, *Contributions to the Philosophy of Technology*, Reidel, Dordrecht and Boston, 1974, pp. 64–8; reprinted in *Science in Flux*, 1975.)

'The Correspondence Principle Revisited' (B. L. Van der Waerden, *Sources of Quantum Mechanics*), *Science* **157** (1967), 794–5.

Review of T. S. Kuhn *et al., Sources of the History of Quantum Physics, Science* **156** (1967), 1589.

I. C. Jarvie and J. A., 'The Rationality of Magic', *Brit. J. Sociology* **18** (1967), 55–74. (Reprinted in B. Wilson, *Rationality*, Oxford and New York, 1970, pp. 172–93; German transl. in Hans G. Kippenberg and Brigitte Luchesi (eds.), *Magie etc.*, Suhrkamp, Frankfurt, 1979, pp. 120–49.)

1968

The Continuing Revolution: A History of Physics from the Greeks to Einstein, in collaboration with Aaron Agassi (McGraw Hill, New York, 1968), 222 pp. (Hebrew transl., Dvir, Tel Aviv, 1976; Italian transl., Armando, Rome, 1979.)

'The Novelty of Popper's Philosophy of Science', *International Phil. Quarterly* 8 (1968), 442–63. (Reprinted in *Science in Flux*, 1975; Italian transl., *Epistemologia etc.*, 1978.)

'Science in Flux: Footnotes to Popper', in R. S. Cohen and M. W. Wartofsky (eds.), *Boston Studies in the Philosophy of Science*, Vol. 3 (Reidel and Humanities, Dordrecht and New York, 1968), pp. 293–323. (Reprinted in *Science in Flux*, 1975; Italian transl., *Epistemologia etc.*, 1978; Russian transl. in a selection from *Boston Studies*, Progress, Moscow, 1978.)

'Precision in Theory and Measurement', *Phil. Sci.* **35** (1968), 287–90. (Reprinted in *Science in Flux*, 1975.)

'The Logic of Technological Development', *Akten des XIV Internationalen Kongress für Philosophie* (Herder, Wien, 1968), pp. 483–88. (Reprinted in *Science in Flux*, 1975.)
'On the Limits of Scientific Explanation: Hempel and Evans-Pritchard', *Philosophical Forum* **1** (1968), 171–83.
'Anthropomorphism', in P. P. Wiener (ed.), *Dictionary of the History of Ideas* (Scribners, New York, 1968), pp. 87–91.
'Logical Positivism' and 'Philosophy and Science' (in Hebrew), D. Knaani (ed.), *Encyclopedia of the Social Sciences* **4** (Tel Aviv, 1968).
'No More Discovery in Physics?' (Schlegel, *Completeness in Physics*), *Synthese* **18** (1968), 103–8.
'Changing Our Background Knowledge' (Bunge, *Scientific Research*, 2 vols.), *Synthese* **19** (1968–9), 453–64.
Editor's Notes, *Philosophical Forum*, Volume 1 (1968), 3–5 and 123–6.

1969

'Popper on Learning from Experience', *American Phil. Quarterly*, Monograph Series, No. 3 (1969), 162–70. (Reprinted in *Science in Flux*, 1975.)
'Can Religion Go Beyond Reason?' *Zygon* **4** (1969), 128–68. (Reprinted in *Science in Flux*, 1975.)
'Privileged Access', *Inquiry* **12** (1969), 420–6. (Reprinted in *Science in Flux*, 1975.)
'Sir John Herschel's Philosophy of Success', *Historical Studies in the Physical Sciences* **1** (1969), 1–36. (Reprinted in *Science and Society*, 1981.)
'Leibniz's Place in the History of Physics', *J. Hist. Ideas* **30** (1969), 331–44.
Review of C. H. Dunhof, *Government Contracting and Technological Change, Physics Today* (1969), p. 95.
'The Concept of Scientific Theory as Illustrated by the Practice of Bloodletting', *Medical Opinion and Review* **5** (1969), 156–69.
'Unity and Diversity in Science', in R. S. Cohen and M. W. Wartofsky (eds.), *Boston Studies in the Philosophy of Science*, Vol. 4 (Reidel, Dordrecht, 1969), pp. 463–522. (Reprinted in *Science in Flux*, 1975.)
'Comments: Theoretical Entities versus Theories', *ibid.*, Vol. 5 (1969), pp. 457–9.
'Fisica' (in Spanish), *Diccionario Enciclopédia Salvat Universal*, Vol. 11, pp. 191–4.
I. C. Jarvie (ed.), in consultation with J. A., *Hong Kong: A Society in Transition* (Routledge and Praeger, London and New York, 1969.)
J. A. and I. C. Jarvie, 'A Study in Westernization', *ibid.*, 129–63.

1970

'Philosophy as Literature: The Case of Borges' (*Other Inquisitions*), *Mind* **39** (1970), 287–94.
'Positive Evidence as a Social Institution', *Philosophia* **1** (1970), 143–57. (Reprinted in *Science in Flux*, 1975.)
'Positive Evidence in Science and Technology', *Phil. Sci.* **37** (1970), 261–70. (Reprinted in *Science in Flux*, 1975.)
'Can We Learn from History?' Suzanne Delorm (ed.), *Proceedings of the 12th International Congress for the History of Science* (Paris, 1970), Volume 2, pp. 5–8.

'Duhem's Instrumentalism and Autonomism', *Ratio* **12** (1970), 148–50. (Also in German edition of *Ratio*; reprinted in *Science in Flux*, 1975.)

'The Origins of the Royal Society' (Margery Purver), *Organon* 7 (1970), 117–35. (Reprinted in *Science and Society*, 1981.)

'The Preaching of John Holt' (*The Underachieving School*), *Interchange* **1** (1970), 115–18.

J. A., I. C. Jarvie and Tom Settle, 'The Ground of Reason', *Philosophy* **45** (1970), 43–50.

1971

Faraday as a Natural Philosopher (Chicago University Press, Chicago and London, 1971), 332 pp.

'Kant's Program', *Synthese* **23** (1971), 18–23. (Excerpt from the above.)

'Qualifying Exams, Do They Qualify?', *Educational Forum* **35** (1971), 156–66.

'Tautology and Testability in Economics', *Philosophy of the Social Sciences* **1** (1971), 49–63.

'What is a Natural Law?', *Studium Generale* **24** (1971), 1051–66. (Reprinted in *Science in Flux*, 1975.)

'The Standard Misinterpretation of Skepticism', *Philosophical Studies* **22** (1971), 49–50. (Reprinted in *Science in Flux*, 1975.)

'The Aims of Higher Education' (in Hebrew), *Keshet* **13** (1971), 62–73.

'Agassi's Alleged Arbitrariness', *Studies in the History and Philosophy of Science* **2** (1971), 157–65.

'On Explaining the Trial of Galileo' (Koestler, *Sleepwalkers*), *Organon* 8 (1971), 138–66. (Reprinted in *Science and Society*, 1981.)

'Tristram Shandy, Pierre Menard, and All That: Comments on *Criticism and the Growth of Knowledge*' (Lakatos and Musgrave), *Inquiry* **14** (1971), 152–64.

Review of H. Reichenbach, *Axiomatics of Special Relativity, Physics Today* (1971), pp. 49–50.

1972

'Sociologism in Philosophy of Science', *Metaphilosophy* **3** (1972), 103–22. (Reprinted in *Science and Society*, 1981.)

'Imperfect Knowledge', *Philosophy and Phenomenological Research* **32** (1972), 465–77. (Reprinted in *Science in Flux*, 1975.)

'The Twisting of the I.Q. Test', *Philosophical Forum* **3** (1972), 260–72. (Finnish transl., *Paradoksi* **3** (1976), 164–85.)

'Scientific and Dogmatic Approaches in the History of Science' (in Hebrew), *Keshet* **14** (1972), 122–35.

'Dimensional Analysis' (in Hebrew), *Hebrew Encyclopedia*, Vol. 23, pp. 799–802.

'The Interface of Philosophy and Physics' (Mario Bunge, ed., *The Delaware Seminar*, 2 vols.), *Phil. Sci.* **39** (1972), 367–8.

'Listening in the Lull' (Borger and Cioffi, *Explanation in the Behavioral Sciences*), *Phil. Soc. Sci.* **2** (1972), 317–32.

Review of Theodore Mischel (ed.), *Cognitive Development and Epistemology, Phil. Soc. Sci.* **2** (1972), 367–8.

1973

'Continuity and Discontinuity in the History of Science', *J. Hist. Ideas* **34** (1973), 609–26. (Reprinted in *Science and Society*, 1981.)

'Rationality and the *Tu Quoque* Argument', *Inquiry* **16** (1973), 395–406. (Reprinted in *Science and Society*, 1981.)

'When Should We Ignore Evidence in Favour of a Hypothesis?', *Ratio* **15** (1973), 183–205. (Also in German edition of *Ratio*; reprinted in *Science in Flux*, 1975.)

'Random versus Unsystematic Observations', *Ratio* **15** (1973), 11–13. (Also in German edition of *Ratio*; reprinted in *Science in Flux*, 1975.)

'Did Agnon Learn from Kafka?' (H. Barzel, *Between Agnon and Kafka*) (in Hebrew), *Keshet* **15** (1973), 74–86.

J. A. and I. C. Jarvie, 'Magic and Rationality Again', *Brit. J. Soc.* **24** (1973), 236–45.

I. C. Jarvie and J. A., editors, selected papers by Ernest Gellner, *Cause and Meaning in the Social Sciences* (1973, Routledge, London and Boston); *Contemporary Thought and Politics* (1974, same publisher); *The Devil in Modern Philosophy* (1975, same publisher); *Spectacles*, 1980 (Cambridge University Press, London and New York).

1974

'Conventions of Knowledge in Talmudic Law', in Bernard Jackson (ed.), *Studies in Jewish Legal History, in Honor of David Daube* (Jewish Chronicle Publications, London 1974); also published as a special issue of the *Journal of Jewish Studies* **25** (1974), 16–34.

'The Logic of Science and Metaphysics', *Philosophical Forum* **5** (1974), 406–16. (Reprinted in *Science in Flux*, 1975.)

'On Pursuing the Unattainable', in R. S. Cohen and M. W. Wartofsky (eds.), *Boston Studies in the Philosophy of Science*, Vol. 11 (Reidel, Dordrecht, 1974), pp. 249–57. (Reprinted in *Science and Society*, 1981.)

'The Logic of Scientific Inquiry', *Synthese* **26** (1974), 498–514. (Reprinted in *Science and Society*, 1981.)

'Criteria for Plausible Argument', *Mind* **83** (1974), 406–16. (Reprinted in *Science in Flux*, 1975.)

'Modified Conventionalism is More Comprehensive than Modified Essentialism', in P. A. Schilpp (ed.), *The Philosophy of Sir Karl Popper* (Open Court, LaSalle, Ill.), pp. 693–6. (An extract from a paper first published in *Science in Flux*, 1975.)

'The Last Refuge of the Scoundrel', *Philosophia* 4 (1974), 315–17.

'Postscript: on the Futility of Fighting the Philistines: Karl Popper's Objective Knowledge', *Philosophia* 4 (1974), 163–201.

T. Settle, J. A., and I. C. Jarvie, 'Towards a Theory of Openness to Criticism', *Phil. Soc. Sci.* 4 (1974), 83–90.

1975

Science in Flux. Collected Essays, *Boston Studies in the Philosophy of Science*, Vol. 28 (Reidel, Dordrecht and Boston, 1975), 523 pp. The volume contains previously unpublished material: 'Prologue: On Stability and Flux'; 'Towards a Theory of 'Ad

Hoc' Hypotheses'; 'The Traditional Ad Hoc Use of Instrumentalism'; 'Modified Conventionalism'; and more. Appendices on Kant and on Buber.

'Three Views of the Renaissance of Science', *Physis* **17** (1975), 1–21. (Reprinted in *Science and Society*, 1981.)

'The Present State of the Philosophy of Science', *Philosophica* **15** (1975), 5–20. (Reprinted in *Science and Society*, 1981.)

'Institutional Individualism', *Brit. J. Soc.* **26** (1975), 144–55.

'Field Theory in De La Rive's Treatise', *Organon* **11** (1975), 285–301.

'Determinism: Metaphysical versus Scientific', *Memoirs*, 5th International Congress of Logic, Methodology and Philosophy of Science (London, Canada, 1975), VI–4.

'Scientists as Sleepwalkers', in Y. Elkana (ed.), *The Interaction Between Science and Philosophy* (Humanities, N. Y., 1975), pp. 391–405. Also discussion notes, pp. 191–3, 284–6, 291, 439–43. (Reprinted in *Science and Society*, 1981.)

'Genius in Science', *Phil. Soc. Sci.* **5** (1975), 145–61. (Hebrew transl., *Keshet* **16** (1974), 135–42. (Reprinted in *Science and Society*, 1981.)

'Between Metaphysics and Methodology', *Poznań Studies in Phil. Sci. and Humanities* **1** (1975), 1–8. (Reprinted in *Science and Society*, 1981.)

'Spontaneity in the Arts' (in Hebrew), *Close-Up* **3–4** (1975), 1–21. (English transl., *Poznań Studies* **2** (1976), 54–64.)

'Subjectivism: From Infantile Disease to Chronic Illness', *Synthese* **30** (1975), 3–14.

'Replies to Critics', *Synthese* **30** (1975), 33–8.

'Verisimilitude: Comments on David Miller', *Synthese* **30** (1975), 197–204.

'In Search of the Zeitgeist' (Lewis Feuer, *Einstein and the Generation of Science*), *Phil. Soc. Sci.* **5** (1975), 339–42.

'The Future of Berkeley's Instrumentalism' (R. J. Brook, *Berkeley's Philosophy of Science*), *International Studies in Philosophy* **7** (1975), 167–78.

J. A. and Paul T. Sagal, 'The Problem of Universals', *Philosophical Studies* **28** (1975), 289–94.

R. S. Cohen and J. A., 'Dinosaurs and Horses', *Synthese* **32** (1975), 233–47.

1976

'Causality and Medicine', *J. of Medicine and Philosophy* **1** (1976), 301–17.

'Medicine: Art or Science' (in Hebrew), *Koroth*, Quarterly for the History of Medicine 7 (1976), 50–61.

'On the Philosophy of Technology', *Methodology and Science* **9** (1976), 41–50.

'Metaphysics as Regulative Ideas for Science', *Science et Metaphysique*, Proceedings of the Freiburg Meeting of the International Academy for the Philosophy of Science (Brussels, 1976), 33–46.

'Can Adults Become Genuinely Bilingual?', in A. Kasher (ed.), *Language in Focus*, Bar-Hillel Memorial Volume, *Boston Studies in the Philosophy of Science*, Vol. 43 (Reidel Dordrecht, 1976), pp. 473–84.

'The Lakatosian Revolution', in R. S. Cohen, P. K. Feyerabend, and M. W. Wartofsky (eds.), Essays in Memory of Imre Lakatos', *Boston Studies in the Philosophy of Science*, Vol. 39 (Reidel, Dordrecht, 1976), pp. 9–21.

'Assurance and Agnosticism', in A. C. Michalos and R. S. Cohen (eds.), *PSA, 1974, Boston Studies in the Philosophy of Science*, Vol. 32 (Reidel, Dordrecht, 1976), 449–54. (Reprinted in *Science in Flux*, 1975.)

'Verisimilitude: Popper, Miller, and Hattiangadi', in M. Przełecki, *et al.* (eds.), *Formal Methods in the Methodology of the Empirical Sciences* (Reidel, Dordrecht and Boston, 1976), pp. 335–72.

'Justification by Society versus Justificationism', *Phil. Forum* 7 (1976), 364–6.

'Comments on Peirce's Review Essay: Philosophy', *Signs* 2 (1976), 512.

Review of James Stephenson, *Francis Bacon and the Style of Science, Philosophy and Rhetorics* 9 (1976), 251–4.

Review of Joseph Weizenbaum, *Computer Power, Technology and Culture* 17 (1976), 813–16.

Review of Paul Feyerabend, *Against Method, Philosophia* 6 (1976), 165–177.

'Reply to Professor Feyerabend', *Philosophia* 6 (1976), 190–1.

A contribution to a symposium on Freud in the 70's (in Hebrew), *Keshet* 18 (1976), 54–5.

Y. Fried and J. A., *Paranoia: A Study in Diagnosis, Boston Studies*, Vol. 50 (Reidel, Dordrecht, 1976), 200 pp.

1977

Towards a Rational Philosophical Anthropology (Martinus Nijhoff, The Hague), 370 pp.

'Who Discovered Boyle's Law?', *Studies in the History and Philosophy of Science* 8 (1977), 189–250.

'Robert Boyle's Anonymous Writings', *Isis* 68 (1977), 284–7.

'The Methodology of Research Projects – a Sketch', *Zeitschrift für Allgemeine Wissenschaftstheorie* 8 (1977), 30–8. (Reprinted in *Science and Society*, 1981.)

'Tradition and Revolution' (in Hebrew), in A. Kasher and J. Levinger (eds.), *The Isiah Leibovitch Book* (Univ. Pbln., Tel Aviv, 1977), pp. 79–89.

'Between Clarity and Rationality' (in Hebrew), in J. Meltser (ed.), Bar-Hillel Memorial Volume (Hebrew Univ., Jerusalem, 1977); also published as *Iyyun* 27 (1976–7), 147–52; English summary, 359.

'More Against the Principle of Loyalty' (in Hebrew), in M. Dascal (ed.), *The Just and the Unjust* (Univ. Pbln., Tel Aviv, 1977), pp. 39–42.

'The Zeitgeist and Professor Feuer' and 'Second Reply to Professor Feuer', *Phil. Soc. Sci.* 7 (1977), 251–3, 263.

Review of the Georgescu-Roegen *Festschrift, Technology and Culture* 18 (1974), 577–8.

1978

'Liberal Forensic Medicine', *J. of Medicine and Philosophy* 3 (1978), 226–41.

'Externalism', *Manuscrito* 2 (1978), 65–78. (Reprinted in *Science and Society*, 1981.)

'The Ideological Import of Newton', *Vistas in Astronomy* 22 (1979), 419–30. (Reprinted in *Science and Society*, 1981.)

'Technology, Mass Movement, and Rapid Social Change', *Research in Philosophy and Technology* 1 (1978), 53–64.

'Shifting from Physical to Social Technology', *Technology* 1 (1978), 199–212.

'In Defence of Standardized On Demand Publications', in Miriam Balaban (ed.), *Scientific Information Transfer: The Editor's Role* (Reidel, Dordrecht and Boston, 1978), pp. 133–9. (Reprinted in *Science and Society*, 1981, as 'Storage and Communication of Knowledge'.)

'Glaube hat nichts mit Rationalität zu tun', in O. Molden (ed.), *Konflikt und Ordnung* (Verlag Molden, Wien, 1978), pp. 58–64; (English version 'Faith Has Nothing to Do With Rationality', in *Science and Society*, 1981.)

'Logic and Logic of' *Poznań Studies* 4 (1978), 1–11.

'Sex and Violence in the Cinema' (in Hebrew), *Close-Up* 5 (1978), 74–82.

'Williams Dodges Agassi's Criticism', *Brit. J. Phil. Sci.* 29 (1978), 248–52.

'Movies Seen Many Times' (Review of three books), *Phil. Soc. Sci.* 8 (1978), 398–405.

'Wittgenstein's Heritage' (Godfrey Vesey, *Understanding Wittgenstein*), *Erkenntnis* **13** (1978), 305–26.

'Mach's Trial and Error' (*Knowledge and Error*), *Philosophia* 8 (1978), 305–26.

Epistemologia, metafiscia, e storia della scienza, translations (Armando, Rome, 1978).

La filosofia dell'uomo libero: verso una storiografia della scienza, translation (Armando, Rome, 1978).

John R. Wettersten and J. A., 'Rationality, Problems, Choice', *Philosophica* **22** (1978), 5–22.

1979

'The Philosophy and the Sciences of Man', *Specificitê des sciences humaines et tant des sciences*, Proceedings of the Trento meeting of the International Academy for the Philosophy of Science, Brussels, 1979, also published as *Epistemologia* 2 (1979), *Special Issue*, pp. 155–66.

'The Choice of Scientific Problems', in J. Bärmark (ed.), *Perspectives in Metascience*, Törnebohm *Festschrift* (Acta Reg. Soc. Sc. et Litt. Gothob, *Interdisciplinaria* 2) pp. 13–25. (Reprinted in *Science and Society*, 1981.)

'Die Legitimation der Erkenntnis', in O. Molden, *Wissen und Macht* (Verlag Molden, Wien, 1979), pp. 237–46; English version, 'The Legitimation of Science', *Dialogos* **35** (1980), 27–35. (Reprinted in *Science and Society*, 1981.)

'Art and Science', *Scientia* **73** (1979), 127–140. (Italian translation, *Scientia* **73** (1979), 141–52.)

'Wissenschaft und Metaphysik', *Grazer Philosophische Studien* 9 (1979), 97–106.

'Quanta in Context', in *Einstein Symposium, Lecture Notes in Physics* (Springer, Berlin), Vol. 100, pp. 180–203.

'The Functions of Intellectual Rubbish', *Research in the Sociology of Knowledge, Science and Art* **2** (1979), 209–27.

'The Whole and Its Parts', *Nature and Systems* **1** (1979), 32–6.

'The Philosophy of Hans Albert' (*Traktat über Rationale Praxis*), *Soziologische Revue* **2** (1979), 241–9.

'The Legacy of Lakatos' (S. Latsis, *Method and Appraisal in Economics*), *Phil. Soc. Sci.* 9 (1979), 316–26.

'Towards a Rational Theory of Superstition' (*Recent Advances in Natal Astrology*), *Zetetic Scholar* **3/4** (1979), 107–20.

'Rejoinder', *Zetetic Scholar* **5** (1979), 85–8.

Dialogo senza fine, translation (Armando, Rome, 1979).

J. A. and I. C. Jarvie, 'The Rationality of Dogmatism', in T. Geraets (ed.), *Rationality Today* (University of Ottawa Press, 1979), pp. 353–62.

1980

'Rights and Reason', *Israel Yearbook of Human Rights* 9 (1980), 9–22.

'Between Science and Technology', *Phil. Sci.* 47 (1980), 82–99.

'The Problem of Scientific Validation', in M. D. Grmek, R. S. Cohen and G. Cimino (eds.), *Scientific Discovery, Boston Studies in the Philosophy of Science*, Vol. 34 (Reidel, Dordrecht, 1980), pp. 103–114.

'The Rationality of Discovery', in Ths. Nickles (ed.), *Scientific Discovery, Logic, and Rationality, Boston Studies in the Philosophy of Science*, Vol. 56 (Reidel, Dordrecht, 1980), pp. 185–199.

'Gehirnwasche', *Unter dem Pflaster liegt der Strand* 7 (1980), 179–191.

'On Mathematics Education: The Lakatosian Revolution', *For The Learning of Mathematics* 1 (1980), 39–41.

'Wie es Euch Gefällt', in H. P. Duerr (ed.), *Versuchungen. Aufsätze zur Philosophie Paul Feyerabends* (Suhrkamp, 1980), pp. 147–157.

'Comments on Stewart Guthry', *Current Anthropology* 21 (1980), 194.

'Comments on Professor Hyman's Paper', *Zetetic Scholar* 6 (1980), 39–41.

J. A., and I. C. Jarvie, 'The Rationality of Irrationalism', *Metaphilosophy* 11 (1980), 127–133.

J. A., and Charles M. Sawyer, 'Was Lakatos an Elitist?', *Ratio* 22 (1980), 61–3. (Also in German edition of *Ratio*.)

J. A., and John R. Wettersten, 'Stegmüller Squared', *Ztsch. für Allg. Wissenschaftstheorie* 11 (1980), 86–94.

INDEX OF NAMES

INDEX OF SUBJECTS

BOSTON STUDIES IN THE PHILOSOPHY OF SCIENCE

Editors:
ROBERT S. COHEN and MARX W. WARTOFSKY
(Boston University)

1. Marx W. Wartofsky (ed.), *Proceedings of the Boston Colloquium for the Philosophy of Science 1961-1962.* 1963.
2. Robert S. Cohen and Marx W. Wartofsky (eds.), *In Honor of Philipp Frank.* 1965.
3. Robert S. Cohen and Marx W. Wartofsky (eds.), *Proceedings of the Boston Colloquium for the Philosophy of Science 1964-1966. In Memory of Norwood Russell Hanson.* 1967.
4. Robert S. Cohen and Marx W. Wartofsky (eds.), *Proceedings of the Boston Colloquium for the Philosophy of Science 1966-1968.* 1969.
5. Robert S. Cohen and Marx W. Wartofsky (eds.), *Proceedings of the Boston Colloquium for the Philosophy of Science 1966-1968.* 1969.
6. Robert S. Cohen and Raymond J. Seeger (eds.), *Ernst Mach: Physicist and Philosopher.* 1970.
7. Milic Capek, *Bergson and Modern Physics.* 1971.
8. Roger C. Buck and Robert S. Cohen (eds.), *PSA 1970. In Memory of Rudolf Carnap.* 1971.
9. A. A. Zinov'ev, *Foundations of the Logical Theory of Scientific Knowledge (Complex Logic).* (Revised and enlarged English edition with an appendix by G. A. Smirnov, E. A. Sidorenka, A. M. Fedina, and L. A. Bobrova.) 1973.
10. Ladislav Tondl, *Scientific Procedures.* 1973.
11. R. J. Seeger and Robert S. Cohen (eds.), *Philosophical Foundations of Science.* 1974.
12. Adolf Grünbaum, *Philosophical Problems of Space and Time.* (Second, enlarged edition.) 1973.
13. Robert S. Cohen and Marx W. Wartofsky (eds.), *Logical and Epistemological Studies in Contemporary Physics.* 1973.
14. Robert S. Cohen and Marx W. Wartofsky (eds.), *Methodological and Historical Essays in the Natural and Social Sciences. Proceedings of the Boston Colloquium for the Philosophy of Science 1969-1972.* 1974.
15. Robert S. Cohen, J. J. Stachel and Marx W. Wartofsky (eds.), *For Dirk Struik. Scientific, Historical and Political Essays in Honor of Dirk Struik.* 1974.
16. Norman Geschwind, *Selected Papers on Language and the Brain.* 1974.
18. Peter Mittelstaedt, *Philosophical Problems of Modern Physics.* 1976.
19. Henry Mehlberg, *Time, Causality, and the Quantum Theory* (2 vols.). 1980.
20. Kenneth F. Schaffner and Robert S. Cohen (eds.), *Proceedings of the 1972 Biennial Meeting, Philosophy of Science Association.* 1974.
21. R. S. Cohen and J. J. Stachel (eds.), *Selected Papers of Léon Rosenfeld.* 1978.
22. Milic Capek (ed.), *The Concepts of Space and Time. Their Structure and Their Development.* 1976.
23. Marjorie Grene, *The Understanding of Nature. Essays in the Philosophy of Biology.* 1974.

24. Don Ihde, *Technics and Praxis. A Philosophy of Technology.* 1978.
25. Jaakko Hintikka and Unto Remes, *The Method of Analysis. Its Geometrical Origin and Its General Significance.* 1974.
26. John Emery Murdoch and Edith Dudley Sylla, *The Cultural Context of Medieval Learning.* 1975.
27. Marjorie Grene and Everett Mendelsohn (eds.), *Topics in the Philosophy of Biology.* 1976.
28. Joseph Agassi, *Science in Flux.* 1975.
29. Jerzy J. Wiatr (ed.), *Polish Essays in the Methodology of the Social Sciences.* 1979.
32. R. S. Cohen, C. A. Hooker, A. C. Michalos, and J. W. van Evra (eds.), *PSA 1974: Proceedings of the 1974 Biennial Meeting of the Philosophy of Science Association.* 1976.
33. Gerald Holton and William Blanpied (eds.), *Science and Its Public: The Changing Relationship.* 1976.
34. Mirko D. Grmek (ed.), *On Scientific Discovery.* 1980.
35. Stefan Amsterdamski, *Between Experience and Metaphysics. Philosophical Problems of the Evolution of Science.* 1975.
36. Mihailo Marković and Gajo Petrović (eds.), *Praxis. Yugoslav Essays in the Philosophy and Methodology of the Social Sciences.* 1979.
37. Hermann von Helmholtz: *Epistemological Writings. The Paul Hertz/Moritz Schlick Centenary Edition of 1921 with Notes and Commentary by the Editors.* (Newly translated by Malcolm F. Lowe. Edited, with an Introduction and Bibliography, by Robert S. Cohen and Yehuda Elkana.) 1977.
38. R. M. Martin, *Pragmatics, Truth, and Language.* 1979.
39. R. S. Cohen, P. K. Feyerabend, and M. W. Wartofsky (eds.), *Essays in Memory of Imre Lakatos.* 1976.
42. Humberto R. Maturana and Francisco J. Varela, *Autopoiesis and Cognition. The Realization of the Living.* 1980.
43. A. Kasher (ed.), *Language in Focus: Foundations, Methods and Systems. Essays Dedicated to Yehoshua Bar-Hillel.* 1976.
46. Peter L. Kapitza, *Experiment, Theory, Practice.* 1980.
47. Maria L. Dalla Chiara (ed.), *Italian Studies in the Philosophy of Science.* 1980.
48. Marx W. Wartofsky, *Models: Representation and the Scientific Understanding.* 1979.
50. Yehuda Fried and Joseph Agassi, *Paranoia: A Study in Diagnosis.* 1976.
51. Kurt H. Wolff, *Surrender and Catch: Experience and Inquiry Today.* 1976.
52. Karel Kosík, *Dialectics of the Concrete.* 1976.
53. Nelson Goodman, *The Structure of Appearance.* (Third edition.) 1977.
54. Herbert A. Simon, *Models of Discovery and Other Topics in the Methods of Science.* 1977.
55. Morris Lazerowitz, *The Language of Philosophy. Freud and Wittgenstein.* 1977.
56. Thomas Nickles (ed.), *Scientific Discovery, Logic, and Rationality.* 1980.
57. Joseph Margolis, *Persons and Minds. The Prospects of Nonreductive Materialism.* 1977.
58. Gerard Radnitzky and Gunnar Andersson (eds.), *Progress and Rationality in Science.* 1978.

59. Gerard Radnitzky and Gunnar Andersson (eds.), *The Structure and Development of Science.* 1979.
60. Thomas Nickles (ed.), *Scientific Discovery: Case Studies.* 1980.
61. Maurice A. Finocchiaro, *Galileo and the Art of Reasoning.* 1980.